Stem Cell Therapy: Medical Frontiers

Stem Cell Therapy: Medical Frontiers

Editor: Mark Walters

FA FOSTER
ACADEMICS

www.fosteracademics.com

www.fosteracademics.com

F A
FOSTER
ACADEMICS

Cataloging-in-Publication Data

Stem cell therapy : medical frontiers / edited by Mark Walters.
     p. cm.
Includes bibliographical references and index.
ISBN 978-1-63242-569-0
1. Stem cells--Therapeutic use. 2. Stem cells. I. Walters, Mark.
QH588.S83 S74 2018
616.027 74--dc23

Foster Academics,
118-35 Queens Blvd., Suite 400,
Forest Hills, NY 11375, USA

ISBN 978-1-63242-569-0 (Hardback)

Contents

Preface

Stem cell therapy involves the transplantation of stem cells for the rebuilding of hematopoietic cells and tissues. Medically, stem cells are used in the treatment of cancers of the blood and bone marrow. The procedure may be autologous, where the patient provides the cells to be transplanted or allogenic, where there is a healthy host and a patient that receives the stem cells. Transplant of stem cells provide a complete cure or long-term remedy to otherwise fatal diseases. Modern medical research has ensured a high success rate of this procedure. This book, with its detailed analyses and data, will prove immensely beneficial to medical professionals and students involved in this area at various levels. Different approaches, evaluations, methodologies and advanced studies on stem cell therapy have been included in this book.

Significant researches are present in this book. Intensive efforts have been employed by authors to make this book an outstanding discourse. This book contains the enlightening chapters which have been written on the basis of significant researches done by the experts.

Finally, I would also like to thank all the members involved in this book for being a team and meeting all the deadlines for the submission of their respective works. I would also like to thank my friends and family for being supportive in my efforts.

Editor

Mesenchymal Stem Cells from Fetal Heart Attenuate Myocardial Injury after Infarction: An *In Vivo* Serial Pinhole Gated SPECT-CT Study in Rats

Venkata Naga Srikanth Garikipati[1], Sachin Jadhav[1], Lily Pal[2], Prem Prakash[3], Madhu Dikshit[3], Soniya Nityanand[1]*

1 Stem Cell Research Facility, Department of Hematology, Sanjay Gandhi Post-Graduate Institute of Medical Sciences, Lucknow, India, 2 Department of Pathology, Sanjay Gandhi Post-Graduate Institute of Medical Sciences, Lucknow, India, 3 Cardio-vascular unit, Division of Pharmacology, Central Drug Research Institute, Lucknow, India

Abstract

Mesenchymal stem cells (MSC) have emerged as a potential stem cell type for cardiac regeneration after myocardial infarction (MI). Recently, we isolated and characterized mesenchymal stem cells derived from rat fetal heart (fC-MSC), which exhibited potential to differentiate into cardiomyocytes, endothelial cells and smooth muscle cells *in vitro*. In the present study, we investigated the therapeutic efficacy of intravenously injected fC-MSC in a rat model of MI using multi-pinhole gated SPECT-CT system. fC-MSC were isolated from the hearts of Sprague Dawley (SD) rat fetuses at gestation day 16 and expanded *ex vivo*. One week after induction of MI, 2×10^6 fC-MSC labeled with PKH26 dye (n = 6) or saline alone (n = 6) were injected through the tail vein of the rats. Initial *in vivo* tracking of 99mTc-labeled fC-MSC revealed a focal uptake of cells in the anterior mid-ventricular region of the heart. At 4 weeks of fC-MSC administration, the cells labeled with PKH26 were located in abundance in infarct/peri-infarct region and the fC-MSC treated hearts showed a significant increase in left ventricular ejection fraction and a significant decrease in the end diastolic volume, end systolic volume and left ventricular myo-mass in comparison to the saline treated group. In addition, fC-MSC treated hearts had a significantly better myocardial perfusion and attenuation in the infarct size, in comparison to the saline treated hearts. The engrafted PKH26-fC-MSC expressed cardiac troponin T, endothelial CD31 and smooth muscle sm-MHC, suggesting their differentiation into all major cells of cardiovascular lineage. The fC-MSC treated hearts demonstrated an up-regulation of cardio-protective growth factors, anti-fibrotic and anti-apoptotic molecules, highlighting that the observed left ventricular functional recovery may be due to secretion of paracrine factors by fC-MSC. Taken together, our results suggest that fC-MSC therapy may be a new therapeutic strategy for MI and multi-pinhole gated SPECT-CT system may be a useful tool to evaluate cardiac perfusion, function and cell tracking after stem cell therapy in acute myocardial injury setting.

Editor: Rajasingh Johnson, University of Kansas Medical Center, United States of America

Funding: This work was supported by an extramural grant sanctioned by Department of Biotechnology, Government of India, to Dr. S. Nityanand (BT/PR6519/MED/14/826/2005). http://dbtindia.nic.in/index.asp. The funders had no role in study design, data collection and analysis, decision to publish, or preparation of the manuscript.

Competing Interests: The authors have declared that no competing interests exist.

* Email: soniya@sgpgi.ac.in

Introduction

Cellular cardiomyoplasty has emerged as a potential therapeutic strategy for patients with acute myocardial infarction (MI). MI results in loss of cardiomyocytes, ventricular remodeling, scar formation, fibrosis and subsequently heart failure [1]. The ultimate goal of any regenerative therapy for ischemic myocardium is to regenerate lost cardiomyocytes and facilitate cardiovascular neovascularization, in order to lead to clinical improvement in cardiac functions. An array of adult stem cell types including skeletal myoblasts, bone marrow derived stem cells, endothelial progenitor as well as cardiac stem cells have been shown to lead to functional benefit in animal models of infarction [2–5], but clinical trials have generated mixed results [6–8]. Hence, a search for a novel stem cell type that is capable of restoring cardiac function is of paramount importance.

Mesenchymal stem cells (MSC) due to their characteristic properties such as ease of isolation, extensive *ex vivo* expansion capacity and multi-lineage differentiation potential are considered to be one of the potential stem cells for cardiac repair and regeneration after MI in both experimental animals [9], and clinical studies [10]. Although originally identified in bone marrow, MSC have also been isolated from many adult organs as well as fetal-stage tissues [11]. Recently it has been suggested that the developmental stage of donor tissues not only affects the ability of MSC to differentiate into cardiomyocyte, but also their capacity to undergo smooth muscle and endothelial differentiation [12]. Moreover, it has been shown that tissue specific MSC possess unique properties with inherent potential of differentiation in to cell lineages of their tissue of origin [13]. In this context, we recently isolated and characterized MSC derived from rat fetal heart and described these cells as fetal cardiac mesenchymal stem cells (fC-MSC). They exhibited the potential to differentiate in to cardiomyocytes, endothelial cells and smooth muscle cells over successive passages, while maintaining expression of TERT and a normal karyotype [14].

Because of the enormous potential of cardiac stem cell therapy, it is being rapidly translated into clinical trials, and thus has left many issues unresolved, and emphasizes the need for concurrent techniques that provide more insights in to the mechanisms involved [15]. Molecular imaging is likely to play an important role in the better understanding of the fate of stem cells and their contribution in recovery of cardiac function [16]. Myocardial gated SPECT/CT is widely accepted as a gold standard for clinical measurement of cardiac functions [17]. With use of pinhole collimators and the advances in data processing, gated SPECT/CT has recently been adapted for small animal cardiovascular molecular imaging [18].

Taken together, we designed the present study to investigate the therapeutic efficacy of intravenously injected fC-MSC in a clinically most relevant rat model of MI (cardiac ischemia-reperfusion (IR) injury), using multi-pinhole gated SPECT/CT system. We also sought the cellular and molecular mechanisms underlying the beneficial effects of fC-MSC therapy.

Materials and Methods

Animals

Adult Sprague-Dawley (SD) rats, aged 8–12 weeks, weighing 180–250 g, were used in all experiments. Animals were housed at a constant temperature and humidity, with a 12:12-h light-dark cycle, and had free access to a standard diet and water. All the procedures were performed as per guidelines of Institutional Animal Ethics Committee and Committee for Purpose of Control and Supervision of Experiments on Animals (CPCSEA), India. The Committee on the Ethics of Animal Experiments of Sanjay Gandhi Post Graduate Institute of Medical Sciences, Lucknow, India, approved the protocol.

Isolation, Culture and Characterization of rat fC-MSC

fC-MSC were isolated and cultured from the hearts of SD rat fetuses at gestation day 16, in accordance with the methods described previously [14]. In brief, fetal hearts (n = 10) were minced, digested with 1 mg/mL collagenase type-IV (Worthington Biochemical, USA) and cultured in 25 cm^2 tissue culture flasks (Becton, Dickinson; USA) using complete culture medium consisting of α-MEM medium, 2 mg/mL of Glutamax (Gibco-Invitrogen), 16.5% fetal bovine serum (Hyclone, USA) and bacteriostatic level of penicillin-streptomycin (Gibco-Invitrogen). The MSC characteristics of the cells between 3rd–5th passages were confirmed by flow cytometric positivity for CD29, CD44, CD73, CD90 and CD105 and negativity for CD31, CD45 and MHC-II or isotype-identical antibodies (IgG) were used as negative controls and differentiation into adipogenic and osteo-genic cells using induction kits (Chemicon, USA).

Rat model of MI

The left anterior descending (LAD) coronary artery of male SD rats (n = 12) was occluded using the method previously described [19], with slight modifications. In brief, all rats were anesthetized with 80 mg/kg ketamine and 10 mg/kg xylazine, injected intra-peritoneally, and then mechanically ventilated. The heart was exposed through a left thoracotomy at the 3rd–4th intercostal space, and the left anterior descending coronary artery (LAD) was ligated with a 6–0 polyester suture and reperfused after 30 min of ligature.

On day 7 of MI induction, rats were randomized into two groups: (i) fC-MSC group (n = 6), and (ii) saline group (n = 6). A total of 2×10^6 fC-MSC labeled with PKH26 dye or saline alone (150 μL each) were injected intravenously through tail vein in the two groups, respectively. In 06 additional SD rats, MI was induced and after 1 week of induction of MI, fC-MSC labeled with Tc99-HMPAO were injected and tracked after 6 h of injection.

The ischemic area and the left ventricle (LV) functions were monitored in all treated and untreated animals by serial 99Tc-sestamibi pinhole gated SPECT performed before fC-MSC therapy (1 week post MI induction) and 4 weeks after fC-MSC therapy. The animals were sacrificed 4 weeks after fC-MSC therapy, and the hearts were removed, rinsed and used to perform the assays described below.

Cell Labeling

For initial tracking of fC-MSC after 6 h of injection, the cultured cells were trypsinized and incubated in a concentration of 2×10^6 cells per mL at 37°C with 2 mCi of Tc-99 with HMPAO linker for a 10-min period, and then the labeling process was stopped by a 5-min centrifugation at 950 g. This 10-min incubation period was found to result in sufficient labeling efficiency (74%) without significant deterioration of cell viability (92%). The 99 m Tc-HMPAO labeled cells were conditioned in an insulin syringe (2×10^6 cells per 50 μL), and a single injection was performed through tail vein.

For tracking of the cells 4 weeks after injection, fC-MSC were labeled with PKH26 (Sigma Aldrich, USA, kind gift from Dr Yucheng Dai, Molecular Medicine and Stem Cell Center of Second Affiliated Hospital of Jiangxi Medical Collage, Nanchang, PR China) according to manufacturer's protocol. In brief, 2×10^6 fC-MSC were taken in a 15 mL cone-shaped tube, washed with 10 mL PBS, centrifuged (200×g, 15 minutes) and a 25 μL of cell pellet was obtained. To the pellet, 1 mL of diluent C and 4×10^{-6} M of PKH26 staining reagent was added to obtain a final volume of 2 mL of 2×10^6 fC-MSC and incubated at room temperature for 5 min. The same amount of 10% FBS DMEM (2 mL) was used to stop the staining reaction. Labeling efficiency was 95% as validated by observation of cells in fluorescence microscope. Cells were conditioned to an insulin syringe (2×10^6 per 150 μL) to inject in animals.

Pinhole Gated SPECT Acquisitions

The animals were sedated by intraperitoneal injection of 10 mg/kg xylazine and 80 mg/kg ketamine. 99mTc-sestamibi (2 mCi in a 0.3- to 0.5 mL volume) was injected intravenously 40–60 min before starting gated SPECT acquisition. During the acquisition, the animals were kept in prone position and were connected to a standard electrocardiogram monitor by 3 electrodes placed on the inner surfaces of limb extremities. Pinhole gated SPECT acquisitions were performed using a dual head γ-camera (Bioscan, USA) equipped with a 3-mm pinhole collimator (195-mm focal length; 43-mm radius of rotation). A number of 24 projections of 60 seconds per step were acquired on a 360° rotation and with 8 frames per cardiac cycle. Additional acquisition parameters were as follows: 64×64 matrix, 2.0 zoom, 126- to 154-keV energy window, beat acceptance window set to ±20% of averaged R-R interval. Total acquisition time was 24 min.

Reconstruction and Analysis of Pinhole Gated SPECT Images

Using the HISPECT-NG software myocardial gated SPECT images were reconstructed. FlowQuant software (Ottawa Heart Institute, Canada) was used to generate myocardial perfusion polar maps and to determine LV end-diastolic volume (EDV) and end-systolic volume (ESV), as well as LV ejection fraction (EF) on the contiguous gated short-axis slices obtained from serial 99mTc-sestamibi gated SPECT. Further, polar maps were used to quantify the viable myocardium using the threshold value of 50% uptake of 99mTc-sestamibi [20].

Immunohistochemistry

Heart tissues fixed in 4% paraformaldehyde were sectioned (5 μm) and incubated with antibodies specific for the detection of the following proteins; Troponin-T (1:500; Serotec, UK), CD31 (1:500; Santa Cruz), sm-MHC (1:1000, Serotec), in 5% normal sheep serum overnight at 4°C. Next day, after washing with PBS, the sections were incubated with fluorochrome tagged respective secondary antibodies for 1 hour in dark. The sections were counterstained with Hoechst 33258 (Molecular probes) nuclear stain and mounted in antifade mountant. Heart sections stained with non-immune serum or IgG instead of the primary antibodies was used as negative control. The pictures were taken using fluorescent microscope (Nikon 80i, Japan).

Real Time PCR

Heart tissues were harvested at 4 weeks post fC-MSC therapy, pulverized in liquid nitrogen, and homogenized in 1 mL Trizol reagent (Invitrogen). Total RNA was extracted with 0.2 ml of chloroform and precipitated with 0.5 ml 80% (vol/vol) isopropanol. After the removal of the supernatant, the RNA pellet was washed with 70% (vol/vol) ethanol, air-dried, and dissolved in DNase RNase-free water. Complementary DNA was synthesized from 1 μg of RNA using First-Strand cDNA Synthesis Kit (USB, USA) at 44°C for 60 minutes, 92°C for 10 minutes. Real-time PCR with relative quantification of target gene copy numbers in relation to β-actin transcripts was carried out using the following primers given in Table 1. Relative fold expression values were determined applying the $\Delta\Delta$ cycle threshold (Ct) method [21].

TUNEL Assay

In situ detection of apoptosis was performed on heart tissue sections (5 μm) fixed in 4% paraformaldehyde using terminal deoxynucleotidetransferase (TdT) - mediated dUTP nick end labeling (TUNEL) kit (Roche, Germany) and positive cells were counted in five random fields. Relative TUNEL positivity was expressed as number of TUNEL positive cells/100 nuclei counterstained using Hoechst dye. Image-Pro plus 5.1 software was used for image capturing and cell counting (Media Cybernetics Inc., USA).

Western Blotting

Heart tissues of saline treated and fC-MSC treated rats were homogenized in RIPA buffer containing 1 mmol/L phenylmethanesulphonyl fluoride (PMSF) and 1% protease inhibitor cocktail (Sigma-Aldrich, MO, USA). Tissue homogenate was centrifuge at 10000 rpm for 10 min and supernatant was stored. 40 mg proteins were loaded and separated by 10% SDS-PAGE. After electrophoresis, separated proteins were transferred to polyvinylidenedifluoride (PVDF) membranes. The membrane was blocked for 1 hour in 5% BSA at room temp followed by overnight incubation at 4°C with primary antibodies, Bax and Bcl-2 (Cell Signaling Technology, MA, USA). β- Actin antibody was used as an internal control. Primary antibodies were detected by corresponding horseradish peroxidase (HRP)-conjugated secondary antibodies using super signal west pico chemiluminescent substrate (Thermo scientific, IL, USA).

Masson's trichome Staining

Hearts tissues fixed in 10% formalin were cut into sections of 5 μm and stained with Masson's Trichome stain (Glaxo Smith Kline, UK) to evaluate fibrosis.

Statistical Analysis

Values were expressed as means ± SEM. Comparisons between saline and stem cell treated groups were made with the use of paired Student's t-tests. $P<0.05$ was considered significant.

Results

Mesenchymal stem cell characteristics of fC-MSC

The fC-MSC grew as plastic adherent cells having trigonal or spindle shaped morphology from primary culture up to the 21st passage. Flow cytometric analysis showed a typical mesenchymal phenotype of fC-MSC with expression of CD29 (96.54±0.44%), CD44 (95.96±0.10%), CD73 (95.10±0.16%), CD90 (98.90±0.18%), and CD105 (98.00±0.12%) and absence of CD31 (0.50±0.02%), CD45 (0.64±0.02%) and MHC-II (0.30±0.08%) markers. Treatment of fC-MSC with adipogenic and osteogenic induction media resulted in their differentiation into adipocytes and osteocytes as demonstrated by Oil red-O and Alizarin red staining, respectively (Fig. 1).

Effect of fC-MSC on LV function and perfusion after MI

Myocardial gated SPECT was used to measure LV function and cardiac perfusion 1 week after MI and 4 weeks after fC-MSC therapy. One week after MI, no significant differences in the ejection fraction (EF), end diastolic volume (EDV), end systolic volume (ESV) and left ventricular (LV) myo-mass were observed between saline and fC-MSC treated hearts (Table 2 and Fig. 2).

Table 1. Primers used for the Real Time PCR study.

Gene	Primer Sequence (5'-3')	Accession No.
GAPDH	F-CCTCTCTCTTGCTCTCAGTAT/R-GTATCCGTTGTGGATCTGACA	NM_017008.3
VEGF-A	f-Tgtgaatgcagaccaaagaaa/R-ctgaacaaggctcacagtgaat	AY702972.1
HGF	F-CCAGCTAGAAACAAAGACTTGAAAGA/R-GAAATGTTTAAGATCTGTTTGCGTT	NM_017017.2
bFGF	F-tcttcctgcgcatccatc/R-gcttggagctgtagtttgacg	X61697.1
IGF-1	F-atgcccaagactcagaagga/R-gtggcattttctgttcctc	X06043.1
TGF-alpha	F-gtattgtttccatgggacctg/R-cgtacccagagtggcagac	NM_012671.2

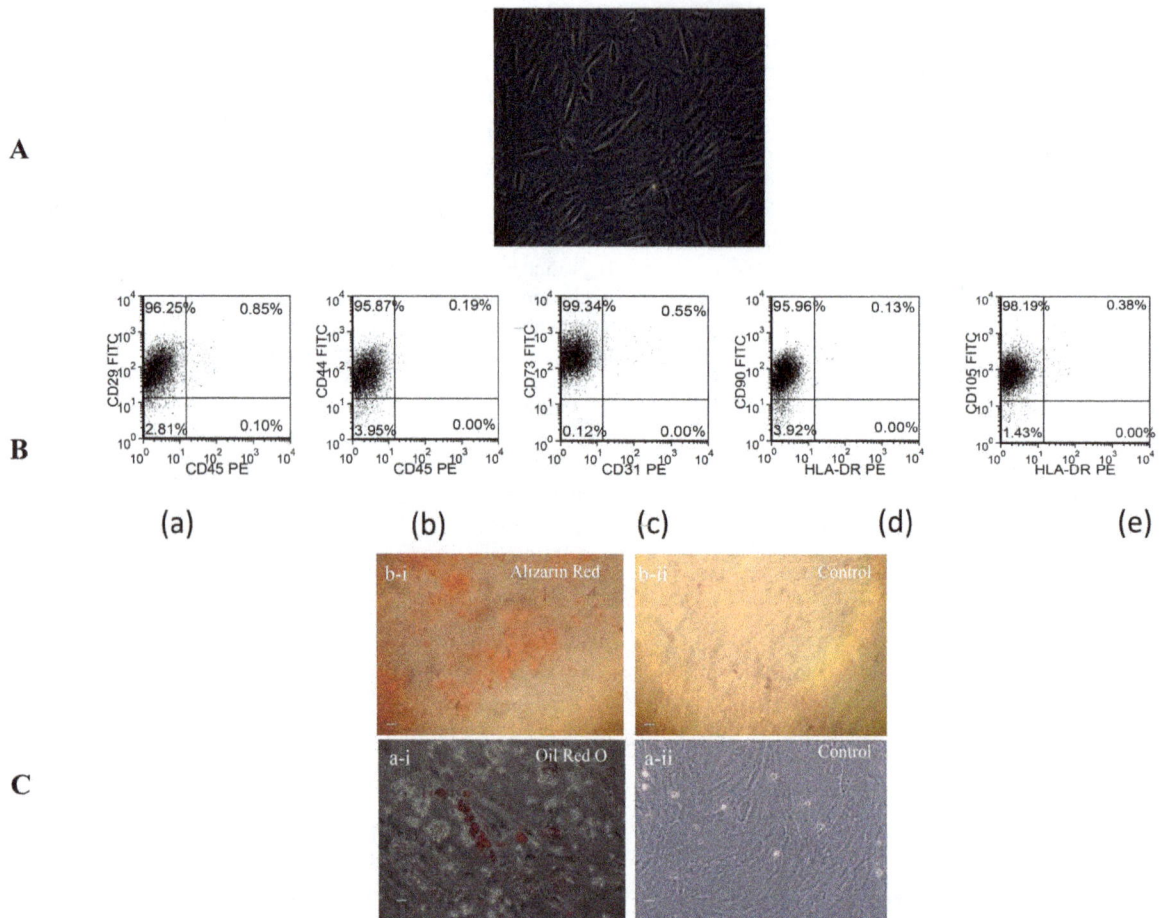

Figure 1. Morphology and characterization of fC-MSC (A) Representative photomicrograph (10X, 20 μm) of fC-MSC in culture showing spindle shaped morphology. (B) Representative flow cytometric dot-plots showing that fC-MSC are (a) CD29+/CD45−; (b) CD44+/CD45−; (c) CD73+/CD31−; (d) CD90+/HLA-DR−; (e) CD105+/HLA-DR−. (C) Representative photomicrographs (10X, 20 μm) showing differentiation of fC-MSC into Osteoblasts (a-i: differentiated cells positive for Alizarin red stain, and a-ii: control cells negative for Alizarin red stain) and Adipocytes (b-i: differentiated cells positive for oil red O stain, and b-ii: control cells negative for oil red O stain).

After 4 weeks of cell/saline therapy, the fC-MSC treated group showed a significant increase in EF from the EF observed 1 week after MI and before cell therapy (48.57 ± 1.5 vs 40.2 ± 2, $p < 0.001$), whereas EF remained significantly depressed in the saline treated animals (33.1 ± 1.2 vs 39.3 ± 1.9, $p < 0.01$). Furthermore, there was a significant difference in EF between fC-MSC and saline treated groups, 4 weeks after cell/saline therapy (48.57 ± 1.5 vs 33.1 ± 1.2, $p < 0.001$). Although both saline and fC-MSC treated hearts showed cardiac remodeling at 1 week after MI, the EDV, ESV and LV myo-mass of fC-MSC treated hearts showed a significant decrease at 4 weeks after therapy as compared to saline treated hearts (EDV 364.5 ± 18 vs 618.5 ± 17.6 μl, $p < 0.001$; ESV 391.66 ± 30.2 vs 179.5 ± 8.7 μl, $p < 0.001$; LV myo-mass 0.24 ± 0.01 vs 0.42 ± 0.02 g, $p < 0.01$) (Table 2 and Fig. 2).

The extent of ischemic region and its myocardial perfusion were evaluated at 4 weeks after fC-MSC administration. As can be seen in Fig. 3, the hearts treated with fC-MSC demonstrated a smaller ischemic lesion (region deficit, dark area) and a significantly greater 99mTc-sestamibi uptake than saline treated group ($31.4 \pm 2.04\%$ vs. $16.0 \pm 2.14\%$; $p < 0.001$).

In vivo tracking and engraftment of fC-MSC

SPECT/CT imaging of MI rats 6 h after intravenous infusion of 99mTc-labeled fC-MSC revealed significant uptake of 99mTc-labeled cells in the lungs, with focal uptake in the anterior mid-ventricular region of the heart (Fig. 4A). Furthermore, to confirm the engraftment and retention of fC-MSC in follow up period, cells were labeled with PKH26. At 4 weeks of fC-MSC administration, cells labeled with PKH26 were located in the fibrotic region of the scar and more rarely in remote myocardium (Fig. 4B).

Cardiomyogenic, endothelial and smooth muscle cell differentiation of fC-MSC in vivo

Regenerating myocardial cells were detected by co-localization of immunostaining for cTnT (cardiomyocytes), sm-MHC (smooth muscle cells) and CD31 (endothelial cells) with PKH26 labelling. PKH 26 -labeled cells in the infarct/peri-infarct region also expressed cTnT, CD31, and sm-MHC indicating ability of fC-MSC to differentiate into cardiomyocytes, endothelial and smooth muscle cells (Fig. 5).

Table 2. Gated SPECT analysis of fC-MSC therapy in rats with MI.

Groups	Ejection Fraction (%)	End Diastolic Volume (µl)	End Systolic Volume (µl)	LV Mass (g)
Healthy Control (Before MI)	59.83±1.19	261.3±22.11	104.66±8.88	0.21±0.008
MI baseline (1 Week after MI)	39.3±1.9***	455.6±16.1***	255.66±17.51**	0.32±0.01***
MI+ Saline (5 weeks post MI)	33.1±1.2†	618.5±17.6†††	391.66±30.2††	0.42±0.02††
MI+ fC-MSC (5 Weeks post MI)	48.57±1.5‡‡ §§§	364.5±18.7‡‡ §§§	179.5±8.7‡‡ §§§	0.24±0.01‡‡ §§

***$P<0.001$,
**$P<0.01$ = Healthy Control Vs MI Baseline;
††$P<0.01$ = MI Baseline Vs MI+Saline;
†††$P<0.001$ ‡‡$P<0.01$,
‡$P<0.05$ = MI baseline Vs MI+fC-MSC;
§§§$P<0.001$,
§§$P<0.01$ = MI+Saline Vs MI+fC-MSC.

Paracrine, anti-apoptotic and anti-fibrotic effects of fC-MSC

We screened hearts of MI rats for gene expression of growth factors by real time PCR and protein expression of pro/anti-apoptotic molecules by western blot. fC-MSC treated hearts showed significant up regulation in gene expression of various growth factors VEGF (40.2 ± 0.9 vs 6.54 ± 0.48 fold; $p<0.001$), βFGF (2.3 ± 0.25 vs 1.3 ± 0.1 fold, $p<0.05$), IGF-1 (61.6 ± 4.4 vs 2.3 ± 0.7 fold, $p<0.001$) and HGF-1 (5.18 ± 0.5 vs 1.96 ± 0.11 fold, $p<0.001$), compared to saline treated group (Fig. 6). We also observed a significant decrease in protein expression of pro-apoptotic molecule Bax (10.50 ± 1.20 v/s 22.15 ± 2.30, $p<0.01$) and increase in expression of anti-apoptotic molecule Bcl2 (52.75 ± 2.57 v/s 33.4 ± 1.75, $p<0.001$) in fC-MSC treated hearts as compared to saline treated hearts (Fig. 7A and B). TUNEL assay also revealed a significant decrease in apoptotic cells in fC-

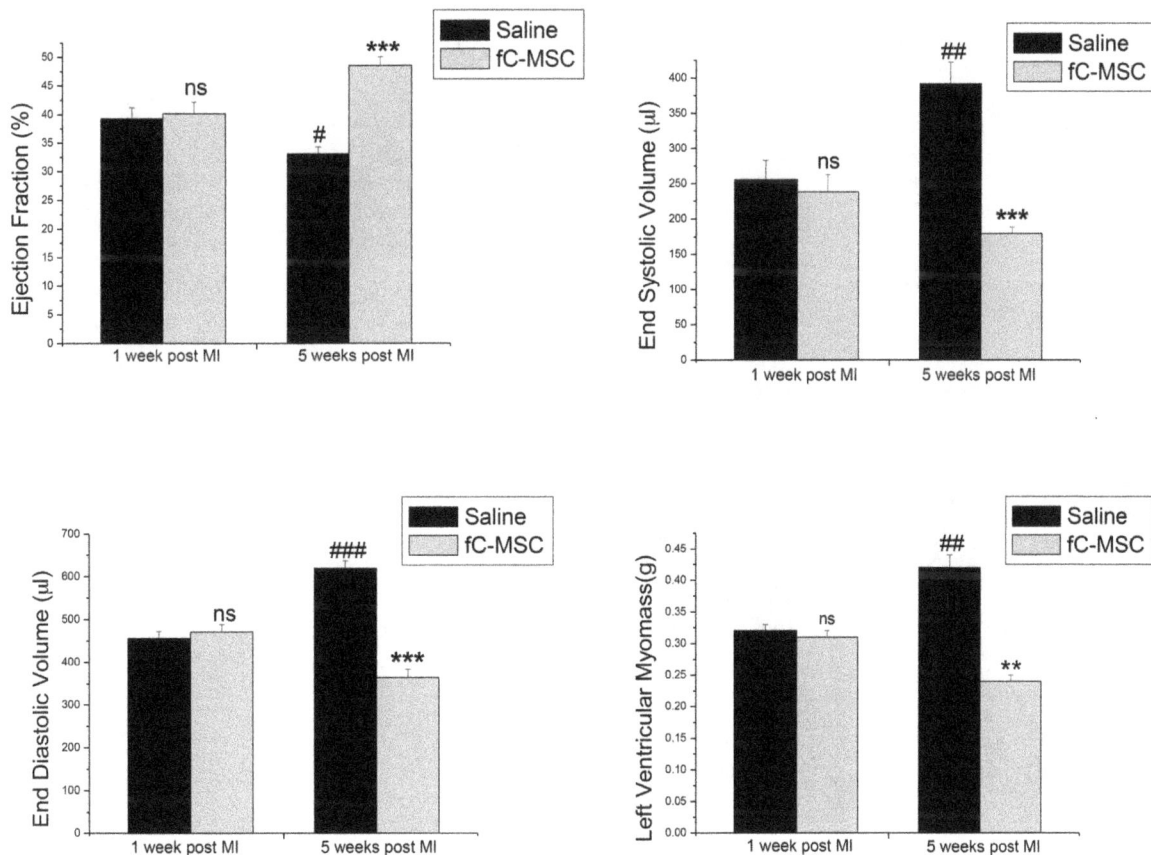

Figure 2. Effect of fC-MSC on LV functions: Bar diagrams showing ejection fraction, end systolic volume, end diastolic volume, and left ventricular myo-mass, measured at 1 week after MI (before fC-MSC therapy) and 4 weeks after fC-MSC therapy using gated SPECT analysis. Values shown are mean ± SEM (n = 6); *P<0.05, *P<0.01, *P<0.001 saline group after cell therapy vs before cell therapy (within the group); #P<0.05, #P<0.01, #P<0.001 fC-MSC group after cell therapy vs before cell therapy (within the group); †P<0.05, †P<0.01, †P<0.001 fC-MSC group (after cell therapy) vs saline group (after cell therapy).

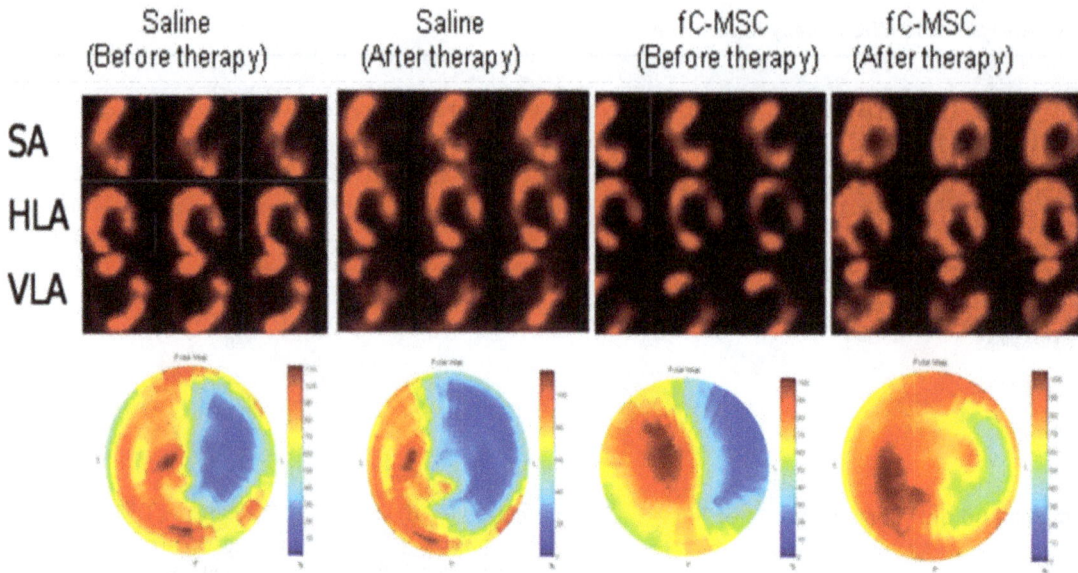

Figure 3. Effect of fC-MSC on LV perfusion: Representative SPECT perfusion images and polar-maps obtained at 1 week after MI (before fC-MSC therapy) and 4 weeks after fC-MSC therapy. (A) Serial 99mTc-sestamibi perfusion images obtained in SPECT short axis (SA), horizontal long axis (HLA), vertical long axis (VLA) of MI hearts treated with saline and fC-MSC (B) Corresponding polar-maps. The perfusion images (A) and the polar-maps (B) show a myocardial-flow defect in the anterolateral wall of left ventricle in both the groups. However, the hearts treated with fC-MSC demonstrate a smaller ischemic lesion (region of deficit) and a better perfusion in the MI segments.

MSC treated group as compared to saline treated group (22 ± 3 vs 11 ± 2, $p < 0.001$) (Fig. 7C and D).

Masson's Trichome staining of tissue slices of MI hearts demonstrated an attenuation of myocardial fibrosis in the fC-MSC treated group, in comparison to the saline treated group (27 ± 1.4 vs $12 \pm 1.15\%$, $p < 0.001$; Fig. 8A and B).

Discussion

The main objective of the present study was to determine the therapeutic capacity of fC-MSC as a novel stem cell type to treat myocardial infarction. We have demonstrated a significant improvement in the cardiac functions following fC-MSC therapy in a rat model of MI using the multi-pinhole SPECT/CT technology. We further demonstrated that intravenously administered fC-MSC were capable of engraftment in the ischemic myocardium and that the engrafted fC-MSC differentiated into all major cells of cardiovascular lineage. In addition, we also documented paracrine, anti-fibrotic and anti-apoptotic effects of fC-MSC.

We have isolated fC-MSC from rat fetal heart, which showed morphologic, phenotypic and differentiation characteristics similar

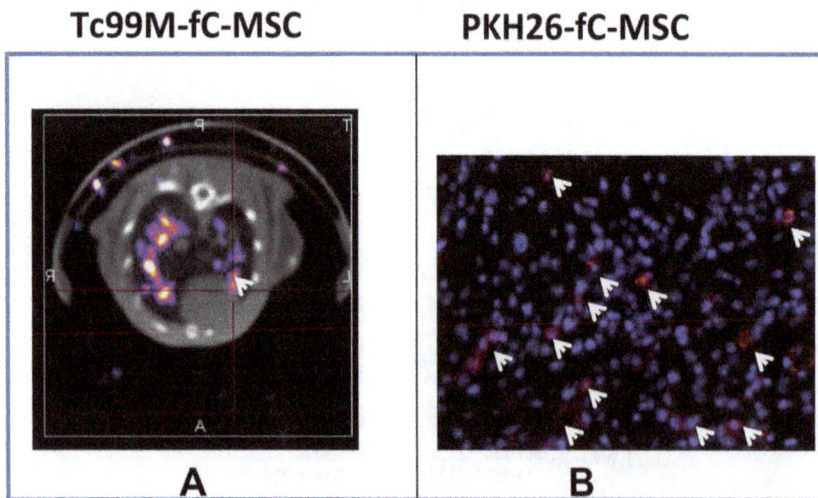

Figure 4. *In vivo* tracking and engraftment of fC-MSC: (A) Representative SPECT/CT images of fC-MSC treated hearts showing focal uptake of fC-MSC labelled with Tc-99 min the anterior mid-ventricular region of the heart6 hoursafter fC-MSC therapy. (B) Representative immunofluorescence photomicrographs of fC-MSC treated hearts showing PKH26 labelled fC-MSC in the injured hearts 4 weeks after fC-MSC therapy.

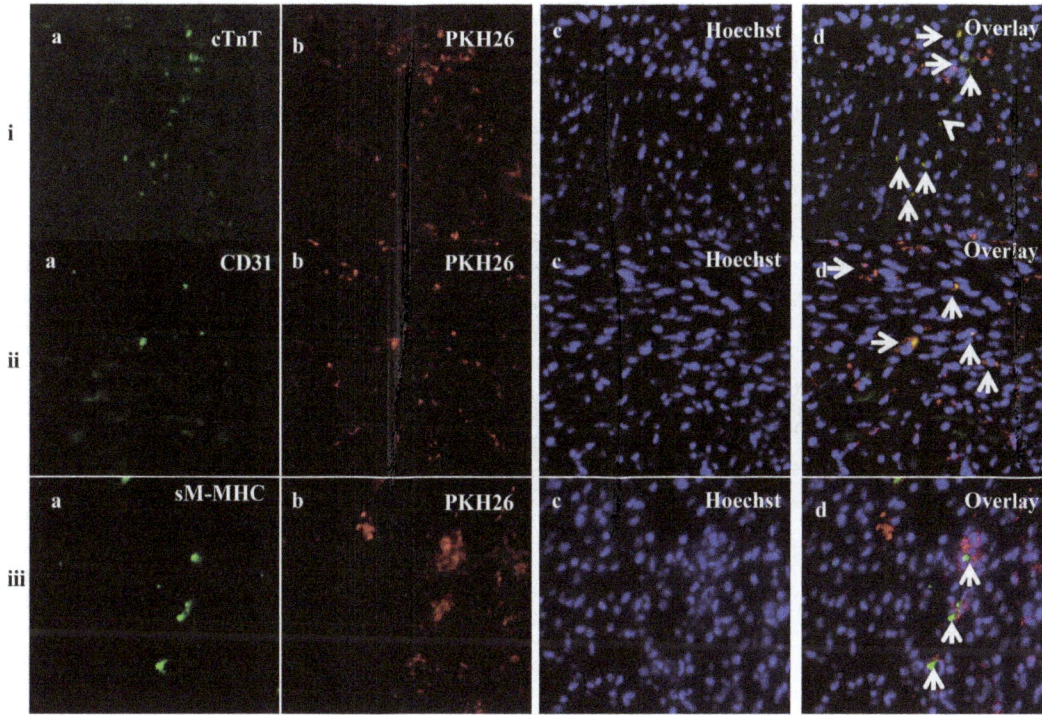

Figure 5. *In vivo* differentiation of fC-MSC in to cardiomyocytes, smooth muscle and endothelial cells. Representative immunofluorescence photomicrographs (40X; 20 μm) of fC-MSC differentiation into cardiovascular lineage cells observed in MI rat models 4 weeks after fC-MSC administration. Row (i): cardiomyocyte showing immunofluorescence staining for (a) Troponin-T; (b) PKH26 dye; (c) Hoechst dye; (d) Overlay of a, b & c images. Row (ii): endothelial cells showing immunofluorescence staining for (a) CD31; (b) PKH26 dye; (c) Hoechst dye; (d) Overlay of a, b & c images. Row (iii): smooth muscle cells showing immunofluorescence staining for (a) SM-MHC; (b) PKH26 dye; (c) Hoechst dye; (d) Overlay of a, b & c images.

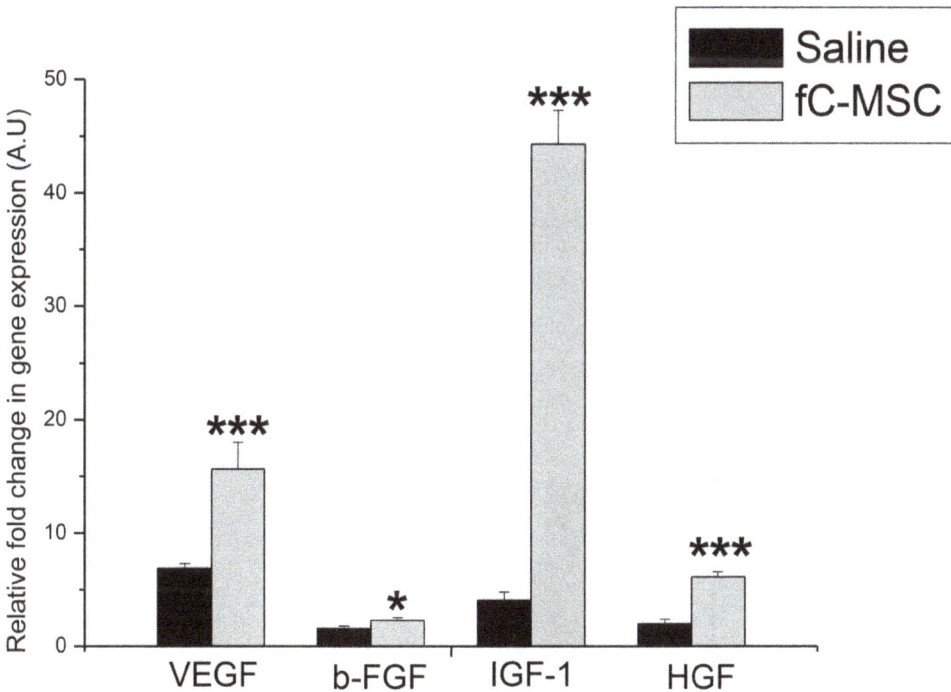

Figure 6. Effect of fC-MSC on gene expression of growth factors in rats with acute myocardial injury. Bar diagrams showing fold change expression of VEGF, b-FGF, IGF-1, and HGF in saline treated and fC-MSC treated ischemic hearts 4 weeks after fC-MSC administration. Values shown are mean ± SEM of 6 experiments, *P<0.05, **P<0.01, ***P<0.001: fC-MSC vs saline treatment.

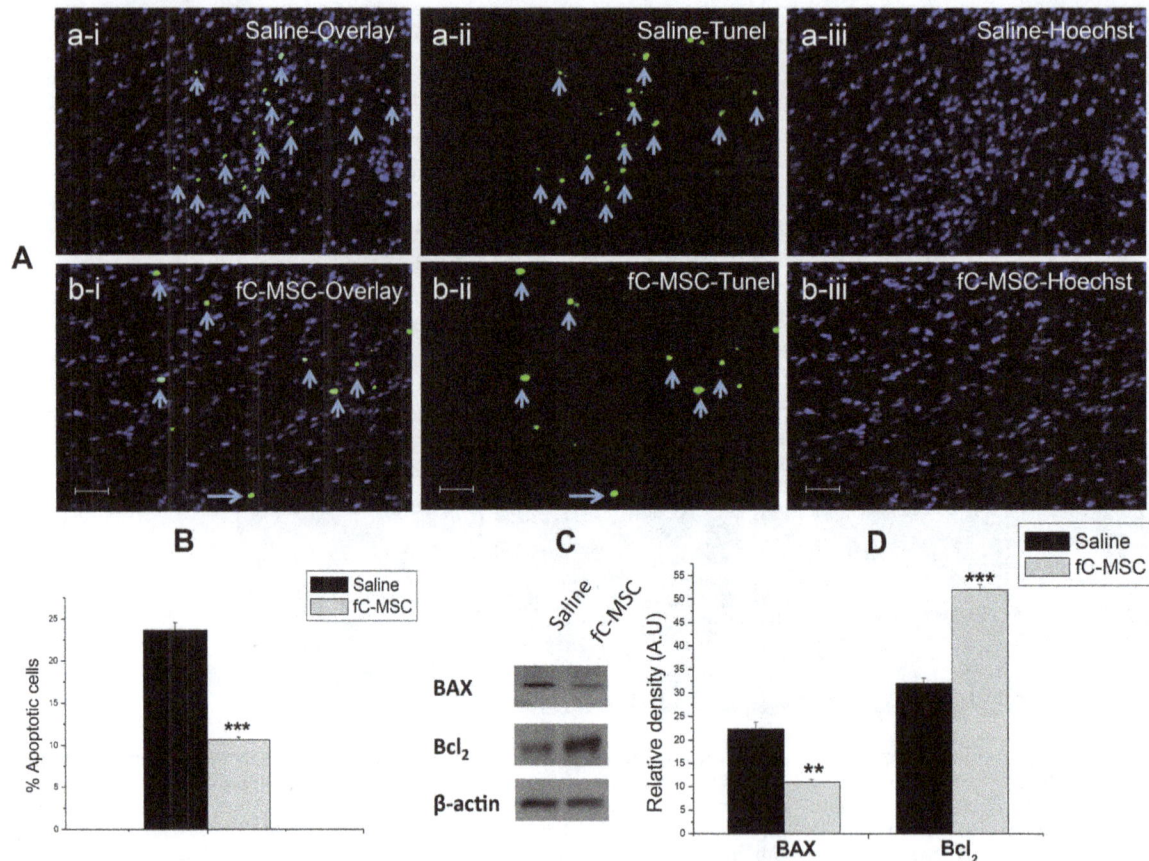

Figure 7. fC-MSC inhibit the apoptosis in rats with acute myocardial injury. (a) Representative immunofluorescence photomicrographs of saline and fC-MSC treated hearts 4 weeks after fC-MSC therapy showing (a-i & b-i) Overlay; (a-ii & b-ii) Tunel positive cells (green dye) counter stain with (a-iii & b-iii) Hoechst dye respectively (B) TUNEL apoptotic index showing significant decrease in apoptotic cells in fC-MSC treated compared to saline treated hearts. Values shown are mean ± SEM (n = 6). **$P<0.01$ fC-MSC treated vs saline treated hearts. (C) Representative immune-blots showing expression of BAX and BCL2 in saline and fC-MSC treated rats and (D) their relative density. Densitometric analysis was applied for comparison of relative protein expression. Values expressed Mean ± SE (n = 6), *$P<0.05$, **$P<0.01$, ***$P<0.001$: fC-MSC treated vs. saline treated group.

to those of typical MSC. We recently reported that fC-MSC are capable of differentiation into all the three cell types of the cardiovascular tri-lineage over successive passages [14]. fC-MSC we identified also displayed cardiovascular transcription signature and could be induced to differentiate into all major cell types of cardiovascular lineage including cardiomyocytes, endothelial cells and smooth muscle cells *in vitro*. In addition these cells exhibited embryonal markers and extensive expansion potential in an undifferentiated state while maintaining expression of TERT and a normal karyotype. The tissue specific commitment and the

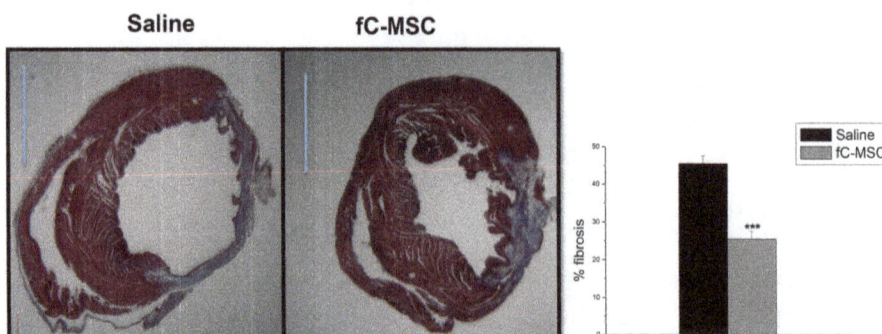

Figure 8. fC-MSC attenuate the myocardial fibrosis in rats with acute myocardial injury (A) Representative heart sections stained with Massons Trichome showing decrease in myocardial fibrosis in fC-MSC treated compared to saline treated hearts 4 weeks after fC-MSC administration. (B) Bar diagram showing percentage of fibrosis in saline treated and fC-MSC treated hearts. Values shown are mean ± SEM (n = 6). **$P<0.05$, **$P<0.01$, ***$P<0.001$: fC-MSC treated vs saline treated hearts.

primitive characteristics of fC-MSC together suggest their potential therapeutic value in cardiovascular regenerative medicine [14,22]. Therefore to further explore the therapeutic effects of fC-MSC, we developed a rat model of cardiac ischemia-reperfusion (IR) injury by transient occlusion of descending coronary artery (LAD). Our model characteristically provides a large trans mural infarction extending from left ventricular apex to the lateral wall and recapitulates the phenotype of ischemic cardiomyopathy. Since acute inflammatory response immediately following reperfusion may result in clearance of administered cells from the infarcted region [23], we decided to inject fC-MSC only after this period of intense inflammation.

The success of any cell-based therapy for myocardial infarction injury is eventually moderated by the improvement of the cardiac functions. In this regard, echocardiography has been widely used to evaluate cardiac function after cell transplantation as it provides a safe, noninvasive, and inexpensive method. However, methods to image heart perfusion and *in vivo* tracking of stem cells in cardiovascular disease using small animal SPECT/CT is an exciting new area of research and has been less understudied. Therefore, in this study we performed multi pinhole 99m Tc-sestamibi gated SPECT/CT for non-invasive evaluation of cardiac perfusion and function in the rat heart at 1 week after MI and 4 weeks after the administration of fC-MSC. Moreover, this technology was also used for initial tracking of fC-MSC 6 h after administration. The systemic administration of fC-MSC attenuated post-infarction LV remodeling, as indicated by a significant improvement in heart function. The functional improvement observed with fC-MSC therapy was comparable with those observed with BMSC [24–26], and cardiac stem cells (CSC) [27], indicating that fC-MSC to be a potential novel cell types for cardiac repair. A recent study has shown adult heart cardiac stem cells expressing c-kit have an enhanced cardiopoietic potential compared to bone marrow-derived MSCs [28]. The same group has recently shown that co-transplantation of MSCs with CSCs reduces scar size and improves heart function post-MI, suggesting that combination therapy represent an effective and robust cell therapeutic strategy [29]. However, further studies to compare the therapeutic efficacy of C-MSC with other cell types are warranted to establish the superiority of fC-MSC over other cell types.

The initial tracking of the administered fC-MSC labeled with Tc99m revealed cells migration to the infarcted myocardium. Similar to our findings Barbash *et al* demonstrated cell migration to the heart in the first hours of systemic administration of MSC labeled with Tc99m in an open chest MI induced rat model. Because of short half-life of radionuclide Tc99m used for labeling of fC-MSC, the fC-MSC can only be tracked up to 6 hours following their injection [30]. To determine homing and retention of administered fC-MSC during follow up period, the cells were labeled with florescence dye, PKH-26. The preferential homing and retention of fC-MSC in the ischemic area of the myocardium might have contributed to efficient healing of the infracted heart. Although the underlying mechanisms remain unclear, ischemic tissue may express specific receptors or ligands to facilitate trafficking, adhesion, and infiltration of fC-MSC to ischemic sites. In the present study, some of engrafted fC-MSC expressed Troponin T, CD31 and sm-MHC, suggesting the potential of fC-MSC to differentiate into cardiomyocytes, endothelial and smooth muscle cells. fC-MSC represent a more primitive progenitor cell population than adult tissue-derived MSC and thus have the differentiation capability along cardiomyogenic lineages. These findings suggest that fC-MSC improved LV functions likely by increasing the local perfusion and also by myocardial regeneration.

Recent cell-based clinical studies suggest paracrine mechanisms for the observed changes in infarct remodeling or functional recovery. VEGF, HGF, and IGF-I have been shown to play important roles in cardiac repair [31]. Moreover, it has been recently reported that local administration of these growth factors in a porcine model of acute myocardial infarction resulted in reduction in the infarct size, increase in the capillary density and improvement in the cardiac contractile function [32]. In agreement with others [33], the present study also showed significant up-regulation in expression of these growth factors in infarcted rat hearts treated with fC-MSC compared to saline treated hearts. All these findings suggest that paracrine factors account for cardio-protective effects of fC-MSC therapy. Furthermore, fC-MSC were found to decrease fibrosis in ischemic myocardium as demonstrated by Masson's Trichome staining. This result is consistent with the previous studies demonstrating that stem cell based therapies in ischemic myocardium improved cardiac function at least in part through an anti-fibrotic effect [34,35]. In addition, we also found that fC-MSC protected cardiomyocyte from I/R induced apoptosis, as shown by a significant down-regulation in expression of pro-apoptotic molecule Bax and up-regulation in expression of anti-apoptotic molecule Bcl2 in infarcted heart following fC-MSC therapy. Detection of TUNEL positive cells also supports these findings. It has been shown that ratio of Bcl-2/Bax could be a key factor in the initiation of apoptosis through a mitochondrial-mediated pathway [36], Thus, fC-MSC may reduce injury through increasing ratio of Bcl-2/Bax.

In conclusion, we have demonstrated that fC-MSC engraft, differentiate into cardiac lineage cells and improve cardiac functions in rat model of MI. In addition, we demonstrate that paracrine, anti-fibrotic and ant-apoptotic effects of fC-MSC may be responsible for observed changes in post-infarct LV remodeling and functional recovery. Taken together, our results suggest that fC-MSC might hold substantial promise to develop a new therapeutic strategy for repair of myocardium after infarction. Finally, in acute myocardial infarction settings, pinhole gated SPECT/CT may be a useful tool to evaluate improvement in cardiac functions and track the cells fate after cell therapy.

Acknowledgments

This work was supported by an extramural grant sanctioned by Department of Biotechnology, Government of India, to Dr S Nityanand (BT/PR6519/MED/14/826/2005). We would like to express gratitude to Dr Sanjay Gambhir, Prof Nuclear Medicine Department, SGPGIMS, Lucknow, for providing the isotope. The authors would also like to extend sincere thanks to Dr. M.M.Godbole, Prof., Department of Endocrinology, SGPGIMS, Lucknow, and his research group, for extending laboratory facilities to perform immunocytochemistry experiments.

Author Contributions

Conceived and designed the experiments: VNSG SJ LP PP MD SN. Performed the experiments: VNSG SJ LP PP. Analyzed the data: VNSG SJ LP SN. Contributed reagents/materials/analysis tools: MD SN. Wrote the paper: VNSG SJ LP PP MD SN.

References

1. Tang XL, Rokosh DG, Guo Y, Bolli R (2010) Cardiac progenitor cells and bone marrow-derived very small embryonic-like stem cells for cardiac repair after myocardial infarction. Circ J 74: 390–404.

2. Hagege AA, Carrion C, Menasche P, Vilquin JT, Duboc D, et al. (2003) Viability and differentiation of autologous skeletal myoblast grafts in ischaemic cardiomyopathy. Lancet 361: 491–492.

3. Kajstura J, Rota M, Whang B, Cascapera S, Hosoda T, et al. (2005) Bone marrow cells differentiate in cardiac cell lineages after infarction independently of cell fusion. Circ Res 96: 127–137.

4. Schuh A, Liehn EA, Sasse A, Hristov M, Sobota R, et al. (2008) Transplantation of endothelial progenitor cells improves neovascularization and left ventricular function after myocardial infarction in a rat model. Basic Res Cardiol 103: 69–77.

5. Rota M, Padin-Iruegas ME, Misao Y, De Angelis A, Maestroni S, et al. (2008) Local activation or implantation of cardiac progenitor cells rescues scarred infarcted myocardium improving cardiac function. Circ Res 103: 107–116.

6. Meyer GP, Wollert KC, Lotz J, Steffens J, Lippolt P, et al. (2006) Intracoronary bone marrow cell transfer after myocardial infarction: eighteen months' follow-up data from the randomized, controlled BOOST (BOne marrOw transfer to enhance ST-elevation infarct regeneration) trial. Circulation 113: 1287–1294.

7. Martin-Rendon E, Brunskill SJ, Hyde CJ, Stanworth SJ, Mathur A, et al. (2008) Autologous bone marrow stem cells to treat acute myocardial infarction: a systematic review. Eur Heart J 29: 1807–1818.

8. Eisen HJ (2008) Skeletal myoblast transplantation: no MAGIC bullet for ischemic cardiomyopathy. Nat Clin Pract Cardiovasc Med 5: 520–521.

9. Silva GV, Litovsky S, Assad JA, Sousa AL, Martin BJ, et al. (2005) Mesenchymal stem cells differentiate into an endothelial phenotype, enhance vascular density, and improve heart function in a canine chronic ischemia model. Circulation 111: 150–156.

10. Hare JM, Traverse JH, Henry TD, Dib N, Strumpf RK, et al. (2009) A randomized, double-blind, placebo-controlled, dose-escalation study of intravenous adult human mesenchymal stem cells (prochymal) after acute myocardial infarction. J Am Coll Cardiol 54: 2277–2286.

11. Chen PM, Yen ML, Liu KJ, Sytwu HK, Yen BL (2011) Immunomodulatory properties of human adult and fetal multipotent mesenchymal stem cells. J Biomed Sci 18: 49.

12. Ramkisoensing AA, Pijnappels DA, Askar SF, Passier R, Swildens J, et al. (2011) Human embryonic and fetal mesenchymal stem cells differentiate toward three different cardiac lineages in contrast to their adult counterparts. PLoS One 6: e24164.

13. Pelekanos RA, Li J, Gongora M, Chandrakanthan V, Scown J, et al. (2012) Comprehensive transcriptome and immunophenotype analysis of renal and cardiac MSC-like populations supports strong congruence with bone marrow MSC despite maintenance of distinct identities. Stem Cell Res 8: 58–73.

14. Srikanth GV, Tripathy NK, Nityanand S (2013) Fetal cardiac mesenchymal stem cells express embryonal markers and exhibit differentiation into cells of all three germ layers. World J Stem Cells 5: 26–33.

15. Beeres SL, Bengel FM, Bartunek J, Atsma DE, Hill JM, et al. (2007) Role of imaging in cardiac stem cell therapy. J Am Coll Cardiol 49: 1137–1148.

16. Morrison AR, Sinusas AJ (2009) New molecular imaging targets to characterize myocardial biology. Cardiol Clin 27: 329–344, Table of Contents.

17. Wu JC, Abraham MR, Kraitchman DL (2010) Current perspectives on imaging cardiac stem cell therapy. J Nucl Med 51 Suppl 1: 128S–136S.

18. Tsui BM, Kraitchman DL (2009) Recent advances in small-animal cardiovascular imaging. J Nucl Med 50: 667–670.

19. Garikipati Venkata Naga Srikanth PP, Naresh Kumar Tripathy, Madhu Dikshit, Soniya Nityanand (2009) Establishment of a rat model of myocardial infarction with a high survival rate: A suitable model for evaluation of efficacy of stem cell therapy Journal of stem cells and regenerative medicine 5: 30–36.

20. Tran N, Li Y, Maskali F, Antunes L, Maureira P, et al. (2006) Short-term heart retention and distribution of intramyocardial delivered mesenchymal cells within necrotic or intact myocardium. Cell Transplant 15: 351–358.

21. Dussault AA, Pouliot M (2006) Rapid and simple comparison of messenger RNA levels using real-time PCR. Biol Proced Online 8: 1–10.

22. Garikipati Venkata Naga Srikanth NKT, Soniya Nityanand (2012) Isolation and characterization of cardiac mesenchymal stem cells from rat foetal hearts. International Journal of Regenerative Medicine 1: 1–8.

23. Krishnamurthy P, Thal M, Verma S, Hoxha E, Lambers E, et al. (2011) Interleukin-10 deficiency impairs bone marrow-derived endothelial progenitor cell survival and function in ischemic myocardium. Circ Res 109: 1280–1289.

24. Berry MF, Engler AJ, Woo YJ, Pirolli TJ, Bish LT, et al. (2006) Mesenchymal stem cell injection after myocardial infarction improves myocardial compliance. Am J Physiol Heart Circ Physiol 290: H2196–2203.

25. Schuleri KH, Amado LC, Boyle AJ, Centola M, Saliaris AP, et al. (2008) Early improvement in cardiac tissue perfusion due to mesenchymal stem cells. Am J Physiol Heart Circ Physiol 294: H2002–2011.

26. Halkos ME, Zhao ZQ, Kerendi F, Wang NP, Jiang R, et al. (2008) Intravenous infusion of mesenchymal stem cells enhances regional perfusion and improves ventricular function in a porcine model of myocardial infarction. Basic Res Cardiol 103: 525–536.

27. Dawn B, Stein AB, Urbanek K, Rota M, Whang B, et al. (2005) Cardiac stem cells delivered intravascularly traverse the vessel barrier, regenerate infarcted myocardium, and improve cardiac function. Proc Natl Acad Sci U S A 102: 3766–3771.

28. Oskouei BN, Lamirault G, Joseph C, Treuer AV, Landa S, et al. (2012) Increased potency of cardiac stem cells compared with bone marrow mesenchymal stem cells in cardiac repair. Stem Cells Transl Med 1: 116–124.

29. Williams AR, Hatzistergos KE, Addicott B, McCall F, Carvalho D, et al. (2013) Enhanced effect of combining human cardiac stem cells and bone marrow mesenchymal stem cells to reduce infarct size and to restore cardiac function after myocardial infarction. Circulation 127: 213–223.

30. Barbash IM, Chouraqui P, Baron J, Feinberg MS, Etzion S, et al. (2003) Systemic delivery of bone marrow-derived mesenchymal stem cells to the infarcted myocardium: feasibility, cell migration, and body distribution. Circulation 108: 863–868.

31. Huikuri HV, Kervinen K, Niemela M, Ylitalo K, Saily M, et al. (2008) Effects of intracoronary injection of mononuclear bone marrow cells on left ventricular function, arrhythmia risk profile, and restenosis after thrombolytic therapy of acute myocardial infarction. Eur Heart J 29: 2723–2732.

32. Yoon YS, Wecker A, Heyd L, Park JS, Tkebuchava T, et al. (2005) Clonally expanded novel multipotent stem cells from human bone marrow regenerate myocardium after myocardial infarction. J Clin Invest 115: 326–338.

33. Iso Y, Spees JL, Serrano C, Bakondi B, Pochampally R, et al. (2007) Multipotent human stromal cells improve cardiac function after myocardial infarction in mice without long-term engraftment. Biochem Biophys Res Commun 354: 700–706.

34. Christoforou N, Oskouei BN, Esteso P, Hill CM, Zimmet JM, et al. (2010) Implantation of mouse embryonic stem cell-derived cardiac progenitor cells preserves function of infarcted murine hearts. PLoS One 5: e11536.

35. Mias C, Lairez O, Trouche E, Roncalli J, Calise D, et al. (2009) Mesenchymal stem cells promote matrix metalloproteinase secretion by cardiac fibroblasts and reduce cardiac ventricular fibrosis after myocardial infarction. Stem Cells 27: 2734–2743.

36. Fehlberg S, Gregel CM, Goke A, Goke R (2003) Bisphenol A diglycidyl ether-induced apoptosis involves Bax/Bid-dependent mitochondrial release of apoptosis-inducing factor (AIF), cytochrome c and Smac/DIABLO. Br J Pharmacol 139: 495–500.

2

Photodynamic Therapy with 5-Aminolevulinic acid (ALA) Impairs Tumor Initiating and Chemo-Resistance Property in Head and Neck Cancer-Derived Cancer Stem Cells

Chuan-Hang Yu[1,2]*, Cheng-Chia Yu[1,2,3]*

1 School of Dentistry, Chung Shan Medical University, Taichung, Taiwan, 2 Department of Dentistry, Chung Shan Medical University Hospital, Taichung, Taiwan, 3 Institute of Oral Sciences, Chung Shan Medical University, Taichung, Taiwan

Abstract

Background: Head and neck cancer (HNC) ranks the fourth leading malignancy and cancer death in male population in Taiwan. Despite recent therapeutic advances, the prognosis for HNC patients is still dismal. New strategies are urgently needed to improve the chemosensitization to conventional chemotherapeutic drugs and clinical responses of HNC patients. Studies have demonstrated that topical 5-aminolevulinic acid-mediated photodynamic therapy (ALA-PDT) is being used in the treatment of various human premalignant and malignant lesions with some encouraging clinical outcomes. However, the molecular mechanisms of ALA-PDT in the therapeutic effect in HNC tumorigenesis and whether ALA-PDT as chemosensitizer for HNC treatment remain unclear. Accumulating data support cancer stem cells (CSCs) contributes chemo-resistance in HNC. Based on the previous studies, the purpose of the study is to investigate the effect of ALA-PDT on CSCs and chemosensitization property in HNC.

Methodology/Principal Finding: CSCs marker ALDH1 activity of HNC cells with ALA-PDT treatment as assessed by the Aldefluor assay flow cytometry analysis. Secondary Sphere-forming self-renewal, stemness markers expression, and invasiveness of HNC-CSCs with ALA-PDT treatment were presented. We observed that the treatment of ALA-PDT significantly down-regulated the ALDH1 activity and CD44 positivity of HNC-CSCs. Moreover, ALA-PDT reduced self-renewal property and stemness signatures expression (Oct4 and Nanog) in sphere-forming HNC-CSCs. ALA-PDT sensitized highly tumorigenic HNC-CSCs to conventional chemotherapies. Lastly, synergistic effect of ALA-PDT and Cisplatin treatment attenuated invasiveness/colongenicity property in HNC-CSCs.

Conclusion/Significance: Our results provide insights into the clinical prospect of ALA-PDT as a potential chemo-adjuvant therapy against head and neck cancer through eliminating CSCs property.

Editor: Shree Ram Singh, National Cancer Institute, United States of America

Funding: This study is supported by research grant from National Science Council (100-2632-B-040-001-MY3, 101-2314-B-040-016, 102-2628-B-040 -001 -MY3).The funders had no role in study design, data collection and analysis, decision to publish, or preparation of the manuscript.

Competing Interests: The authors have declared that no competing interests exist.

* E-mail: d94422004@ntu.edu.tw (C-HY); ccyu@csmu.edu.tw (C-CY)

Introduction

Head and neck cancer (HNC) is one of the most common cancers in the world and one of causes of cancer-related death due to frequent recurrence after chemotherapy resistance [1]. Despite improvements in the diagnosis and management of HNC, long-term survival rates have improved only marginally over the past decade [1]. New drugs or strategies are urgently needed to improve the chemosensitization to conventional chemotherapeutic drugs and clinical responses of HNC patients [2].

Recent studies have pointed out resistance to conventional radiotherapies/chemotherapies treatment is a major clinical criteria for characterizing cancer stem cells (CSCs) with self-renewal and highly tumorigenenic capacity in malignant tumors including HNC[2–4] Both intracellular ALDH1+ and CD44+ cells have been proposed to exhibit CSCs properties and have been used as CSCs functional markers for head and neck cancer-derived cancer stem cells (HNC-CSCs)[5–7]. Nevertheless, searching an effective approach targeting HNC-CSCs to improve HNC -related malignancies is urgently required [7].

Photodynamic therapy (PDT) involves two individually non-toxic components, light and photosensitizer [8]. PDT is a new treatment and holds considerable promise for many solid tumors[9]. Studies have demonstrated that topical 5-aminolevulinic acid-mediated PDT (ALA-PDT) is being used in the treatment of various human premalignant and malignant lesions with some encouraging clinical outcomes [10–13]. PDT might have the potential of inhibiting the metastasis of incompletely treated HNC [14]. In a mice Lewis lung carcinoma model, ALA-PDT decreased the metastasis of cancer cells in vivo[15]. In addition, ALA-PDT increases apoptotic ability of oral cancer cells through NF-κB/JNK signaling [16]. ALA-PDT also abrogated migration capacity of oral cancer cells by down-regulation of FAK and ERK [17].

Based on the previous studies, the purpose of the study is to investigate the effect of ALA-PDT on chemosensitization and CSCs property in HNC. Overall, we first demonstrated ALA-PDT effectively reduce CSC-like property including ALDH1 activity, CD44 positivity, self-renewal, invasion, and enhanced the chemosensitivity in HNC. ALA-PDT would be a potential chemo-adjuvant therapy and may be beneficial for anti-CSCs treatment in HNC.

Materials and Methods

Primary cultivated cells from HNC tissues

Surgical tissue specimens from HNC patients were collected after obtaining written informed consent and this study was approved by The Institutional Review Board in Chung Shan Medical University Hospital (CSMUH No: CS10249). Primary HNC cells were established as previously described [6,18]. In brief, after surgical removal of the HNC tissues, the tissues were washed 3 times in glucose containing HBSS, then the samples were sliced at a thickness of 300 μm and the sliced tissues were immersed in 0.1% (w/w) collagenase containing glucose containing HBSS for 15 minutes at 37°C and subjected to rotation shaker shaking at 125 rpm. HNC primary culture were cultured in DMEM (Invitrogen, Carlsbad, CA, USA) with 10% fetal bovine serum, supplemented with 1 mM sodium pyruvate, non-essential amino acids, 2 mM L-glutamine, 100 units/mL penicillin, and 100 μg/mL streptomycin under standard culture conditions (37°C, 95% humidified air, 5% CO_2).

Chemicals

5-ALA was obtained purchased from Sigma (St. Louis, MO, USA). Just before use, 5-ALA was further diluted in culture medium to appropriate final concentrations.

HNC cell lines cultivation and 5-ALA-based photodynamic therapy

For ALA-PDT studies, the cells were incubated with 5-ALA for 3 hours, and then irradiated with red light of 635±5 nm at various doses.

Enrichment of sphere-forming HNC cancer stem cells (HNC-CSCs)

Spheres from HNC cells were isolated in medium consisting of serum-free DMEM/F12 medium (GIBCO), N2 supplement (GIBCO), 10 ng/mL human recombinant basic fibroblast growth factor (basic FGF), and 10 ng/mL epidermal growth factor (EGF) (R&D Systems, Minneapolis, MN). Cells were plated at a density of 10^4 live cells/10-mm low attachment dishes, and the medium was changed every other day until the tumor sphere formation was observed in about 1 week [2].

Aldefluor assay

Aldefluor assay kit is purchased from StemCell Technologies, Inc. (Vancouver, BC, Canada) 1×10^5 cells will be suspended in 50 μl of assay buffer and added Aldefluor to final concentration of 1 μM. For ALDH1 inhibitor control, DEAB will be added to final concentration of 150 μM. Cells will be then incubated at 37°C for 45 min and stained with 7-AAD on ice for further 5 min. After washed with PBS, green fluorescence positive cells in live cells (7AAD-) will be analyzed by flow cytometry (FACSCalibur™, BD Bioscience) by comparing the fluorescence intensity of DEAB treated sample and these cells will be represented as cells with high ALDH activity (ALDH+ cells)[2].

SYBR Real-time reverse transcription-polymerase chain reaction (RT-PCR)

Total RNA of cells was purified using Trizol reagent (Invitrogen, Carlsbad, CA, USA) according to the manufacturer's protocol. Briefly, the total RNA (1 μg) of each sample was reversely transcribed by Superscript II RT (Invitrogen, Carlsbad, CA, USA). Then, the amplification was carried out in a total volume of 20 μl containing 0.5 μM of each primer, 4 mM $MgCl_2$, 2 μl LightCycler™–FastStart DNA Master SYBR green I (Roche Molecular Systems, Alameda, CA, USA) and 2 μl of 1:10 diluted cDNA. The *GAPDH* housekeeping gene was amplified as a reference standard. *GAPDH* primers were designed: *GAPDH* (forward): GGGCCAAAAGGGTCATCATC (nt 414–434, GenBank accession no. BC059110.1), *GAPDH* (reverse): AT-GACCTTGCCCACAGCCTT (nt 713–733). PCR reactions were prepared in duplicate and heated to 95°C for 10 minutes followed by 40 cycles of denaturation at 95°C for 10 seconds, annealing at 55°C for 5 seconds, and extension at 72°C for 20 seconds. All PCR reactions were performed in triplicate. Standard curves (cycle threshold values versus template concentration) were prepared for each target gene and for the endogenous reference (*GAPDH*) in each sample. To confirm the specificity of the PCR reaction, PCR products were electrophoresed on a 1.2% agrose gel [3]. Primer sequences are listed in Table 1[3].

Western blot

The extraction of proteins from cells and western blot analysis were performed as described. Samples (15 μL) were boiled at 95°C for 5 min and separated by 10% SDS-PAGE. The proteins were wet-transferred to Hybond-ECL nitrocellulose paper (Amersham, Arlington Heights, IL, USA). The following primary antibodies were used: rabbit anti–human Oct4, rabbit anti–human Nanog (Santa Cruz Biotechnology, Santa Cruz, CA, USA); rabbit anti-GAPDH (MDBio, Inc., Taipei, Taiwan); and mouse anti–β-actin (Novus Biologicals, Littleton, CO, USA). Immunoreactive protein bands were detected by the ECL detection system (Amersham Biosciences Co., Piscataway, NJ, USA).

In vitro cell invasion Assay

For transwell migration assays, 2×10^5 cells were plated into the top chamber of a transwell (Corning, Acton, MA, USA) with a porous membrane (8.0 μm pore size). Cells were plated in medium with lower serum (0.5% FBS), and medium supplemented with higher serum (10% FBS) was used as a chemoattractant in the lower chamber. The cells were incubated for 24 hour at 37°C and cells that did not migrate through the pores were removed by a cotton swab. Cells on the lower surface of the membrane were stained with Hoechst 33258 (Sigma-Aldrich Co., St. Louis, MO, USA) to show the nuclei; fluorescence was detected at a magnification of 100x using a fluorescence microscope (Carl Zeiss, Oberkochen, Germany). The number of fluorescent cells in a total of five randomly selected fields was counted. In vitro cell invasion analysis was as described previously [19].

Soft agar colony forming assay

Each well (35 mm) of a six-well culture dish was coated with 2 ml bottom agar (Sigma-Aldrich Co., St. Louis, MO, USA) mixture (DMEM, 10% (v/v) FCS, 0.6% (w/v) agar). After the bottom layer was solidified, 2 ml top agar-medium mixture (DMEM, 10% (v/v) FCS, 0.3% (w/v) agar) containing 2×10^4 cells was added, and the dishes were incubated at 37°C for 4 weeks. Plates were stained with 0.005% Crystal Violet then the colonies were counted. The number of total colonies with a

Table 1. The sequences of the primers for quantitative RT-PCR.

Gene (Accession No.)	Primer Sequence (5' to 3')	Product size (bp)	Tm (°C)
Oct4 (NM_002701)	F: GTGGAGAGCAACTCCGATG R: TGCTCCAGCTTCTCCTTCTC	86	60
Nanog (NM_024865)	F: ATTCAGGACAGCCCTGATTCTTC R: TTTTTGCGACACTCTTCTCTGC	76	60
GAPDH (NM_002046)	F: CATCATCCCTGCCTCTACTG R: GCCTGCTTCACCACCTTC	180	60

diameter ≥100 µm was counted over five fields per well for a total of 15 fields in triplicate experiments [3].

Statistical analysis

Statistical Package of Social Sciences software (version 13.0) (SPSS, Inc., Chicago, IL) was used for statistical analysis. Student's t test was used to determine statistical significance of the differences between experimental groups; p values less than 0.05 were considered statistically significant. The level of statistical significance was set at 0.05 for all tests.

Results

Inhibition of ALDH1 activity in HNC cells under ALA-PDT treatment

ALDH1, a cytosolic isoenzyme that is responsible for the oxidation of retinol to retinoic acid during early stem cell differentiation have both been shown to be CSC markers in head and neck cancer[20,21]. First, we used the in vitro cell-based ALDH activity assay system, which has been demonstrated as a HNC-CSCs marker, to examine the effects of ALA-PDT on ALDH1 activity in HNC cell lines and primary cultivated HNC cells by flow cytometry described in material and methods. Our

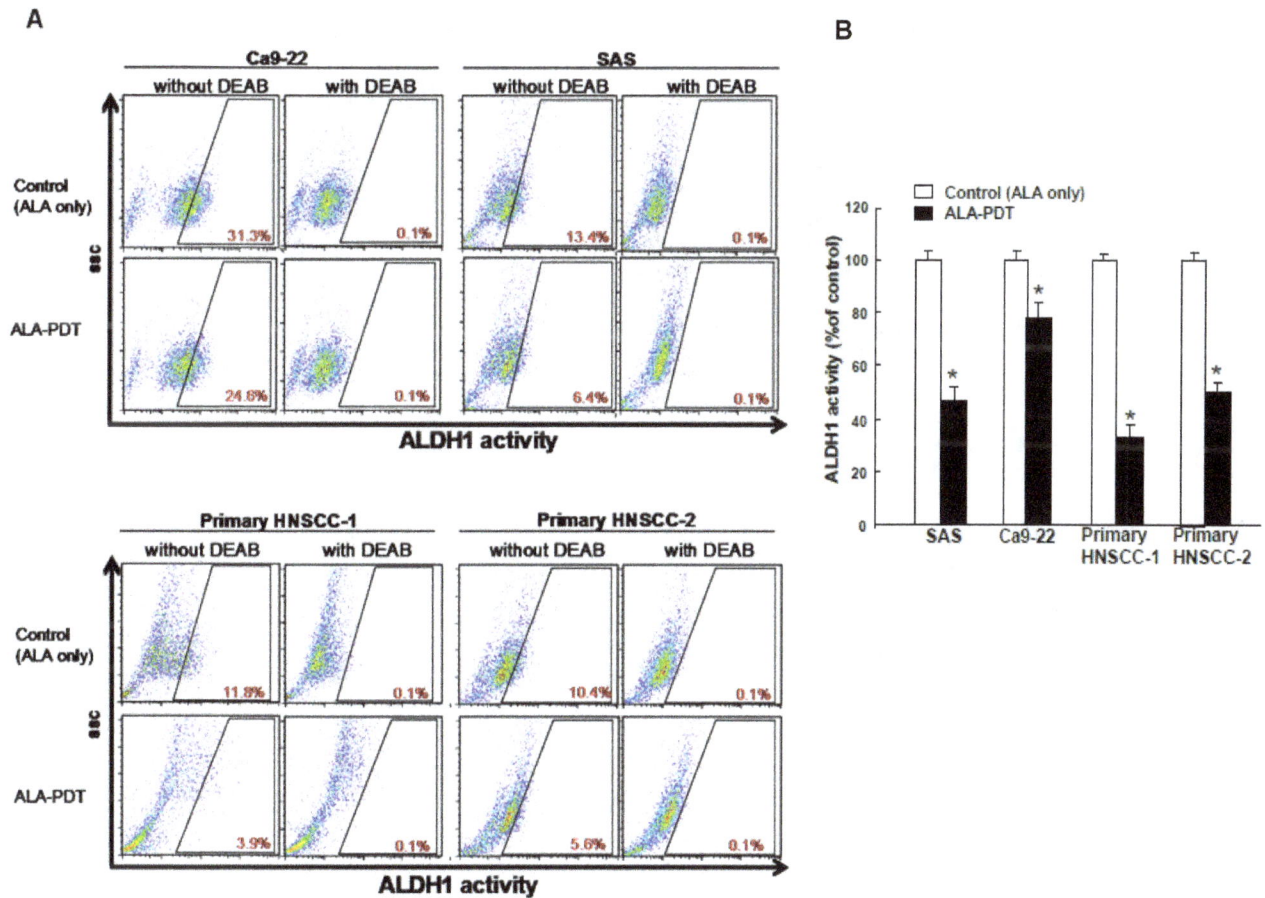

Figure 1. Effects of 5-aminolevulinic acid-mediated photodynamic therapy (ALA-PDT) on ALDH1 activity in HNSCC cell lines and primary HNC cells. (A) HNC cell lines (SAS and Ca9-22) and primary HNC cells (HNC-1 and HNC-2) were seeded as 1×10⁶ cells/well in 6-well-plate and then pretreated 1 mM ALA for 3 hr followed by PDT with red light at a dose of 4 J/cm2 or ALA only treatment control group. ALDH+ cells were determined by Alderfluor assay and viable cells (7-AAD negative) were used for analysis. DEAB-treated cells were used as negative control. (B) The quantification results were shown in bar graph. The experiments were repeated three times and representative results were shown. Results are presented as means ± SD *, p<0.05.

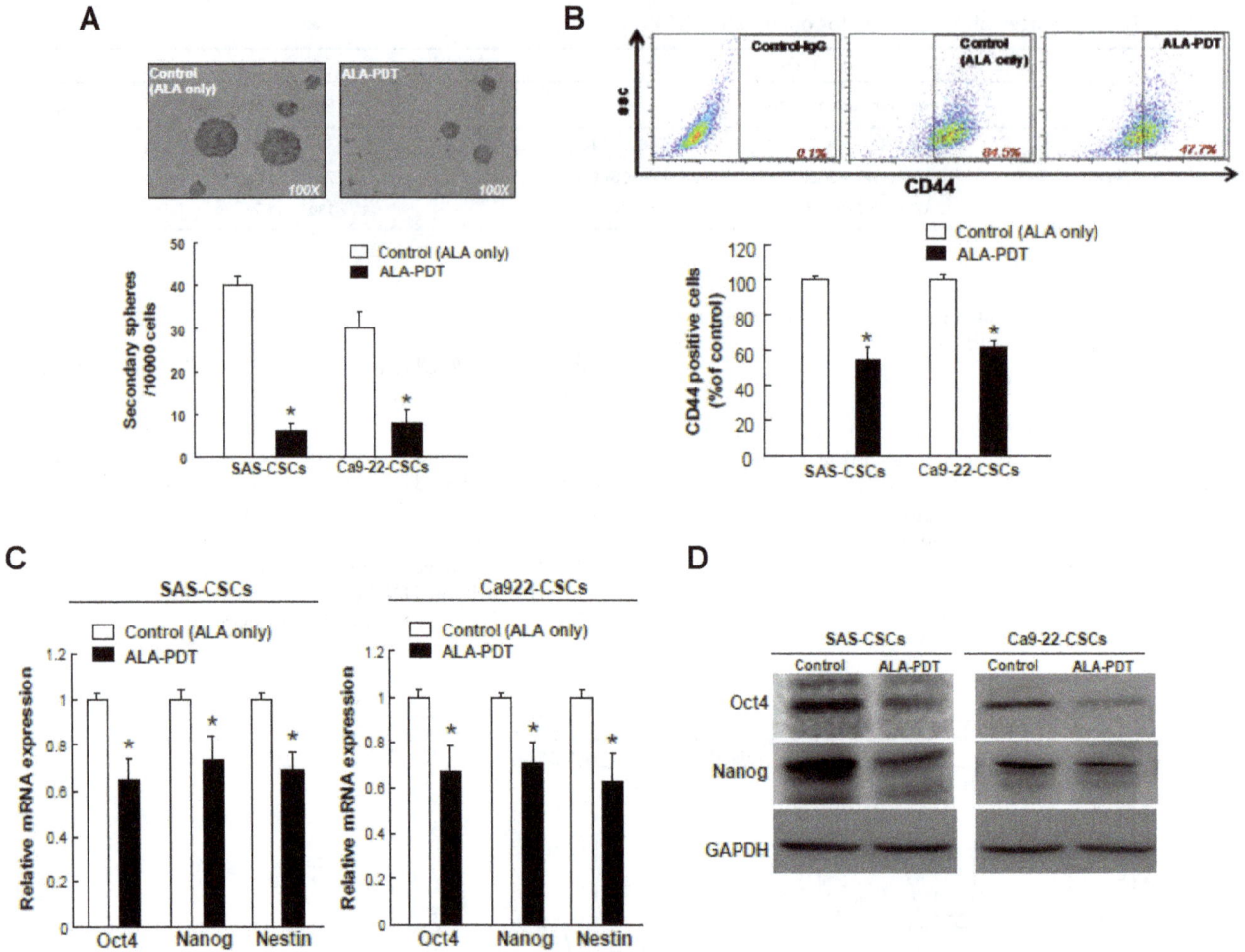

Figure 2. ALA-PDT suppressed secondary sphere-forming capability, CD44 activity, and stemness marker expression in HNC-CSCs. (A) For self-renewal analysis, single cell suspension was obtained from primary HNC-CSCs spheres by accutase digestion and secondary sphere formation capacity was determined with primary sphere culture procedure except the plating cell density as 10000 cells/ml. (B) The expression of CD44 positivity of control and ALA-PDT-treated HNC-CSCs was determined by flow cytometry analysis. Data shown here are the mean ± SD of three independent experiments. *, p<0.05 vs. Control. SAS or Ca9-22-derived HNC-CSCs treated with control or ALA-PDT and analyzed transcripts and protein level of Oct-4 and Nanog by real-time RT-PCR (C) and immunoblotting analysis (D), respectively. *, p<0.05 vs. Control.

data suggested ALA-PDT treatment significantly decreases ALDH1 activity of HNC cells (Fig. 1).

ALA-PDT effectively eliminates self-renewal capacity, CD44 positivity, and stemness signatures in HNC-CSCs

Previously, we examined the successful sphere formation over serial passages of culture, one of indexes for evaluating the persistent self-renewal property of cancer stem cells, and showed that the self-renewal capacity of all HNC-CSCs[22]. To investigate the effect of ALA-PDT in maintaining self-renewal HNC-CSCs, we evaluated the secondary sphere-forming ability with ALA-PDT treatment in HNC cells. Treatment of HNC cells with ALA-PDT interfered with spheres body size and numbers of HNC-CSCs (Fig. 2A). Flow cytometry analysis of CD44 expression indicated that the ALA-PDT treatment reduced the percentage of both CD44+ cells in HNC-CSCs (Fig. 2B). To further determine whether the reduction in tumor sphere formation efficiency with ALA-PDT treatment was due to decreased stemness markers expression, stemness genes (Oct-4 and Nanog) of sphere-forming HNC-CSCs with control and ALA-

PDT treatment were determined by western blot analysis. The results confirmed that ALA-PDT -treated HNC-CSCs markedly reduced the expression transcript (Fig. 2C) and protein level (Fig. 2D) of stemness genes.

ALA-PDT treatment delivery enhances the efficacy of chemotherapy in HNC-CSCs

CSCs are relatively resistant to chemotherapy[6]. The findings that ALA-PDT treatment regulated CSCs properties suggested a role for ALA-PDT as a potential chemo-adjuvant therapy. Cell viability assays showed that the cytotoxic effect on HNC-CSCs to cisplatin and fluorouracil (5-FU) was significantly increased with the ALA-PDT combination treatment (Fig. 3A). Flow cytometry analysis of multidrug resistance (MDR) genes indicated that the decreased chemoresistance by ALA-PDT treatment could be attributed to the reduced expression of ABCG2 (Fig. 3B). These data suggest that ALA-PDT treatment ameliorated the drug resistance of HNC-CSCs to cisplatin and fluorouracil (5-FU) treatment through ABCG2 down-regulation.

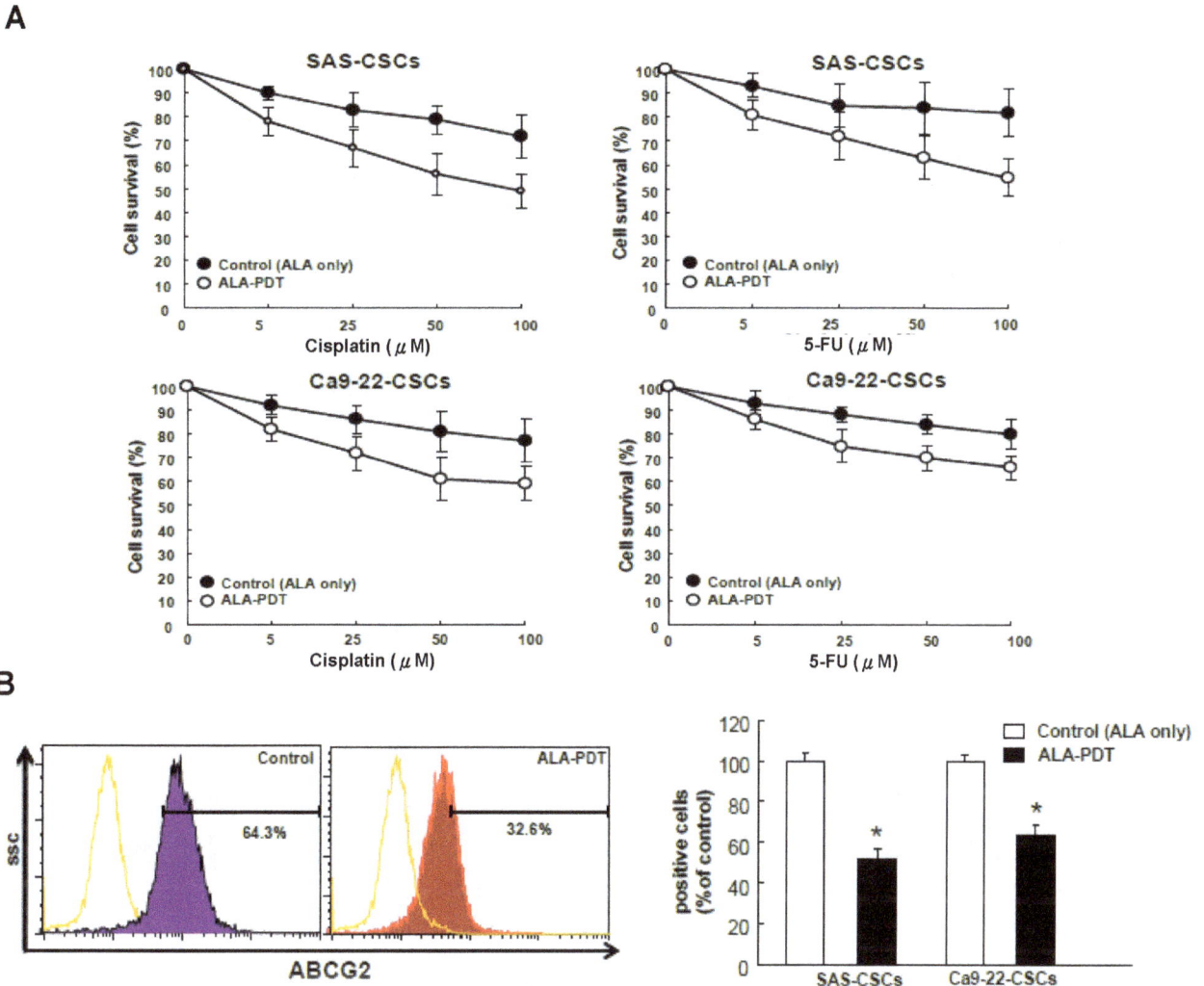

Figure 3. ALA-PDT treatment enhances the efficacy of chemotherapy. (A) HNC-CSCs with ALA-PDT or control treatment were subjected to treatment with different concentrations of cisplatin or 5-FU. Cell viability was determined by MTT assay. (B) Flow cytometry analysis of the expression level of drug-resistant marker ABCG2 expression in the HNC-CSCs as indicated.

Co-administration of ALA-PDT treatment and cisplatin abrogated invasiveness and clonogenicity of HNC-CSCs

Since CSCs appear to play a significant role in promoting tumor initiating activity[3], we sought to measure the synergistic effects of ALA-PDT combined with cisplatin treatment on invasion/clonogenicity of HNC-CSCs. Single cell suspension of ALA-PDT-treated HNC-CSCs were treated with or without cisplatin treatment was used for analysis of their invasion/clonogenicity *in vitro* as described in Materials and Methods section. Treatment with cisplatin alone did not affect the invasion ability in HNC-CSCs (Fig. 4A), the combination of ALA-PDT and cisplatin treatment enhanced the efficacy of these treatments (Fig. 4A). Meanwhile, similar synergistic effect of ALA-PDT treatment and chemo-treatment was also observed in colony formation assay (Fig. 4B). The combination treatment showed a synergistic effect in abrogating clonogenicity in HNC-CSCs. These data indicate that the effectiveness of cisplatin chemotherapy treatment on HNC-CSCs can be improved with ALA-PDT treatment.

Discussion

CSCs are considered to be responsible for the initiation, propagation and metastasis of tumors[23]. Importantly, the existence of CSCs might explain cancer recurrences, even after clinical treatment with either radiotherapy or chemotherapy on cancer patients[24]. Therefore, searching the novel treatment strategy targeting CSCs hopefully provide us with new therapeutic approaches. In the present report, we firstly showed ALA-PDT provided a therapeutic effect in HNC-CSCs by inhibiting the CSCs-like properties of head and neck cancer, such as the stemness signature, migration ability, and chemoresistance. To the best of our knowledge, this is the first study to demonstrate the critical role of an ALA-PDT-based therapy in targeting HNC-derived CSC-like cells and in blocking HNC-CSCs-mediated tumor initiating activity.

Epithelial-mesenchymal transition (EMT) is a process critical for appropriate embryonic development, and it is also re-engaged in adults during tumorigenesis[25]. EMT is widely accepted as one of the CSCs properties, [26] and Oct4/Nanog signaling has been demonstrated to be involved in the regulation of EMT and

A

B

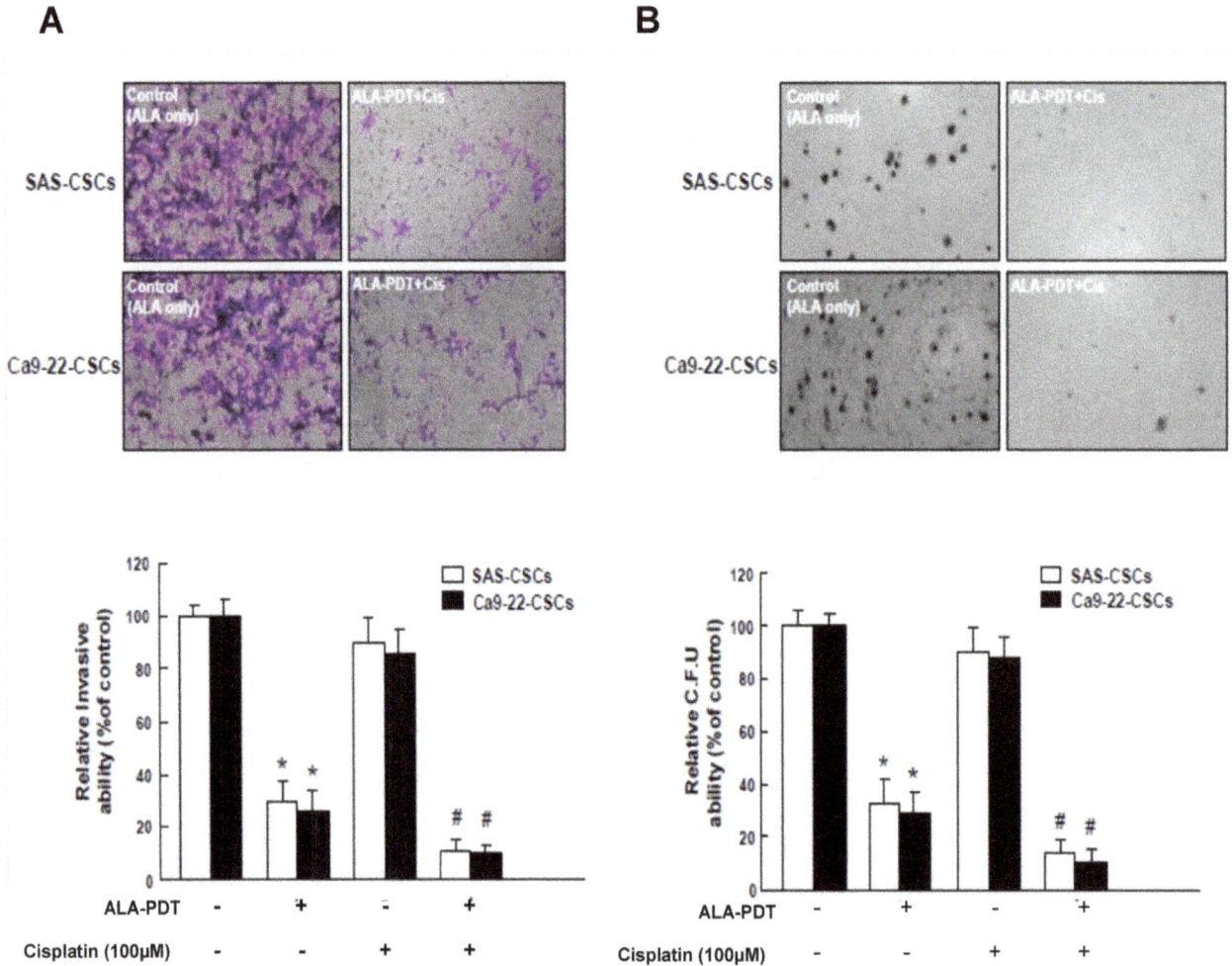

Figure 4. Reduced oncogenic properties in HNC-CSCs by ALA-PDT combined with cisplatin treatment. (A) Invasion ability and (B) colony-forming ability in HNC-CSCs was examined after treatment with either ALA-PDT or cisplatin chemotherapy or both. *, p<0.05 ALA-PDT vs. Control; #, p<0.05 ALA-PDT +Cisplatin vs. ALA-PDT alone.

metastasis [27]. Oral cancer epithelial cells can acquire mesenchymal traits which facilitate migration and invasion through EMT process.[28] It is known that EMT can give rise to cells with stem cell, and cancer initiating stem cells properties that have undergone EMT are therefore more motile and metastasized.[26]. Since we have found that the effect of ALA-PDT on invasion ability in HNC-CSCs, exploring whether the ALA-PDT - mediated CSCs and invasion capabilities depending on EMT pathway will be investigated in the future.

Chemotherapy is the current platform for treating HNC patients with metastasis[29]; however, the chemotherapeutic effect is limited, and its side effects largely interfere with the quality of life of patients. CSCs are clinically characterized by resistance to chemotherapy[29]. The presence of CSCs results in the low efficacy of anti-cancer therapies and the failure of tumor eradication and eventually leads to tumor recurrence and metastasis[30]. The HNC-CSCs were highly resistant to chemotherapy[18], and the chemoresistance of these cells was signifi-

cantly inhibited when they were treated with ALA-PDT. It is noteworthy that ALA-PDT treatment resulted in reduced MDR protein levels in HNC-derived CSCs after chemotreatment. The detailed mechanisms of the regulatory network between ALA-PDT treatment on drug-resistant genes requires further investigation.

In conclusion, the present study demonstrated the inhibitory effects of ALA-PDT on stem-like properties and chemoresistance in HNC. Notably, here is great need to unravel the underlying mechanisms of the ALA-PDT-mediated pathway in HNC-CSCs and to further evaluate the therapeutic possibilities of ALA-PDT treatment clinically.

Author Contributions

Conceived and designed the experiments: CHY. Performed the experiments: CHY. Analyzed the data: CCY. Contributed reagents/materials/analysis tools: CHY CCY. Wrote the paper: CHY CCY.

References

1. Siegel R, Naishadham D, Jemal A (2013) Cancer statistics, 2013. CA Cancer J Clin 63: 11–30.

2. Hu FW, Tsai LL, Yu CH, Chen PN, Chou MY, et al. (2012) Impairment of tumor-initiating stem-like property and reversal of epithelial-mesenchymal

transdifferentiation in head and neck cancer by resveratrol treatment. Mol Nutr Food Res 56: 1247–1258.

3. Yu CC, Tsai LL, Wang ML, Yu CH, Lo WL, et al. (2013) miR145 targets the SOX9/ADAM17 axis to inhibit tumor-initiating cells and IL-6-mediated paracrine effects in head and neck cancer. Cancer Res 73: 3425–3440.

4. Wu MJ, Jan CI, Tsay YG, Yu YH, Huang CY, et al. (2010) Elimination of head and neck cancer initiating cells through targeting glucose regulated protein78 signaling. Mol Cancer 9: 283.

5. Prince ME, Sivanandan R, Kaczorowski A, Wolf GT, Kaplan MJ, et al. (2007) Identification of a subpopulation of cells with cancer stem cell properties in head and neck squamous cell carcinoma. Proc Natl Acad Sci U S A 104: 973–978.

6. Yu CC, Chen YW, Chiou GY, Tsai LL, Huang PI, et al. (2011) MicroRNA let-7a represses chemoresistance and tumourigenicity in head and neck cancer via stem-like properties ablation. Oral Oncol 47: 202–210.

7. Chen YW, Chen KH, Huang PI, Chen YC, Chiou GY, et al. (2010) Cucurbitacin I suppressed stem-like property and enhanced radiation-induced apoptosis in head and neck squamous carcinoma–derived CD44(+)ALDH1(+) cells. Mol Cancer Ther 9: 2879–2892.

8. Green B, Cobb AR, Hopper C (2013) Photodynamic therapy in the management of lesions of the head and neck. Br J Oral Maxillofac Surg 51: 283–287.

9. Dolmans DE, Fukumura D, Jain RK (2003) Photodynamic therapy for cancer. Nat Rev Cancer 3: 380–387.

10. Chen HM, Chen CT, Yang H, Kuo MY, Kuo YS, et al. (2004) Successful treatment of oral verrucous hyperplasia with topical 5-aminolevulinic acid-mediated photodynamic therapy. Oral Oncol 40: 630–637.

11. Chen HM, Yu CH, Tu PC, Yeh CY, Tsai T, et al. (2005) Successful treatment of oral verrucous hyperplasia and oral leukoplakia with topical 5-aminolevulinic acid-mediated photodynamic therapy. Lasers Surg Med 37: 114–122.

12. Yu CH, Chen HM, Hung HY, Cheng SJ, Tsai T, et al. (2008) Photodynamic therapy outcome for oral verrucous hyperplasia depends on the clinical appearance, size, color, epithelial dysplasia, and surface keratin thickness of the lesion. Oral Oncol 44: 595–600.

13. Yu CH, Chen HM, Lin HP, Chiang CP (2013) Expression of Bak and Bak/Mcl-1 ratio can predict photodynamic therapy outcome for oral verrucous hyperplasia and leukoplakia. J Oral Pathol Med 42: 257–262.

14. Lou PJ, Jager HR, Jones L, Theodossy T, Bown SG, et al. (2004) Interstitial photodynamic therapy as salvage treatment for recurrent head and neck cancer. Br J Cancer 91: 441–446.

15. Lisnjak IO, Kutsenok VV, Polyschuk LZ, Gorobets OB, Gamaleia NF (2005) Effect of photodynamic therapy on tumor angiogenesis and metastasis in mice bearing Lewis lung carcinoma. Exp Oncol 27: 333–335.

16. Chen HM, Liu CM, Yang H, Chou HY, Chiang CP, et al. (2011) 5-aminolevulinic acid induce apoptosis via NF-kappaB/JNK pathway in human oral cancer Ca9-22 cells. J Oral Pathol Med 40: 483–489.

17. Yang TH, Chen CT, Wang CP, Lou PJ (2007) Photodynamic therapy suppresses the migration and invasion of head and neck cancer cells in vitro. Oral Oncol 43: 358–365.

18. Lo WL, Yu CC, Chiou GY, Chen YW, Huang PI, et al. (2011) MicroRNA-200c attenuates tumour growth and metastasis of presumptive head and neck squamous cell carcinoma stem cells. J Pathol 223: 482–495.

19. Chiou SH, Yu CC, Huang CY, Lin SC, Liu CJ, et al. (2008) Positive correlations of Oct-4 and Nanog in oral cancer stem-like cells and high-grade oral squamous cell carcinoma. Clin Cancer Res 14: 4085–4095.

20. Yu CC, Lo WL, Chen YW, Huang PI, Hsu HS, et al. (2011) Bmi-1 Regulates Snail Expression and Promotes Metastasis Ability in Head and Neck Squamous Cancer-Derived ALDH1 Positive Cells. J Oncol 2011.

21. Chen YC, Chang CJ, Hsu HS, Chen YW, Tai LK, et al. (2010) Inhibition of tumorigenicity and enhancement of radiochemosensitivity in head and neck squamous cell cancer-derived ALDH1-positive cells by knockdown of Bmi-1. Oral Oncol 46: 158–165.

22. Lo JF, Yu CC, Chiou SH, Huang CY, Jan CI, et al. (2011) The epithelial-mesenchymal transition mediator S100A4 maintains cancer-initiating cells in head and neck cancers. Cancer Res 71: 1912–1923.

23. Visvader JE, Lindeman GJ (2008) Cancer stem cells in solid tumours: accumulating evidence and unresolved questions. Nat Rev Cancer 8: 755–768.

24. Korkaya H, Wicha MS (2010) Cancer stem cells: nature versus nurture. Nat Cell Biol 12: 419–421.

25. Polyak K, Weinberg RA (2009) Transitions between epithelial and mesenchymal states: acquisition of malignant and stem cell traits. Nat Rev Cancer 9: 265–273.

26. Mani SA, Guo W, Liao MJ, Eaton EN, Ayyanan A, et al. (2008) The epithelial-mesenchymal transition generates cells with properties of stem cells. Cell 133: 704–715.

27. Chiou SH, Wang ML, Chou YT, Chen CJ, Hong CF, et al. (2010) Coexpression of Oct4 and Nanog enhances malignancy in lung adenocarcinoma by inducing cancer stem-like properties and epithelial-mesenchymal transdifferentiation. Cancer Res 70: 10433–10444.

28. Yang MH, Wu MZ, Chiou SH, Chen PM, Chang SY, et al. (2008) Direct regulation of TWIST by HIF-1alpha promotes metastasis. Nat Cell Biol 10: 295–305.

29. Haddad RI, Shin DM (2008) Recent advances in head and neck cancer. N Engl J Med 359: 1143–1154.

30. Clarke MF, Dick JE, Dirks PB, Eaves CJ, Jamieson CH, et al. (2006) Cancer stem cells–perspectives on current status and future directions: AACR Workshop on cancer stem cells. Cancer Res 66: 9339–9344.

3

Secretome of Peripheral Blood Mononuclear Cells Enhances Wound Healing

Michael Mildner[1,9], Stefan Hacker[2,3,9], Thomas Haider[3,4], Maria Gschwandtner[1], Gregor Werba[5], Caterina Barresi[1], Matthias Zimmermann[3,4], Bahar Golabi[3,4], Erwin Tschachler[1,6], Hendrik Jan Ankersmit[3,4]*

1 Department of Dermatology, Medical University Vienna, Vienna, Austria, 2 Department of Plastic Surgery, Medical University Vienna, Vienna, Austria, 3 Christian Doppler Laboratory for Cardiac and Thoracic Diagnosis and Regeneration, Vienna, Austria, 4 Department of Thoracic Surgery, Medical University Vienna, Vienna, Austria, 5 Department of Surgery, Medical University Vienna, Vienna, Austria, 6 Centre de Recherches et dInvestigations Epidermiques et Sensorielles (CE.R.I.E.S.), Neuilly, France

Abstract

Non-healing skin ulcers are often resistant to most common therapies. Treatment with growth factors has been demonstrated to improve closure of chronic wounds. Here we investigate whether lyophilized culture supernatant of freshly isolated peripheral blood mononuclear cells (PBMC) is able to enhance wound healing. PBMC from healthy human individuals were prepared and cultured for 24 hours. Supernatants were collected, dialyzed and lyophilized (SEC^{PBMC}). Six mm punch biopsy wounds were set on the backs of C57BL/6J-mice and SEC^{PBMC} containing emulsion or controls were applied daily for three days. Morphology and neo-angiogenesis were analyzed by H&E-staining and CD31 immuno-staining, respectively. In vitro effects on diverse skin cells were investigated by migration assays, cell cycle analysis, and tube formation assay. Signaling pathways were analyzed by Western blot analysis. Application of SEC^{PBMC} on 6 mm punch biopsy wounds significantly enhanced wound closure. H&E staining of the wounds after 6 days revealed that wound healing was more advanced after application of SEC^{PBMC} containing emulsion. Furthermore, there was a massive increase in CD31 positive cells, indicating enhanced neo-angiogenesis. In primary human fibroblasts (FB) and keratinocytes (KC) migration but not proliferation was induced. In endothelial cells (EC) SEC^{PBMC} induced proliferation and tube-formation in a matrigel-assay. In addition, SEC^{PBMC} treatment of skin cells led to the induction of multiple signaling pathways involved in cell migration, proliferation and survival. In summary, we could show that emulsions containing the secretome of PBMC derived from healthy individuals accelerates wound healing in a mouse model and induce wound healing associated mechanisms in human primary skin cells. The formulation and use of such emulsions might therefore represent a possible novel option for the treatment of non-healing skin ulcers.

Editor: Johanna M. Brandner, University Hospital Hamburg-Eppendorf, Germany

Funding: This study was funded by the Christian Doppler Research Association, APOSIENCE AG, and the Medical University of Vienna. Maria Gschwandtner was supported by a grant from the Austrian Science Fund FWF: T545-B19. The funders had no role in study design, data collection and analysis, decision to publish, or preparation of the manuscript.

Competing Interests: The authors have read the journal's policy and have the following conflicts: The Medical University of Vienna has claimed financial interest (Patent number: PCT/EP09/67534, filed 18 Dec 2008; Patent name: PHARMACEUTICAL PREPARATION COMPRISING SUPERNATANT OF BLOOD MONONUCLEAR CELL CULTURE). Hendrik Jan Ankersmit is a shareholder of APOSIENCE AG, which owns the rights to commercialize SECPBMC for therapeutic use. All other authors declare that they have no conflict of interest. APOSIENCE AG is a funder of this study.

* E-mail: hendrik.ankersmit@meduniwien.ac.at

9 These authors contributed equally to this work.

Introduction

The skin is the largest organ of the human body. It covers and protects the underlying organs from ultraviolet radiation, mechanical and chemical damage, invading microorganisms and excessive water loss [1]. Due to this essential function, skin wounds need to be efficiently repaired within a very short time frame. Optimum healing of cutaneous wounds requires a well-orchestrated integration of complex biological and molecular events, including cell migration and proliferation, extracellular matrix deposition and remodeling as well as neo-angiogenesis [2–4]. Unfortunately, adequate healing of skin wounds, in particular in the elderly population or in diabetic patients, is often impaired, leading to increased morbidity [5]. The seminal work by

Holzinger et al. has shown that autologous transplantation of mononuclear cells effectively initiates and improves granulation and epithelialization of skin ulcers [6]. In addition, topic application of a mixture of peripheral blood mononuclear cells (PBMC) together with basic fibroblast growth factor also resulted in a dramatic improvement in the treatment of a diabetic gangrene [7]. Several reports showed that transplantation of stem cells accelerates wound re-epithelialization and neovascularization in various models [8–10]. Recently, it has been demonstrated that the efficacy of a stem cell based therapy is dependent on soluble factors produced by these cells, since the secretome of stem cell cultures is sufficient to accelerate cutaneous wound healing [11–13].

The idea of using conditioned medium as a therapeutic agent evolved in the field of stem cell research and originated from cell based therapy of myocardial infarction. Many of the regenerative effects seen after administration of stem cells were shown to be rather mediated via a paracrine signaling than by direct cellular interactions [14,15]. Conditioned culture medium, containing the secretome of mesenchymal stem cells, is rich in angiogenic and chemotactic factors [16,17]. In addition, first evidence emerges that stem cell conditioned medium has immunosuppressive properties [18,19]. We have previously shown that infusion of cultured irradiated apoptotic PBMC suspensions in a rat acute myocardial infarction (AMI) model restored long-term cardiac function [20,21]. In a further study, we demonstrated that the regenerative function of these cells was solely mediated via soluble factors produced by the cells [22]. These highly active culture supernatants not only showed positive effects in a rodent and porcine AMI model, but also displayed cytoprotective functions on human primary cardiomyocytes [22].

Based on these beneficial pathophysiological concepts we speculated whether an emulsion containing lyophilized culture supernatants of freshly isolated unstimulated PBMC (SEC^{PBMC}) is able to enhance skin wound healing in a mouse model *in vivo* and whether it is able to activate wound healing associated mechanisms in primary human skin cells *in vitro*.

Methods

Ethics statement

Animal experiments were approved by the committee for animal research, Medical University of Vienna (vote: 66.009/0108-II/10b/2009).

Human peripheral blood mononuclear cells (PBMC) were obtained from young healthy volunteers. The study was approved by the local ethics committee (Medical University of Vienna, vote: EK-Nr 2010/034), and informed written consent was obtained from all volunteers.

Mice

C57BL/6J mice were purchased (Taconic, Ry, Denmark) and kept at the Division of Biomedical Research at the Medical University of Vienna. Three months old mice were used for the wounding experiments.

Wounding protocol

Mice were anesthetized using Ketalar/Rompun solution intraperitoneally (50 mg Ketalar and 16 mg Rompun per kg body weight) and placed on a heating plate prior to surgery. One six mm full-thickness punch biopsy wound was set on the back of each mouse by folding the back skin and punching through 2 thicknesses of skin. Wounds were immediately treated with emulsions containing SEC^{PBMC}, lyophilized medium or physiological NaCl-solution (15 mice per group). Wounds were left uncovered and emulsions were applied once a day for 3 consecutive days. Wound area was measured in two dimensions with handyman's calipers for the first 3 days until a crust was formed and photographed at the beginning and the end of the experiment. Mice were sacrificed on day 7 and wounds were analyzed with regard to their morphology by Hematoxylin-Eosin (H&E)-staining, and the expression of Ki67 and CD31 was analyzed by specific immuno-stainings.

Cell culture

Human peripheral blood mononuclear cells (PBMC) from young healthy volunteers were cultured in serum-free UltraCul-

ture Medium (Lonza, Basel, Switzerland). Human dermal Fibroblasts (FB, Lonza) were cultured in DMEM (Gibco BRL, Gaithersburg, USA) medium, supplemented with 10% fetal bovine serum (FBS, PAA, Linz, Austria), 25 mM L-glutamine (Gibco) and 1% penicillin/streptomycin (Gibco). Human primary Keratinocytes (KC) were cultured in KC-growth medium (KGM, Lonza). Endothelial cells (EC) were cultured in EC-growth medium (EGM2 MV, Lonza). All cells were cultured at 37°C, 5% CO$_2$ and at 95% relative humidity.

Generation of lyophilized cell culture supernatant derived from PBMC (SEC^{PBMC})

Fifty ml of venous blood were drawn from 10 healthy volunteers. PBMC were separated by Ficoll-Paque (GE Health-care Bio-Sciences AB, Stockholm, Sweden) density gradient centrifugation. Cells were washed twice with PBS (Gibco), resuspended in serum-free UltraCulture Medium (Lonza) and cultured for 24 h at a cell density of 2.5×10^6 cells per ml. After 24 h supernatants were collected by spinning cell-suspensions at 500 g. All supernatants were pooled and dialyzed against ammonium acetate (50 mM, Sigma, Vienna, Austria) for 24 h at 4°C. For dialysis a cut-off of 6–8 kDa of the dialysis-membranes (Spectrum Medical Industries Inc., Los Angeles, CA, USA) was used. The obtained liquid was sterile filtered (Whatman Filter 0.22 μm, Dassel, Germany), frozen at −80°C and lyophilized overnight (Lyophilizator Christ alpha 1–4, Martin Christ Gefriertrocknungsanlagen GmbH, Osterode am Harz, Germany). Lyophilization was performed at −20°C and a pressure of 0.1 mbar. The so obtained powder was reconstituted with the respective media for *in vitro* use or further processed to generate an emulsion for *in vivo* applications.

Preparation of SEC^{PBMC} emulsions

Lyophilized SEC^{PBMC} was resolved in physiological NaCl-solution (Braun, Melsungen, Germany) to obtain a solution corresponding to an equivalent of the secretome from 2.5×10^7 cells per ml. As control we used lyophilized UltraCulture medium resolved in physiological NaCl-solution and physiological NaCl-solution alone. One part of the solutions was mixed with one part of a hydrophilic cream base (Ultrasicc®). 100 μl of these emulsions (containing SEC^{PBMC} of 1.25×10^6 PBMC) were applied to one skin wound per day for the first 3 days. Initially we performed the experiments with the dose used mentioned above and a lower dose (1/5). In this pre-experiment we found similar effects with the higher dose, but the lower dose was not effective. The frequency of application of the emulsion was chosen in agreement with our veterinary.

Antibodies

All antibodies used are listed in Table 1.

Histochemistry and immunohistochemistry

For analyses, the complete wounds including 2 mm of the epithelial margins were excised, bisected and immediately processed. Skin wounds were either fixed in 4% formalin overnight and embedded in paraffin, or immediately embedded in optimal cutting temperature compound (TissueTek, Sakura Finetek, Zoeterwoude, The Netherlands), snap frozen in liquid nitrogen and stored at −80°C until further processing. H&E-staining was done on 5 μm thick paraffin sections.

Staining of paraffin-sections

Ki67 staining was performed on paraffin embedded tissues after antigen retrieval by boiling in a microwave for 5 minutes in citrate-

Table 1. List of Antibodies.

Antigen	Company	No.	Dilution	
			Western blot	Immuno-staining
CD31	BD Biosciences[1]	553710	n.d.	1:50
Ki67	Novus Biologicals[2]	NB500-170	n.d.	1:50
Ki67	Abcam[3]	AB15580	n.d.	1:1000
c-Jun	NEB[4]	9165	1:100	n.d.
phospho-c-Jun	NEB[4]	9261	1:100	n.d.
CREB	NEB[4]	9197	1:100	n.d.
phospho-CREB	NEB[4]	9198	1:100	n.d.
Akt	NEB[4]	2938	1:100	n.d.
phospho-Akt	NEB[4]	9271	1:100	n.d.
Erk1/2	NEB[4]	4695	1:100	n.d.
phospho-Erk1/2	NEB[4]	4376	1:100	n.d.
Hsp27	NEB[4]	2402	1:100	n.d.
phospho-Hsp27	NEB[4]	2404	1:100	n.d.
GAPDH	HyTest[5]	5G4	1:2000	n.d.
mouse IgG HRP	Amersham[6]	NA931V	1:10000	n.d.
rabbit IgG HRP	Thermo Fisher[7]	31460	1:10000	n.d.
rabbit Fluor546	Alexa[8]	A-11035	n.d.	1:500
rat igG HRP	Amersham[6]	NA9350	n.d.	1:500

Company addresses:
[1]Bedford, MA, USA; [2]Littleton, CO, USA; [3]Cambridge, UK; [4]NEB: New England Biolabs, Beverly, MA, USA; [5]Turku Finland; [6]Buckinghamshire, UK; [7]Rockford, IL, USA; [8]Eugene, OR, USA. n.d.: not done.

buffer (pH = 6, Dakocytomation, Glostrup, Denmark). Non-specific staining was blocked by incubation with 10% normal goat serum for 1 h. The slides were subsequently incubated overnight in a humidified chamber at 4°C with either a Ki67 antibody or isotype-matched control antibodies (Table 1) diluted in PBS containing 2% bovine serum albumin (BSA) and 10% goat serum. To visualize Ki67, sections were incubated with a rabbit Fluor546 antibody (Alexa) for 60 min. After washing nuclei were stained with Hoechst (Dakocytomation) and the slides were mounted with Fluoprep (bioMerieux, Marcy l'Etoile, France).

Staining of cryo-sections

CD31 immunostaining was performed on 10 μm thick sections of frozen tissues after fixation in aceton/methanol (1:1) for 10 min. The staining was performed as described for paraffin-sections without boiling in a microwave. The CD31 antibody and the second step antibody used are listed in Table 1. After washing with PBS, slides were incubated sequentially with an HRP-linked second step antibody in PBS containing 2% BSA and 10% normal goat serum for 1 h, followed by incubation with DAB Chromogen tablets (DAKO, Carpinteria, CA). After washing, nuclear staining was performed by incubation with hematoxylin for 10 sec. Slides were mounted with Fluoprep (bioMérieux).

In vitro scratch assay

3×10^5 FB and KC were seeded in 6-well plates. After reaching 100% confluence, cells were washed twice with PBS and scratched using a pipette-tip. The resulting scratch was investigated under the microscope and the area for photographs was marked using a pen on the bottom of the well. Immediately after scratching, the first photograph was taken (initial wound size) exactly at the marked area. Lyophilized SEC^{PBMC} (2.5×10^6 per ml) as well as lyophilized medium were resolved in the respective medium without additional growth factors and added to the scratch-wounds. After 18 h the same area of the scratch wounds were photographed again and the reduction in wound width was measured using ImageJ 1.45 s software (National Institutes of Health, Bethesda, MD, USA).

Cell cycle analysis

1×10^5 FB, KC and EC were seeded in 6-well plates and cultured overnight in their respective growth medium. To synchronize cells they were cultivated in basic medium for 18 hours. Lyophilized SEC^{PBMC} (2.5×10^6 per ml) and lyophilized control medium were resolved in either DMEM, basic keratinocyte medium (KBM, Lonza) or basic endothelial cell medium (EBM, Lonza) - all without serum and growth factors - and added to the respective cell types for 24 h. Cell cycle analysis was performed according to the manufacturer's instructions (BD Biosciences, Franklin Lakes, NJ, USA). Briefly, after 24 h of treatment, cells were incubated with BrdU (10 μM final concentration, kit-component) for 2 h. After fixation (BD Cytofix/Cytoperm^{TM}, kit-component) cells were stained with a fluorescein isothiocyanate (FITC) conjugated anti-BrdU antibody (BD Biosciences, kit-component) for 30 min. After washing cells were stained with 7-AAD (BD Biosciences, kit-component) and immediately analyzed on a FACS-Calibur (BD Biosciences). Gates were set according to the manufacturer's instructions and data were evaluated using the FlowJo software (Tree Star, Ashland, OR, USA).

Tube formation assay

Twenty-four-well plates were coated with 300 μl of Matrigel (BD Biosciences) and incubated for 1 h at 37°C. 1×10^5 EC in 1 ml EBM containing 10% FBS with either lyophilized SEC^{PBMC} (derived from 2.5×10^6 per ml) or lyophilized medium were seeded into duplicate wells and observed 24 h after incubation. EC treated with 100 ng/ml vascular endothelial cell growth factor (VEGF, R&D-systems, Minneapolis, USA) served as positive control.

Western blot analysis

3×10^6 FB, KC and EC were seeded in 6-well plates and cultured overnight in their respective growth medium. After removal of the medium cells were washed twice with PBS (Gibco) and cultured in their respective basal medium without growth factors for 3 h. Aliquots of lyophilized SEC^{PBMC} and medium were resolved in the different basal media at a 10-fold concentration. One tenth of this solution was then directly added to the cell cultures. After 1 h cells were washed with PBS and lyzed in SDS-PAGE loading buffer. After sonication and centrifugation, proteins were size fractionated by SDS-PAGE through an 8 to 18% gradient gel (Amersham Pharmacia Biotech, Uppsala, Sweden) and transferred to nitrocellulose membranes (BioRad, Hercules, CA, USA). Immunodetection was performed with anti-c-Jun, anti-phospho-c-Jun, anti-CREB, anti-phospho-CREB, anti-Akt, anti-phospho-AKT, anti-Erk1/2, anti-phospho-Erk1/2, anti-Hsp27, anti-phospho-Hsp27 (Ser15) and anti-GAPDH followed by a HRP-conjugated goat anti-mouse IgG antiserum or a goat anti-rabbit IgG antiserum (all antibodies are listed in Table 1). Reaction products were detected by chemiluminescence with the ChemiGlow reagent (Biozyme Laboratories Limited, South Wales, U.K.) according to the manufacturer's instructions. Bands from 3 independent experiments were quantified using ImageJ software and the mean increase in phosphorylation was calculated in relation to the expression of the respective non-phosphorylated proteins.

Quantification of blood vessel density

To quantify neo-angiogenesis, the granulation tissue underneath the wounds of each CD31 stained wound from 15 animals in each group was evaluated. CD31 positive blood vessels were counted and expressed in absolute numbers. In addition, wound angiogenesis was analyzed by measuring the percentage of the CD31$^+$ area using ImageJ software. The percentage of the CD31$^+$ area in relation to the granulation tissue was calculated.

Quantification of Ki67 positive keratinocytes

High-power images of Ki67-stained sections were used to quantify the number of proliferating KC. Digital images of Ki67-stained slides were taken from the regenerated epidermis at the wound edges of each sample. Ki67-positive KC within the full width of the captured field were counted. Regenerated epidermis was defined by hyperthickening and lack of hair follicles.

Statistical methods

Statistical analysis was performed using the program Graph-Pad Prism version 5 (GraphPad Software Inc., San Diego, CA, USA). Statistical comparison of two means was performed by using Student's t-test, and comparison of more than two means was performed using analysis of variance (ANOVA) with Bonferroni multiple comparisons post-test. For Figure 1A Kruskal-Wallis testing was used. A p-value <0.05 was defined as the level of significance and is depicted with *.

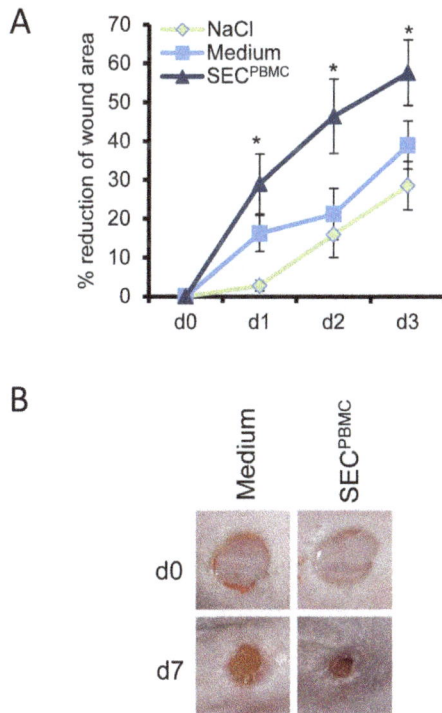

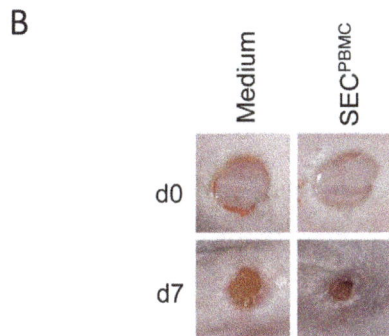

Results

SEC^{PBMC} containing emulsions enhance wound healing in mice in vivo

To study the capacity of SEC^{PBMC} to promote skin wound healing we used the full thickness punch wound model in C57BL/6J mice. We generated emulsions containing either lyophilized SEC^{PBMC}, lyophilized UltraCulture medium or physiological NaCl-solution. Six mm punch biopsies were set on the backs of C57BL/6J mice and treated daily with the different emulsions for 3 days. Wound areas were analyzed throughout the healing process until a crust had formed. Compared to an emulsion containing lyophilized medium as well as an emulsion containing NaCl-solution, application of an SEC^{PBMC}-containing emulsion strongly enhanced wound closure during the first 3 days (Figure 1A). This advantage in wound closure was still visible at day 7, as shown by representative photographs of the wounds (Figure 1B). H&E staining of the wounds after 7 days revealed that wound healing was more advanced after application of SEC^{PBMC} containing emulsion (Figure 1C). Microscopic examination revealed that the wound gap distance was smaller in the SEC^{PBMC}-treated mice and re-epithelialization was markedly enhanced in these animals (Figure 1C).

SEC^{PBMC} induces migratory capacity of dermal fibroblasts and epidermal keratinocytes

To investigate the capacity of SEC^{PBMC} to induce cell migration on human primary FB and KC the cells were grown to confluency, scratched and treated with lyophilized control medium or

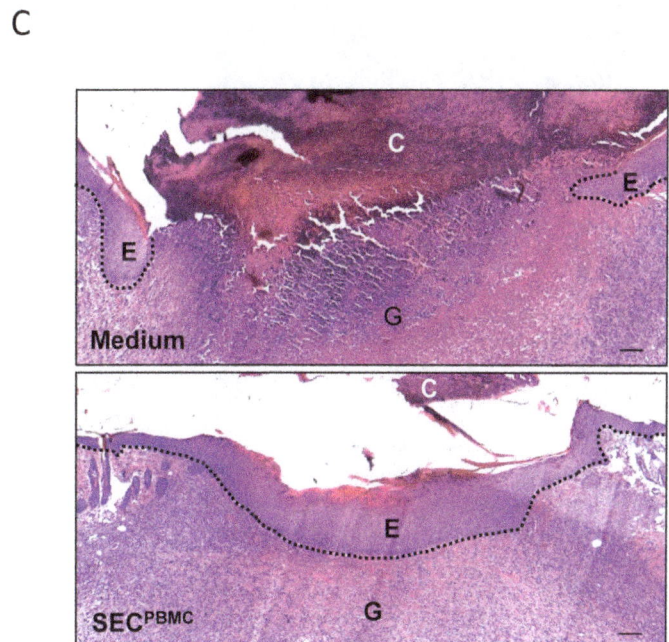

Figure 1. SEC^{PBMC} leads to enhanced wound closure and re-epithelialization. (A) Wound areas were measured during the first 3 days after wounding. Treatment with SEC^{PBMC} significantly enhanced wound closure. Error bars represent one standard deviation calculated from 15 animals for each set of values (*: p$<$0.05). (B) Representative photographs from mouse wounds (n = 15 from each group) immediately after wounding and at day 7 after wounding are shown. (C) H&E staining of wounds treated with medium or SEC^{PBMC} 7 days after wounding is shown. While medium treated wounds still show a thick crust and little re-epithelialization, SEC^{PBMC} treated wounds are fully re-epithelialized. C = crust, E = newly formed epidermis, G = granulation tissue. Scale bars: 100 μm. One representative animal of 15 is shown.

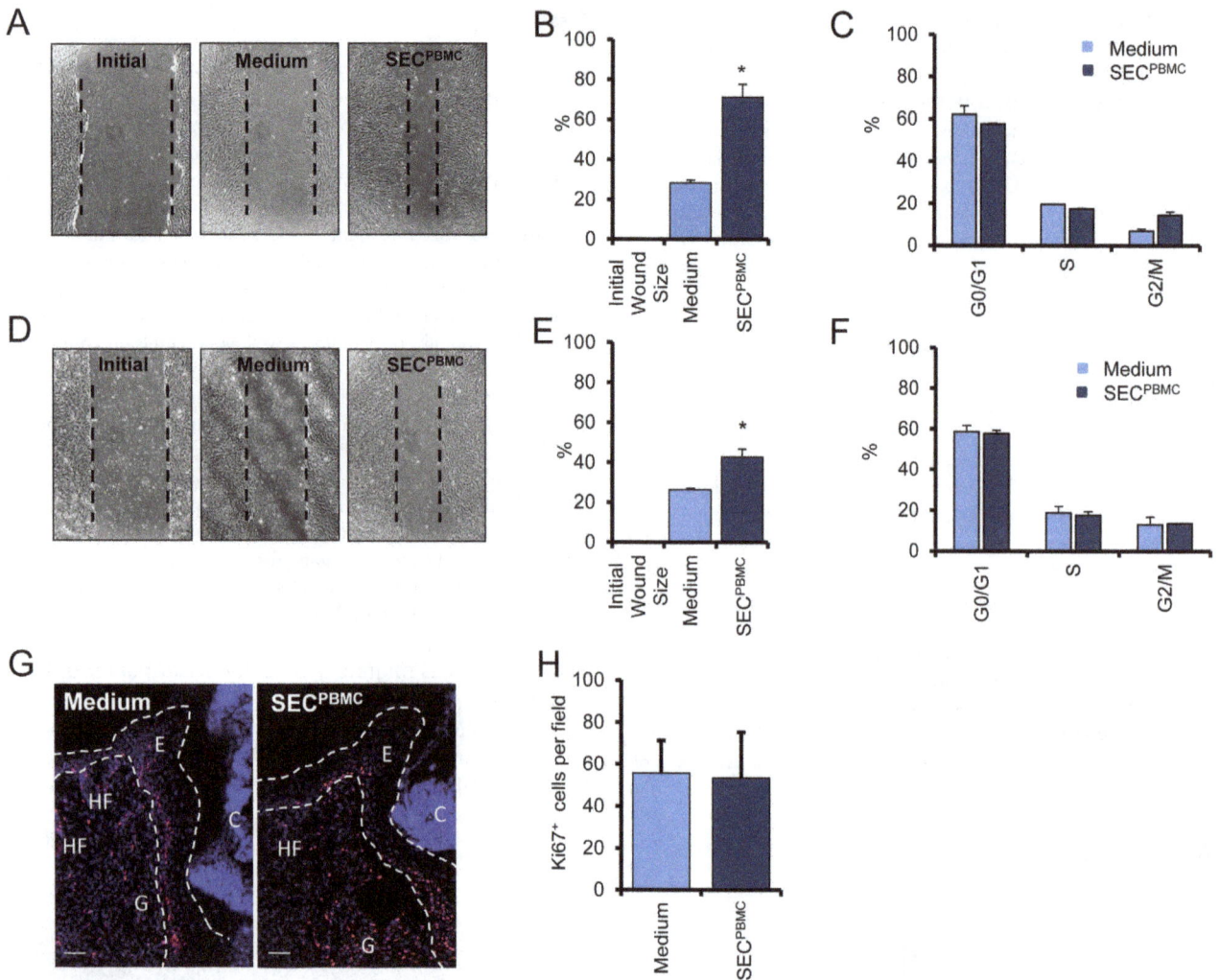

Figure 2. SEC^PBMC induces migration of human primary fibroblasts and keratinocytes. Scratch wounds of FB (**A**) and KC (**D**) are shown. One representative experiment of three each done in triplicates is shown. The mean-width of the gaps of nine scratch wounds after 18 h was measured and the percentage of closure for FB (**B**) and KC (**E**) was calculated. (*: p<0.01). Cells cycle analyses revealed no significant differences in FB (**C**) and in KC (**F**). One representative experiment of three each done in triplicates is shown. (**G**) Ki67 staining of medium and SEC^PBMC treated wounds showed no significant alterations in proliferating cells. Photographs were taken at the wound-edge of wounds 7 days after wounding. Ki67 staining is shown in red and nuclear staining is shown in blue. C = crust, E = newly formed epidermis, G = granulation tissue HF = hair follicle. One representative animal of 15 is shown. Scale bars: 50 μm. (**H**) The mean from 15 animals per group of Ki67 positive cells at the wound edge of one high power image was calculated.

SEC^PBMC for 18 hours. As shown in figure 2, SEC^PBMC significantly induced cell migration in both, dermal FB (Figure 2A,B) and epidermal KC (Figure 2D,E). By contrast, cell cycle analysis revealed no significant changes in cell cycle progression after SEC^PBMC treatment in both cell types (Figure 2C,F). This finding was consistent with our *in vivo* observation, where we could not detect a significant increase in proliferating cells in the regenerated epidermis of the skin wounds treated with SEC^PBMC containing emulsions, as demonstrated by Ki67 staining (Figure 2G,H).

SEC^PBMC shows strong angiogenic properties in vivo and in vitro

Since a crucial event during wound healing is the sprouting of newly formed blood vessels into the wounded tissue, we examined the efficacy of SEC^PBMC to induce neo-angiogenesis *in vivo*. Skin wounds, harvested seven days after wounding, were stained for

CD31, a specific marker for blood vessels. As shown in figure 3, we found a massive increase in CD31 positive cells in SEC^PBMC treated wounds, indicating enhanced neo-angiogenesis (Figure 3A). Blood vessel density in SEC^PBMC treated wounds increased from 52 ± 10.7 per high power field in medium treated wounds to 132 ± 31 in SEC^PBMC treated wounds (Figure 3B). In addition, morphometric analyses revealed that the percentage of the granulation tissue taken up by blood vessels was markedly increased in the SEC^PBMC treated animals compared to the medium treated group (Figure 3C). We further investigated the effect of SEC^PBMC on primary microvascular skin EC *in vitro*. In contrast to FB and KC, treatment of EC with SEC^PBMC strongly induced proliferation in these cells (Figure 3D). The observed increase in the proliferation rate after SEC^PBMC-treatment in EC was even higher than that observed after incubation with VEGF (100 ng/ml), which is a known strong promotor of EC-proliferation. In addition to enhanced cell growth, newly formed EC need

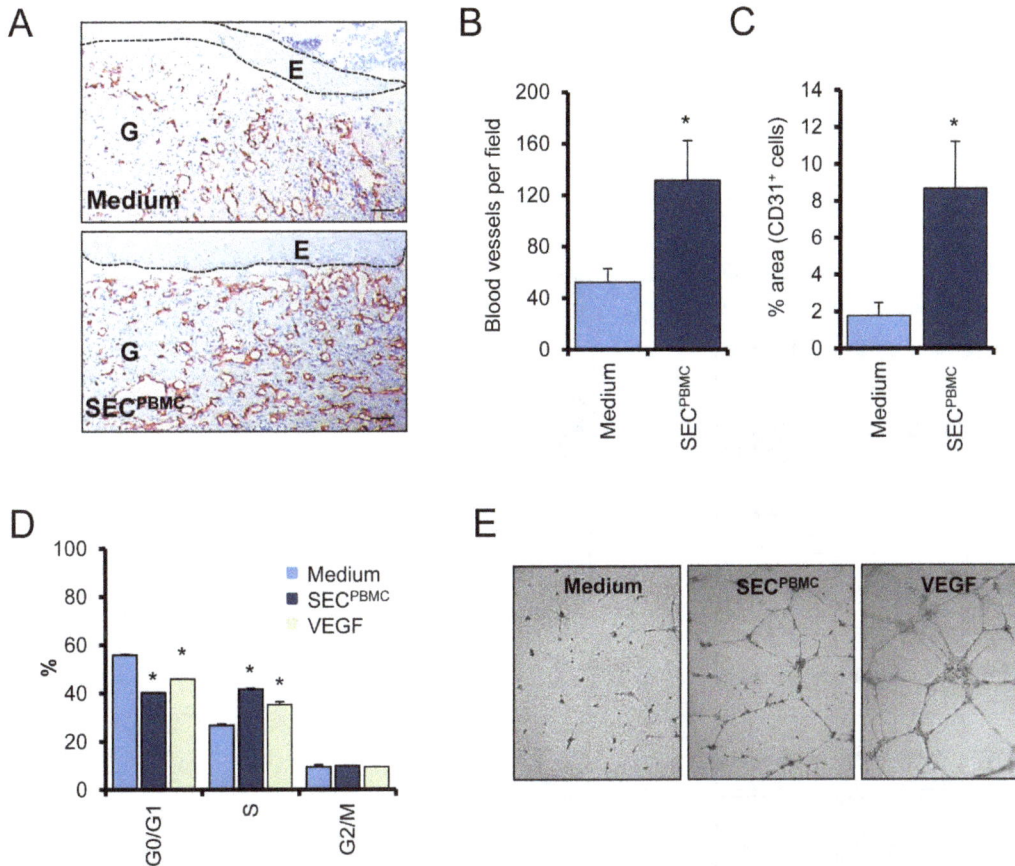

Figure 3. SEC^PBMC induces formation of new blood vessels. (**A**) Representative CD31 stainings of wounds underneath the original wound edge are shown. Scale bars: 50 µm (**B**) The numbers of CD31 positive cells underneath the newly formed epidermis were evaluated. The graph represents the mean of 15 animals in each group (*: p<0.01). (**C**) The area of the granulation tissue taken up by CD31+ cells was evaluated. The graph represents the mean of 15 animals in each group. (**D**) Cell cycle analysis of SEC^PBMC treated microvascular EC shows a strong increase in proliferating cells. VEGF treatment served as positive control. One representative experiment of three each done in triplicates is shown (*: p<0.01). (**E**) A tube formation assay is shown. Compared to medium alone SEC^PBMC strongly induced tube formation in a matrigel assay. One representative experiment of three is shown.

to reorganize into a three-dimensional tubular structure in a wound healing scenario. We therefore investigated the ability of SEC^PBMC to induce tube formation in a matrigel assay system. As shown in Figure 3E, cultivation of EC together with SEC^PBMC indeed led to the formation of tubular structures. In comparison, no tube formation was observed in EC cultured with control medium (Figure 3E).

SEC^PBMC treatment leads to activation of signaling cascades involved in cell migration, proliferation and survival

To get more information about the underlying mechanisms, we analyzed a variety of signaling cascades involved in cell migration, proliferation and survival. In human primary KC SEC^PBMC led to a rapid activation of CREB, Erk1/2, c-Jun, Akt and Hsp27, in human primary dermal FB Erk1/2, c-Jun, Akt and Hsp27 were activated and in dermal microvascular EC SEC^PBMC led to the activation of CREB, c-Jun and Hsp27 (Figure 4).

Discussion

The skin protects the organism against environmental aggressions and microbial pathogens and forms an inside-out barrier preventing fluid loss [1]. Loss of its integrity as a result of injury or

illness may cause chronic skin ulcers leading to major disability or even death. Chronic wounds are often difficult to treat, encouraging investigations for new innovative therapeutics that enhance wound healing and tissue regeneration. Recently, cell based therapies have been suggested to be of great advantage in a wound healing scenario. Most of these studies showed positive effects of directly applied highly purified cells to the wounds, which however, are often difficult to isolate and not easily applicable. Increasing evidence is surfacing that all observed therapeutic effects rely on their ability to secrete a cocktail of factors that enhance tissue regeneration, modulate the local environment and stimulate proliferation, migration, differentiation, survival and functional recovery of resident cells [23,24]. Recent publications have demonstrated that progenitor cells secrete soluble proteins and induce regenerative mechanisms in a paracrine manner [25–27]. In addition, also mesenchymal stem cells have been shown to augment the proliferative phase in wound healing that is characterized by angiogenesis, granulation tissue formation, epithelialization and wound contraction [9,10,28,29]. In the present study we demonstrated that lyophilized supernatants of unstimulated cultured PBMC (SEC^PBMC) are able to enhance wound healing in a mouse model *in vivo*, and induce characteristics of wound healing in human cells *in vitro*.

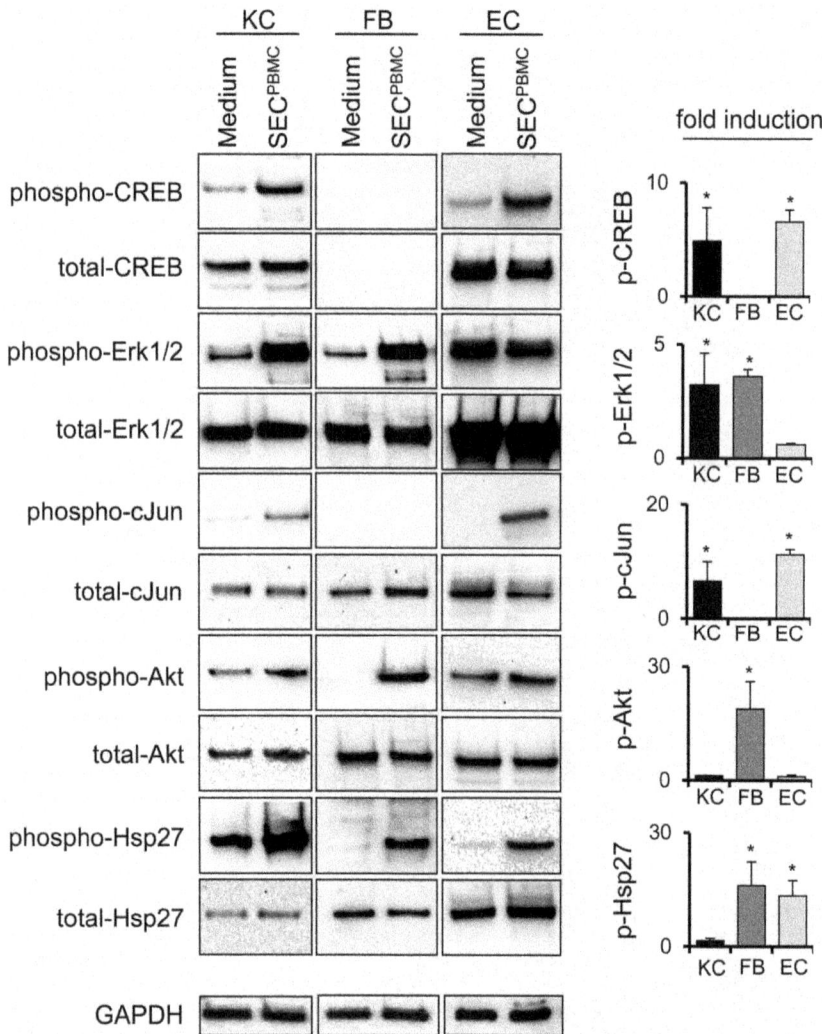

Figure 4. SEC^PBMC leads to activation of several signaling cascades. KC, FB and EC were treated for 1 h with medium or SEC^PBMC. Western blot analyses of several signaling factors are shown. One representative experiment of three is shown. The graphs in the right panel represent the mean band intensity of all three experiments. The increase in expression of the phosphorylated proteins was calculated in relation to the respective non-phosphorylated proteins (*: $p < 0.01$).

In a different experimental setting we could previously show that infusion of apoptotic PBMC suspensions in an experimental model of acute myocardial infarction prevented myocardial damage and tissue remodeling [12,20], and that this protective effect was also conferred by only applying the secretome of these cells [22]. However, in contrast to stem cells isolated by complicated and time consuming protocols, we used easily obtainable PBMC and showed that one single infusion of PBMC derived "paracrine factors" prevented myocardial damage. We analyzed the secretome of both, apoptotic and living PBMC, and found a variety of highly expressed factors, which have been associated with cytoprotection and tissue regeneration [22]. In line with the work of Di Santo et al., we were not able to block these *in vitro* effects by IL-8, ENA-78, VEGF and MMP9 blocking antibodies [22]. Thus, our data strongly suggest that for tissue protection and regeneration an interaction of several factors is necessary for its full regenerative capacity.

Cell migration and proliferation are rate limiting events in skin wound healing. Here we could show that SEC^PBMC not only improved cutaneous wound healing in a murine model, but also induced migration and proliferation of primary human skin cells. Importantly, SEC^PBMC treatment of mouse wounds led to a massive sprouting of newly formed blood vessels. Similar angiogenic effects were also found on human dermal microvascular EC *in vitro*, since SEC^PBMC strongly induced proliferation and tube formation in these cells. In a further set of experiments we could identify multiple signaling cascades, which were induced after SEC^PBMC treatment of different human primary skin cells. Nevertheless, a clear conclusion on the signaling cascades and the initiating factors that might be responsible for the diverse effects in the different cell types cannot be drawn at the present time. Further studies attempting to unravel the factors involved in these processes are ongoing. Currently, we can just speculate on the mechanisms initiated after the treatment with SEC^PBMC. We have previously determined high amounts of angiogenic factors present in SEC^PBMC (VEGF, PDGF, IL-8, ENA-78 and others) [22]. In addition, factors that have been associated with enhanced skin re-epithelialization or that promote wound healing as a chemoattractant to cells of the immune system (eg. MCP-1, RANTES, IL-8) are found in high concentrations [22]. In a recent publication by

Lin and coworkers it was demonstrated that toll-like receptor 3 ligand, polyinosinic:polycytidylic acid (Poly (I:C)), promotes wound healing in a similar mouse model. They further showed that treatment with Poly(I:C) led to a massive overproduction of chemokines such as MIP-2/CXCL2, MIP-1α/CCL3, MCP-1/CCL2, and RANTES/CCL5, which were responsible for the observed enhanced re-epithelialization [30]. The data by Lin et al. suggest that in our setting these factors might also play an important role for the re-epithelialization of skin wounds. However, in contrast to the study by Lin et al., where a significant effect on wound closure was shown several days after Poly(I:C) treatment, SEC^{PBMC} induces an immediate wound closure. This might be explained by the fact that SEC^{PBMC} already contains high amounts of cytokines and chemokines, in contrast to other treatments, where production of these factors has to be stimulated first. Together we showed a series of events, all contributing to an improved wound healing. The exact mechanisms however, as well as the responsible factor(s) are still not known. Thus, our data provide the basis for further investigations which will address these points in future studies.

In summary, we have shown for the first time that cell culture supernatants obtained from PBMC (SEC^{PBMC}) enhance wound healing in a mouse model and lead to activation of several wound healing associated events in human skin cells. We think that SEC^{PBMC} shows several advantages for the clinical use: 1) The raw material (PBMC) is easily obtainable, 2) It shows minimal or no antigenicity due to cell-free content and 3) it allows "off the shelf" utilization in the clinical setting. Therefore, the formulation and use of SEC^{PBMC} containing emulsions might represent an interesting new option in the treatment of non-healing skin ulcers.

Acknowledgments

We thank Heidemarie Rossiter for essential technical support.

Author Contributions

Conceived and designed the experiments: MM SH ET HJA. Performed the experiments: MM SH TH MG GW CB MZ BG. Analyzed the data: MM SH MG. Wrote the paper: MM SH HJA.

References

1. Elias PM (2005) Stratum corneum defensive functions: An integrated view. Journal of Investigative Dermatology 125: 183–200.
2. Martin P (1997) Wound healing - Aiming for perfect skin regeneration. Science 276: 75–81.
3. Falanga V (2005) Wound healing and its impairment in the diabetic foot. Lancet 366: 1736–1743.
4. Singer AJ, Clark RA (1999) Cutaneous wound healing. N Engl J Med 341: 738–746. doi:10.1056/NEJM199909023411006.
5. Menke NB, Ward KR, Witten TM, Bonchev DG, Diegelmann RF (2007) Impaired wound healing. Clin Dermatol 25: 19–25. doi:10.1016/j.clindermatol.2006.12.005.
6. Holzinger C, Zuckermann A, Kopp C, Schollhammer A, Imhof M, et al. (1994) Treatment of Nonhealing Skin Ulcers with Autologous Activated Mononuclear-Cells. European Journal of Vascular Surgery 8: 351–356.
7. Asai J, Takenaka H, Ichihashi K, Ueda E, Katoh N, et al. (2006) Successful treatment of diabetic gangrene with topical application of a mixture of peripheral blood mononuclear cells and basic fibroblast growth factor. Journal of Dermatology 33: 349–352.
8. Sander AL, Jakob H, Henrich D, Powerski M, Witt H, et al. (2011) Systemic Transplantation of Progenitor Cells Accelerates Wound Epithelialization and Neovascularization in the Hairless Mouse Ear Wound Model. Journal of Surgical Research 165: 165–170.
9. Wu YJ, Chen L, Scott PG, Tredget EE (2007) Mesenchymal stem cells enhance wound healing through differentiation and angiogenesis. Stem Cells 25: 2648–2659.
10. Barcelos LS, Duplaa C, Krankel N, Graiani G, Invernici G, et al. (2009) Human CD133(+) Progenitor Cells Promote the Healing of Diabetic Ischemic Ulcers by Paracrine Stimulation of Angiogenesis and Activation of Wnt Signaling. Circulation Research 104: 1095–U199.
11. Walter MNM, Wright KT, Fuller HR, MacNeil S, Johnson WEB (2010) Mesenchymal stem cell-conditioned medium accelerates skin wound healing: An in vitro study of fibroblast and keratinocyte scratch assays. Experimental Cell Research 316: 1271–1281.
12. Watson SL, Marcal H, Sarris M, Di Girolamo N, Coroneo MTC, et al. (2010) The effect of mesenchymal stem cell conditioned media on corneal stromal fibroblast wound healing activities. British Journal of Ophthalmology 94: 1067–1073.
13. Yew TL, Hung YT, Li HY, Chen HW, Chen LL, et al. (2011) Enhancement of Wound Healing by Human Multipotent Stromal Cell Conditioned Medium: The Paracrine Factors and p38 MAPK Activation. Cell Transplantation 20: 693–706.
14. Gnecchi M, He HM, Liang OD, Melo LG, Morello F, et al. (2005) Paracrine action accounts for marked protection of ischemic heart by Akt-modified mesenchymal stem cells. Nature Medicine 11: 367–368.
15. Angoulvant D, Ivanes F, Ferrera R, Matthews PG, Nataf S, et al. (2011) Mesenchymal stem cell conditioned media attenuates in vitro and ex vivo myocardial reperfusion injury. Journal of Heart and Lung Transplantation 30: 95–102.
16. Hsiao ST, Asgari A, Lokmic Z, Sinclair R, Dusting GJ, et al. (2012) Comparative Analysis of Paracrine Factor Expression in Human Adult Mesenchymal Stem Cells Derived from Bone Marrow, Adipose, and Dermal Tissue. Stem Cells Dev. doi:10.1089/scd.2011.0674.
17. Horn AP, Frozza RL, Grudzinski PB, Gerhardt D, Hoppe JB, et al. (2009) Conditioned medium from mesenchymal stem cells induces cell death in organotypic cultures of rat hippocampus and aggravates lesion in a model of oxygen and glucose deprivation. Neuroscience Research 63: 35–41.
18. Tasso R, Ilengo C, Quarto R, Cancedda R, Caspi RR, et al. (2012) Mesenchymal stem cells induce functionally active T-regulatory lymphocytes in a paracrine fashion and ameliorate experimental autoimmune uveitis. Invest Ophthalmol Vis Sci 53: 786–793. doi:10.1167/iovs.11-8211.
19. Demircan PC, Sariboyaci AE, Unal ZS, Gacar G, Subasi C, et al. (2011) Immunoregulatory effects of human dental pulp-derived stem cells on T cells: comparison of transwell co-culture and mixed lymphocyte reaction systems. Cytotherapy 13: 1205–1220. doi:10.3109/14653249.2011.605351.
20. Ankersmit HJ, Hoetzenecker K, Dietl W, Soleiman A, Horvat R, et al. (2009) Irradiated cultured apoptotic peripheral blood mononuclear cells regenerate infarcted myocardium. European Journal of Clinical Investigation 39: 445–456.
21. Lichtenauer M, Mildner M, Baumgartner A, Hasun M, Werba G, et al. (2011) Intravenous and intramyocardial injection of apoptotic white blood cell suspensions prevents ventricular remodelling by increasing elastin expression in cardiac scar tissue after myocardial infarction. Basic Research in Cardiology 106: 645–655.
22. Lichtenauer M, Mildner M, Hoetzenecker K, Zimmermann M, Podesser BK, et al. (2011) Secretome of apoptotic peripheral blood cells (APOSEC) confers cytoprotection to cardiomyocytes and inhibits tissue remodelling after acute myocardial infarction: a preclinical study. Basic Research in Cardiology 106: 1283–1297.
23. Zhang Y, Klassen HJ, Tucker BA, Perez MTR, Young MJ (2007) CNS progenitor cells promote a permissive environment for neurite outgrowth via a matrix metalloproteinase-2-dependent mechanism. Journal of Neuroscience 27: 4499–4506.
24. Togel F, Westenfelder C (2007) Adult bone marrow-derived stem cells for organ regeneration and repair. Developmental Dynamics 236: 3321–3331.
25. Di Santo S, Yang ZJ, von Ballmoos MW, Voelzmann J, Diehm N, et al. (2009) Novel Cell-Free Strategy for Therapeutic Angiogenesis: In Vitro Generated Conditioned Medium Can Replace Progenitor Cell Transplantation. Plos One 4.
26. Gnecchi M, Zhang ZP, Ni AG, Dzau VJ (2008) Paracrine Mechanisms in Adult Stem Cell Signaling and Therapy. Circulation Research 103: 1204–1219.
27. Korf-Klingebiel M, Kempf T, Sauer T, Brinkmann E, Fischer P, et al. (2008) Bone marrow cells are a rich source of growth factors and cytokines: implications for cell therapy trials after myocardial infarction. European Heart Journal 29: 2851–2858.
28. Smith AN, Willis E, Chan VT, Muffley LA, Isik FF, et al. (2010) Mesenchymal stem cells induce dermal fibroblast responses to injury. Experimental Cell Research 316: 48–54.
29. Yoon BS, Moon JH, Jun EK, Kim J, Maeng I, et al. (2010) Secretory Profiles and Wound Healing Effects of Human Amniotic Fluid-Derived Mesenchymal Stem Cells. Stem Cells and Development 19: 887–902.
30. Lin Q, Wang L, Lin Y, Liu X, Ren X, et al. (2012) Toll-Like Receptor 3 Ligand Polyinosinic:Polycytidylic Acid Promotes Wound Healing in Human and Murine Skin. J Invest Dermatol 132: 2085–92.

PDGF, NT-3 and IGF-2 in Combination Induced Transdifferentiation of Muscle-Derived Stem Cells into Schwann Cell-Like Cells

Yi Tang[1,2⑨], **Hua He**[3⑨], **Ning Cheng**[4⑨], **Yanling Song**[1], **Weijin Ding**[1], **Yingfan Zhang**[1], **Wenhao Zhang**[5], **Jie Zhang**[1], **Heng Peng**[6], **Hua Jiang**[1]*

1 Department of Plastic Surgery, Changzheng Hospital, Second Military Medical University, Shanghai, China, 2 Department of Plastic Surgery, No. 411 Hospital of CPLA, Shanghai, China, 3 Department of Neurosurgery, Changzheng Hospital, Second Military Medical University, Shanghai, China, 4 Department of Transfusion, Changhai Hospital, Second Military Medical University, Shanghai, China, 5 Department of Hematology, XinHua Hospital, Affiliated to Shanghai Jiao Tong University (SJTU) School of Medicine, Shanghai, China, 6 Department of Mathematics, Hong Kong Baptist University, Kowloon, Hong Kong

Abstract

Muscle-derived stem cells (MDSCs) are multipotent stem cells with a remarkable long-term self-renewal and regeneration capacity. Here, we show that postnatal MDSCs could be transdifferentiated into Schwann cell-like cells upon the combined treatment of three neurotrophic factors (PDGF, NT-3 and IGF-2). The transdifferentiation of MDSCs was initially induced by Schwann cell (SC) conditioned medium. MDSCs adopted a spindle-like morphology similar to SCs after the transdifferentiation. Immunocytochemistry and immunoblot showed clearly that the SC markers S100, GFAP and p75 were expressed highly only after the transdifferentiation. Flow cytometry assay showed that the portion of S100 expressed cells was more than 60 percent and over one fourth of the transdifferentiated cells expressed all the three SC markers, indicating an efficient transdifferentiation. We then tested neurotrophic factors in the conditioned medium and found it was PDGF, NT-3 and IGF-2 in combination that conducted the transdifferentiation. Our findings demonstrate that it is possible to use specific neurotrophic factors to transdifferentiate MDSCs into Schwann cell-like cells, which might be therapeutically useful for clinical applications.

Editor: Joao carlos Bettencourt de Medeiros Relvas, IBMC - Institute for Molecular and Cell Biology, Portugal

Funding: This work was supported by National Natural Science Foundation of China (Grant No. 81100950. Website http://www.nsfc.gov.cn/. The funders had no role in study design, data collection and analysis, decision to publish, or preparation of the manuscript.

Competing Interests: The authors have declared that no competing interests exist.

* E-mail: lmiao.simm@gmail.com

⑨ These authors contributed equally to this work.

Introduction

Schwann cells (SCs) play a crucial role in peripheral nerve development and regeneration, and are thus an attractive therapeutic target in peripheral nerve injuries [1–3]. It is reported that cultured SCs could induced neuronal sprouting and regrowth in cell culture experiments and improve peripheral nerve regeneration in vivo [4,5]. SCs can be obtained from nerve biopsies for autologous transplantation and will not elicit an intense immune response. However, it's difficult to culture sufficient numbers of autologous SCs because of their restricted mitotic activity, and there are also other disadvantages such as limitations in the supply of nerve material [6,7]. Use of allogeneic cells would need subsequent clinical immunosuppression [4]. Stem cells may be an alternative source for SCs. However, the clinical application of embryonic stem cells is limited because of ethical problems and their carcinogenic potential [8]. Increase evidence shows that adult stem cells may be promising candidate sources of cells [9,10].

Skeletal muscle may represent a convenient and valuable source of stem cells for stem cell-mediated gene therapy. Previous evidence supports the existence of MDSCs that exhibits both multipotentiality and self-renewal capabilities and therefore can be used for tissue engineering and regenerative therapy [11,12]. MDSCs have the ability to differentiate, upon stimulation with defined media, into multiple types of cells, including myogenic, hematopoietic, osteogenic, adipogenic, and chondrogenic-like cells [13]. The apparent advantages of MDSCs have led us to investigate whether they could be transdifferentiated to a Schwann cell phenotype.

Our aim was to assess the phenotypic and bioassay characteristics of MDSCs transdifferentiated to SC-like cells. Importantly, we also sought to determine the neurotrophic factors which directed the transdifferentiation.

Materials and Methods

Ethics Statement

All animal experiments were approved by the Administrative Committee of Experimental Animal Care and Use of Second Military Medical University (SMMU, Licence No. 2011023), and conformed to the National Institute of Health guidelines on the ethical use of animals.

Isolation and culture of mouse MDSCs

Primary muscle cultures were prepared from newborn (3–5 d) normal C57BL/6 mice, and the MDSCs were purified from the primary culture using a previously described preplate technique [11]. Skeletal muscle was dissected under a light microscopy followed by an enzymatic dissociation. The muscle cells were centrifuged, resuspended and cultured. Different populations of muscle-derived cells were isolated based on their adhesion characteristics. The muscle cells were plated on collagen-coated flasks for 24 h (pp1). The nonadherent cells were then transferred to other flasks (pp2). After 24 h, the floating cells in pp2 were collected, centrifuged, and plated on new flasks (pp3). These procedures were repeated at 24-h intervals until serial preplates (pp4–6) were obtained.

Cell viability and growth assay of mouse MDSCs

Cell viability of mouse MDSCs were measured by trypan blue dye exclusion method. Trypan blue is a dye that cann't enter cells with an intact membrane and therefore stains only the cells with membrane disruption. The MDSCs were stained with 0.025% Trypan blue in PBS. The number of cells was counted using a hemocytometer.

Cell growth curves of mouse MDSCs (PP6 cells) were estimated by counting cell numbers every 24 hours. Cell numbers at individual time points were normalized to those at 0 day.

Conditioned medium of mouse SCs

Primarily cultured mouse SCs were obtained as previously reported [6]. The sciatic nerve and dorsal root ganglia of six mice were isolated and dissected followed by an enzymatic dissociation. The cells were centrifuged, resuspended and cultured. Cytarabine (10 μM) was added in the medium to suppress fibroblast growth. The medium was changed each two days. After 3–4 days, half of the medium was collected and replaced with fresh medium. The conditioned medium of SCs was obtained by repeating the collection three times.

Transdifferentiation of MDSCs to SC-like cells

Cultured MDSCs were stimulated with SC conditioned medium or medium containing different combinations of neurotrophic factors for three days. The morphological changes of the cells were studied using a light microscopy. Expression of SC markers S100, GFAP and p75 in the cells were analyzed by immunocytochemistry, flow cytometry and immunoblot.

Immunocytochemistry

The primary antibodies used in this study were rabbit anti-desmin (1:50, Cell Signaling), rat anti-Sca-1 (1:40, Sigma-Aldrich), mouse anti-S100 (1:40, Invitrogen). Cultured mouse MDSCs, SCs and tMDSCs were fixed and stained according to standard procedures [14].

Flow cytometry

The percentages of Sca-1 and desmin positive MDSCs were analyzed by flow cytometry. All antibodies used in this assay were from eBioscience. Live cell events were collected and analyzed on a FACSCalibur flow cytometer using Cell Quest software.

S100, GFAP and p75 positive SCs or tMDSCs were analyzed using a similar protocol.

Immunoblot

Total proteins were extracted from cells (Cultured MDSCs, SCs, and tMDSCs) using sodium dodecyl sulfate lysis buffer. The protein were electrophoresed on SDS/PAGE and transferred to polyvinyldifluoridine membranes. The membranes were incubated with the primary antibodies followed by the horseradish peroxidase–conjugated anti-rabbit or anti-mouse secondary antibodies. The protein bands were visualized using the ECL system and scanned.

Enzyme-linked immunosorbent assay (ELISA)

The quantities of neurotrophic factors in the SC conditioned medium were measured using ELISA kit from Invitrogen. The tested medium was incubated in plates coated with capture antibodies (anti-BDNF, anti-PDGF, anti-NT-3, anti-IGF-2, anti-NGF, anti-GDNF, anti-FGF). After that, plates were incubated with secondary antibodies and then with peroxidase-conjugated anti-mouse IgG. Soluble colorimetric product was measured.

Statistical analysis

Statistical differences between two groups were determined by two-tailed Student's t test. Multiple group comparisons were made by ANOVA test, using a significance level of 95%. Data were presented as means $\pm$ standard error of the mean.

Results

Isolation and Characterization of mouse MDSCs

The pre-plate technique was used to isolate different populations of muscle cells on the basis of their adherence to collagen-coated flasks [11,15]. Consistent with the previous reports, the cells adhered early on (1st preplate passage, PP1) were mostly fibroblasts without significant directional growth (Fig. S1). Myoblasts adhered at PP3 and satellite cells adhered at approximately PP5 (Fig. S1). The cells of PP6 had a more rounded shape like stem cells (Fig. 1 A). Cell viability determined by trypan blue staining demonstrated that over 95% of the PP6 cells were viable (Data not shown). Cell growth assay showed that the cells of PP6 had a quiescent slow-cycling phenotype which was also a feature of stem cells (Fig. 1 B). To further demonstrate that the PP6 cells were MDSCs, we investigated the expression of Sca1 which was a well-defined marker for putative MDSCs in PP6 cells. Immunocytochemistry showed that the PP6 are Sca 1 positive and are also desmin positive which indicated they were muscle-derived (Fig. 1 C). Flow cytometry assay demonstrated that 93.23±0.93% of the PP6 cells were Sca 1 positive, and 94.18±0.38% were desmin positive. 90.1±1.28% were double positive cells, indicating that most of the PP6 cells were MDSCs (Fig. 1 D). These results suggested that the PP6 cells isolated from mouse skeletal muscles were highly purified MDSCs.

Transdifferentiation of mouse MDSCs to a Schwann cell phenotype

To obtain conditioned medium of SC which would be used for differentiation of MDSCs, we isolated SCs from mouse sciatic nerve and dorsal root ganglia. The isolated cells had typical spindle-shaped SC morphology (Fig. 2 A). The isolated cells were assessed for expression of the SC markers S100, GFAP, and p75. Immunocytochemistry with anti-S100 demonstrated they highly expressed S100 (Fig. 2 B). In addition, flow cytometry assay showed that most of the cells were S100 (96.77±1.46%), GFAP (92.92±4.94%), and p75 (93.38±0.90%) positive, and 86.12±1.53% of the cells expressed all of the three proteins (Fig. 2 C). Thus, the isolated cells were SCs with high purity.

Next, the conditioned medium of isolated mouse SCs was added to the cultured MDSCs to induce their differentiation. 72 hours after the induction, morphology of MDSCs was changed to

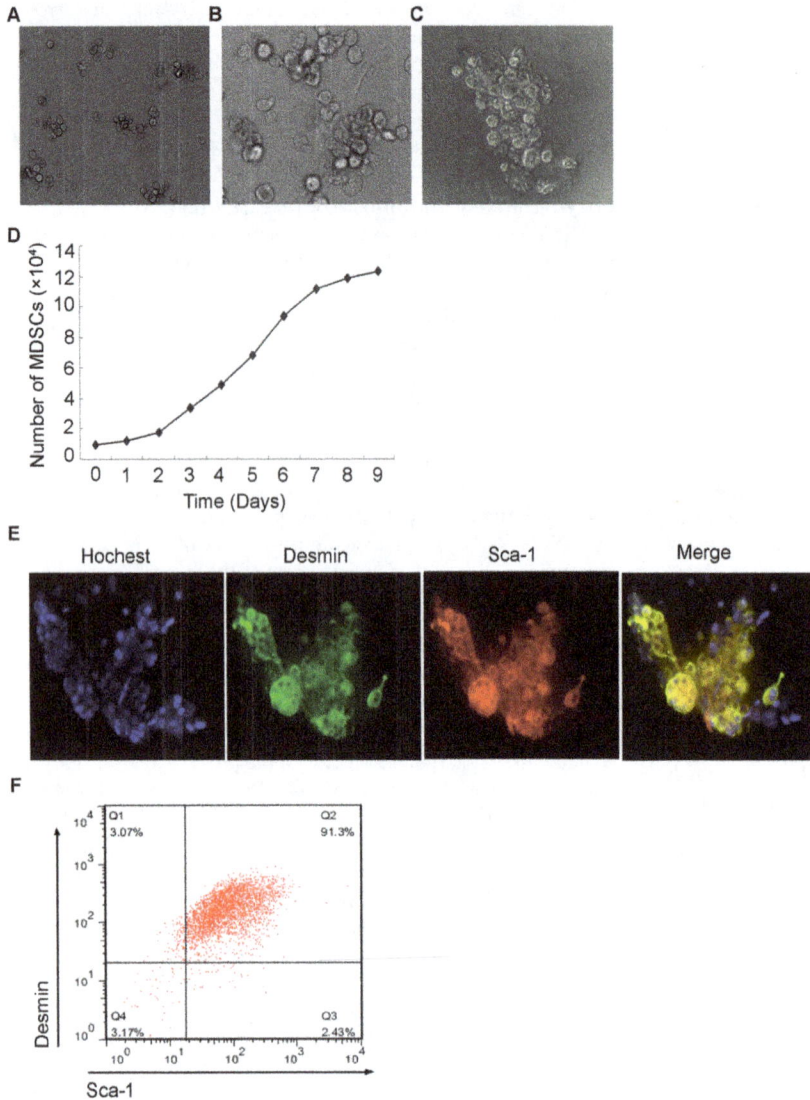

Figure 1. Isolation and Characterisation of mouse MDSCs. (A) Phase-contrast micrographs of undifferentiated PP6 cells isolated from mouse skeletal muscle. The cells had a rounded shape like stem cells. (B) Growth curve of PP6 cells showed a slow-cycling phenotype like stem cells. (C) Immunofluorescence staining showed the PP6 cells were desmin and Sca-1 positive which indicated they were MDSCs. (D) Flow cytometry assay demonstrated that $93.23\pm0.93\%$ of the PP6 cells were Sca 1 positive, and $94.18\pm0.38\%$ were desmin positive. $90.1\pm1.28\%$ were double positive cells (data are mean % cells$\pm$S.E.M.). All experiments were repeated at least three times in triplicates.

spindle-like shape with processes, which was a typical morphology of SC-like cells (Fig. 3 A). The transdifferentiated SC-like cells (tMDSCs) were assessed for expression of the SC markers S100, GFAP, and p75 to study evidence of phenotypic progression to a SC lineage. Immunocytochemisty showed that the tMDSCs highly expressed the SC marker S100 protein (Fig. 3 B). Moreover, flow cytometry assay demonstrated that the portion of S100, GFAP, and p75 positive cells were $65.48\pm6.20\%$, $39.84\pm1.66\%$ and $41.08\pm0.78\%$, while $25.86\pm5.37\%$ of the cells expressed all of the three proteins (Fig. 3 C). Immunoblot assay of the S100, GFAP and p75 expression showed an accord with flow cytometry (Fig. 3 D). The MDSCs expressed the SC markers highly only after transdifferentiation, indicating the tMDSCs were progressed along a SC lineage.

Neurotrophic factors essential for the transdifferentiation

It has been reported that neurotrophic factors secreted by SCs support the survival of neurons cultured in vitro and in vivo after peripheral nerve injury [16,17]. In this study, the conditioned medium of isolated mouse SC could induce the transdifferentiation of MDSCs to SC-like cells. We sought to determine the neurotrophic factors involved in this process.

First, we detected the quantity of seven neurotrophic factors in the conditioned medium, including nerve growth factor (NGF), brain-derived neurotrophic factor (BDNF), neurotrophin-3 (NT-3), glial cell line-derived neurotrophic factor (GDNF), platelet-derived growth factor (PDGF), fibroblast growth factor (FGF) and insulin-like growth factor-2 (IGF-2). Quantitative method of ELISA was conducted and the results showed that there were high levels of BDNF, PDGF, NT-3 and IGF-2 in the conditioned medium (Fig. 4 A). The levels of NGF were relatively low, while GDNF and FGF were undetectable (Fig. 4 A). It seemed that

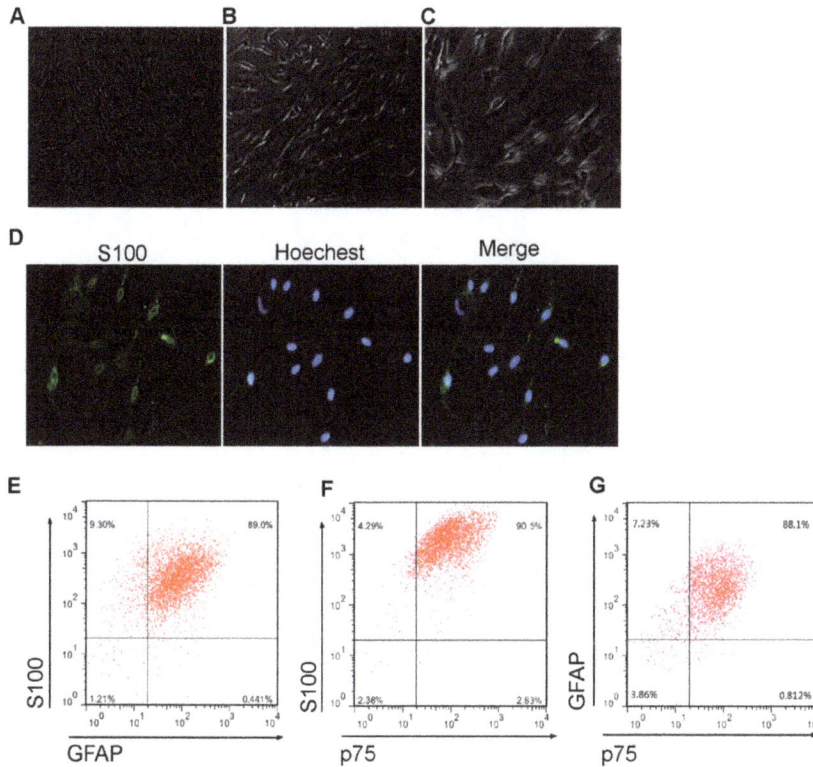

Figure 2. Isolation and Characterisation of mouse SCs to obtain the conditioned medium. (A) Phase-contrast micrographs of SCs isolated from mouse sciatic nerve and dorsal root ganglia. The cells had typical spindle-shaped SC morphology. (B) Immunofluorescence staining demonstrated the isolated cells expressed SC marker S100 protein. (C) Flow cytometry assay showed that most of the cells were S100 (96.77±1.46%), GFAP (92.92±4.94%), and p75 (93.38±0.90%) positive, and 86.12±1.53% of the cells expressed all of the three SC markers (data are mean % cells±S.E.M.).

BDNF, PDGF, NT-3, IGF-2 and NGF might be involved in the transdifferentiation of MDSCs to SC-like cells.

To determine which neurotrophic factors directed the transdifferentiation, we treated the MDSCs with one alone, two in combination or three in combination of the five neurotrophic factors and detect the morphological changes of the cells. None of the five neurotrophic factors could induce morphological changes of MDSCs alone, and the results were similar when MDSCs were treated with two factors in combination (data not shown). In the treatments of three neurotrophic factors in combination, only the combination of PDGF (1000 pg/ml), NT-3 (500 pg/ml) and IGF-2 (200 pg/ml) could induce the morphological change of MDSCs to SC-like cells (Fig. 4 B). Moreover, immunocytochemistry showed that PDGF, NT-3 and IGF-2 in combination also induced expression of SC marker S100 protein in MDSCs (Fig. 4 C). The results of flow cytometry were consistent with the SC conditioned medium induced transdifferentiation (Fig. 3 C), as the portion of S100, GFAP, and p75 positive cells were 58.64±4.38%, 47.38±0.84% and 44.33±2.39%, while 27.89±5.98% of the cells expressed all of the three proteins (Fig. 4 D). In addition, immunoblot assay demonstrated that treatment of PDGF, NT-3 and IGF-2 in combination could induced high expression of S100, GFAP, and p75 in transdifferentiated MDSCs (tMDSCs) (Fig. 4 E). These results indicated the effects of SC conditioned medium on MDSCs transdifferentiation might be mediated by PDGF, NT-3 and IGF-2 in combination.

Discussion

In this study, we showed that primarily cultured mouse MDSCs could be transdifferentiated to Schwann cell phenotype. The transdifferentiation could be induced by Schwann cell conditioned medium or by the combined treatment of three neurotrophic factors. Several kinds of evidence support the success of transdifferentiation. First, we observed that morphology of MDSCs was changed along a typical SC-like spindle-like shape with processes after the transdifferentiation. Second, we showed that the SC marker S100 was highly expressed in the tMDSCs using immunocytochemistry assay. Third, the immunoblot demonstrated clearly that the SC markers S100, GFAP and p75 were expressed highly only after the transdifferentiation. Finally, flow cytometry assay showed that the portion of S100 expressed cells was about 60 percent and over one fourth of the transdifferented cells expressed all the three SC markers, indicating an efficient transdifferentiation.

MDSCs are a potentially new type of undifferentiated cell isolated from skeletal muscle without myogenic restrictions. MDSCs have been reported differentiate into different types of cells, including myogenic, hematopoietic, osteogenic, adipogenic, and chondrogenic-like cells [13]. MDSCs also have a remarkable long-term self-renewal and regeneration capacity [18]. Skeletal muscle is also a convenient source which could not induce the problem of clinically immunosupression. These properties form the basis of potential clinical use of MDSCs in the therapies of degenerative diseases. In this study, we have successfully induced the transdifferentiation of MDSCs towards SC like phenotype,

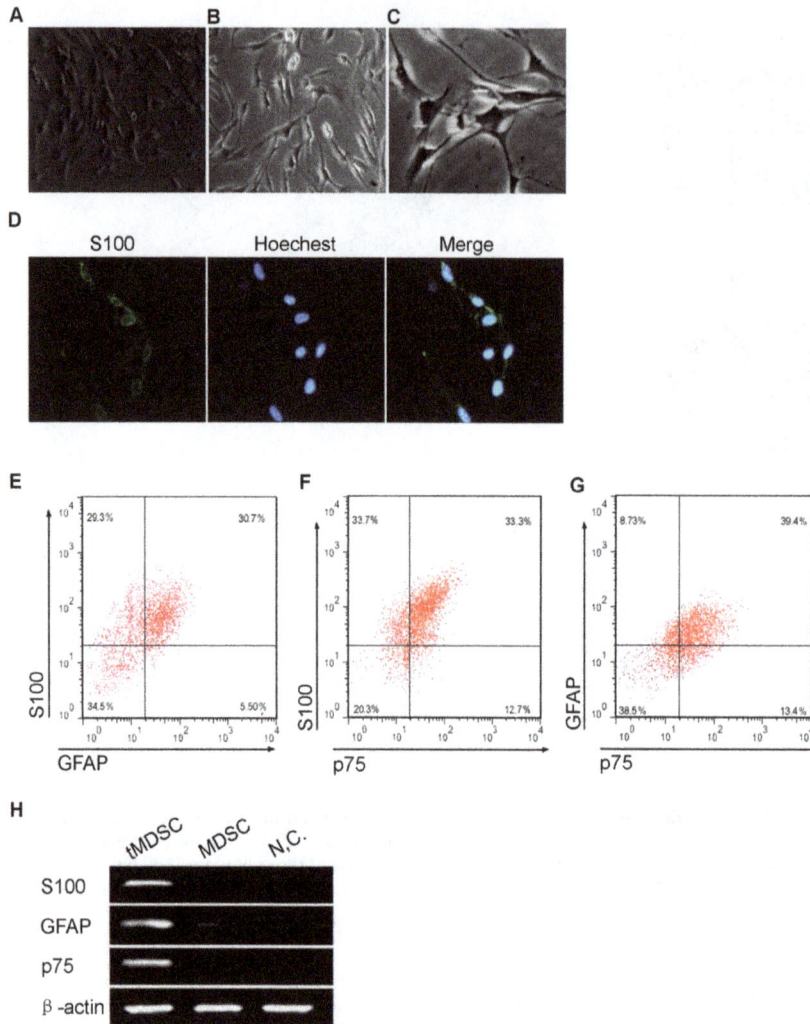

Figure 3. Transdifferentiation of mouse MDSCs to a Schwann cell phenotype. (A) Phase-contrast micrographs showed that MDSCs adopted a spindle-like shape with processes which was a characteristic of Schwann cells after treatment with SC conditioned medium. (B) Immunocytochemisty showed that the tMDSCs highly expressed the SC marker S100 protein. (C) Flow cytometry assay demonstrated that the portion of S100, GFAP, and p75 positive cells in tMDSCs were $65.48\pm6.20\%$, $39.84\pm1.66\%$ and $41.08\pm0.78\%$, and $25.86\pm5.37\%$ of the tMDSCs expressed all of the three SC markers (data are mean % cells$\pm$S.E.M.). (D) Immunoblot assay showed the MDSCs expressed of S100, GFAP and p75 only after transdifferentiation. NC, negative control. β-actin served as loading control. Right, bar graph of qualitative data in statistics. S100, GFAP and p75 protein levels were normalized to that of β-actin, shown as mean$\pm$S.E.M. **, $p<0.01$ vs. MDSC.

however, the true function of these SC like cells remains to be fully investigated.

Previous studies have shown that growth factors affect differentiation directly in stem cell populations [19]. However, there are not many reports about single factor which directs differentiation exclusively to one cell type. SCs developed through stages known as SC precursor cells, early SC and mature myelinating or non-myelinating SC. Several growth factors, such as bFGF, PDGF, neuregulin-1 (NRG-1) and its isoforms, neurotrophin-3 and IGF-1, have been reported involved in the development from SC precursor cells into early SC [20,21]. Adipose-derived stem cells treated with a mixture of glial growth factors (GGF-2, bFGF, PDGF and forskolin) adopted a phenotype similar to Schwann cells [22]. Combinations of bFGF, PDGF-AA and Her-β have proved to successfully induce the differentiation of bone marrow stromal cells to SC like cells [23]. In this study, the combinations of PDGF, NT-3 and IGF-2 successfully induced transdifferentiation of MDSCs along SC like phenotype. From the

neurotrophic factors detected in the Schwann cell conditioned medium, only this combination could induce the transdifferentiation. We also used the reported growth factors (GGF-2, bFGF, PDGF, forskolin, PDGF-AA and Her-β) for the transdifferentiation, but none of these growth factors or the reported combination had positive effects on MDSCs transdifferentiation. These results suggested that specific growth factors conduct the differentiation of different types of stem cells, even differentiation towards the same cell types.

Supporting Information

Figure S1   Morphology of cells in preplate method.

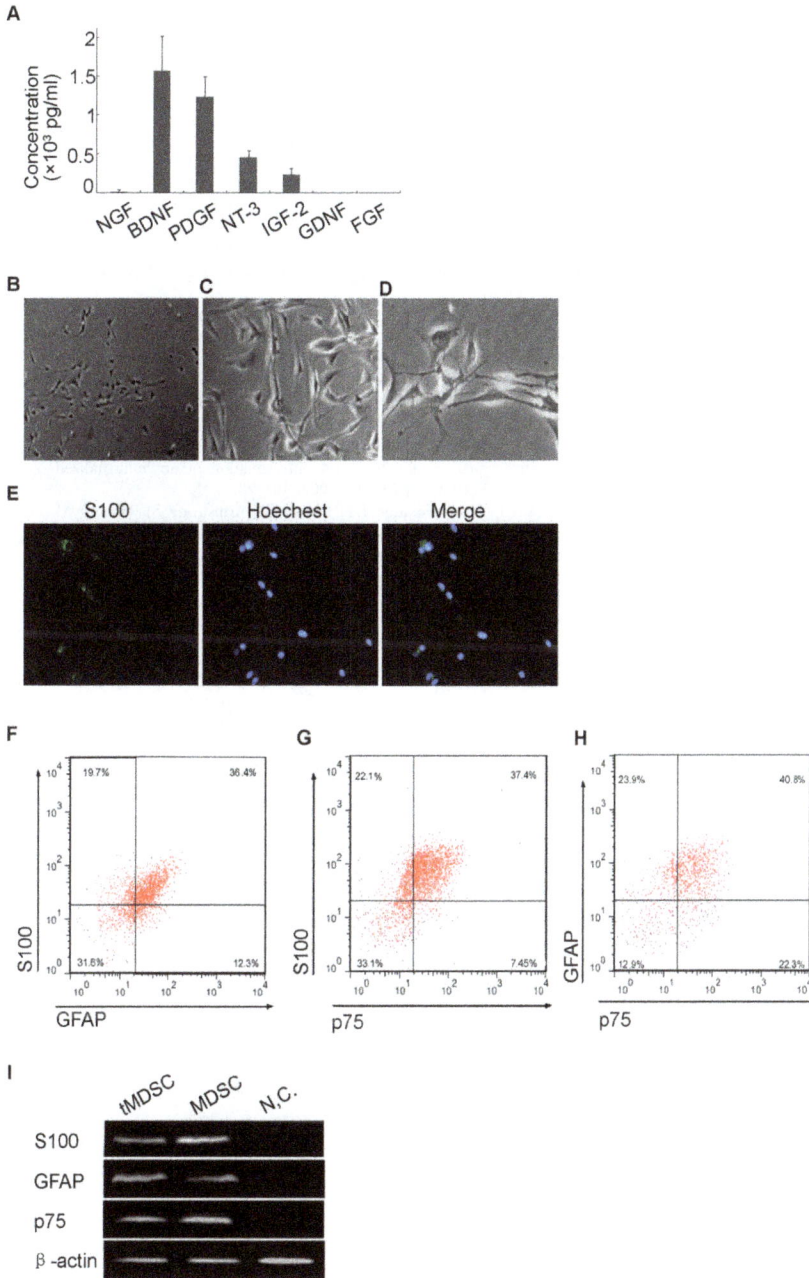

Figure 4. Neurotrophic factors which direct the transdifferentiation. (A) ELISA assay showed the quantity of seven neurotrophic factors in the SC conditioned medium. There were high levels of BDNF, PDGF, NT-3 and IGF-2. The levels of NGF were relatively low, while GDNF and FGF were undetectable. (B) Phase-contrast micrographs showed that MDSCs adopted a SC-like shape upon treatment with PDGF (1000 pg/ml), NT-3 (500 pg/ml) and IGF-2 (200 pg/ml) in combination. (C) Immunocytochemisty showed that the tMDSCs highly expressed the SC marker S100 protein. (D) Flow cytometry assay demonstrated that the portion of S100, GFAP, and p75 positive cells in tMDSCs were $58.64 \pm 4.38\%$, $47.38 \pm 0.84\%$ and $44.33 \pm 2.39\%$, while $27.89 \pm 5.98\%$ of the tMDSCs expressed all of the three SC markers (data are mean % cells $\pm$ S.E.M.). (E) Immunoblot assay showed the MDSCs expressed of S100, GFAP and p75 only after transdifferentiation. β-actin served as loading control. Right, bar graph of qualitative data in statistics. S100, GFAP and p75 protein levels were normalized to that of β-actin, shown as mean $\pm$ S.E.M. **, $p < 0.01$ vs. MDSC.

Acknowledgments

We thank Department of Neurobiology, Second Military Medical University directed by Professor Cheng He for their technical assistance.

References

1. Bunge RP (1994) The role of the Schwann cell in trophic support and regeneration. J Neurol 242: 19–21.

2. Ide C (1996) Peripheral nerve regeneration. Neurosci Res 25: 101–21.

Author Contributions

Conceived and designed the experiments: HJ YT HH. Performed the experiments: YT HH NC YS. Analyzed the data: YT HH WZ HP. Contributed reagents/materials/analysis tools: WD YZ JZ. Wrote the paper: YT HH.

3. Jessen KR, Mirsky R (1999) Schwann cells and their precursors emerge as major regulators of nerve development. Trends Neurosci 22: 402–10.

4. Guenard V, Kleitman N, Morrissey TK, Bunge RP, Aebischer P (1992) Syngenic Schwann cells derived from adult nerves seeded in semi-permeable guidance channels enhance peripheral nerve regeneration. J Neurosci 12: 3310–20.

5. Mosahebi A, Woodward B, Wiberg M, Martin R, Terenghi G (2001) Retroviral labelling of Schwann cells: In vitro charactarisation and in vivo transplantation to improve peripheral nerve regeneration. Glia 34: 8–17.

6. Keilhoff G, Fansa H, Schneider W, Wolf G (1999) In vivo predegeneration of peripheral nerves: an effective technique to obtain activated Schwann cells for nerve conduits. J Neurosci Methods 89: 8917–24.

7. Keilhoff G, Fansa H, Smalla KH, Schneider W, Wolf G (2000) Neuroma: a donor-age independent source of human Schwann cells for tissue engineered nerve grafts. Neuroreport 11: 3805–9.

8. Brustle O, Jones KN, Learish RD, Karram K, Choudhary K, et al. (1999) Embryonic stem cell-derived glial precursors: a source of myelinating transplants. Science 285: 754–6.

9. Grompe M (2012) Tissue stem cells: new tools and functional diversity. Cell Stem Cell 10(6): 685–9.

10. Zhao J, He H, Zhou K, Ren Y, Shi Z, et al. (2012) Neuronal transcription factors induce conversion of human glioma cells to neurons and inhibit tumorigenesis. PLoS One 7(7): e41506.

11. Qu-Petersen Z, Deasy B, Jankowski R, Ikezawa M, Cummins J, et al. (2002) Identification of a novel population of muscle stem cells in mice: potential for muscle regeneration. J Cell Biol 157(5): 851–64.

12. Usas A, Huard J (2007) Muscle-derived stem cells for tissue engineering and regenerative therapy. Biomaterials 28: 5401–6.

13. Deasy BM, Jankowski RJ, and Huard J (2001) Muscle-Derived Stem Cells: Characterization and Potential for Cell-Mediated Therapy. Blood Cells, Molecules, and Diseases 27(5): 924–33.

14. Lee JY, Qu-Petersen Z, Cao B, Kimura S, Jankowski R, et al. (2000) Clonal isolation of muscle-derived cells capable of enhancing muscle regeneration and bone healing. J Cell Biol 150: 1085–100.

15. Gharaibeh B, Lu A, Tebbets J, Zheng B, Feduska J, et al. (2008) Isolation of a slowly adhering cell fraction containing stem cells from murine skeletal muscle by the preplate technique. Nat Protoc 3: 1501–9.

16. Frostick SP, Yin Q, Kemp GJ (1998) Schwann cells, neurotrophic factors, and peripheral nerve regeneration. Microsurgery 18(7): 397–405.

17. Li Q, Ping P, Jiang H, Liu K (2006) Nerve conduit filled with GDNF genemodified Schwann cells enhances regeneration of the peripheral nerve. Microsurgery 26(2): 116–21.

18. Deasy BM, Gharaibeh BM, PoHett JB, Jones MM, Lucas MA, et al. (2005) Long-term self-renewal of postnatal Muscle-derived Stem Cells. Mol Biol Cell 16(7): 3323–33.

19. Schuldiner M, Yanuka O, Itskovitz-Elder J, Melton DA, Benvenisty N (2000) Effects of eight growth factors on the differentiation of cells derived from human embryonic stem cells. Proc Natl Acad Sci USA 97: 11307–12.

20. Cheng HL, Shy M, Feldman EL (1999) Regulation of insulin-like growth factor-binding protein-5 expression during Schwann cell differentiation. Endocrinology 140: 4478–85.

21. Cohen RI, McKay R, Almazan G (1999) Cyclic AMP regulates PDGF-stimulated signal transduction and differentiation of an immortalized optic-nerve-derived cell line. J Exp Biol 202: 461–73.

22. Kingham PJ, Kalbermatten DF, Mahay D, Armstrong SJ, Wiberg M, et al. (2007) Adipose-derived stem cells differentiate into a Schwann cell phenotype and promote neurite outgrowth in vitro. Exp Neurol 207: 267–74.

23. Keilhoff G, Goihl A, Langnase K, Fansa H, Wolf G (2006) Transdifferentiation of mesenchymal stem cells into Schwann cell-likemyelinating cells. Eur J Cell Biol 85: 11–24.

5

Over-Expression of HSP47 Augments Mouse Embryonic Stem Cell Smooth Muscle Differentiation and Chemotaxis

Mei Mei Wong, Xiaoke Yin, Claire Potter, Baoqi Yu, Hao Cai, Elisabetta Di Bernardini, Qingbo Xu*

Cardiovascular Division, King's College London, British Heart Foundation Centre, London, United Kingdom

Abstract

In the recent decade, embryonic stem cells (ESC) have emerged as an attractive cell source of smooth muscle cells (SMC) for vascular tissue engineering owing to their unlimited self-renewal and differentiation capacities. Despite their promise in therapy, their efficacy is still hampered by the lack of definitive SMC differentiation mechanisms and difficulties in successful trafficking of the ESC towards a site of injury or target tissue. Heat shock protein 47 (HSP47) is a 47-kDa molecular chaperone that is required for the maturation of various types of collagen and has been shown to be a critical modulator of different pathological and physiological processes. To date, the role of HSP47 on ESC to SMC differentiation or ESC chemotaxis is not known and may represent a potential molecular approach by which ESC can be manipulated to increase their efficacy in clinic. We provide evidence that HSP47 is highly expressed during ESC differentiation into the SMC lineage and that HSP47 reduction results in an attenuation of the differentiation. Our experiments using a HSP47 plasmid transfection system show that gene over-expression is sufficient to induce ESC-SMC differentiation, even in the absence of exogenous stimuli. Furthermore, HSP47 over-expression in ESC also increases their chemotaxis and migratory responses towards a panel of chemokines, likely via the upregulation of chemokine receptors. Our findings provide direct evidence of induced ESC migration and differentiation into SMC via the over-expression of HSP47, thus identifying a novel approach of molecular manipulation that can potentially be exploited to improve stem cell therapy for vascular repair and regeneration.

Editor: Qingzhong Xiao, William Harvey Research Institute, Barts and The London School of Medicine and Dentistry, Queen Mary University of London, United Kingdom

Funding: Grant Number: BHFPG/04/29/British Heart Foundation/United Kingdom. FunderWebsite: http://www.bhf.org.uk/. The funders had no role in study design, data collection and analysis, decision to publish, or preparation of the manuscript.

Competing Interests: The authors have declared that no competing interests exist.

* E-mail: qingbo.xu@kcl.ac.uk

Introduction

The construction of vascular tissues to replace damaged and injured vessels is a fundamental strategy in development of new treatments for cardiovascular disease. The use of mature endothelial and smooth muscle cells (SMC) is ideal for the repair of diseased vessels. SMC, in particular, are the most abundant cell type within blood vessels and are known to play a key role in maintenance of the vasculature [1]. Nevertheless, the exploitation of these vascular cells for therapy is often limited, due to insufficient supply of the cells, as they can only divide for a finite number of times before undergoing growth arrest and senescence.

In recent years, stem cells have emerged as an attractive cell source for vascular tissue engineering owing to their unlimited self-renewal and differentiation capacities. Indeed cell types such as embryonic stem cells (ESC), induced (or partially induced) pluripotent stem cells (iPS or PiPs) have been documented for their capacity to differentiate into functional SMC [2,3,4]. Furthermore, several novel mechanisms have been postulated to be involved in driving the differentiation process i.e. extracellular signal-regulated kinase (ERK)/β-catenin [5], dickkopf-related protein-3 (DKK-3) [2], reactive oxygen species [6], histone deacetylases 7 [7], microRNA (miRNA)-221 and miRNA-25 [3]. Despite the existing data, a clear consensus has not been reached

and an abundance of studies are still ongoing with the aim to improve the efficacy of stem cells in translational and clinical studies.

Although stem cells are promising tools in the cardiovascular regeneration field, their efficacy is very often hampered by the lack of successful trafficking of the cells towards a site of injury or target tissue [8]. Furthermore, effective transdifferentiation of stem cells is likely to be highly dependent on their regulated chemotaxis, whether via the administration of exogenous factors or molecular programming of the stem cells. Chemokine receptors and their corresponding ligands have been established as key mediators of migration and trafficking in many cell types [9,10]. For instance, the CXCR4/stromal-derived factor-1 (SDF-1) axis has been documented to play a critical role in initiating stem cell movement, in particular that of adventitial stem cells [11], endothelial progenitor cells (EPC) and haematopoietic stem cells (HSC) [12]. To date the chemokine receptor expression profile, of the aforementioned stem cell types, with respect to cardiovascular regeneration remains poorly defined. Identification of the precise means by which chemokine receptor expression can be manipulated in stem cells is likely to improve their delivery towards target tissues.

Heat shock proteins (HSP) are a class of evolutionarily conserved proteins that can be expressed under the influence of various types of stress, thus allowing them to be critical modulators of different pathological and physiological processes [13]. HSP47, or SERPINH1, is a 47-kDa HSP that acts as a molecular chaperone required for proper assembly of triple-helical procollagen molecules, which are eventually transported into extracellular space via the Golgi apparatus [14]. While previous studies show increased expression of HSP47 under several pathological conditions, such as idiopathic pulmonary fibrosis [15], renal scarring [16], neointima formation [17] and glomerulosclerosis [18], the protein has also been found to act as therapeutic mediators that suppress cytotoxicity by reducing abnormally aggregated or misfolded procollagen molecules [19]. Presently, the molecular and functional roles of HSP47 in stem cell differentiation and migration remain unknown.

In this study, we provide novel evidence that endoplasmic reticulum (ER)-resident HSP47 is highly expressed during mouse ESC differentiation into SMC and that lack of HSP47 results in a complete attenuation of differentiation. Furthermore, the overexpression of HSP47 is sufficient to drive ESC-SMC differentiation in the absence of any exogenous stimulation. More intriguingly, our study provides the first evidence that the overexpression of HSP47 also increases ESC chemotaxis and their responses towards chemokines, likely via the upregulation of chemokine receptors. Together, our study suggests a putative role of HSP47 in enhancing the chemotaxis and differentiation of ESC into SMC, thus indicating a novel mechanism of molecular manipulation that can be exploited to improve stem cell therapy for vascular repair and regeneration.

Experimental Procedures

Materials

Cell culture media, serum and cell culture supplements were purchased from ATCC, Millipore, Gibco and PAA. Antibodies against Calponin, Smooth Muscle 22-α (SM-22α) and Smooth Muscle-Myosin Heavy Chain II (SM-MyHCII) were purchased from Abcam. Antibodies against Smooth Muscle-α Actin (SM-αA) and Glyceraldehyde 3-phosphate dehydrogenase (GAPDH) was purchased from Sigma Aldrich. Antibodies against HSP47 were purchased from MBL International and Millipore. Secondary antibodies for immunostaining anti-mouse Alexa488, anti-rabbit Alexa488 and anti-rabbit Alexa594 were purchased from Invitrogen, whereas secondary antibodies for Western Blotting were purchased from Dako.

Embryonic Stem Cell Culture and Differentiation

Mouse ESC were purchased as a cell line (ES-D3, ATCC) and cultured in gelatin-coated flasks (Sigma-Aldrich) in DMEM (ATCC) medium with 10% FBS, leukemia inhibitory factor (10 ng/ml) and 0.1 mM 2-mercaptoethanol. The ESC were differentiated into SMC by culturing on Collagen IV for different lengths of time as stated in α-MEM (Gibco) with 10% FBS and 0.5 mM 2-mercapotoethanol.

Quantitative RT-PCR

Total RNA was isolated from ESC using an RNeasy Mini kit (QIAGEN Inc.) according to manufacturer's instructions. In brief, 2 μg RNA were reverse transcribed into cDNA with random primers by MMLV reverse transcriptase (RT) (Promega) and real time RT-PCR was performed using 2 ng of cDNA per sample with a SYBR Green Master Mix (Promega) in a 25-μl reaction. For each sample, Ct values were measured using ABI PRISM 7000 Sequence Detector (Applied Biosystems) and 18 S ribosomal RNA was used as an endogenous control for normalizing the

amounts of RNA. Sequences of mouse SMC and chemokine receptor primer sets used were as previously described by our laboratory [5,20]. Sequences of the mouse HSP47 primer set are as follows. HSP47: forward5′>GCAGCAGCAAGCAACACTA-CAACT<3′,reverse5′>AGAACATGGCGTTCACAAGCAGT G<3′.

Western Blot Analysis

Embryonic stem cells were harvested and lysed with IP-A buffer (25 mM Tris-HCl pH 7.5, 150 mM NaCl, 1 mM EDTA pH 8.0, 1%Triton X-100 plus protease inhibitors) and proteins were sequentially measured using the Bradford method. ESC lysates (40 μg) were applied to an acrylamide gel using SDS-PAGE and transferred to a nitrocellulose membrane (Amersham Biosciences), followed by a standard western blotting procedure. Polyclonal antibodies against SM-22α, calponin (both purchased from Abcam, UK) GAPDH, SM-αA (both from Sigma Alrich) and HSP47 (MBL, International Corporation, Woburn MA) were used to detect the respective proteins. A HRP-conjugated secondary antibody and an ECL detection system (Amersham Biosciences) were used to detect bound primary antibodies.

Immunofluorescence Staining

Cultures of ESC were fixed with 4% paraformaldehyde, permeabilized with 0.1% Triton X-100 in PBS and blocked with 10% normal swine serum (Dako). Incubation of cell or tissue samples with primary antibodies was performed at 4°C overnight, followed by incubation with secondary antibodies for 30 mins at 37°C after three thorough washes with PBS. Subsequently, the cells were counterstained with DAPI (1:1000 in PBS) for 3 mins at room temperature and mounted with fluorescent mounting media (Dako) before image acquisition using an Axio Imager. M2 microscope and AxioVision Digital Imaging System (Carl Zeiss Ltd.). Primary antibodies used were Calponin, SM-22α (both from Abcam), SM-αA (Sigma Aldrich), HSP47 (Abcam) and DAPI (Sigma Aldrich). The appropriate fluorescent-conjugated (Alexa 488 and Alexa 546) IgG antibodies were used as secondary antibodies (Invitrogen).

Transwell Chemotaxis Assay

Chemotaxis assays were performed with transwell inserts of 5.0 micron membrane pore size (Becton Dickinson Labware, USA). ESC were harvested using trypsin-EDTA and subsequently loaded onto the upper chamber at 1×10^5 cells in serum free media, whereas the bottom chamber contained serum free media with either FBS (20%), eotaxin, KC, MCP-1, Rantes or SDF-1 (all chemokines at 10 ng/ml). After an overnight incubation, nonmigrating ESC on the upper side of the filters were removed using a cotton tip applicator, whereas ESC on the underside of the membrane were fixed with 3% PFA for 10 mins before staining with 1% crystal violet solution (diluted with dH_2O) at room temperature for 15 mins. Data was expressed as the mean number of migrated ESC in 5 random fields of view (at 20×). Chemotaxis assays were also carried out on cells following HSP47 overexpression.

HSP47 Over-expression

For transient over-expression of HSP47, 1.0 μg per 1×10^6 ESC of pCMV6-HSP47 (Origene) expression plasmid was introduced into ESC by nucleofector II (Amaxa, Germany) with a mouse ESC nucleofection kit (Amaxa, VPH-1001). The program A-23 was used according to manufacturer's instructions. An empty vector (Addgene) plasmid was included as a negative control (mock).

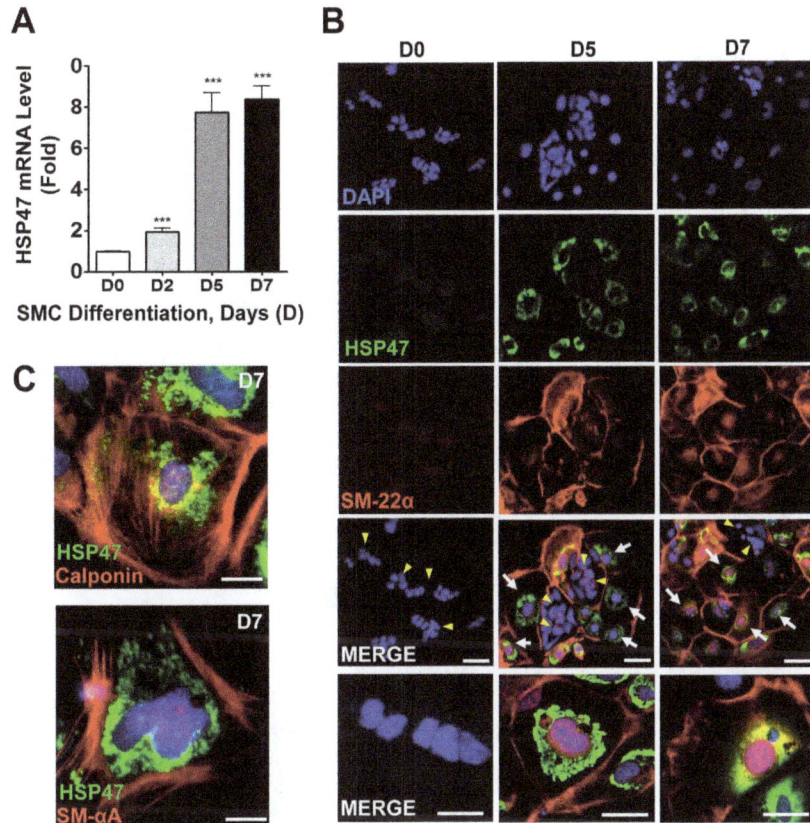

Figure 1. HSP47 is expressed during SMC differentiation of ESC. A–C, HSP47 is increased during ESC-SMC differentiation. Murine ESC were differentiated into SMC in the presence of Collagen IV for the time indicated, followed by quantitative RT-PCR analysis of HSP47 mRNA (A), immunofluorescent staining with HSP47 (green) and SM-22α (red) antibodies (B), calponin (red) and SM-αA (red) (C). DAPI was included to counterstain the nucleus (blue). Scale bar, 30 μm. Images shown are representative of at least three separate experiments, whereas graphs are shown as mean ± SEM of at least three independent experiments, ***$P < 0.005$ compared with Day 0.

Total ESC RNA and proteins were harvested after gene over-expression and subjected to real time RT-PCR or western blot analysis, respectively.

HSP47 Knockdown

Gene suppression of HSP47 (NM_009825.1) was carried out using MISSION short hairpin RNA (shRNA) lentiviral plasmids transfer as previously described [21]. The non-targeting vector, SHC002 was used as a negative control. All shRNA lentiviral plasmids were purchased from Sigma Aldrich UK. Total RNA or proteins were harvested after gene ablation and subjected to real time RT-PCR or western blot analysis, respectively.

Statistical Analysis

Data in this study represent mean and standard error of the mean (S.E.M.) of at least three separate experiments. Statistical analysis was performed using Graphpad Prism V.4 (GraphPad Software, San Diego CA) with analysis of variance (ANOVA) followed by Dunnett's multiple comparison tests. Significance was considered when $p < 0.05$.

Results

HSP47 is expressed during SMC differentiation

Firstly, we wanted to investigate if HSP47 was expressed during the process of SMC differentiation from ESC. Using a previously

established differentiation protocol [22], mouse ESC were differentiated into functional SMC in the presence of Collagen IV. Over the course of differentiation, we found a time-dependent increase of the HSP47 gene that peaked at day 7 (Figure 1A). The upregulation of HSP47 was also confirmed at protein level by immunofluorescence staining with HSP47 (green) and SM-22α (red) antibodies (Figure 1B). Experiments also showed that expression of HSP47 increases in a time dependent manner and that it is localized specifically in the endoplasmic reticulum of the ESC (Figure 1). Intriguingly, HSP47 was only found on cells that also expressed the SMC marker (SM-22α) (indicated by white arrows), but not within clusters of undifferentiated ESC (indicated by yellow arrow heads). Consistently, HSP47 was also expressed concomitantly with other SMC markers such as calponin and SM-αA following 7 days of differentiation (Figure 1C). Thus, these data suggest that HSP47 is likely to play a role in mediating the differentiation of ESC into functional SMCs.

Over-expression of HSP47 can induce stem cell-SMC differentiation

To confirm the role of HSP47 in mediating ESC-SMC differentiation, the gene was ablated in ESC using lentiviral shRNA knockdown prior to differentiation. Figure 2A and 2B show that knockdown of HSP47 results in a marked and significant ablation of SMC markers in response to Collagen IV-induced differentiation.

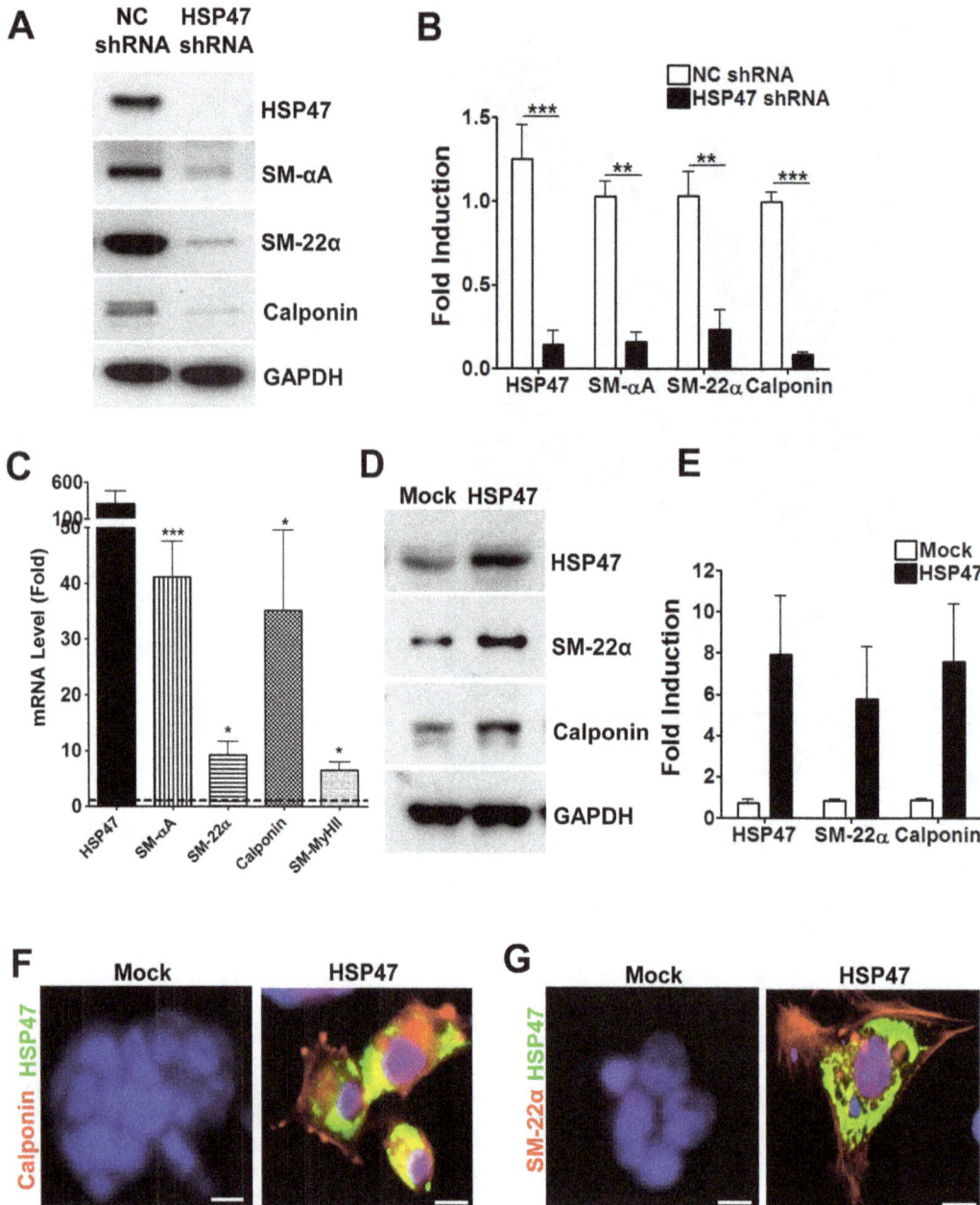

Figure 2. HSP47 can regulate ESC-SMC differentiation. A and B, knockdown of HSP47 abolishes HSP47 expression and attenuates ESC-SMC differentiation. ESCs were infected with non-coding (NC) or HSP47 shRNA lentiviruses for 24 h prior to SMC differentiation, followed by Western blot analysis (A) and quantification (B). C–E, over-expression of HSP47 induces SMC differentiation of ESC. ESCs were transfected either with a pCMV6-HSP47 or mock plasmid (1 μg/1×10⁶ cells) via electroporation, followed by real time RT-PCR (C), Western blotting (D, E) and Immunofluorescent staining with HSP47 (green), calponin (red) (F) and SM-22α (red) (G) antibodies. DAPI was included to counterstain the nucleus (blue). Scale bar, 30 μm. Images and blots shown are representative of at least three separate experiments, whereas graphs are shown as mean ± SEM of at least three independent experiments, *P<0.05, **P<0.01, ***P<0.005 fold increase compared with mock control, dotted line.

While experiments thus far involved SMC differentiation in response to an exogenous stimulus, we wondered whether over-expression of HSP47 itself is sufficient to induce ESC-SMC differentiation. For induction of transient HSP47 over-expression, a HSP47 expression plasmid was introduced into the ESC via nucleofection and maintained in normal culture conditions (without Collagen IV). ESC were also transfected with an empty vector plasmid (mock) as a control. Results show that HSP47-overexpressing ESC had significantly higher gene expression levels of a panel of SMC markers including SM-αA, SM-22α, calponin and SM-MyHCII (Figure 2C). The upregulation of SMC marker expression in HSP47-overexpressing cells was also evident at the protein level, as shown with western blotting (Figures 2D and 2E) and immunofluorescence staining (Figures 2F and 2G). These results show that the over-expression of HSP47 is sufficient to drive SMC differentiation of ESC, albeit the absence of exogenous stimulants such as Collagen IV.

A

B

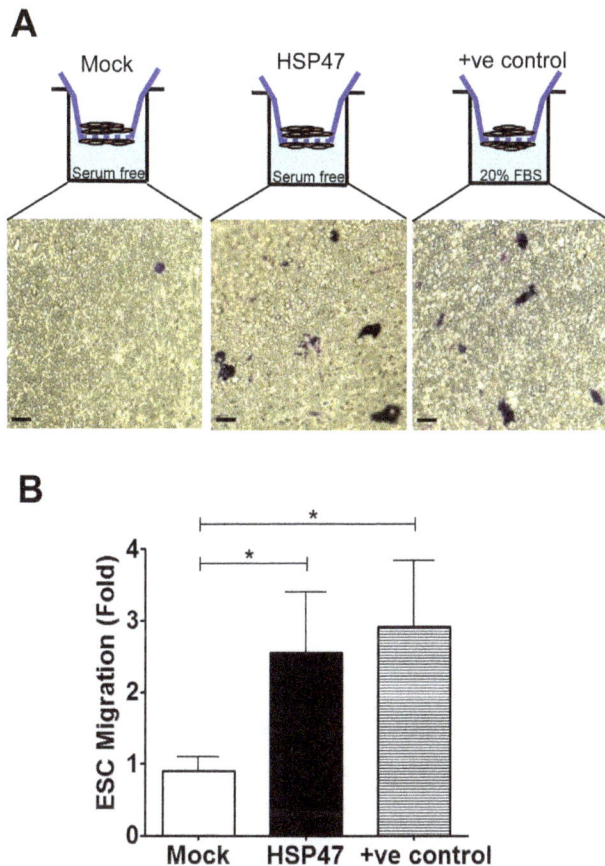

Figure 3. HSP47 over-expression induces ESC chemotaxis. A and B, over-expression of HSP47 induces chemotaxis of ESC. ESCs were transfected with pCMV6-HSP47 plasmid (1 µg/1×10^6 cells) via electroporation and subjected to 5.0micron transwell chemotaxis assays 48 hours later. Chemotaxis of ESCs (either HSP47 or mock) towards serum free media in the transwells was documented following 1% crystal violet staining. (Scale bars, 30 µm) Chemotaxis of ESCs towards media containing 20% FBS was used as a positive control. Chemotaxis index was defined as the mean number of ESCs counted per 10 random fields of view with 20× objective and presented as fold increase compared to the mock control. Graphs are shown as mean ± SEM of three independent experiments. *p<0.05 compared with mock control.

Over-expression of HSP47 induces stem cell chemotaxis

The efficacy of ESC in therapy largely depends on their capacity of homing and localizing within tissues of interest. for repair and regeneration [8]. This concern prompted us to investigate whether, in light of its promising role in inducing the differentiation of ESC into potentially functional SMC, HSP47 could also play a role in inducing ESC migratory capacity. Therefore, we compared the chemotaxis of HSP47-stem cells to mock-stem cells using a 5.0 micron transwell system. Our data confirmed that HSP47-overexpressing ESC had markedly increased chemotaxis capacities compared to mock controls (Figure 3A). Furthermore, the increment that was seen was comparable to stem cell chemotaxis towards 20% FBS (established positive control) (Figure 3A). Quantitative analysis of these transwell assays revealed a significant increase in ESC chemotaxis following HSP47 over-expression (Figure 3B).

HSP47 over-expression induces chemokine-mediated stem cell chemotaxis via induction of chemokine receptor expression

Chemokines and their corresponding receptors play critical roles in regulating the mobilization of many cell types [9,10,23], depending on the appropriate patho/physiological conditions that are present. Therefore, we wondered if the HSP47-induced migratory capacity of ESC could also enhance their tropism towards various chemokines. Notably, we found that in the absence of any stimulus or genetic manipulation, ESC were able to respond to a panel of chemokines at variable affinities; most of which were relatively low when compared to the positive control i.e. 20% FBS (Figure 4A). Intriguingly, we found that over-expression of HSP47 resulted in a super-induction of their migratory responses towards all the chemokines (eotaxin, keratinocyte chemokine (KC), monocyte chemotactic protein-1 (MCP-1), Rantes and SDF-1) and 20% FBS (Figure 4B). Interestingly, the enhancement of HSP47-stem cell chemotaxis was most apparent in response to MCP-1 and SDF-1, both of which have been shown to induce vascular stem/progenitor cell recruitment [11,24].

While evaluating stem cell chemotaxis in response to various chemokines, we also considered the potential effects of HSP47 over-expression on stem cell chemokine receptor expression. Using ESC from 3 separate preparations (in the absence of any stimulus or genetic manipulation), we observed that the cells expressed variable gene levels of chemokine receptors; CCR1-5, CCR7, CXCR1 and CX3CR1 were detectable (Figure 4C). Following the over-expression of HSP47, we found a marked increase in the gene expression levels of most, if not all, chemokine receptors (Figure 4D). Interestingly, chemokine receptors CCR3, 4, 6 and 9, CXCR4 and CX3CR1 were significantly upregulated by at least 3 folds as compared to the mock controls (Figure 4D). Together, our study provides first evidence that HSP47 can play a putative role in driving stem cell-SMC differentiation and enhancing their chemotactic capacities through the induction of chemokine receptors (Figure 5A). These data indicate a potential avenue in which stem cell therapy for vascular diseases could be exploited and improved for clinical use.

Discussion

Vascular SMC are critical for the maintenance of homeostasis within normal blood vessels and are often compromised as a result of insult from vascular diseases. In the recent decades, many investigators have strived to identify numerous therapeutic strategies to engineer vascular tissues that can efficiently replace the lost cells [2]. While the exploitation of ESC as a cell source for functional SMC is extremely attractive [6,7,22,25,26], there remains the need for improvement of their efficacy in clinic. In our study, we provide the first evidence that HSP47 gene expression is significantly upregulated during the differentiation of ESC into SMC. Although predominantly known as an ER-expressed protein, HSP47 has also been reported to be expressed on the cell surface of several epidermoid carcinoma lines [27]. Interestingly, HSP47 was also detected in the peripheral blood of patients with rheumatic autoimmune disease [28], suggesting that it is also expressed as a secreted protein. In our study, we found a distinct pattern of HSP47 expression that was restricted within the ER of differentiating ESCs, and not on the cell surface. Furthermore, our study confirms that the lack of functional HSP47 results in an attenuation of ESC-SMC differentiation altogether.

Subsequent experiments demonstrated that the over-expression of HSP47 alone (i.e. in the absence of exogenous stimulants) was

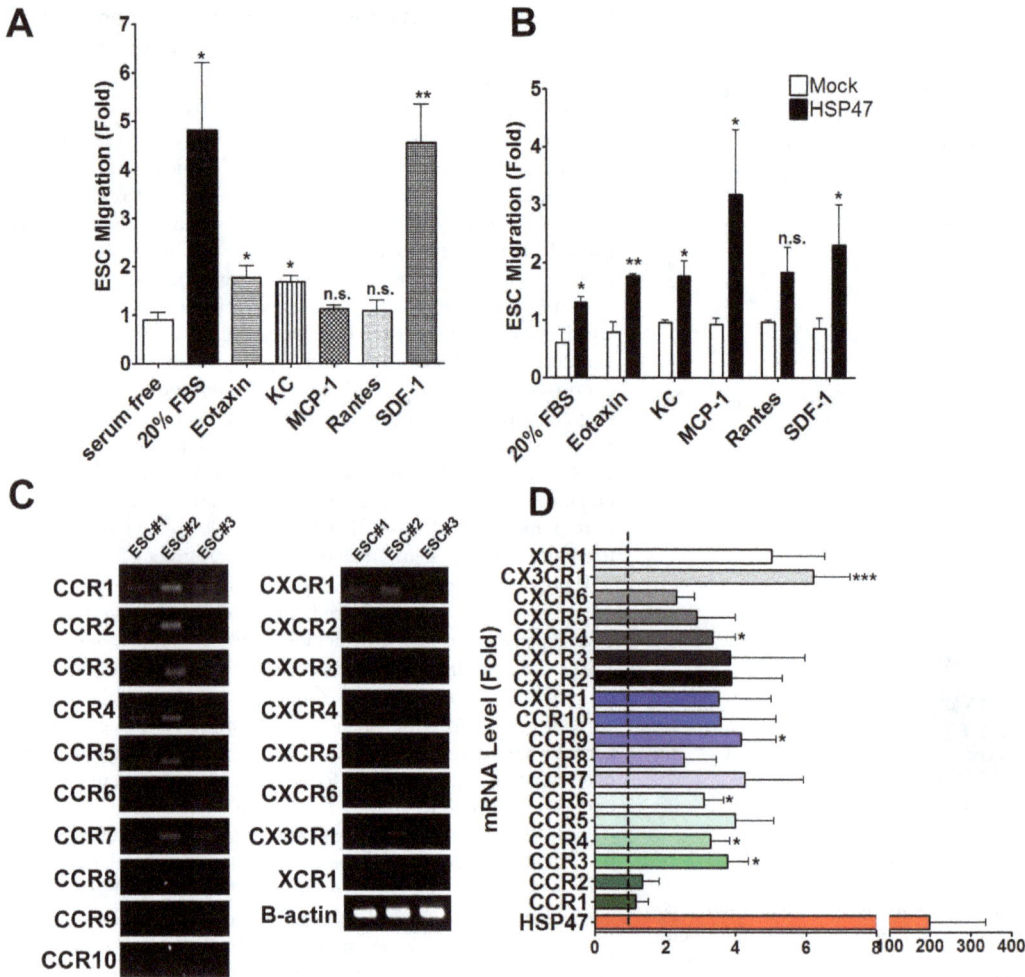

Figure 4. HSP47 over-expression enhances chemokine-mediated ESC chemotaxis and chemokine receptor expression. ESCs migrate in variable affinities in response to a restricted set of chemokines. Chemotaxis of ESCs towards serum free media containing either 20% FBS, eotaxin, KC, MCP-1, Rantes and SDF-1 (all chemokines at 10 ng/ml) in the transwells was documented following 1% crystal violet staining (A). Over-expression of HSP47 induces chemokine-mediated chemotaxis of ESC. Chemotaxis of ESCs (either HSP47 or mock) towards serum free media containing either 20% FBS, eotaxin, KC, MCP-1, Rantes and SDF-1 (all chemokines at 10 ng/ml) was documented following 1% crystal violet staining and represented graphically as a fold increase over mock ESC controls (B). *P<0.05, **P<0.01 compared with mock control. ESCs express a restricted set of chemokine receptor. ESCs cultured under normal culture conditions were harvested and subjected to routine RT-PCR for analysis of chemokine receptor expression (C). Over-expression of HSP47 induces ESC chemokine receptor expression. ESCs were transfected either with a pCMV6-HSP47 or mock plasmid (1 µg/1×10^6 cells) via electroporation, followed by real time RT-PCR (D). Graphs are shown as mean ± SEM of at least three independent experiments, *P<0.05, ***P<0.005 fold increase compared with mock control, dotted line.

sufficient to induce a marked increase in ESC-SMC differentiation. Although the precise mechanisms that drive ESC-SMC differentiation via HSP47 over-expression remain to be elucidated in forthcoming studies, it is likely that it is facilitated by an increase in the production and secretion of procollagens [29]. Indeed, collagen (types I and IV) have been shown to induce SMC differentiation of ESC via integrin (α1/β1/αV) signaling [22]. Furthermore, Matsuoka et al. demonstrated that ESC derived from HSP47-null mice resulted in dysfunctional type I and type IV collagens that lead to abnormal structural formation of embryoid bodies [30].

Transdifferentiation of stem cells is often coupled with efficient trafficking and directed migration of the cells towards a specific target tissue to ensure that a successful initiation of tissue repair and regeneration can be carried out. Moreover, the capacity of stem cell chemotaxis is also crucial for functional integration during which the cells are injected directly into a site of injury.

Chemokine-mediated signalling is fundamental in the migratory behaviour of many cell types and is upregulated especially in response to tissue injury or pathological conditions [9,10]. In the absence of exogenous manipulation, we showed that ESC can migrate towards the bottom of 8.0 µm transwells in response to a restricted set of chemokines such as eotaxin, KC and SDF-1. It is noteworthy that, except for SDF-1, the migration towards the chemokines was either relatively weak or undetectable when compared to controls. The weak affinity of their migratory response was not surprising owing to the variable and low expression of chemokine receptors that the ESCs express, namely CCR1-5, CCR7, CXCR1 and CX3CR1. These data provide potential evidence as to why translational and clinical studies using stem cells as therapeutic agents remain variable and controversial [31,32].

To date, our data comprise the first indication that the over-expression of HSP47 results in a significant induction of ESC

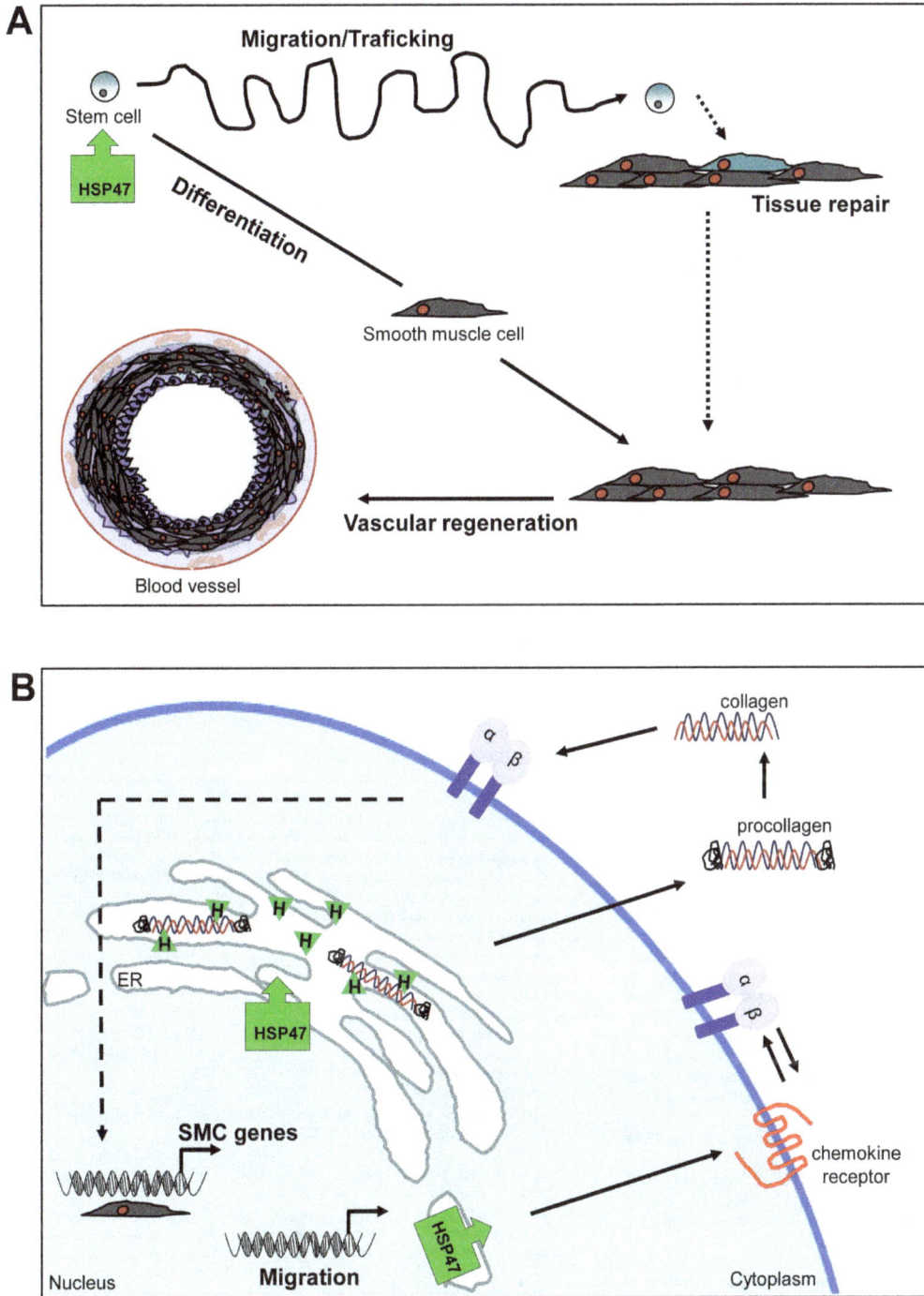

Figure 5. Schematic illustration of the role of HSP47 in enhancing ESC-SMC differentiation and chemotaxis. The over-expression of HSP47 results in an increase of both ESC chemotaxis and SMC differentiation that can lead to tissue repair and vascular regeneration (A). The over-expression of the ER-resident HSP47 (green block arrow) induces SMC differentiation by preferentially binding to procollagen molecules, assists in their functional maturation and subsequent release into the extracellular space of ESC. Increased secretion of collagen can in turn induce SMC differentiation by the activation of SMC specific genes via integrin stimulated signal pathways. The over-expression of HSP47 (green block arrow) also indirectly induces a panel of chemokine receptors, potentially through the co-operation with integrin signaling pathways (B).

chemotaxis. More importantly, the over-expression further induced ESC response towards the panel of chemokines and this was concomitant with an upregulation of most chemokine receptors, except for CCR1 and CCR2, by at least 2 folds compared to the controls. Indeed, the over-expression of chemokine receptors such as CXCR4 [33] and CCR1[34] in mesenchymal stem cells (MSC) was shown to result in increased cellular migration and survival. Mechanistically, the induction of ESC chemokine receptor expression is unlikely to be a direct result of HSP47 over-expression because HSP47 is known to bind solely to collagens of at least types I-V in the ER [14,35]. It is plausible that the increase in ESC chemotaxis is mediated by a co-operation between the chemokine receptor and integrin signaling pathways [36,37] (Figure 5B). The precise mechanisms involved remain to be elucidated.

In summary, the present study provides fundamental evidence that HSP47 is a ER-resident protein that is highly expressed and required during SMC differentiation of ESC. Furthermore, our data provides evidence that the expression of HSP47 can be manipulated in ESC to induce their active chemotaxis and differentiation towards the SMC lineage in the absence of exogenous stimulation. Our simple but concise study postulates a novel and promising approach to improve stem cell therapy for vascular repair and regeneration.

Author Contributions

Conceived and designed the experiments: MMW QX. Performed the experiments: MMW XY CP BY HC EDB. Analyzed the data: MMW XY CP QX. Contributed reagents/materials/analysis tools: QX. Wrote the paper: MMW QX.

References

1. van der Loop FT, Gabbiani G, Kohnen G, Ramaekers FC, van Eys GJ (1997) Differentiation of smooth muscle cells in human blood vessels as defined by smoothelin, a novel marker for the contractile phenotype. Arterioscler Thromb Vasc Biol 17: 665–671.

2. Karamariti E, Margariti A, Winkler B, Wang X, Hong X, et al. (2013) Smooth muscle cells differentiated from reprogrammed embryonic lung fibroblasts through DKK3 signaling are potent for tissue engineering of vascular grafts. Circ Res 112: 1433–1443.

3. Xiao Q, Wang G, Luo Z, Xu Q (2010) The mechanism of stem cell differentiation into smooth muscle cells. Thromb Haemost 104: 440–448.

4. Xie C, Hu J, Ma H, Zhang J, Chang LJ, et al. (2011) Three-dimensional growth of iPS cell-derived smooth muscle cells on nanofibrous scaffolds. Biomaterials 32: 4369–4375.

5. Wong MM, Winkler B, Karamariti E, Wang X, Yu B, et al. (2013) Sirolimus Stimulates Vascular Stem/Progenitor Cell Migration and Differentiation Into Smooth Muscle Cells via Epidermal Growth Factor Receptor/Extracellular Signal-Regulated Kinase/beta-Catenin Signaling Pathway. Arterioscler Thromb Vasc Biol 33: 2397–2406.

6. Xiao Q, Luo Z, Pepe AE, Margariti A, Zeng L, et al. (2009) Embryonic stem cell differentiation into smooth muscle cells is mediated by Nox4-produced H2O2. Am J Physiol Cell Physiol 296: C711–723.

7. Zhang L, Jin M, Margariti A, Wang G, Luo Z, et al. (2010) Sp1-dependent activation of HDAC7 is required for platelet-derived growth factor-BB-induced smooth muscle cell differentiation from stem cells. J Biol Chem 285: 38463–38472.

8. Smart N, Riley PR (2008) The stem cell movement. Circ Res 102: 1155–1168.

9. DeVries ME, Ran L, Kelvin DJ (1999) On the edge: the physiological and pathophysiological role of chemokines during inflammatory and immunological responses. Semin Immunol 11: 95–104.

10. Kiefer F, Siekmann AF (2011) The role of chemokines and their receptors in angiogenesis. Cell Mol Life Sci 68: 2811–2830.

11. Chen Y, Wong MM, Campagnolo P, Simpson R, Winkler B, et al. (2013) Adventitial stem cells in vein grafts display multilineage potential that contributes to neointimal formation. Arterioscler Thromb Vasc Biol 33: 1844–1851.

12. Moore MA, Hattori K, Heissig B, Shieh JH, Dias S, et al. (2001) Mobilization of endothelial and hematopoietic stem and progenitor cells by adenovector-mediated elevation of serum levels of SDF-1, VEGF, and angiopoietin-1. Ann N Y Acad Sci 938: 36–45; discussion 45–37.

13. Minowada G, Welch WJ (1995) Clinical implications of the stress response. J Clin Invest 95: 3–12.

14. Widmer C, Gebauer JM, Brunstein E, Rosenbaum S, Zaucke F, et al. (2012) Molecular basis for the action of the collagen-specific chaperone Hsp47/SERPINH1 and its structure-specific client recognition. Proc Natl Acad Sci U S A 109: 13243–13247.

15. Iwashita T, Kadota J, Naito S, Kaida H, Ishimatsu Y, et al. (2000) Involvement of collagen-binding heat shock protein 47 and procollagen type I synthesis in idiopathic pulmonary fibrosis: contribution of type II pneumocytes to fibrosis. Hum Pathol 31: 1498–1505.

16. Razzaque MS, Ahsan N, Taguchi T (2000) Heat shock protein 47 in renal scarring. Nephron 86: 339–341.

17. Murakami S, Toda Y, Seki T, Munetomo E, Kondo Y, et al. (2001) Heat shock protein (HSP) 47 and collagen are upregulated during neointimal formation in the balloon-injured rat carotid artery. Atherosclerosis 157: 361–368.

18. Moriyama T, Kawada N, Ando A, Yamauchi A, Horio M, et al. (1998) Up-regulation of HSP47 in the mouse kidneys with unilateral ureteral obstruction. Kidney Int 54: 110–119.

19. Barral JM, Broadley SA, Schaffar G, Hartl FU (2004) Roles of molecular chaperones in protein misfolding diseases. Semin Cell Dev Biol 15: 17–29.

20. Margariti A, Xiao Q, Zampetaki A, Zhang Z, Li H, et al. (2009) Splicing of HDAC7 modulates the SRF-myocardin complex during stem-cell differentiation towards smooth muscle cells. J Cell Sci 122: 460–470.

21. Margariti A, Winkler B, Karamariti E, Zampetaki A, Tsai TN, et al. (2012) Direct reprogramming of fibroblasts into endothelial cells capable of angiogenesis and reendothelialization in tissue-engineered vessels. Proc Natl Acad Sci U S A 109: 13793–13798.

22. Xiao Q, Zeng L, Zhang Z, Hu Y, Xu Q (2007) Stem cell-derived Sca-1+ progenitors differentiate into smooth muscle cells, which is mediated by collagen IV-integrin alpha1/beta1/alphav and PDGF receptor pathways. Am J Physiol Cell Physiol 292: C342–352.

23. Cho C, Miller RJ (2002) Chemokine receptors and neural function. J Neurovirol 8: 573–584.

24. Grudzinska MK, Kurzejamska E, Bojakowski K, Soin J, Lehmann MH, et al. (2013) Monocyte chemoattractant protein 1-mediated migration of mesenchymal stem cells is a source of intimal hyperplasia. Arterioscler Thromb Vasc Biol 33: 1271–1279.

25. Pepe AE, Xiao Q, Zampetaki A, Zhang Z, Kobayashi A, et al. (2010) Crucial role of nrf3 in smooth muscle cell differentiation from stem cells. Circ Res 106: 870–879.

26. Xiao Q, Pepe AE, Wang G, Luo Z, Zhang L, et al. (2012) Nrf3-Pla2g7 interaction plays an essential role in smooth muscle differentiation from stem cells. Arterioscler Thromb Vasc Biol 32: 730–744.

27. Hebert C, Norris K, Della Coletta R, Reynolds M, Ordonez J, et al. (1999) Cell surface colligin/Hsp47 associates with tetraspanin protein CD9 in epidermoid carcinoma cell lines. J Cell Biochem 73: 248–258.

28. Yokota S, Kubota H, Matsuoka Y, Naitoh M, Hirata D, et al. (2003) Prevalence of HSP47 antigen and autoantibodies to HSP47 in the sera of patients with mixed connective tissue disease. Biochem Biophys Res Commun 303: 413–418.

29. Rocnik EF, van der Veer E, Cao H, Hegele RA, Pickering JG (2002) Functional linkage between the endoplasmic reticulum protein Hsp47 and procollagen expression in human vascular smooth muscle cells. J Biol Chem 277: 38571–38578.

30. Matsuoka Y, Kubota H, Adachi E, Nagai N, Marutani T, et al. (2004) Insufficient folding of type IV collagen and formation of abnormal basement membrane-like structure in embryoid bodies derived from Hsp47-null embryonic stem cells. Mol Biol Cell 15: 4467–4475.

31. Murry CE, Soonpaa MH, Reinecke H, Nakajima H, Nakajima HO, et al. (2004) Haematopoietic stem cells do not transdifferentiate into cardiac myocytes in myocardial infarcts. Nature 428: 664–668.

32. Abdel-Latif A, Bolli R, Tleyjeh IM, Montori VM, Perin EC, et al. (2007) Adult bone marrow-derived cells for cardiac repair: a systematic review and meta-analysis. Arch Intern Med 167: 989–997.

33. Du Z, Wei C, Yan J, Han B, Zhang M, et al. (2013) Mesenchymal stem cells overexpressing C-X-C chemokine receptor type 4 improve early liver regeneration of small-for-size liver grafts. Liver Transpl 19: 215–225.

34. Huang J, Zhang Z, Guo J, Ni A, Deb A, et al. (2010) Genetic modification of mesenchymal stem cells overexpressing CCR1 increases cell viability, migration, engraftment, and capillary density in the injured myocardium. Circ Res 106: 1753–1762.

35. Natsume T, Koide T, Yokota S, Hirayoshi K, Nagata K (1994) Interactions between collagen-binding stress protein HSP47 and collagen. Analysis of kinetic parameters by surface plasmon resonance biosensor. J Biol Chem 269: 31224–31228.

36. Hartmann TN, Burger JA, Glodek A, Fujii N, Burger M (2005) CXCR4 chemokine receptor and integrin signaling co-operate in mediating adhesion and chemoresistance in small cell lung cancer (SCLC) cells. Oncogene 24: 4462–4471.

37. Umehara H, Bloom E, Okazaki T, Domae N, Imai T (2001) Fractalkine and vascular injury. Trends Immunol 22: 602–607.

6

In Vitro Epigenetic Reprogramming of Human Cardiac Mesenchymal Stromal Cells into Functionally Competent Cardiovascular Precursors

Matteo Vecellio[1,2], Viviana Meraviglia[2], Simona Nanni[3], Andrea Barbuti[4], Angela Scavone[4], Dario DiFrancesco[4], Antonella Farsetti[5,6], Giulio Pompilio[1,2], Gualtiero I. Colombo[7], Maurizio C. Capogrossi[8], Carlo Gaetano[9*❦], Alessandra Rossini[1*❦]

1 Dipartimento di Scienze Cliniche e di Comunità, Università di Milano, Milano, Italy, 2 Laboratorio di Biologia Vascolare e Medicina Rigenerativa, Centro Cardiologico Monzino, IRCCS, Milano, Italy, 3 Istituto di Patologia Medica, Università Cattolica del Sacro Cuore, Roma, Italy, 4 Dipartimento di Bioscienze, The PaceLab, Università di Milano, Milano, Italy, 5 Istituto Nazionale Tumori Regina Elena, Roma, Italy, 6 Istituto di Neurobiologia e Medicina Molecolare, Consiglio Nazionale delle Ricerche (CNR), Roma, Italy, 7 Laboratorio di Immunologia e Genomica Funzionale, Centro Cardiologico Monzino, IRCCS, Milano, Italy, 8 Laboratorio di Patologia Vascolare, Istituto Dermopatico dell'Immacolata - IRCCS, Roma, Italy, 9 Division of Cardiovascular Epigenetics, Department of Cardiology, Goethe University, Frankfurt am Main, Germany

Abstract

Adult human cardiac mesenchymal-like stromal cells (CStC) represent a relatively accessible cell type useful for therapy. In this light, their conversion into cardiovascular precursors represents a potential successful strategy for cardiac repair. The aim of the present work was to reprogram CStC into functionally competent cardiovascular precursors using epigenetically active small molecules. CStC were exposed to low serum (5% FBS) in the presence of 5 µM all-trans Retinoic Acid (ATRA), 5 µM Phenyl Butyrate (PB), and 200 µM diethylenetriamine/nitric oxide (DETA/NO), to create a novel epigenetically active cocktail (EpiC). Upon treatment the expression of markers typical of cardiac resident stem cells such as c-Kit and MDR-1 were up-regulated, together with the expression of a number of cardiovascular-associated genes including KDR, GATA6, Nkx2.5, GATA4, HCN4, NaV1.5, and α-MHC. In addition, profiling analysis revealed that a significant number of microRNA involved in cardiomyocyte biology and cell differentiation/proliferation, including miR 133a, 210 and 34a, were up-regulated. Remarkably, almost 45% of EpiC-treated cells exhibited a TTX-sensitive sodium current and, to a lower extent in a few cells, also the pacemaker I_f current. Mechanistically, the exposure to EpiC treatment introduced global histone modifications, characterized by increased levels of H3K4Me3 and H4K16Ac, as well as reduced H4K20Me3 and H3s10P, a pattern compatible with reduced proliferation and chromatin relaxation. Consistently, ChIP experiments performed with H3K4me3 or H3s10P histone modifications revealed the presence of a specific EpiC-dependent pattern in c-Kit, MDR-1, and Nkx2.5 promoter regions, possibly contributing to their modified expression. Taken together, these data indicate that CStC may be epigenetically reprogrammed to acquire molecular and biological properties associated with competent cardiovascular precursors.

Editor: Toru Hosoda, Tokai University, Japan

Funding: The present study was supported by the Italian Ministry of Health (RC 2010-2011) and by the Italian Ministry of Education, University and Research (FIRB-MIUR RBFR087JMZ to A.R. and A.F; FIRB-MIUR RBFR10URHP_001 to S.N; PRIN-MIUR 2008NY72S to A.F; PRIN-MIUR 2008ETWBTW to DD). C.G. is supported by the LOEWE-CGT centre of excellence. The funders had no role in study design, data collection and analysis, decision to publish, or preparation of the manuscript.

Competing Interests: The authors have declared that no competing interests exist.

* E-mail: Gaetano@em.uni-frankfurt.de (CG); alessandra.rossini@unimi.it (AR)

❦ These authors contributed equally to this work.

Introduction

Cellular cardiomyoplasty is a promising therapy to reconstitute injured hearts. Cell based interventions aimed at structurally regenerating the heart imply that transplanted cells graft in the host tissue and adopt the phenotype of resident cardiomyocytes, endothelial cells and smooth muscle cells. In this light, cells possessing pluripotency, such as embryonic stem (ES) cells and the so-called induced-pluripotent stem cells (iPS) [1] may be considered a good candidate. Yet, although several attempts have been made to simplify iPS cell generation methods avoiding the undesired effect of neoplastic transformation [2], their use still raises safety concerns. Consequently, much effort has been put

into promoting cardiovascular differentiation of adult cells. Intriguingly, a recent work has shown the possibility of directly converting neonatal or embryonic mouse fibroblasts into cardiomyocytes by a transcription-factor based reprogramming strategy [3]. In spite of its potential practical relevance, the efficiency of this procedure is low and genetic manipulation of target cells is still required. In this context, chemical strategies based on the use of small active molecules represent an easier, more effective and safer alternative to genetic methods [4]. Further, achieving terminal differentiation of adult somatic or stem cells into cardiomyocytes may not be the right therapeutic approach, as multiple cell types (i.e. cardiomyocytes, vascular cells, and fibroblasts) should be generated to rebuild damaged heart tissue. In this perspective, reprogramming of adult cardiac cells into progenitors, which are less de-differentiated than iPS cells and exhibit lineage commitment restricted to the cell types of interest, may represent a successful strategy.

In addition to adult cardiac stem cells, a different category of heart cells, namely cardiac mesenchymal stromal cells (CStC) deriving from the cardiac parenchyma, have been recently isolated and characterized by our group [5]. CStC are c-Kit negative, reveal positivity for both pericytes (i.e. CD146) and fibroblast markers (i.e. vimentin and human fibroblast surface antigen) and share similarities with syngeneic bone marrow mesenchymal stromal cells (BMStC), showing comparable morphology and expression of mesenchymal-associated antigens (i.e. CD105, CD73, CD29, and CD44). Despite their similarities, significant differences between CStC and BMStC emerged. In fact, CStC may be identified by the expression of a specific microRNA signature and exhibit a residual plasticity toward the cardiovascular lineage, possessing the ability to contribute new adult-like cardiomyocytes after heart ischemia with higher efficiency than BMStC [5]. Of note, CStC are easily obtained from small biopsyspecimens and may be efficiently grown *in vitro*.

A large number of evidences indicate that biological response modifiers, including epigenetically active small molecules, such as histone deacetylase inhibitors (HDACi), may facilitate the redirection of adult cellular functions toward stemness [4]. So far, however, no reportsdescribed the *in vitro* enhancement of adult cardiac precursors via a defined cocktail of drugs.

In this report, we describe the properties of CStC chemically converted into functional cardiovascular precursors by means of nutrients deprivation in the presence of retinoids, phenyl butyrate and nitric oxide drugs. Remarkably, these drugs have been used in the past to stimulate *in vitro* cardiomyocyte production in different experimental contexts [6,7,8], but they were never combined together on cells isolated from human adult heart. These compounds have different mechanisms of action including nuclear receptor activation (retinoic acid) [9] and inhibition of histone deacetylases (phenyl butyrate). Further, nitric oxide donors such as DETA/NO, have been shown to play a role in the prevention of apoptosis, microRNA up-regulation [10] and cardiac commitment [11]. Therefore it is possible that these drugs may synergize in determining the generation of functionally competent cardiovascular precursor-like cells.

Materials and Methods

Ethics Statement

Cardiac stromal cells were obtained from right auricle samples of donor patients undergoing valve substitution or by-pass surgery after signed informed consent and approval by the Centro Cardiologico Monzino (Milano, Italy) Ethical Committee. Investigations were conducted according to the principles expressed in the Declaration of Helsinki. Data were analyzed anonymously.

CStC Isolation and Culture

CStC were isolated from right auricles and cultured in growth medium (GM) as previously described [5]. CStC at passages 4–8 were incubated for 7 days with an Epigenetic cocktail (EpiC), composed by Iscove's Modified Dulbecco's Medium (IMDM), 5% Foetal Bovine Serum (FBS), 10.000 U/ml Penicillin/Streptomicin, 10 μg/ml L-Glutamine, 5 μM All trans Retinoic Acid (ATRA), 5 μM Phenyl Butyrate (PB), and 200 μM diethylenetriamine/nitric oxide (DETA/NO), all purchased from Sigma Aldrich. The EpiC medium was changed every 48 hrs.

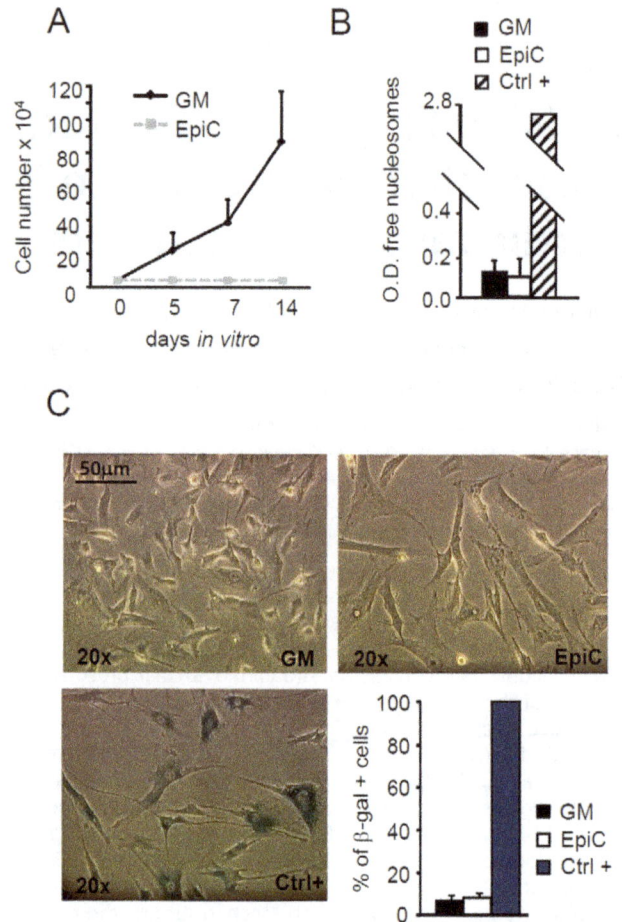

Figure 1. Effect of EpiC treatment on CStC growth, viability and senescence. (A) Growth curve of CStC cultured in GM or EpiC for 5, 7, and 14 days (n=7). (B) Quantification of free nucleosomes, used as markers of apoptosis, in CStC exposed to GM or EpiC for7 days. Ctrl+ = positive control supplied by the manufacturer. (C) Staining for senescence-related acidic β-galactosidase (β-gal) performed on CStC grown either in GM or in EpiC medium. Positivity for β-gal is indicated by the presence of a dark grey stain within the cytoplasm. Ctrl+ = primary bone marrow mesenchymal stromal cells at passage 10 (replicative senescence). Bar Graph shows average percentage of β-gal positive cells (n=4). Original magnification: 20×.

Western Blot Analysis

CStC in control and EpiC media were lysed with Laemmli buffer in presence of protease and phosphatase inhibitors (Roche Diagnostic). Proteins were resolved by SDS–PAGE, transferred onto nitrocellulose membranes (Bio-Rad Laboratories), and incubated overnight at 4°C with primary antibodies listed in Supplementary Table S1. Subsequently, the blots were incubated with the appropriate anti-rabbit, anti-mouse, or anti-goat horseradish peroxidase-conjugated secondary antibody (Amersham-GE Healthcare). ECL or, when appropriate, ECL plus (Amersham-GE Healthcare) were used for chemiluminescence detection. Each filter was also probed with anti-β-actin, anti-β-tubulin, anti-H3, or anti-H4, to verify equal protein loading. Densitometric analyses were performed by NIH ImageJ software, version 1.4.3.67.

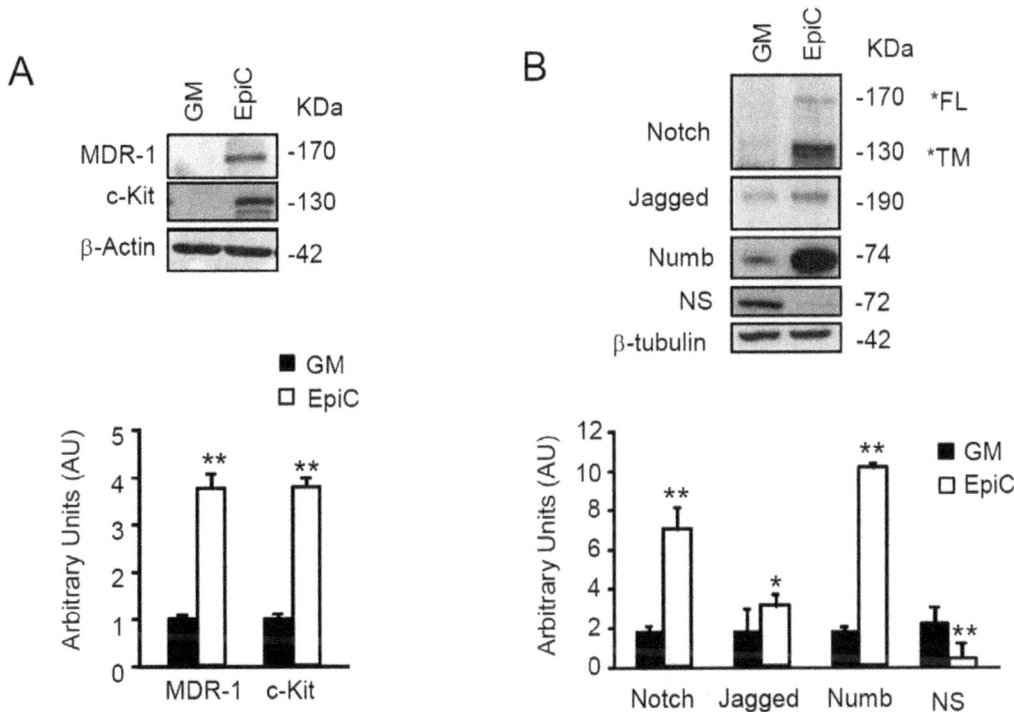

Figure 2. Effect of EpiC treatment on stem cell marker expression. Western blotting analysis of (A) adult cardiac stem cell marker c-Kit and MDR-1 (n = 4) and (B) Notch, Jagged-1, Numb and nucleostemin (NS) in CStC, grown either in GM or in EpiC (n = 6). Densitometry is shown in the bar graphs at the bottom of each panel (**P≤0.01, *P≤0.05 vs GM). FL = full length protein; TM = Transmembrane domain.

Real-time Reverse Transcription–polymerase Chain Reaction Analysis

Total RNA was extracted from cells using the TRIzol reagent and 500 ng of RNA were reverse-transcribed using Superscript III reverse transcriptase (Invitrogen). cDNA was amplified by SYBR-GREEN quantitative PCR in an iQ5 Cycler (Bio-Rad Laboratories). Primer sequences are reported in Supplementary Table S2. Relative expression was estimated using the DeltaCt (ΔCt) method. $\Delta\Delta$Ct were calculated using average ΔCt for each gene expression in GM. Fold changes in gene expression were estimated as $2^{(-\Delta\Delta Ct)}$. ΔCt = 25 was arbitrarily assigned to non-expressed genes.

microRNA Profiling Analysis

Total RNA was extracted from cells using TRIzol reagent (Invitrogen), according to the manufacturer's instructions. The concentration and purity of RNA were determined using a NanoDrop 1000 spectrophotometer (Thermo Scientific), and only highly pure preparations (ratio of 260/280>1.8 and 260/230>1.8) were used. The integrity of total RNA was assessed using an Experion electrophoresis system and the RNA high sense Analysis Kit (Bio-Rad Laboratories) and only highly quality RNA (RQI >9.5/10) was subjected to subsequent analysis. Comparative microRNA expression profiling was carried out using TaqMan Low Density Arrays Human MicroRNA A+B Cards Set v3.0 (Applied Biosystems). All procedures were performed according to the manufacturer's instructions, on a 7900HT Real-Time PCR System (Applied Biosystems).

Quality control and low level analysis of TaqMan Arrays were performed using the software ABI Prism SDS v2.4. Optimal baseline and Ct (threshold cycle) were determined automatically by the software algorithm. All Ct values reported as greater than

35 or as not detected were changed to 35 and considered negative calls. Raw expression intensities of target microRNAs were normalized for differences in the amount of total RNA added to each reaction using the mean expression value of all expressed microRNAs in a given sample, following the method described by Mestdagh and co-workers [12]. Relative quantitation of micro-RNA expression was performed using the comparative Ct method (ΔCt). ΔCt values were defined as the difference between the Ct of any microRNA in the calibrator sample (the sample with the highest expression, i.e. lowest Ct value) and the Ct of the same microRNA in experimental sample. The Ct values were transformed to raw quantities using the formula $\eta^{\Delta Ct}$, where amplification efficiency (η) was set arbitrarily to 2 (100%). The normalized relative expression quantity of each microRNA was calculated by dividing its raw quantity by the normalization factor. MicroRNAs that had missing values in greater than 50% of the samples (i.e. that were not present at least in all cells on one treatment) were deemed as uninformative and removed from the dataset.

Chromatin Immunoprecipitation (ChIP) Assay

ChIP assays were performed and DNA fragments analyzed by quantitative real-time PCR (qPCR) as previously described [13]. Briefly, standard curves were generated by serially diluting the input DNA (5-log dilutions in triplicate) and qPCR was done on an ABI Prism 7500 PCR instrument (Applied Biosystems), using a SYBR Master mix (Applied Biosystems) and evaluating the dissociation curves. The qPCR analyses were performed in duplicate and the data obtained were normalized to the corresponding DNA input. Data are represented as relative enrichment of specific histone modifications in EpiC-treated cells compared to growth medium (GM). Primers for human promoters

A

B

C

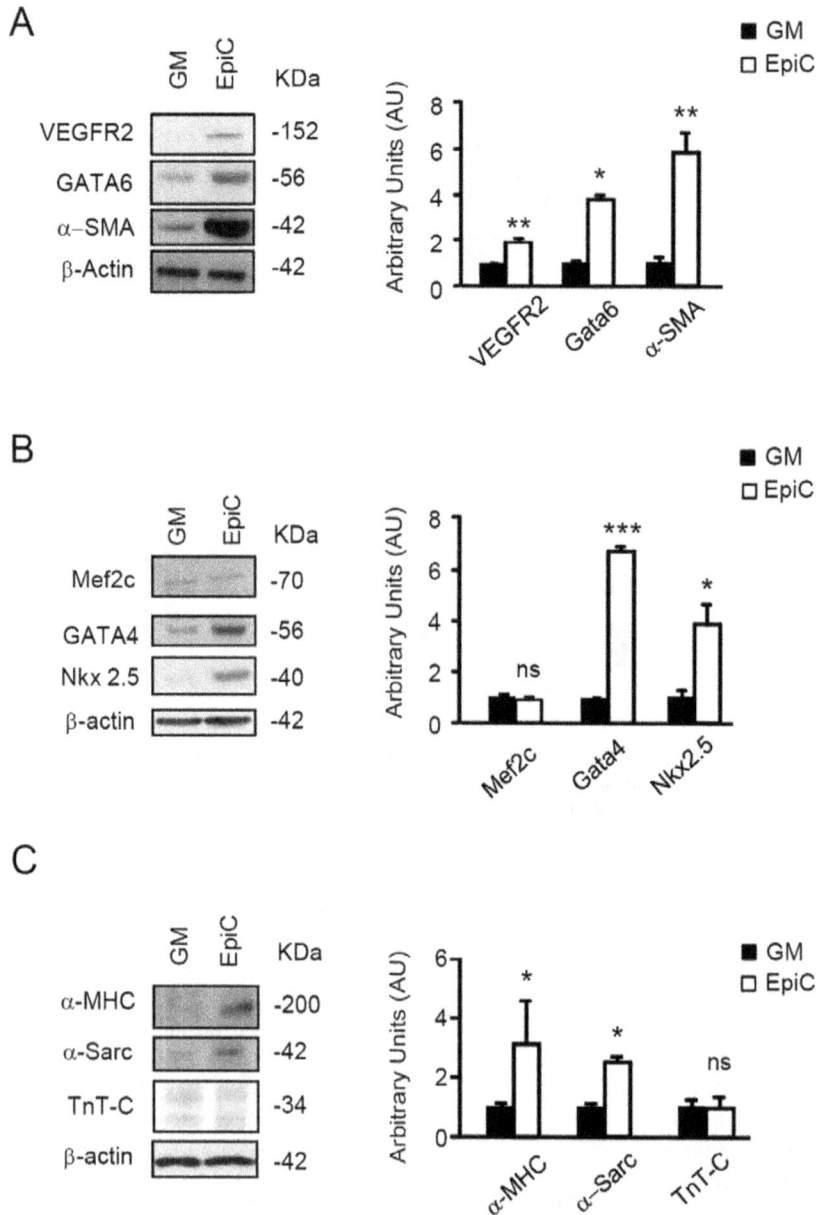

Figure 3. Effect of EpiC treatment of cardiovascular marker expression. Immunoblots showing expression analyses of (A) vascular markers: VEGFR-2, GATA6, and α-smooth muscle actin (α-SMA); (B) early cardiomyogenic markers: Mef2c, GATA4, and Nkx2.5; (C) late cardiomyogenic markers α-MHC, α-Sarc, and Tn-TC in EpiC-treated CStC compared to cells in GM. Bar graphs represent average results, normalized to β-actin, of western blot densitometric analyses (*P≤0.05, **P<0.01, and ***P≤0.005, vs GM). The data represent the mean ± SD of 4 independent experiments performed in duplicate.

were (position from transcriptional starting site, TSS): c-Kit fw: GAGCAGAAACAATTAGCGAAACC (−560 bp); c-Kit rev: GGAAATTGAGCCCCGACATT (−468 bp); Nkx2.5 fw: TGACTCTGCATGCCTCTGGTA (−198 bp); Nkx2.5 rev: TGCAGCCTGCGTTTGCT (−138 bp); MDR-1 fw: TTCCTCCACCCAAACTTATCCTT (-93 bp); MDR-1 rev: CCCAGTACCAGAGGAGGAGCTA (−2 bp); hGNL3 fw: GAGTTTGTGTCGAACCGTCAAG (−563 bp); hGNL3 rev: TCCCTCAGTCCCCAATACCA (−457 bp).

Electrophysiology

Patch-clamp analysis was performed on CStC perfused with a normal Tyrode solution containing (mM): 140 NaCl, 5.4 KCl, 1.8 CaCl$_2$, 5.5 D-glucose, 5 Hepes-NaOH; pH 7.4. Patch pipettes were filled with a solution containing (mM): 130 K-Aspartate, 10 NaCl, 5 EGTA-KOH, 2 CaCl$_2$, 2 MgCl$_2$, 2 ATP (Na-salt), 5 creatine phosphate, 0.1 GTP (Na-salt), 10 Hepes-KOH; pH 7.2 and had resistances of 2 to 4 MOhm. Experiments were carried out at room temperature.

The fast Na+ current (I$_{Na}$) was activated by 50 ms steps to the range −80/+30 mV from a holding potential of −90 mV. Peak I-V relations were constructed by plotting the normalized peak

Table 1. Differentially expressed microRNAs.

microRNA	GM	EpiC	FC	P-value	FDR
miR-133a	0.07±0.07	0.86±0.43	11.88	0.0488	0.13
miR-34a	0.26±0.10	1.23±0.13	4.78	0.0266	0.12
miR-664	0.30±0.09	1.44±0.34	4.75	0.0035	0.06
miR-34a#	0.31±0.13	1.33±0.61	4.27	0.0209	0.10
miR-210	0.25±0.10	1.04±0.29	4.19	0.0266	0.11
miR-200b	0.33±0.16	1.34±0.63	4.06	0.0492	0.13
miR-146b-5p	0.29±0.04	1.00±0.31	3.42	0.0442	0.14
miR-145	0.31±0.23	0.95±0.47	3.09	0.0346	0.13
miR-362-3p	0.48±0.10	1.50±0.35	3.09	0.0002	0.02
miR-1300	0.31±0.18	0.93±0.38	3.02	0.0126	0.09
miR-204	0.38±0.03	1.14±0.16	2.99	0.0124	0.09
miR-452	0.42±0.13	1.20±0.17	2.87	0.0113	0.10
miR-30a-5p	0.43±0.05	1.13±0.23	2.63	0.0056	0.08
miR-21	0.40±0.19	1.02±0.29	2.52	0.0478	0.13
miR-140-5p	0.52±0.07	1.28±0.05	2.46	0.0114	0.09
miR-30e-3p	0.55±0.12	1.31±0.27	2.37	0.0181	0.10
miR-26b#	0.49±0.07	1.09±0.18	2.22	0.0423	0.14
miR-22#	0.64±0.25	1.40±0.31	2.19	0.0162	0.09
miR-30a-3p	0.56±0.19	1.14±0.25	2.03	0.0140	0.09
miR-574-3p	0.60±0.11	1.21±0.16	2.02	0.0445	0.13
miR-663B	1.32±0.34	0.63±0.02	−2.08	0.0397	0.14
miR-20b	1.28±0.34	0.58±0.20	−2.21	0.0455	0.13
miR-708	1.07±0.51	0.46±0.19	−2.34	0.0020	0.04
miR-92a	1.58±0.27	0.68±0.07	−2.34	0.0303	0.12
miR-376a	1.15±0.70	0.43±0.23	−2.71	0.0093	0.09
miR-942	1.38±0.92	0.36±0.22	−3.82	0.0009	0.04
miR-18a	1.16±0.79	0.30±0.19	−3.91	0.0072	0.08
miR-223	1.35±0.38	0.32±0.12	−4.17	0.0017	0.05
miR-29b-1#	1.51±0.82	0.35±0.09	−4.38	0.0282	0.11
miR-155	1.83±0.43	0.40±0.15	−4.54	0.0206	0.10
miR-15b#	1.53±0.75	0.15±0.04	−10.42	0.0429	0.14

microRNA normalized relative expression levels in GM and EpiC-treated cells are expressed as mean ± SD. FC = fold change. FDR = false discovery rate.

current against test voltages. The time-independent inwardly-rectifying K+ current (I_{K1}) was investigated by applying 4 s voltage-ramps from −100 to 25 mV in Tyrode solution and after addition of Ba^{2+} (2 mM $BaCl_2$), a known blocker of I_{K1}. To record the I_f current, 1 mM $BaCl_2$ and 2 mM $MnCl_2$ were added to normal Tyrode in order to block contaminating currents. I_f was activated by a standard activation protocol [14]. Hyperpolarizing test steps to the range −35/−125 mV were applied from a holding potential of −30 mV, followed by a fully activating step at −125 mV. Each test step was long enough to reach steady-state current activation. Normalized tail currents measured at −125 mV were used to plot activation curves, which were fitted to the Boltzmann distribution function: $y = (1/(1+\exp((V-V_{1/2})/s))$, where V is voltage, y fractional activation, $V_{1/2}$ the half-activation voltage, and s the inverse slope factor. Measured values are reported as mean ± SEM.

Statistics

Statistical analysis of real-time PCR, Western Blot, and HDAC Class I activity data was performed using Student's t-test. A P<0.05 was considered significant.

Statistical analysis of the TaqMan Arrays was performed using the MultiExperiment Viewer (MeV) software v4.8.1 [15]. microRNAs with a fold change between the two treatment groups less than ±2 were filtered out. Normalized expression values were $\log_2$ transformed and differentially expressed microRNAs were identified using a paired t-test computing p-values based on all available permutations with a confidence level of 95% and limiting the false discovery rate (FDR) proportion to <0.15. Differences in microRNA expression were considered statistically significant if their P-value was <0.05. Unsupervised hierarchical cluster analysis was performed to assess whether differential profile discriminates the EpiC from the control treated cells. The similarity of microRNA expression among arrays and probes was assessed by calculating the Pearson's correlation (uncentered) coefficient. Normalized $\log_2$ transformed expression values were mean centered and clustered by correlation average linkage, using leaf order optimization.

Supplementary Methods

An additional Methods section is available as Supporting Information Document S1.

Results

Epigenetic Cocktail (EpiC) Design

Human cardiac stromal cells (CStC) cultured in standard medium for mesenchymal cells (GM) are positive for the mesenchymal markers CD105, CD29 and CD73, but negative for adult cardiac stem cell markers Sca-1, c-Kit and VEGFR2 (Figure S1A).

The level of nutrients (i.e. foetal serum) and the presence of selected drugs can modify cell phenotype and fate inducing functional reprogramming [4]. In light of this, after expansion in a medium routinely used for mesenchymal cell culture (growth medium, GM), CStC were exposed for 7 days to a medium with reduced level of foetal bovine serum (5% FBS) either in the presence or in the absence of 5 μM all-trans retinoic acid (ATRA), 5 μM phenyl butyrate (PB) and 200 μM diethylenetriamine/nitric oxide (DETA/NO), alone or in combination. In all these conditions, the expression of markers associated with resident cardiac stem cells (c-Kit, VEGFR2, and MDR-1) [16,17] has been evaluated. Our findings indicated that, although serum deprivation alone or a single drug exhibited the ability to up-regulate the expression of one or more markers, only the complete formulation, defining a novel "epigenetic cocktail" (EpiC), induced the coincident expression of c-Kit, VEGFR2, and MDR-1 in CStC (Figure S1B). Notably, EpiC treatment, while stopping cell proliferation, did neither induce apoptosis or senescence (Figure 1), nor stimulate CStC to differentiate into other mesodermal cells such as adipocytes or osteoblasts (Figure S1C), nor modulated the expression of Sca-1 and typical mesenchymal markers such as CD105 (not shown).

Effects of EpiC Treatment on Stem and Cardiovascular Precursor Markers

EpiC treatment of CStC strongly up-regulated markers associated to cardiac resident adult stem cell such as c-Kit and MDR-1 (Figure 2A and Figure S2A). In this condition, we were able to demonstrate that the MDR-1 transporter was functionally active as indicated by the rhodamine extrusion assay (Figure S2B).

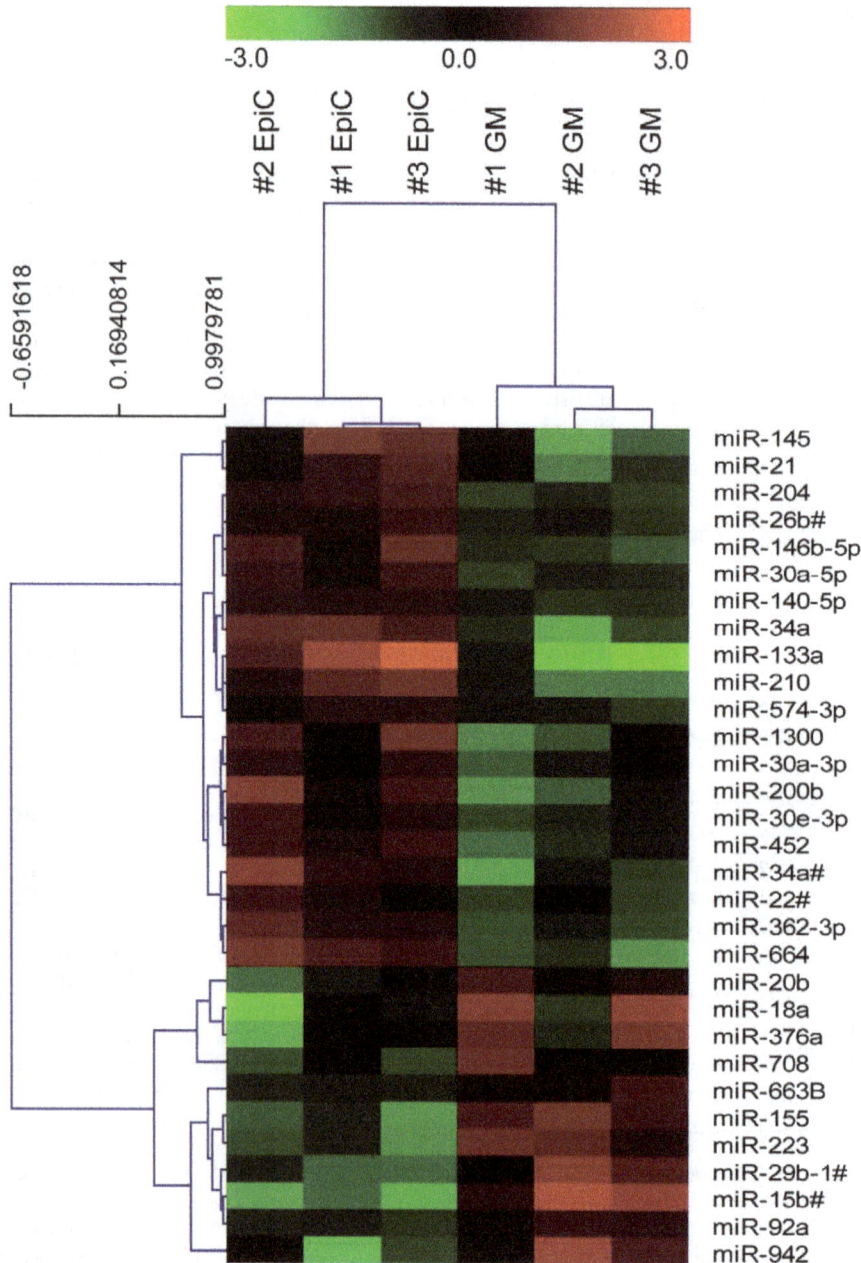

Figure 4. Hierarchical clustering of differentially expressed microRNAs in GM and EpiC-treated CStC. Unsupervised cluster analysis was performed using the 31 microRNAs that showed a significant modulation induced by the EpiC treatment. The dendrogram above shows that the differential expression profile completely discriminates the two treatment groups. The dendrogram on the left shows two distinct clusters of microRNAs up- and down-regulated by EpiC treatment in CStC (respectively at the top and the bottom of the heatmap). The mean centered value of normalized $\log_2$ transformed relative expression level of each microRNA is represented with a green, black, and red color scale (green indicates below mean; black, equal to mean; and red, above mean).

Other proliferation and differentiation markers including Notch, Jagged-1 and Numb [18,19] were also increased (Figure 2B), while the expression of the pluripotency factors Oct4 and Nanog remained negative (not shown). In this condition, EpiC-treated cells were growth arrested (Figure 1A) and nucleostemin (NS), a nucleolar protein present in proliferating stem cells [20], was down-regulated (Figure 2B). Of note, untreated CStC expressed detectable level of GATA6, α-smooth muscle actin and GATA4, whose expression is associated with processes ongoing during

vascular and cardiac commitment [21,22,23]. Interestingly, EpiC treatment increased the expression of markers for both vascular (VEGFR2, GATA6, and α-smooth muscle actin) and cardiomyogenic (GATA4 and Nkx2.5) precursors, while leaving Mef2C expression unaltered (Figure 3A and 3B). However, in spite of the evidence that more mature cardiomyogenic markers could be detected in EpiC-treated CStC, such as α-sarcomeric actin (α -Sarc) and α-myosin heavy chain (α -MHC), neither sarcomere

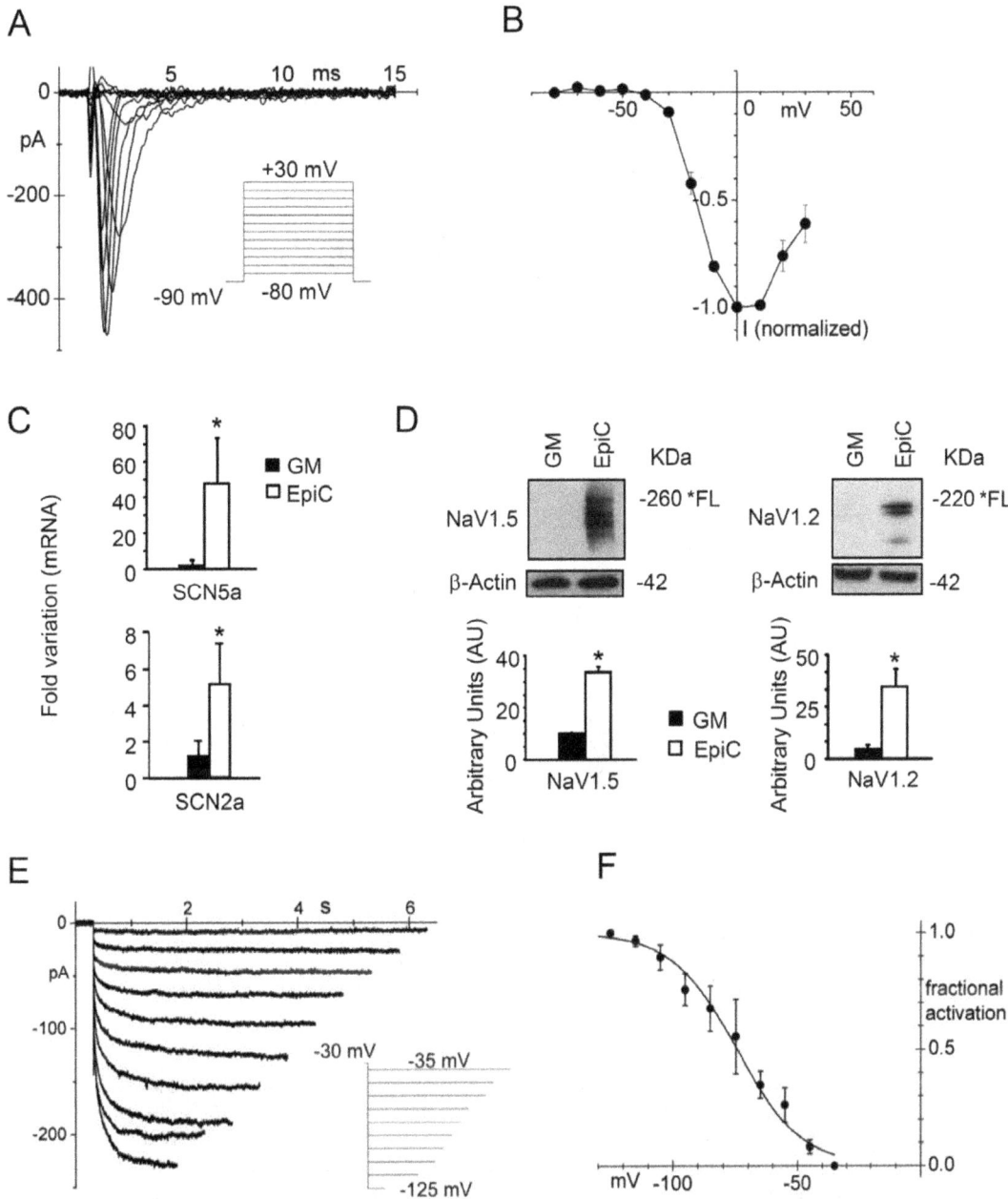

Figure 5. EpiC-treated CStCs express functionally competent ion channels. In panel (A) a family sodium current recorded from a representative EpiC-treated TTX (10 μM) low -sensitive CStC, following the application of a standard depolarizing protocol (see inset), is shown. (B) Mean current-voltage relation of normalized TTX-sensitive currents showing a threshold of activation around −40 mV and a peak around 0 mV. (C) SCN5A and SCN2A genes were up-regulated (n = 3, *P≤0.05) by real-time PCR analysis. (D) Western Blot evidences increase of the type V (NaV1.5) and type II (NaV1.2) voltage-gated sodium channel protein level in CStC cells exposed to EpiC (n = 5, *P≤0.05). FL = full length protein. (E) In few cells, membrane hyperpolarization in the range −35 to −125 mV (see inset) revealed an inward current with the kinetic features of the native pacemaker I_f current. (F) Plot of the mean activation curve obtained from the analysis of the I_f currents recorded in a subset of EpiC-treated CStC.

striation nor increased cardiac troponin (TnT-C) expression were observed (Figure 3C).

Effects of EpiC Treatment on CStC MicroRNA Expression Profile

microRNAs (miR) are 22–23 nucleotide-long, single-stranded ribonucleic molecules usually involved in transcriptional repression and gene silencing [24]. As miR are known to have important roles in cell reprogramming [25] and cell differentiation [26], their expression profile was evaluated in CStC after 7 days of exposure to EpiC. Two hundred and sixty-one microRNAs passed the quality assurance and filtering criteria: their normalized relative expression levels and fold-differences between EpiC and GM-treated cells, along with the raw P-values, are reported in Supplementary Table S3. Results of the differential analysis (Table 1) showed that 31 microRNAs were significantly modulated by EpiC treatment. In particular, 20 microRNAs were up-regulated >2-fold (among which miR-133a, miR-34a, and miR-

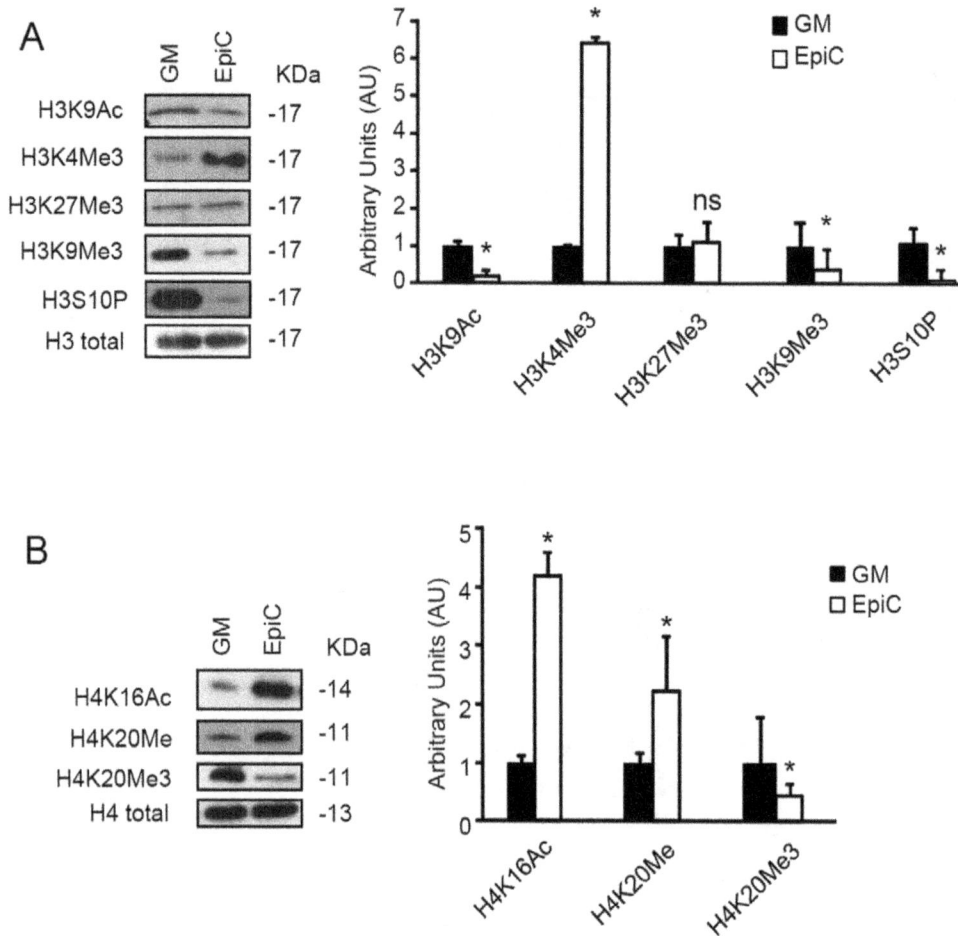

Figure 6. EpiC treatment modifies CStC epigenetic landscape. Western Blot analysis showing histone modification changes in EpiC-treated CStC compared to GM. (A) H3 modifications: H3K9Ac, H3K4Me3, H3K27Me3, H3K9Me3, and H3S10P. (B) H4 modifications: H4K16Ac, H4K20Me, and H4K20Me3. The same filter was probed with anti-total histone H3 or H4, respectively, to control for equal nuclear protein loading. Band densitometric analyses are reported in the bar graphs on the right (n = 3, *P≤0.05).

210) and 10 were down-regulated (such as miR-155). Unsupervised hierarchical cluster analysis was performed using the whole dataset of 261 microRNAs, revealing that the global expression profile well discriminates between the two groups of treatment (Figure S3). Likewise, unsupervised cluster analysis of the differentially expressed microRNAs correctly discriminated between EpiC and control treated CStC, sorting them into two independent groups (Figure 4). In addition, it showed two clusters of microRNAs with highly correlated expression, the first increased and the latter decreased in EpiC-treated cells, which may be deemed as a specific signature of co-regulated microRNAs.

EpiC Treatment Changes CStC Electrophysiological Properties

The exposure to EpiC had a profound impact on CStC function, determining the appearance of electrophysiological features typical of cells committed towards the cardiomyocyte lineage. Specifically, a significant fraction (44.8%, 13 out of 29 cells) of EpiC-treated cells exhibited a fast-activating inward sodium current (Figure 5A), which activated at voltages more positive than −40 mV, peaked around 0 mV, and was completely blocked by TTX (10 µM, Figure 5B). At a lower concentration (50 nM) TTX had variable effects on sodium current. In fact, in a

group of cells, sodium current was not blocked at all (n = 3) while in another group the block ranged from 38 to 94% (mean value: 76.3±9.2%, n = 3, data not shown). These results suggest that EpiC-treated cells may present both TTX-sensitive and TTX-resistant sodium currents. Accordingly, expression analyses showed that, in the presence of EpiC, the type V (NaV1.5) and type II (NaV1.2) voltage-gated sodium channels (known for having different TTX sensitivity and encoded by the SCN5A and the SCN2A genes, respectively) were significantly up-regulated, both at the mRNA (Figure 5C) and at the protein level (Figure 5D). A pacemaker I_f current (Figure 5 E and F) has also been recorded in 5 out of 33 EpiC-treated cells (15.1%), with kinetic properties compatible with those of immature cardiomyocytes and of native pacemaker cells ($V_{1/2}$: −74.4±5.7 mV, n = 3; Figure 5E and F) [27]. Accordingly, EpiC increased the expression of the pacemaker channel subunit HCN4 (Figure S4). Notably, CStC maturation towards the cardiomyocyte lineage was far from being complete, as demonstrated by the absence of the intracellular Ca2+ handling proteins NCX1 and RyR2 (not shown) and the depolarized resting potential characterizing CStC exposed to EpiC (−12.7±2.7 mV, n = 7). This last observation is in accordance with the negligible expression of the inward rectifying I_{K1} current (0.43±0.04 pA/pF

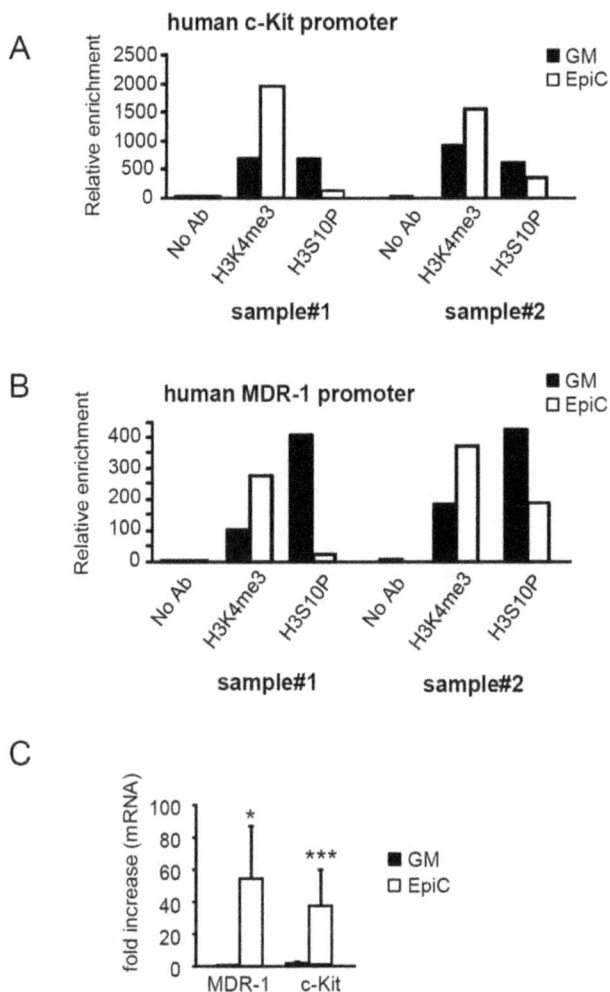

A human c-Kit promoter

B human MDR-1 promoter

C

Figure 7. Effects of EpiC treatment on specific-gene promoters. ChIP experiments, performed on CStC isolated from 2 different patients, show that H3K4Me3 association to (A) c-Kit and (B) MDR-1 promoter is enriched after EpiC treatment, while H3S10P is decreased. Data are expressed as relative enrichment of specific histone modifications i in EpiC-treated cells compared to GM measured by real-time PCR amplification. (C) Real-time RT-PCR analysis of c-Kit and MDR-1 mRNA in EpiC-treated CStC (n = 3, *P≤0.05 and ****P≤0.005).

at −100 mV, n = 5; not shown), physiologically important in setting the resting potential of working cardiomyocytes.

EpiC Treatment Modulated CStC Epigenetic Landscape

As expected by the composition of the EpiC described above, EpiC-treated CStC revealed a significantly lower HDAC Class I activity compared to cells kept in control condition (Figure S5) and changes in a number of genome wide histone modifications could be detected. Specifically, in agreement with its anti-proliferative effect EpiC also induced a significant decrease in Histone H3 phosphorylation at Ser10 (H3S10P, Fig. 6A) [28]. Further, we observed a global decrease in histone H3 lysine 9 trimethylation (H3K9Me3) and H4 lysine 20 trimethylation (H4K20Me3), paralleled by an increase in histone H4K20 monomethylation (H4K20Me, Fig. 6A and B). These modifications are compatible with an open chromatin structure and have been reported in cells undergoing differentiation or cell cycle arrest [29]. On the other hand, an increased density of the permissive marks histone H3K4

trimethylation (H3K4Me3) [30] and histone H4K16 acetylation (H4K16Ac) [31] was observed (Fig. 6 A and B), suggesting the presence of cells with a high developmental potential [32]. Along this line of evidence, repressive markers [33] remained stable (H3K27Me3) or were significantly down-modulated (H3K9Me3 and H4K20Me3), suggesting that EpiC-treated cells underwent a site-selective chromatin remodelling process leading to regulation of gene transcription.

EpiC Introduces Chromatin Changes at Specific-gene Promoters

To validate the hypothesis that Epic may specifically regulate gene expression, a series of chromatin immunoprecipitation (ChIP) experiments were performed in CStC cultured in the presence or the absence of EpiC. Specifically, the highly divergent H3K4Me3 and H3S10P were used to immunoprecipitate chromatin, followed by real-time PCR to detect the relative modulation of these specific histone modifications in the promoter region of c-Kit, MDR-1, Nkx2.5 and nucleostemin. As shown in Figure 7A and B and in Figure S6A, H3K4Me3 association to c-Kit, MDR-1, and Nkx2.5 promoter was increased by EpiC treatment, suggesting that chromatin conformational modifications may account for the increased expression of these genes (Figure 7C). Accordingly, the presence of H3K4Me3 and H3S10P in the NKX-2.5 and nucleostemin (human GNL3 gene product) promoters was reduced, suggesting local structural changes as the basis of their down-regulation (Figure S6B).

Discussion

Resident cardiac stem cells are specialized multipotent cells which, at very low rate, contribute to cardiomyocyte turn-over or cardiac post-ischemic regeneration. They are present in defined cardiac districts, identified as niches, in which these cells are kept quiescent until activation signals come [17]. Specific markers identify a cardiac stem cell, including the presence of c-Kit and the P-pump (MDR-1) gene products [16,17]. Adult cardiac stem cells are neither easily grown ex-vivo nor effectively differentiated into mature cardiomyocytes. Further, the preparation of near-termi-nally differentiated cardiomyocytes may not be the best approach for applications aimed at cardiac regeneration where maximum plasticity is required to regenerate all the cell types forming damaged tissues.

Our group has recently isolated, characterized and efficiently amplified a human population of adult cardiac mesenchymal-like stromal cells (CStC) showing in vitro and in vivo cardiovascular plasticity [5]. Although CStC proneness to acquire cardiomyo-genic markers was higher than that of syngeneic bone marrow cells, CStC efficiency to differentiate into adult cardiovascular cell types remained low, in spite of the expression of detectable mRNA encoding for early cardiovascular markers including c-Kit, GATA4 and GATA6 [5].

In the present work, CStC differentiation potential was improved by using a novel combination of small epigenetically active molecules, defined here as epigenetic cocktail, or "EpiC". This cocktail was designed to modify the CStC chromatin landscape and, thus, to unmask and drive CStC plasticity possibly towards the cardiovascular lineage. In detail, EpiC was made of: (i) all-trans retinoic acid (ATRA), which has genome-wide regulation properties [34] and whose receptors are known to recruit p300 and CBP acetyl-transferases facilitating their action at the histone level of specific gene loci [35]. Relevant for this study, retinoic acid retains well known morphogenetic properties including a profound effect on heart development and regeneration [36]; (ii) phenyl butyrate (PB), a drug belonging to the family of histone deacetylase inhibitors, which are known to enhance mesoderm maturation [37]; (*iii*) diethylenetriamine/nitric oxide (DETA/NO), a nitric oxide donor associated to cell survival growth arrest in vascular cells [38] and increased mesendoderm differentiation in mouse ES cells [10]; (*iv*) a reduced serum content, known to induce spontaneous differentiation in a variety of cell types including C2C12 myoblasts [39] or cardiac mesoangioblasts [40]. Importantly, all drugs used in this study are approved for clinical use or, in case of DETA/NO, currently undergoing clinical trials.

The idea that stem cell fate can be modulated by specific chemicals dates back decades but, recently, an increasing number of studies are showing the potential role of small molecules to promote cardiogenesis in mouse ES cells including different BMP inhibitors such as dorsomorphin [41] and Wnt pathway modulators [42]. On the other hand, similar attempts on adult cells were inefficient and the induction of true cardyomyogenesis is still vigorously debated. In 1999 Makino and colleagues described the appearance of spontaneously beating MHC, MLC-2v, GATA4, Nkx2.5 positive cells following treatment of immortalized murine bone marrow stromal cells with 5′-azacytidine (5-AZA) [43]. Since then, this drug was commonly used to induce cardiomyogenesis in isolated cells [44]. Nevertheless, in some experiments several cells stained positive for adipogenic markers suggesting that this method was not cardiac selective, while other groups were unable to reproduce these findings [45]. Of note, the putative mechanism of 5-AZA induced cardiomyogenesis was firstly attributed to demethylation of cardiac-related genes. However, a study by Cho et al. [46] demonstrated that the effect of 5-AZA was not related to the epigenetic activation of cardio-specific genes, but rather to the transcriptional inhibition of the glycogen synthase kinase (GSK)-3 gene, a major player in the Wnt signaling pathway [46]. In this light, our findings suggest that the EpiC treatment determines global activation-prone changes in chromatin structure, including that at c-Kit, MDR-1 and Nkx2.5 promoters. Importantly, the expression of these genes was induced without apparently altering that of non-cardiac osteogenic and adipogenic differentiation markers. This suggests that CStC response to EpiC may be predominantly cardiovascular oriented [10,47]. In fact, along with the expression of adult cardiac stem markers, EpiC treatment also induced the up-regulation of specific transcription factors associated with the vascular and cardiomyocyte lineage commitment.

It is widely accepted that microRNAs (miRs), are important for cardiac gene expression regulation and differentiation control [48,49]. In the present study, miR expression has been evaluated by profiling analysis. The evidence that unsupervised cluster analysis correctly separated between EpiC-treated and control CStC, further indicates that the EpiC treatment is potentially able to induce a specific transcriptional programme. Specifically, the expression of miR-133a, associated with cardiomyogenic differentiation [50], was strongly up-regulated, together with the expression of miR-210 and mir-34a, involved in stem cell survival [51] and negative growth control [52], respectively. Of note, miR-155 whose expression is associated with proliferation [53] and to the protection of cardiomyocyte precursors from apoptosis [54], was down-modulated in agreement with the reduced proliferation ability observed in EpiC-treated CStC. Interestingly, a recent paper by Anversa and co-workers shows that microRNA typically associated to adult cardiomyocytes (such as miR-1, mir-499 and mir-133) are expressed in cardiovascular precursors, but at lower levels than in adult cardiomyocytes [49]. In this light Epic-treated CStC expressing miR-133, more closely resemble cardiovascular

precursors than control CStC, in which miR-133 expression is very low.

Based on the expression of typical stem and cardiovascular precursor markers and considering previous hypothesis for cardiovascular stem cell differentiation hierarchy in the adult heart [16], it is difficult to establish the differentiation stage to which CStC may belong. The fact that untreated CStC are GATA4, GATA6 and Mef2C positive, but only EpiC-treated CStC are cKit, MDR-1and VEGFR2 positive suggests that our cocktail may induce a cascade of events able to reprogram CStC towards a more immature state, characterized by the expression of cardiovascular stem cell markers [55]. On the other hand, EpiC-treated cells also up-regulated some markers associated with differentiating cardiomyocyte precursors (i.e. Nkx2.5, Gata4, α–sarcomeric actin, α–myosin heavy chain, miR-133a) and exhibited functional, although not yet operational, properties typical of differentiating cardiomyocytes, suggesting that EpiC treatment may induce cardiomyogenic differentiation at least in a fraction of the CStC population. It is thus possible that our approach induced the production of a mixed population composed of cells at different stages of differentiation, or acted on different cell populations present inside the CStC preparations. More experiments are required to elucidate this important point.

Importantly, many cells presented a fast sodium current (normally responsible for the action potential upstroke), which, however, was small in size and with kinetic properties slightly different from those of mature channels. In fact, TTX-sensitivity and expression analyses evidenced that EpiC-treated cells expressed both the TTX-resistant NaV1.5 isoform, the primary cardiac type, and the TTX-sensitiveNaV1.2 isoform, a typical neuronal type which has been also detected, albeit at low levels, in the mouse and human hearts [56,57].

Moreover, few cells displayed the If current, which, although negligible in the adult ventricle, is present in atria and ventricles during the late embryonic and perinatal stage [27,58].

However, due to the lack of inwardly rectifying potassium currents, such as IK1, EpiC-treated cells exhibited a depolarized resting potential keeping both the Na+ and the HCN4 channels inactive, thus preventing the Epic-treated CStC cells from acquiring induced or spontaneous electrical activity.

In conclusion, it is reported here the first evidence that CStC may be chemically reprogrammed to acquire functionally competent cardiac precursor-like features.

Supporting Information

Figure S1   CStC characterization and Epigenetic Cocktail (EpiC) design. (A) Representative FACS analysis of CStC surface markers. (B) Western blot showing MDR-1, c-Kit, and VEGFR-2 expression of CStC cultured in growth medium (GM) or in low serum (LS) with or without epigenetic drugs for 7 days. ATRA = all-trans-retinoic acid; PB = phenyl butyrate; DETA/NO = diethylenetriamine/nitric oxide; EpiC = LS+ATRA+PB+−DETA/NO. (C) Real-Time RT-PCR analysis demonstrates no up-regulation of adipogenic (Adipsin and PPRγ2) and osteogenic (Osteopontin) markers (n = 3). ns = not significant.

Figure S2   Effects of EpiC treatment on c-Kit and MDR-1 expression in CStC. (A) Representative immunofluorescence images for c-Kit and MDR-1 in GM and EpiC-treated CStC. Original magnification: 20×. (B) Rhodamine 123 assay in GM and EpiC treatments (n = 4). Only EpiC-treated CStC were able to extrude Rhodamine through Verapamil sensitive MDR-1 channels.

Figure S3   Hierarchical clustering of microRNAs in GM and EpiC-treated CStC. Unsupervised cluster analysis was performed using the whole dataset of microRNAs that passed the quality assurance and filtering criteria: the global expression profile discriminates treatment groups.

Figure S4   Effect of EpiC treatment on the expression of the pacemaker channel subunit HCN4 in CStC. Representative immunofluorescence images for HCN4 in GM and EpiC-treated CStC. Original magnification: 40×.

Figure S5   Effect of EpiC treatment on HDAC activity in CStC. Bar graphs show Class I HDAC activity in CStC cultured in GM or EpiC for 7 days (n = 4; * P≤0.05).

Figure S6   Effects of EpiC treatment on specific-gene promoters in CStC. (A) and (B) Bar graphs show relative enrichment for H3KMe3 and H3S10P in Nkx2.5 and GNL3 (nucleostemin) promoter.

Table S1   List and working concentrations of primary antibodies used.

Table S2   List of primers for Real-Time RT-PCR.

Table S3   microRNA normalized relative expression levels in cardiac mesenchymal stromal cells (CStC) cultured in growth medium (GM) or Epigenetic Cocktail (EpiC) are expressed as mean ± SD. The dataset includes microRNAs that passed the quality assurance and filtering criteria.

Document S1   Additional Method section.

Author Contributions

Conceived and designed the experiments: AR CG MV AB. Performed the experiments: MV VM SN AB AS. Analyzed the data: MV VM SN GIC. Contributed reagents/materials/analysis tools: GP DD AF MCC. Wrote the paper: AR GC AB GIC.

References

1. Takahashi K, Yamanaka S (2006) Induction of pluripotent stem cells from mouse embryonic and adult fibroblast cultures by defined factors. Cell 126: 663–676.
2. Nussbaum J, Minami E, Laflamme MA, Virag JA, Ware CB, et al. (2007) Transplantation of undifferentiated murine embryonic stem cells in the heart: teratoma formation and immune response. FASEB J 21: 1345–1357.
3. Ieda M, Fu JD, Delgado-Olguin P, Vedantham V, Hayashi Y, et al. (2010) Direct reprogramming of fibroblasts into functional cardiomyocytes by defined factors. Cell 142: 375–386.
4. Anastasia L, Pelissero G, Venerando B, Tettamanti G (2010) Cell reprogramming: expectations and challenges for chemistry in stem cell biology and regenerative medicine. Cell Death Differ 17: 1230–1237.
5. Rossini A, Frati C, Lagrasta C, Graiani G, Scopece A, et al. (2011) Human cardiac and bone marrow stromal cells exhibit distinctive properties related to their origin. Cardiovasc Res 89: 650–660.
6. Ventura C, Cantoni S, Bianchi F, Lionetti V, Cavallini C, et al. (2007) Hyaluronan mixed esters of butyric and retinoic Acid drive cardiac and endothelial fate in term placenta human mesenchymal stem cells and enhance cardiac repair in infarcted rat hearts. J Biol Chem 282: 14243–14252.
7. Ybarra N, del Castillo JR, Troncy E (2011) Involvement of the nitric oxide-soluble guanylyl cyclase pathway in the oxytocin-mediated differentiation of porcine bone marrow stem cells into cardiomyocytes. Nitric Oxide 24: 25–33.
8. Zhang Q, Jiang J, Han P, Yuan Q, Zhang J, et al. (2011) Direct differentiation of atrial and ventricular myocytes from human embryonic stem cells by alternating retinoid signals. Cell Res 21: 579–587.
9. Majumdar A, Petrescu AD, Xiong Y, Noy N (2011) Nuclear translocation of cellular retinoic acid binding protein II is regulated by retinoic acid-controlled SUMOylation. J Biol Chem.
10. Rosati J, Spallotta F, Nanni S, Grasselli A, Antonini A, et al. (2011) Smad-interacting protein-1 and microRNA 200 family define a nitric oxide-dependent molecular circuitry involved in embryonic stem cell mesendoderm differentiation. Arterioscler Thromb Vasc Biol 31: 898–907.
11. Mujoo K, Sharin VG, Bryan NS, Krumenacker JS, Sloan C, et al. (2008) Role of nitric oxide signaling components in differentiation of embryonic stem cells into myocardial cells. Proc Natl Acad Sci U S A 105: 18924–18929.
12. Mestdagh P, Van Vlierberghe P, De Weer A, Muth D, Westermann F, et al. (2009) A novel and universal method for microRNA RT-qPCR data normalization. Genome Biol 10: R64.
13. Nanni S, Benvenuti V, Grasselli A, Priolo C, Aiello A, et al. (2009) Endothelial NOS, estrogen receptor beta, and HIFs cooperate in the activation of a prognostic transcriptional pattern in aggressive human prostate cancer. J Clin Invest 119: 1093–1108.
14. Barbuti A, Crespi A, Capilupo D, Mazzocchi N, Baruscotti M, et al. (2009) Molecular composition and functional properties of f-channels in murine embryonic stem cell-derived pacemaker cells. J Mol Cell Cardiol 46: 343–351.
15. Saeed AI, Bhagabati NK, Braisted JC, Liang W, Sharov V, et al. (2006) TM4 microarray software suite. Methods Enzymol 411: 134–193.
16. Anversa P, Kajstura J, Leri A, Bolli R (2006) Life and death of cardiac stem cells: a paradigm shift in cardiac biology. Circulation 113: 1451–1463.
17. Bearzi C, Rota M, Hosoda T, Tillmanns J, Nascimbene A, et al. (2007) Human cardiac stem cells. Proc Natl Acad Sci U S A 104: 14068–14073.
18. Boni A, Urbanek K, Nascimbene A, Hosoda T, Zheng H, et al. (2008) Notch1 regulates the fate of cardiac progenitor cells. Proc Natl Acad Sci U S A 105: 15529–15534.
19. Cottage CT, Bailey B, Fischer KM, Avitable D, Collins B, et al. (2010) Cardiac progenitor cell cycling stimulated by pim-1 kinase. Circ Res 106: 891–901.
20. Tjwa M, Dimmeler S (2008) A nucleolar weapon in our fight for regenerating adult hearts: nucleostemin and cardiac stem cells. Circ Res 103: 4–6.
21. Maitra M, Schluterman MK, Nichols HA, Richardson JA, Lo CW, et al. (2009) Interaction of Gata4 and Gata6 with Tbx5 is critical for normal cardiac development. Dev Biol 326: 368–377.
22. Abe M, Hasegawa K, Wada H, Morimoto T, Yanazume T, et al. (2003) GATA-6 is involved in PPARgamma-mediated activation of differentiated phenotype in human vascular smooth muscle cells. Arterioscler Thromb Vasc Biol 23: 404–410.
23. Wada H, Hasegawa K, Morimoto T, Kakita T, Yanazume T, et al. (2002) Calcineurin-GATA-6 pathway is involved in smooth muscle-specific transcription. J Cell Biol 156: 983–991.
24. Sevignani C, Calin GA, Siracusa LD, Croce CM (2006) Mammalian microRNAs: a small world for fine-tuning gene expression. Mamm Genome 17: 189–202.
25. Miyoshi N, Ishii H, Nagano H, Haraguchi N, Dewi DL, et al. (2011) Reprogramming of mouse and human cells to pluripotency using mature microRNAs. Cell Stem Cell 8: 633–638.
26. Liu SP, Fu RH, Yu HH, Li KW, Tsai CH, et al. (2009) MicroRNAs regulation modulated self-renewal and lineage differentiation of stem cells. Cell Transplant 18: 1039–1045.
27. Avitabile D, Crespi A, Brioschi C, Parente V, Toietta G, et al. (2011) Human cord blood CD34+ progenitor cells acquire functional cardiac properties through a cell fusion process. Am J Physiol Heart Circ Physiol 300: H1875–1884.
28. McManus KJ, Hendzel MJ (2006) The relationship between histone H3 phosphorylation and acetylation throughout the mammalian cell cycle. Biochem Cell Biol 84: 640–657.
29. Kourmouli N, Jeppesen P, Mahadevhaiah S, Burgoyne P, Wu R, et al. (2004) Heterochromatin and tri-methylated lysine 20 of histone H4 in animals. J Cell Sci 117: 2491–2501.
30. Wu S, Rice JC (2011) A new regulator of the cell cycle: the PR-Set7 histone methyltransferase. Cell Cycle 10: 68–72.
31. Piccolo FM, Pereira CF, Cantone I, Brown K, Tsubouchi T, et al. (2011) Using heterokaryons to understand pluripotency and reprogramming. Philos Trans R Soc Lond B Biol Sci 366: 2260–2265.
32. Pan G, Tian S, Nie J, Yang C, Ruotti V, et al. (2007) Whole-genome analysis of histone H3 lysine 4 and lysine 27 methylation in human embryonic stem cells. Cell Stem Cell 1: 299–312.
33. Mattout A, Biran A, Meshorer E (2011) Global epigenetic changes during somatic cell reprogramming to iPS cells. J Mol Cell Biol.
34. Das S, Foley N, Bryan K, Watters KM, Bray I, et al. (2010) MicroRNA mediates DNA demethylation events triggered by retinoic acid during neuroblastoma cell differentiation. Cancer Res 70: 7874–7881.
35. Dietze EC, Troch MM, Bowie ML, Yee L, Bean GR, et al. (2003) CBP/p300 induction is required for retinoic acid sensitivity in human mammary cells. Biochem Biophys Res Commun 302: 841–848.
36. Pan J, Baker KM (2007) Retinoic acid and the heart. Vitam Horm 75: 257–283.

37. Karamboulas C, Swedani A, Ward C, Al-Madhoun AS, Wilton S, et al. (2006) HDAC activity regulates entry of mesoderm cells into the cardiac muscle lineage. J Cell Sci 119: 4305–4314.

38. Gu M, Lynch J, Brecher P (2000) Nitric oxide increases p21(Waf1/Cip1) expression by a cGMP-dependent pathway that includes activation of extracellular signal-regulated kinase and p70(S6k). J Biol Chem 275: 11389–11396.

39. Goto S, Miyazaki K, Funabiki T, Yasumitsu H (1999) Serum-free culture conditions for analysis of secretory proteinases during myogenic differentiation of mouse C2C12 myoblasts. Anal Biochem 272: 135–142.

40. Galvez BG, Sampaolesi M, Barbuti A, Crespi A, Covarello D, et al. (2008) Cardiac mesoangioblasts are committed, self-renewable progenitors, associated with small vessels of juvenile mouse ventricle. Cell Death Differ 15: 1417–1428.

41. Kudo TA, Kanetaka H, Mizuno K, Ryu Y, Miyamoto Y, et al. (2011) Dorsomorphin stimulates neurite outgrowth in PC12 cells via activation of a protein kinase A-dependent MEK-ERK1/2 signaling pathway. Genes Cells 16: 1121–1132.

42. Meijer L, Skaltsounis AL, Magiatis P, Polychronopoulos P, Knockaert M, et al. (2003) GSK-3-selective inhibitors derived from Tyrian purple indirubins. Chem Biol 10: 1255–1266.

43. Makino S, Fukuda K, Miyoshi S, Konishi F, Kodama H, et al. (1999) Cardiomyocytes can be generated from marrow stromal cells in vitro. J Clin Invest 103: 697–705.

44. Oh H, Bradfute SB, Gallardo TD, Nakamura T, Gaussin V, et al. (2003) Cardiac progenitor cells from adult myocardium: homing, differentiation, and fusion after infarction. Proc Natl Acad Sci U S A 100: 12313–12318.

45. Martin-Rendon E, Sweeney D, Lu F, Girdlestone J, Navarrete C, et al. (2008) 5-Azacytidine-treated human mesenchymal stem/progenitor cells derived from umbilical cord, cord blood and bone marrow do not generate cardiomyocytes in vitro at high frequencies. Vox Sang 95: 137–148.

46. Cho J, Rameshwar P, Sadoshima J (2009) Distinct roles of glycogen synthase kinase (GSK)-3alpha and GSK-3beta in mediating cardiomyocyte differentiation in murine bone marrow-derived mesenchymal stem cells. J Biol Chem 284: 36647–36658.

47. Spallotta F, Rosati J, Straino S, Nanni S, Grasselli A, et al. (2010) Nitric oxide determines mesodermic differentiation of mouse embryonic stem cells by activating class IIa histone deacetylases: potential therapeutic implications in a mouse model of hindlimb ischemia. Stem Cells 28: 431–442.

48. Boettger T, Braun T (2012) A new level of complexity: the role of microRNAs in cardiovascular development. Circ Res 110: 1000–1013.

49. Hosoda T, Zheng H, Cabral-da-Silva M, Sanada F, Ide-Iwata N, et al. (2011) Human cardiac stem cell differentiation is regulated by a mircrine mechanism. Circulation 123: 1287–1296.

50. Liu N, Bezprozvannaya S, Williams AH, Qi X, Richardson JA, et al. (2008) microRNA-133a regulates cardiomyocyte proliferation and suppresses smooth muscle gene expression in the heart. Genes Dev 22: 3242–3254.

51. Kim HW, Haider HK, Jiang S, Ashraf M (2009) Ischemic preconditioning augments survival of stem cells via miR-210 expression by targeting caspase-8-associated protein 2. J Biol Chem 284: 33161–33168.

52. Li L, Yuan L, Luo J, Gao J, Guo J, et al. (2012) MiR-34a inhibits proliferation and migration of breast cancer through down-regulation of Bcl-2 and SIRT1. Clin Exp Med [Epub ahead of print].

53. Sluijter JP, van Mil A, van Vliet P, Metz CH, Liu J, et al. (2010) MicroRNA-1 and -499 regulate differentiation and proliferation in human-derived cardiomyocyte progenitor cells. Arterioscler Thromb Vasc Biol 30: 859–868.

54. Liu J, van Mil A, Vrijsen K, Zhao J, Gao L, et al. (2010) MicroRNA-155 prevents necrotic cell death in human cardiomyocyte progenitor cells via targeting RIP1. J Cell Mol Med 15: 1474–1482.

55. Bearzi C, Leri A, Lo Monaco F, Rota M, Gonzalez A, et al. (2009) Identification of a coronary vascular progenitor cell in the human heart. Proc Natl Acad Sci U S A 106: 15885–15890.

56. Chandler NJ, Greener ID, Tellez JO, Inada S, Musa H, et al. (2009) Molecular architecture of the human sinus node: insights into the function of the cardiac pacemaker. Circulation 119: 1562–1575.

57. Haufe V, Camacho JA, Dumaine R, Gunther B, Bollensdorff C, et al. (2005) Expression pattern of neuronal and skeletal muscle voltage-gated Na+ channels in the developing mouse heart. J Physiol 564: 683–696.

58. Yasui K, Liu W, Opthof T, Kada K, Lee JK, et al. (2001) I(f) current and spontaneous activity in mouse embryonic ventricular myocytes. Circ Res 88: 536–542.

7

Prostaglandin E_2 (PGE$_2$) Exerts Biphasic Effects on Human Tendon Stem Cells

Jianying Zhang, James H-C. Wang*

MechanoBiology Laboratory, Departments of Orthopaedic Surgery, Bioengineering, Mechanical Engineering and Materials Science, and Physical Medicine and Rehabilitation, University of Pittsburgh, Pittsburgh, Pennsylvania, United States of America

Abstract

Prostaglandin E_2 (PGE$_2$) has been reported to exert different effects on tissues at low and high levels. In the present study, cell culture experiments were performed to determine the potential biphasic effects of PGE$_2$ on human tendon stem/progenitor cells (hTSCs). After treatment with PGE$_2$, hTSC proliferation, stemness, and differentiation were analyzed. We found that high concentrations of PGE$_2$ (>1 ng/ml) decreased cell proliferation and induced non-tenocyte differentiation. However, at lower concentrations (<1 ng/ml), PGE$_2$ markedly enhanced hTSC proliferation. The expression levels of stem cell marker genes, specifically SSEA-4 and Stro-1, were more extensive in hTSCs treated with low concentrations of PGE$_2$ than in cells treated with high levels of PGE$_2$. Moreover, high levels of PGE$_2$ induced hTSCs to differentiate aberrantly into non-tenocytes, which was evident by the high levels of PPARγ, collagen type II, and osteocalcin expression in hTSCs treated with PGE$_2$ at concentrations >1 ng/ml. The findings of this study reveal that PGE$_2$ can exhibit biphasic effects on hTSCs, indicating that while high PGE$_2$ concentrations may be detrimental to tendons, low levels of PGE$_2$ may play a vital role in the maintenance of tendon homeostasis *in vivo*.

Editor: Giovanni Camussi, University of Torino, Italy

Funding: This work was supported in part by National Institutes of Health under award numbers AR061395, AR060920, and AR049921. No additional external funding received for this study. The funding agency had no role in study design, data collection and analysis, decision to publish, or preparation of the manuscript.

Competing Interests: The authors have declared that no competing interests exist.

* E-mail: wanghc@pitt.edu

Introduction

Tendons transmit muscular forces to bone and, as a result, they are subjected to large mechanical loads *in vivo*. Consequently, tendons are frequently injured, especially during intense sport activities. Tendon injuries are generally difficult to treat; tendinopathy, a chronic tendon disorder involving tendon inflammation and/or degeneration, is a particularly significant challenge in orthopaedics and sports medicine. Thus far, strategies that stimulate the complete regeneration of tendons after injury have not been developed. To this end, a better understanding of tendon cell biology is essential to devise improved treatment options for tendon injuries such as tendinopathy [1].

One of the major causative factors that contribute to the development of tendinopathy is excessive mechanical loading (or overuse and over-loading) placed on tendons [2,3]. Such excessive mechanical loading has been shown to increase the production of prostaglandin E_2 (PGE$_2$) in cultures of human tendon fibroblasts (tenocytes) *in vitro* [3,4]. In addition, PGE$_2$ production was shown to increase after exercise in the peritendinous space of Achilles tendons *in vivo* [5].

Although PGE$_2$ levels increase after mechanical loading, baseline levels of PGE$_2$ are present in the patellar and Achilles tendons of mice under normal conditions without mechanical loading such as treadmill running [6]. This suggests that PGE$_2$ could have an impact on the tendon stem/progenitor cells (TSCs) that reside in tendons [6–8] and could play an important physiological role in the maintenance of tendon homeostasis.

Therefore, PGE$_2$ may have biphasic effects depending on its concentration. A better understanding of the concentration-dependent effects of PGE$_2$ on tendon cells, particularly TSCs, may shed new light on tendon physiology and pathology. Thus, in this study we hypothesized that lower concentrations of PGE$_2$ increase TSC proliferation and decrease non-tenocyte differentiation of TSCs, while higher concentrations produce the opposite effects. To test this hypothesis, we carried out cell culture experiments by treating human TSCs (hTSCs) with low and high levels of PGE$_2$. We also performed *in vivo* implantation experiments to determine the differentiation fate of hTSCs after treatment with various concentrations of PGE$_2$ *in vitro*.

Materials and Methods

Ethics Statement

The Gift of Hope Organ and Tissue Donor Network (Elmhurst, IL) provided normal human knee tissues, after obtaining written consent from donors' families and approval from the local ethics committee (Gift of Hope Organ and Tissue Donor Network). The University of Pittsburgh IRB also approved the study protocol for using human tendon tissues in the cell culture and animal studies performed in this study. These specimens were used for investigation only and no human subjects were involved in this project. Data obtained for the study was not through intervention or interaction with individuals and does not have any identifiable private information. Further, the University of Pittsburgh IACUC approved the protocol for the use of rats in the *in vivo* implantation experiments.

hTSC Culture

hTSCs were isolated from the patellar tendons of six human donors (20 to 44 years old) using our previously published protocol [8]. Briefly, after removing the paratenons, the core portions of the patellar tendons were cut into small pieces and digested with collagenase type I (3 mg/ml) and dispase (4 mg/ml) at 37°C for 1 hr. After centrifugation at 3,000 rpm for 15 min and removal of the enzyme-containing supernatant, a single-cell suspension was obtained, which was cultured in growth medium (DMEM plus 20% FBS) at 37°C with 5% CO_2. After 8 to 10 days in culture dishes, hTSCs formed colonies. The stem cell colonies were then isolated and cultured in DMEM with 20% FBS. These hTSCs at passage 1 were used in the following experiments.

Verification of the Stemness of hTSCs

The stemness of human tendon stem cells (hTSCs) from the patellar tendon used in this study was verified by immunocytochemical analysis of three stem cell markers, including octamer-binding transcription factor 4 (Oct-4), Nanog, and nucleostemin (NS). hTSCs were first seeded into 12-well plates at a density of 20,000 cells/well with 1.5 ml medium and cultured for 3 days. Then, the hTSCs were fixed in 4% paraformaldehyde in PBS for 20 min at room temperature and washed in 0.5% Triton-X-100 in PBS for 15 min. Subsequently, the fixed cells were incubated with either mouse anti-human Oct-4 (1:500), rabbit anti-human Nanog (1:500), or goat anti-human nucleostemin (1:500) overnight at 4°C. After washing three times with PBS, the cells were again incubated for 2 hrs at room temperature with either Cy3-conjugated goat anti-mouse IgG antibodies (1:1000) for Oct-4, Cy3-conjugated goat anti-rabbit IgG (1:500) for Nanog, or Cy3-conjugated donkey anti-goat IgG antibodies (1:500) for Nucleostemin. Nuclei were stained with Hoechst fluorochrome 33342 (1 μg/ml; Sigma, St. Louis, MO). Stained cells were then examined using fluorescence microscopy. All antibodies were obtained from Chemicon International (Temecula, CA), BD Biosciences (Franklin Lakes, NJ), Neuromics (Edina, MN), or Santa Cruz Biotechnology Inc. (Santa Cruz, CA).

Measurement of Proliferation of hTSCs Treated with PGE$_2$

hTSCs were seeded in 6-well plates (6×10^4/well) and six different concentrations of PGE$_2$ (0, 0.01, 0.1, 1, 10, and 100 ng/ml) were added to the culture. Three replicates were maintained for each concentration. The medium was changed every day and PGE$_2$ was replenished. After 6 days, cell number was measured using a digital cellometer (Nexcelcom Bioscience, Lawrence, MA), and the population doubling time (PDT), which is a measure of cell proliferation, was calculated based on the formula: $\log_2[Nc/N_0]$, where Nc is the total number of cells at confluence, and N_0 is the initial number of cells seeded [8].

Determination of the Effect of PGE$_2$ Treatment on hTSC Stemness

Stemness of hTSCs was determined by immunocytochemistry and FACS analysis. For immunocytochemistry, hTSCs were seeded in 12-well plates (3×10^4/well) and treated with six different PGE$_2$ concentrations ranging from 0 to 100 ng/ml for 5 days, with three replicates for each concentration. The effect of PGE$_2$ treatment on hTSC stemness was then determined by performing immunocytochemistry for stem cell markers SSEA-4 and Stro-1. Briefly, cells were fixed in 4% paraformaldehyde in PBS for

30 min at room temperature. After washing with PBS, the cells were incubated at room temperature with mouse anti-human SSEA-4 (1:350; Invitrogen, Cat. # 414000) for 3 hrs or mouse anti-human Stro-1 (1:200; Invitrogen, Cat. # 398401) for 4 hrs. The cells were then washed three times with PBS, followed by incubation with Cy3-conjugated goat anti-mouse IgG (1:500; Invitrogen, Cat. # A10521) secondary antibody at room temperature for 2 hrs. After a final wash with PBS, the nuclei were stained with Hoechst fluorochrome 33342, as described above. Stained cells were examined and images of cells were obtained using a fluorescence microscope (Nikon eclipse microscope, TE2000-U).

Semi-quantification of Positively-stained hTSCs

For the semi-quantification of stem cell markers *in vitro*, seven random images were captured from each well at a magnification of 20x under the Nikon eclipse microscope. The positively-stained cells in each picture were manually identified and analyzed using SPOT^{TM} imaging software (Diagnostic Instruments, Inc., Sterling Heights, MI). The positive staining percentage was calculated by dividing the number of positively-stained cells by the total number of cells under the microscopic field. The average value of all seven images from each well represented the percentage of positive staining, which indicates the stemness of hTSCs in the respective PGE$_2$ concentrations.

Fluorescence Activated Cell Sorting (FACS) Analysis of hTSCs

To determine the effect of PGE$_2$ treatment on hTSC stemness by FACS analysis, hTSCs (1.5×10^6 in 50 μl PBS) were incubated with 20 μl of the appropriate serum in a centrifuge tube at 4°C for 30 min. Subsequently, 0.4 μg of mouse anti-human SSEA-4 (Cell Signaling, Cat. #4755S) or mouse anti-human Stro-1 (Millipore, Cat. #MAB4315) primary antibody was added and incubated at 4°C overnight. The cells were then washed three times with 2% FBS-PBS, followed by centrifugation at 500 g for 5 min/each time. Then the cells were treated with 1 μg Cy3 conjugated goat anti-mouse IgG secondary antibody at room temperature for 2 hrs. The cells treated with the second antibody only were used as a staining negative control. Finally, the cells were washed twice with PBS and fixed in 1% paraformaldehyde, followed by FACS analysis on a BD LSR II Flow Cytometer (BD Biosciences).

Determination of hTSC Differentiation *in vitro* by qRT-PCR

To determine the effect of PGE$_2$ treatment on the differentiation of hTSCs, we performed quantitative RT-PCR (qRT-PCR) to measure gene expression using a QIAGEN QuantiTect SYBR Green PCR Kit (QIAGEN). Briefly, total RNA was isolated from hTSCs using the RNeasy Mini Kit with an on-column DNase I digest (Qiagen, Valencia, CA). Then first-strand cDNA was reverse transcribed using SuperScript II (Invitrogen, Grand Island, NY) in a 20 μl reaction containing 1 μg total RNA. Conditions for the cDNA synthesis included 65°C for 5 min, 4°C for 1 min, 42°C for 50 min, and finally 72°C for 15 min. qRT-PCR was performed in a 25 μl PCR reaction mixture with 2 μl cDNA (~100 ng RNA) in a Chromo 4 Detector (MJ Research, Maltham, MA) by incubating at 94°C for 5 min, followed by 30 to 60 cycles of a three temperature program consisting of 1 min at 94°C, 40 sec at 57°C, and 40 sec at 72°C. The PCR reaction was terminated after a 10 min extension at 70°C and stored at 4°C until further analysis. Expression of stem cell markers (Oct-4 and Nanog), tenocyte markers (collagen type I and tenascin C), adipocyte marker (PPARγ), chondrocyte marker (Sox9), and osteocyte marker (Runx2) were measured using the primers listed in **Table 1**. GAPDH was used as an internal control. All primers were synthesized by Invitrogen. Expression of each target gene was normalized to GAPDH gene expression and the relative gene expression levels were calculated from the formula $2^{-\Delta\Delta CT}$. Details of the calculation are described in our previous study [9]. The mean and standard deviation (SD) of the CT values were determined from at least three replicates.

Determination of hTSC Differentiation *in vivo* by Implantation

To verify the effect of PGE2 treatment on the differentiation of hTSCs *in vivo*, eight female nude rats (10 weeks old; 200–250 g) were used. hTSCs at passage 2 were seeded into 24-well plates (8×10^6 cells/well) and cultured in DMEM with or without various concentrations of PGE2 for 6 days, with a change of medium every day. For implantation experiments, the cells were trypsinized from each well and mixed with 0.25 ml Matrigel (BD Scientific) to enable gel formation after implantation. These hTSC-Matrigel mixtures were placed subcutaneously in the back of anesthetized rats. Three pieces of hTSC-Matrigel were positioned in three distinct places on each side of each rat's back. Four weeks after implantation, tissue samples were harvested from the implanted area and placed in pre-labeled base molds filled with frozen section medium (Neg 50; Richard-Allan Scientific; Kalamazoo, MI). The tissue blocks were stored at −80°C until histological analysis.

Immunohistochemical and Histological Analyses

Each frozen tissue block was cut into 10 μm thick sections, placed on glass slides, and then allowed to dry overnight at room temperature. The tissue sections were fixed in 4% paraformaldehyde for 30 min and further washed three times with PBS. They were then incubated at room temperature with mouse anti-human PPARγ antibody (Santa Cruz Biotechnology, Inc., Cat. #271392, Santa Cruz, CA) diluted to 1:350 for 2 hrs, mouse anti-collagen type II antibody (1:300; Millipore, Cat. #MAB8887, Temecula,

CA) for 2 hrs, or mouse anti-human osteocalcin antibody (1:300; Santa Cruz Biotechnology, Inc., Cat. #74495, Santa Cruz, CA) for 3 hrs. After washing with PBS, Cy3-conjugated goat anti-mouse IgG (1:500; Santa Cruz Biotechnology) was added as secondary antibody and incubated at room temperature for 1 hr, followed by staining the nuclei with Hoechst fluorochrome 33342 (1 μg/ml; Sigma, St. Louis, MO) at room temperature for 5 min. Additionally, cell morphology and distribution in those tissues that received hTSCs, which had been treated with various concentrations of PGE2 in culture, were assessed by staining with hematoxylin and eosin (H&E). Finally, all tissue sections were examined under a fluorescence microscope.

Semi-quantification of Positively Stained Tissue Sections

Each tissue section was examined under a microscope (Nikon eclipse, TE2000-U) and five random images were taken for the semi-quantification of hTSC differentiation *in vivo*. SPOT™ imaging software (Diagnostic Instruments, Inc., Sterling Heights, MI) was used to process positively stained areas, which were manually identified by examining the images taken. The total area viewed under the microscope was divided by the positively stained area to calculate the proportion of positive staining. Five tissue sections were used for each group and five images were obtained per tissue section. These values were averaged to represent the percentage positive staining in all the groups treated with various PGE2 concentrations, which indicated the extent of cell differentiation.

Statistical Analysis

Data are expressed as mean ± standard deviation (mean ± SD). Unless otherwise indicated, at least three replicates were used for each experimental condition. For statistical analysis of data, one-way ANOVA or a student *t*-test was used wherever appropriate. All comparisons were between each PGE2-treated group and the respective control. A P-value less than 0.05 was considered to indicate statistically-significant differences between the groups compared.

Table 1. Primers used in qRT-PCR for gene expression analysis.

Gene	Primer Sequence	Accession numbers	Reference
Oct-4	Forward 5′-CGC AAG CCC TCA TTT CAC-3′	NM_002701	[32]
	Reverse 5′-CAT CAC CTC CAC CAC CTG-3′		
Nanog	Forward 5′-TCC TCC TCT TCC TCT ATA CTA AC-3′	NM_024865	[33]
	Reverse 5′-CCC ACA ATC ACA GGC ATA C-3′		
Tenascin C	Forward 5′- CGG GGC TAT AGA ACA CCA GT-3′	NM_002160.2	[34]
	Reverse 5′- AAC ATT TAA GTT TCC AAT TTC AGG TT-3′		
Collagen I	Forward 5′-AGG GTG AGA CAG GCG AAC AG-3′	NM_000088	[35]
	Reverse 5′-CTC TTG AGG TGG CTG GGG CA-3′		
PPARγ	Forward 5′- GGC TTC ATG ACA AGG GAG TTT C-3′	NM_13871	[36]
	Reverse 5′- CTT TAT GGA GCC CAA GTT TGA GTT-3′		
Sox9	Forward 5′- CCC CAA CAG ATC GCC TAC AG-3′	NM_000346	[37]
	Reverse 5′- GAG TTC TGG TCG GTG TAG TC-3′		
Runx2	Forward 5′- ATG CTT CAT TCG CCT CAC AAA-3′	NM_001015051	[38]
	Reverse 5′- CCA AAA GAA GTT TTG CTG ACA TGG-3′		
GAPDH	Forward 5′-GCC AAA AGG GTC ATC ATC-3′	NM_002046	[32]
	Reverse 5′-ATG ACC TTG CCC ACA GCC TT-3′		

Results

Verification of the Stemness of hTSCs

Prior to using hTSCs for cell culture experiments in this study, we first verified the stemness of these tendon cells. Microscopic examination of hTSCs revealed the typical cobblestone-shaped morphology of tendon stem cells under phase contrast microcopy (**Fig. 1A**). Further, cells in culture also showed robust expression of all three stem cells markers, Oct-4 (**Fig. 1B**), Nanog (**Fig. 1C**), and nucleostemin (**Fig. 1D**), in immunohistochemical analyses. These characteristics indicated that the cells derived from the human patellar tendons were indeed tendon-specific stem cells.

Effect of PGE$_2$ on the Proliferation of hTSCs

After establishing that the cells in culture were hTSCs, we investigated cell proliferation after PGE$_2$ treatment of hTSCs by determining their population doubling time (PDT). Treatment of hTSCs with a lower concentration (0.01 ng/ml) of PGE$_2$ significantly increased cell proliferation, as evidenced by decreased PDT when compared to the control (**Fig. 2**). PGE$_2$ treatment at a higher concentration (0.1 ng/ml) also induced similar proliferative effects, although to a smaller extent. At concentrations of 1 and 10 ng/ml, the proliferation of hTSCs was not significantly different from the control without PGE$_2$ treatment. At the highest concentration (100 ng/ml), TSC proliferation was significantly decreased.

Effect of PGE$_2$ Treatment on the Stemness of hTSCs

Immunofluorescence assays for stem cell markers revealed that hTSCs treated with a low concentration of PGE$_2$ (0.01 ng/ml) expressed SSEA-4 (**Fig. 3B**) and Stro-1 (**Fig. 4B**) more extensively

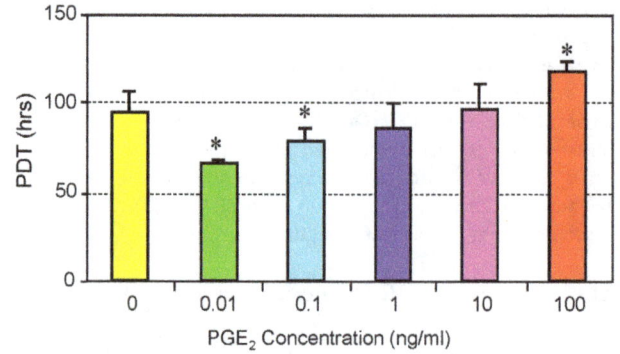

Figure 2. Population doubling time (PDT) of hTSCs treated with various concentrations of PGE$_2$. hTSCs were seeded in 6-well plates and cultured for six days on medium containing six different concentrations of PGE$_2$. PDT increased with increasing concentration of PGE$_2$, meaning that increased PGE$_2$ resulted in decreased cell proliferation (*p<0.05 when compared to control cells without PGE$_2$ treatment).

than controls (without PGE$_2$ treatment) (**Fig. 3A, 4A**) and those treated with higher concentrations of PGE$_2$ (10 or 100 ng/ml) (**Fig. 3E, 3F, 4E, 4F**). Indeed, the expression levels of both stem cell markers were significantly inhibited by higher concentrations of PGE$_2$ (10 or 100 ng/ml) (**Fig. 3, 4**). However, semi-quantification of the staining results revealed that the levels of both SSEA-4 (**Fig. 3G**) and Stro-1 (**Fig. 4G**) were similar between the control hTSCs and hTSCs treated with 0.01 ng/ml PGE$_2$. Consistent with the microscopic observations, higher concentra-

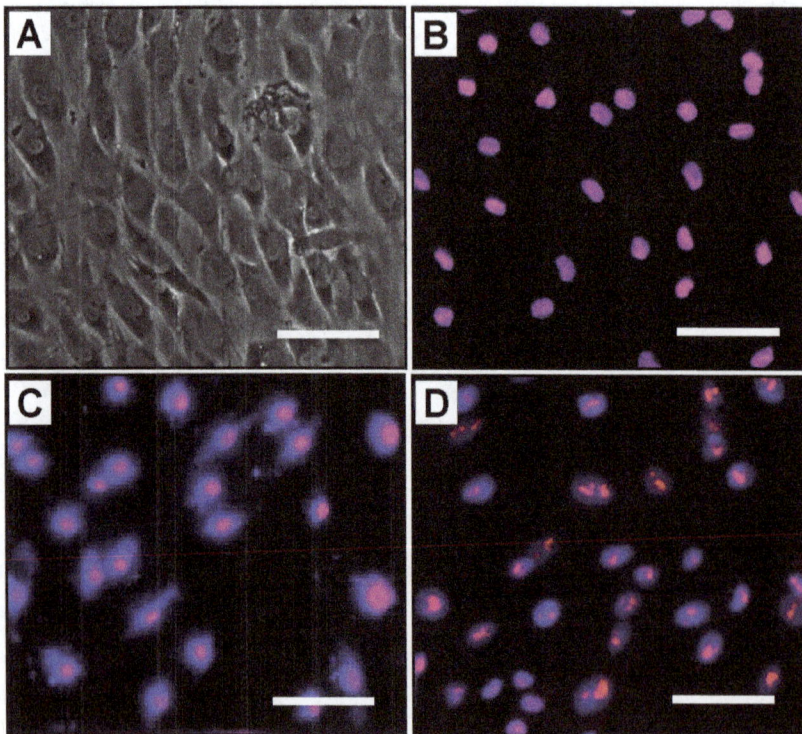

Figure 1. Verification of the stemness of hTSCs. Cobblestone shaped morphology of hTSCs visualized under phase contrast microcopy (**A**). hTSCs also expressed Oct-4 (**B**), Nanog (**C**), and nucleostemin (**D**). Staining for all three stem cell markers was nearly 100% positive with the respective antibodies. Bar = 50 μm.

tions of PGE_2 significantly reduced staining for both stem cell markers. Particularly, the concentration-dependent effect of PGE_2 on Stro-1 was more profound than its effect on SSEA-4 (**Fig. 3G, 4G**), with 81% reduction at 100 ng/ml, 76% at 10 ng/ml, 52% at 1 ng/ml, and 38% at 0.1 ng/ml for Stro-1, and 61% at 100 ng/ml, 40% at 10 ng/ml, 17% at 1 ng/ml, and 12% at 0.1 ng/ml for SSEA-4.

Additionally, FACS analysis of the stem cell markers also corroborated the immunocytochemical findings. Specifically, as PGE_2 concentrations increased from 0 to 0.01 ng/ml, more cells positively stained with SSEA-4 and Stro-1 (**Fig. 5**, blue dots) were evident; however, when PGE_2 concentrations were further increased to 1 and 100 ng/ml, few positively-stained cells were detected. Quantification of the results from two independent FACS experiments also confirmed these observations (**Fig. 6**).

To further characterize the stemness of hTSCs after treatment with PGE_2, we examined the expression of stem cell genes using qRT-PCR. We found that the gene expression levels of Nanog and Oct-4 were significantly ($p < 0.05$) up-regulated in hTSCs treated with lower concentrations (0.01 and 0.1 ng/ml) of PGE_2 (**Fig. 7**). Notably, the expression level of Oct-4 was twice as high as that of Nanog at 0.01 ng/ml PGE_2 concentration. When treated with higher concentrations (1, 10, and 100 ng/ml) of PGE_2, expression levels of both Nanog and Oct-4 were down-regulated and almost reached the levels of controls without PGE_2 treatment.

Effect of PGE_2 on the Differentiation of hTSCs

We next examined the effects of PGE_2 on hTSC differentiation by determining the expression of tenocyte and non-tenocyte related genes. Treatment of hTSCs with lower concentrations

SSEA-4

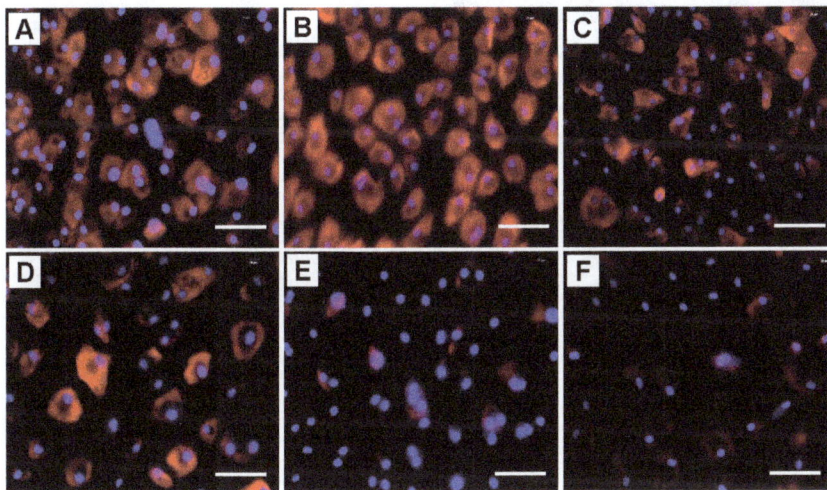

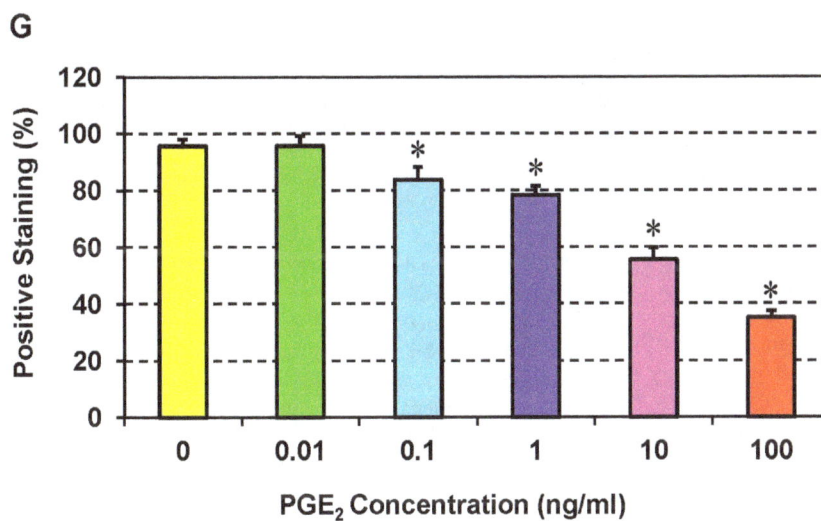

Figure 3. Expression of the stem cell marker SSEA-4 by hTSCs cultured *in vitro* in various concentrations of PGE_2. A: without PGE_2 treatment; **B**: 0.01 ng/ml PGE_2; **C**: 0.1 ng/ml PGE_2; **D**: 1 ng/ml PGE_2; **E**: 10 ng/ml PGE_2; and **F**: 100 ng/ml PGE_2. hTSCs were seeded in 12-well plates, cultured with six different concentrations of PGE_2, incubated with mouse anti-human SSEA-4 primary antibody, and detected with Cy3-conjugated goat anti-mouse IgG. Nuclei were stained with Hoechst (Blue). Expression of SSEA-4 (red) is dose-dependent, with more robust expression seen in hTSCs treated with low levels of PGE_2 (**A–D**) than expression levels seen in those treated with high levels (**E, F**). Positively stained cells were also counted to calculate percentage staining (**G**) (*$p < 0.05$ with respect to hTSCs not treated with PGE_2). Bar: 100 μm.

Stro-1

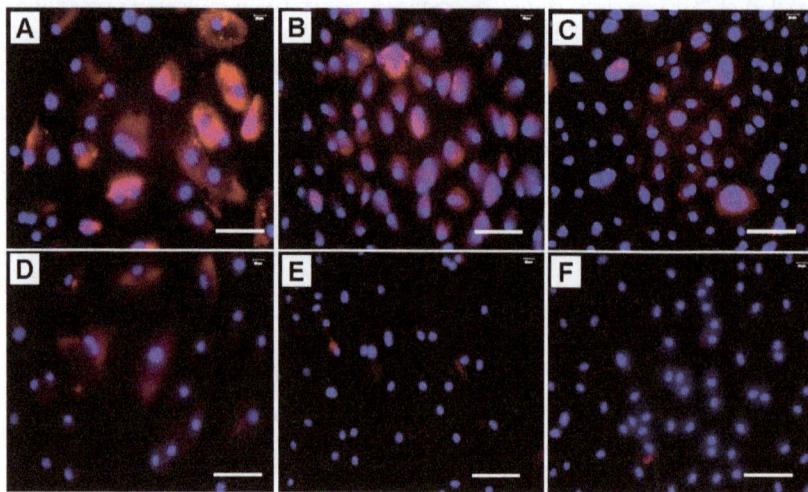

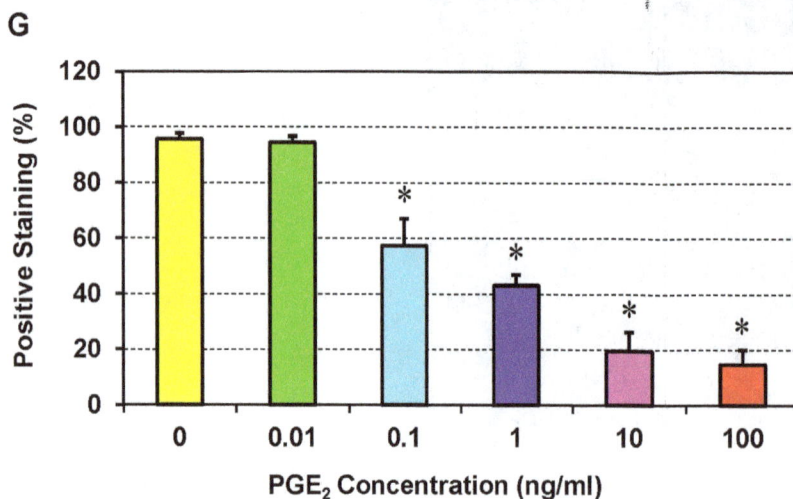

Figure 4. Expression of the stem cell marker Stro-1 by hTSCs cultured *in vitro* in medium containing various concentrations of PGE₂.
A: without PGE$_2$ treatment; **B**: 0.01 ng/ml PGE$_2$; **C**: 0.1 ng/ml PGE$_2$; **D**: 1 ng/ml PGE$_2$; **E**: 10 ng/ml PGE$_2$; and **F**: 100 ng/ml PGE$_2$. hTSCs were seeded in 12-well plates, cultured with six different concentrations of PGE$_2$, incubated with mouse anti-human Stro-1, and detected with Cy3-conjugated goat anti-mouse IgG. Hoechst was used to stain nuclei (blue). Expression of Stro-1 (red) is higher in hTSCs treated with low PGE$_2$ concentrations (**A, B**) than hTSCs treated with high concentrations (**E–F**). Similar to SSEA-4, expression of Stro-1 is also dose-dependent. Positively stained cells were also counted to calculate percentage staining (**G**) (*p<0.05 in comparison with control hTSCs not treated with PGE$_2$). Bar: 100 μm.

(0.01, 0.1, and 1 ng/ml) of PGE$_2$ significantly (p<0.05) enhanced the expression of both collagen type I and tenascin C, two tenocyte-associated genes (**Fig. 8A**). However, at these lower concentrations, the expression levels of non-tenocyte associated genes PPARγ, Sox9, and Runx2 were lower or only marginally higher than the control (**Fig. 8B**). On the other hand, treatment of hTSCs with higher concentrations (10 and 100 ng/ml) of PGE$_2$ significantly (p<0.05) up-regulated PPARγ, Sox9, and Runx2 genes associated with adipogenic, chondrogenic, and osteogenic differentiation, respectively (**Fig. 8B**). This up-regulation corresponded with the down-regulation of collagen type I and tenascin C at 10 and 100 ng/ml PGE$_2$ concentrations (**Fig. 8A**).

Non-tendinous Tissue Formation after Implantation of PGE₂-treated hTSCs

To determine whether PGE$_2$-treated hTSCs underwent non-tenogenic differentiation, we subcutaneously implanted PGE$_2$-treated hTSCs into nude rats. We found that 4 weeks after implantation, non-tenocyte differentiation of hTSCs was more extensive in the cells treated with higher concentrations (10 and 100 ng/ml) of PGE$_2$ (**Fig. 9E–G,** and **Fig. 9I–K**) when compared to the hTSCs that received the lowest concentration of PGE$_2$ (0.1 ng/ml) (**Fig. 9A–C**), as evidenced by higher amounts of PPARγ, collagen type II and osteocalcin (stained in red/pink). It appeared that more cells (black dots) were present in tissues that received hTSCs treated with high PGE$_2$ concentrations (**Fig. 9H–L**) than those that were treated with low PGE$_2$ concentrations (**Fig. 9D**). Specifically, at 100 ng/ml (**Fig. 9L**), numerous cells were concentrated in a specific region (triangle). The immunohis-

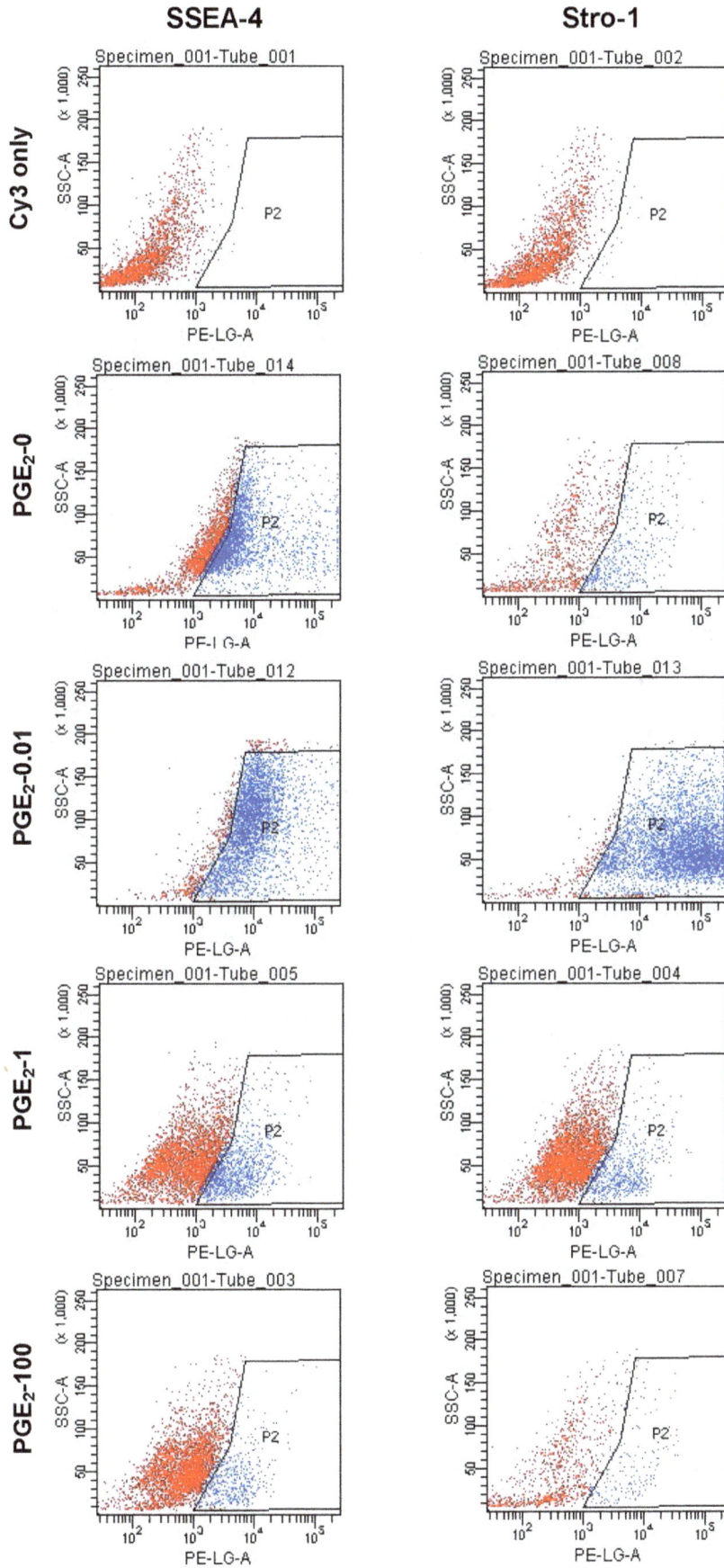

Figure 5. FACS analysis of SSEA-4 and Stro-1 expression in hTSCs treated with various concentrations of PGE$_2$. hTSCs in culture were treated with various concentrations of PGE$_2$. FACS analysis was performed on these cells (for details, see Methods section). It is evident that when cells were treated with 0.01 ng/ml of PGE$_2$ (PGE$_2$-0.01), more cells positively stained with SSEA-4 and Stro-1 were detected (blue dots in the P2 area) compared to control cells without PGE$_2$ treatment (PGE$_2$-0). When PGE$_2$ concentration increased to 1 ng/ml (PGE$_2$-1) and 100 ng/ml (PGE$_2$-100), fewer cells were actually stained with SSEA-4 and Stro-1 when compared to control cells (PGE$_2$-0).

tochemical observations were also confirmed by semi-quantification, which showed a significant (P<0.001) dose-dependent increase in the staining of non-tenocyte associated genes with increasing amounts of PGE$_2$ (**Fig. 9M**). When compared to hTSCs treated with 0.1 ng/ml PGE$_2$, those treated with 10 ng/ml had a ~2 fold increase in PPARγ, collagen type II, and osteocalcin. These increases were higher when hTSCs were treated with 100 ng/ml PGE$_2$ (~3 fold for PPARγ, ~4 fold for collagen type II, and ~4 fold for osteocalcin).

Discussion

PGE$_2$ is one of the most abundant prostaglandins in the body, and an important causative factor of inflammation that results from tissue damage or infection. Since our previous study showed that high levels of PGE$_2$ (1, 10, and 100 ng/ml) decrease proliferation and induce differentiation of mouse TSCs into non-tenocytes [6], in the present study we investigated the effects of comparable and lower doses of PGE$_2$ (0.01 to 100 ng/ml) on hTSC proliferation and differentiation by performing cell culture and cell implantation experiments. Our results revealed a concentration-dependent biphasic effect of PGE$_2$ on the proliferation and differentiation of hTSCs. PGE$_2$ treatment of hTSCs increased cell proliferation at lower concentrations, but decreased it at higher concentrations. In particular, low levels of PGE$_2$ promoted the stemness of hTSCs, as evidenced by the extensive expression of stem cell markers SSEA-4 and Stro-1 in hTSCs treated with low concentrations of PGE$_2$. The range of PGE$_2$ concentrations used in this study also includes the *in vivo* physiological concentrations of PGE$_2$ reported in human Achilles tendons (0.8±0.2 ng/ml, [10] or 54±24 pg/ml [11]). It should be

noted that these values are likely lower due to two reasons: i) patients in these studies were at rest during these measurements and did not undergo intensive exercise, and ii) these values are average microdialysis measurements of PGE$_2$ concentrations over a large portion of the tendon instead of at a local site, where PGE$_2$ concentrations could be much higher.

The biphasic effects of PGE$_2$ on various tissue properties have been reported in previous studies. For example, PGE$_2$ has been shown to exert biphasic effects on vascularity [12]; it elicits vasodilation at low concentrations and reverses this effect at higher concentrations. Similarly, PGE$_2$ treatment reduced proliferation of mesenchymal stem cells (MSCs) in a dose-dependent manner (0.25 μM to 25 μM PGE$_2$, or 88 ng/ml to 8.8 μg/ml), with the two lowest concentrations (0.25 nM and 2.5 nM PGE$_2$, or 88 pg/ml to 880 pg/ml) slightly increasing MSC proliferation over baseline levels [13]. In this study, the authors demonstrated that the biphasic effect of PGE$_2$ was executed by differential activation of two types of protein kinase A (PKA). At low concentrations, PGE$_2$ activated PKA II, leading to a cascade of events that resulted in cell proliferation; at high concentrations, PGE$_2$ caused PKA I activation, resulting in cell cycle arrest which reduced MSC proliferation. In addition, PGE$_2$ was reported to have a biphasic influence on injured esophagus: at low doses PGE$_2$ was protective, but at high doses it damaged the esophagus, with this effect being mediated by the EP1 receptor [14]. Interestingly, the biphasic effects of PGE$_2$ were also reported to depend on the growth state of the tissue type. For example, PGE$_2$ promoted proliferation of quiescent smooth muscle cells indicated by an increase in both DNA and RNA synthesis with increasing levels of PGE$_2$ (10^{-10}–10^{-5}M, or 3.5 ng/ml - 3.5 μg/ml). However, when proliferating smooth muscle cells were treated with the same concentrations of PGE$_2$, DNA synthesis decreased by 48%, indicating that PGE$_2$ had an inhibitory effect [15].

In this study, we established the stemness of hTSCs based on three characteristics described previously for human tendon stem/progenitor cells: a) the ability to form colonies in culture; b)

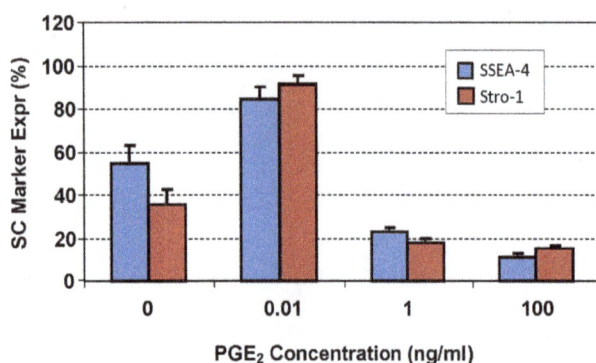

Figure 6. Quantification of SSEA-4 and Stro-1 expression in hTSCs treated with various concentrations of PGE$_2$. Average percentage of cells expressing SSEA-4 and Stro-1 in two independent FACS experiments showed that hTSCs treated with 0.01 ng/ml of PGE$_2$ expressed the most SSEA-4 (84.6±5.4%) and Stro-1 (91.1±4.2%), which was higher than the control cells without PGE$_2$ treatment (54.5±8.6% for SSEA-4, and 35.8±6.5% for Stro-1) and the cells treated with 1 ng/ml (SSEA-4:23.0±2.0%; and Stro-1:17.7±1.7%) or 100 ng/ml PGE$_2$ (SSEA-4:10.6±2.1%, and Stro-1:15.3±1.1%). SC: Stem cell; PGE$_2$-0: PGE$_2$ at 0 concentration; PGE$_2$-0.01: PGE$_2$ concentration at 0.01 ng/ml; PGE$_2$-1: PGE$_2$ concentration at 1 ng/ml; and PGE$_2$-100: PGE$_2$ concentration at 100 ng/ml.

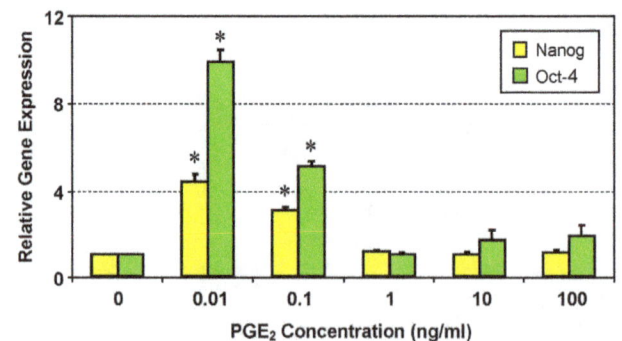

Figure 7. Expression of stem cell markers Nanog and Oct-4 in hTSCs treated with various concentrations of PGE$_2$. Total RNA was collected from various hTSCs cultured with or without PGE$_2$ and used for qRT-PCR. Expression levels of Nanog and Oct-4 are more up-regulated in hTSCs treated with low concentrations of PGE$_2$ (0.01 and 0.1 ng/ml) than in those treated with high concentrations (1, 10, and 100 ng/ml) (*p<0.05 with respect to hTSCs not treated with PGE$_2$).

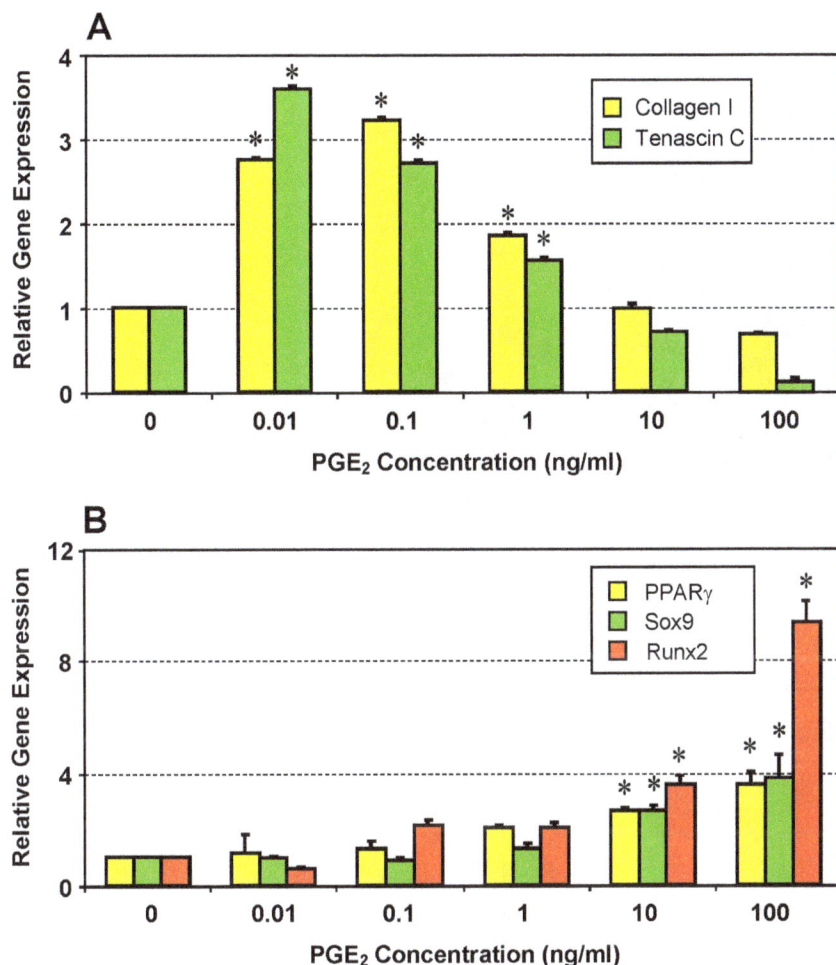

Figure 8. Expression of tenocyte (A) and non-tenocyte (B) related genes in hTSCs treated with various concentrations of PGE₂. qRT-PCR was performed on total RNA collected from cultured hTSCs treated with PGE₂. Expression levels of tenocyte related genes, collagen type I (Collagen I) and Tenascin C, were higher in hTSCs treated with low concentrations of PGE₂ (0.01, 0.1 and 1 ng/ml) than in those treated with high concentrations (10 and 100 ng/ml) (**A**). However, expression levels of non-tenocyte related genes, PPARγ, Sox9, and Runx2, were more reduced in hTSCs treated with low (0.01, 0.1 and 1 ng/ml) than with high concentrations of PGE₂ (10 and 100 ng/ml) (**B**) (*$p < 0.05$ with respect to corresponding controls that did not receive PGE₂ treatment).

expression of stem cell markers Oct-4, Nanog, and nucleostemin; and c) multi-differentiation potential [7,8]. In addition, these hTSCs assumed a cobblestone shape when grown to confluence [8]. Further, we used two stem cell markers, SSEA-4 and Stro-1, to measure the stemness of hTSCs treated with various concentrations of PGE₂. SSEA-4 and Stro-1 are highly expressed in undifferentiated stem cells and therefore are used as markers for stem cell identification. However, after differentiation, SSEA-4 is down-regulated in human embryonic stem cells [16]. Our results showing higher expression of SSEA-4 in cells treated with low levels of PGE₂ indicate that stemness is enhanced in these cells, but not in cells treated with higher levels of PGE₂. Additionally, we also found that cells treated with low levels of PGE₂ produced higher levels of stem cell-related genes (Oct-4 and Nanog) than cells treated with high levels of PGE₂. Oct-4 and Nanog are both required for the self-renewal and maintenance of stem cells in an un-differentiated state [17]. These genes were reported to downregulate the expression and activity of lineage specific factors, thereby promoting pluripotency [18]. Their downregulation, however, increased differentiation and thereby decreased the capacity of mouse embryonic stem cells for self-renewal [19–21].

This study found that higher expression levels of both Nanog and Oct-4 and corresponded low levels of non-tenocyte related genes, particularly in cells treated with low levels of PGE₂ (0.01, and 0.1 ng/ml). The results indicate maintenance of hTSCs in an undifferentiated state, at least in part through Nanog and Oct-4 suppression of adipocyte- (PPARγ), chondrocyte- (Sox9), and osteocyte- (Runx2) related genes. Further, lower expression levels of Nanog and Oct-4, especially in cells treated with high concentrations of PGE₂ (10 and 100 ng/ml), also corresponded to higher expression levels of non-tenocyte related genes. This effect, however, was not observed in the control cells (those without PGE₂ treatment), indicating the role high PGE₂ levels have in promoting non-tenocyte differentiation of hTSCs, which in turn reduces their stemness. Taken together, these results strongly suggest that the beneficial effects of the constitutively maintained low levels of PGE₂ may be critical for the maintenance of homeostasis in tendons *in vivo*.

hTSCs treated with higher concentrations of PGE₂ exhibited extensive expression of non-tenocyte related genes. In the *in vivo* experiment, non-tenocyte proteins PPARγ, collagen type II, and osteocalcin were up-regulated after implantation of hTSCs treated

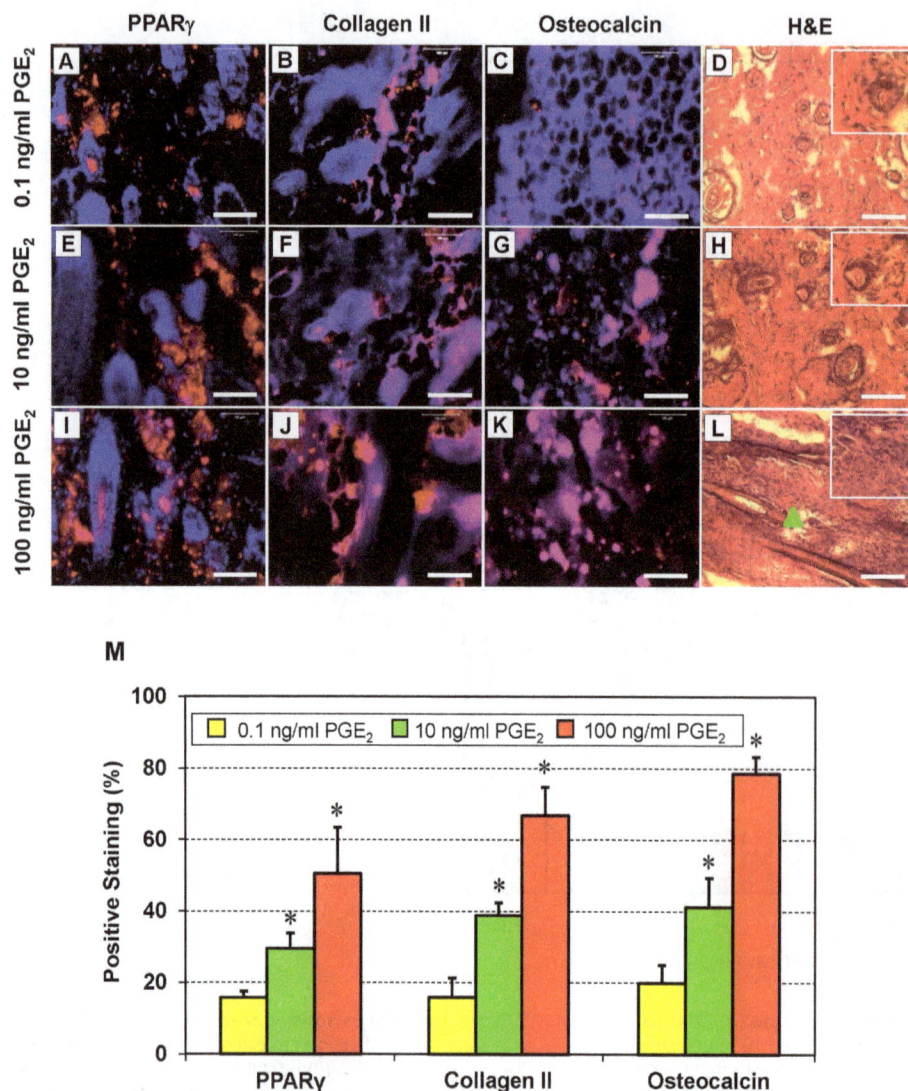

Figure 9. *In vivo* **expression of non-tenocyte markers PPARγ, collagen type II, and osteocalcin in rats implanted with hTSCs treated with various concentrations of PGE$_2$ and their respective hematoxylin and eosin (H&E) stained tissue sections.** hTSCs cultured with three concentrations (0.1, 10, and 100 ng/ml) of PGE$_2$ were implanted subcutaneously into rats; later, immunohistochemical and histological analyses were performed on tissue sections. For the immunohistochemical staining, fixed tissue sections were incubated with mouse anti-human PPARγ antibody, mouse anti-collagen type II (Collagen II) antibody, or mouse anti-human osteocalcin antibody. Cy3-conjugated goat anti-mouse IgG was then used to detect primary binding. Nuclei were stained with Hoechst (blue). Expression levels of PPARγ, collagen type II, and osteocalcin (red) were lower in cells treated with 0.1 ng/ml PGE$_2$ (**A–C**) than those treated with the higher concentrations (10 and 100 ng/ml) of PGE$_2$ (**E–G, I–K**). H&E staining was also performed on tissue sections (**D, H, L**). More cells (black dots; see insets in **D, H,** and **L**) were observed in tissues implanted with hTSCs that had been treated with high concentrations of PGE$_2$ in culture (**H, L**). Specifically, at 100 ng/ml (**L**), cells were concentrated in a specific region (triangle). Additionally, semi-quantification of the stained cells was performed by counting immuno-positive cells and calculating percentage staining (**M**) (*$p < 0.05$ in comparison with control hTSCs not treated with PGE$_2$). Bar: 100 μm.

with high levels of PGE$_2$. These findings suggest that PGE$_2$ at high concentrations could cause differentiation of TSCs into non-tenocytes; this could lead to impaired tendon healing and the formation of non-tendinous tissues in affected tendons, which would consequently reduce tendon strength. Indeed, it has been suggested that PGE$_2$, as a local hormone in tendons, may contribute to the development of tendinopathy [2,22–24]. In addition, prostaglandins (PGs) are known to play a pathophysiological role in the skeletal system, including contributing to the pathology of osteoporosis by enhancing bone resorption [25]. However, in the same milieu, PGs also exert a physiological role by stimulating bone formation through increased osteoblast

proliferation and differentiation. These functions of PGs are consistent with the biphasic effects of PGE$_2$ that maintain tendon homeostasis and lead to tendon pathology or tendinopathy.

It should be noted that when hTSCs were treated with low levels of PGE$_2$, tenocyte-related genes, including collagen type I and tenascin C, were highly expressed (**Fig. 7A**). These results suggest that PGE$_2$ at low concentrations may exert its effects on TSCs in two ways: promoting the stemness of TSCs, and inducing TSCs to differentiate towards tenocytes (or progenitor cells for tenocytes). TSCs in our cultures presumably consisted of two sub-populations of cells: one population consisted of early-stage stem cells expressing stem cell markers, such as Nanog, Oct-4, SSEA-4, and Stro-1, and the other population consisted of progenitor cells, which have differentiated towards tenocytes and expressed collagen type I and tenascin C, as demonstrated in this study. In other words, low levels of PGE_2 not only promote TSC self-renewal, but also promote the differentiation of TSCs into progenitor cells for tenocytes, suggesting that low concentrations of PGE_2 cause hTSCs to undergo asymmetric differentiation. Endogenous PGE_2 has also been shown to stimulate the proliferation of human MSCs [26], protect mouse embryonic stem cells from apoptosis through EP receptor activation [27], and enhance homing, survival, and proliferation of mouse and human hematopoietic stem cells that lead to increased numbers of repopulating cells and units [28]. As tendon-specific stem cells, TSCs play an important role in the repair of injured tendons by proliferating and differentiating *in vivo*. When tendons are injured, more tenocytes are needed, and TSCs must be activated to effectively repair injured tendons. Our results indicate that the constitutive baseline levels of PGE_2, which are low, may be used to effectively expand TSCs for cell therapy of injured tendons by promoting proliferation and maintaining tendon homeostasis.

The beneficial effects of low PGE_2 levels on TSCs have several potential applications in tendon tissue engineering. Since PGE_2 at low levels can promote the stemness of TSCs, it may be used to maintain TSCs in culture. In addition, because low PGE_2 levels can accelerate TSC proliferation, they could be used to quickly expand TSC populations for the use of cell therapy to treat injured tendons. Moreover, *in vivo* tendon injuries could be potentially treated by injecting low levels of PGE_2 at the site of injury. This could enhance the healing of injured tendons because of the ability of low levels of PGE_2 to stimulate self-renewal of TSCs and promote tenogenesis. A recent study showed that low levels of PGE_2 injected into rat patellar tendons enhanced their structural properties (the ultimate load, stiffness, and elastic modulus) [29].

While this is the first study to demonstrate the biphasic effects of PGE_2 on hTSCs, the molecular mechanisms responsible for these biphasic effects are yet to be investigated. PGE_2 is known to exert its diverse biological effects through the EP receptors [14,15,30] and by differential activation of PKA types [13]. Hence, the biphasic response of hTSCs to PGE_2 observed in this study may

also involve multiple EP receptor subtypes and/or differential activation of PKA types. Also, while we have shown the beneficial effects of low PGE_2 levels on hTSCs, one limitation of the study is the use of static culture without mechanical loading applied to hTSCs. However, tendons, and therefore the TSCs *in vivo*, are constantly subjected to mechanical loading, which regulates the expression levels of collagen type I, PPARγ, collagen type II, Sox9, and Runx2 genes. In addition, mechanical loading also increases PGE_2 levels in both patellar and Achilles tendons [9], indicating a potential interaction between mechanical loading and PGE_2. Additional studies are required to reveal the mechanisms behind this interaction. Further, we investigated only the long term effects (up to 6 days) of PGE_2 treatment on hTSCs. It is known that exercise increases PGE_2 levels in human blood only transiently, with maximum levels observed 2 hrs after exercise [31]. Therefore, it would be of interest to study the short term effects of PGE_2 on hTSCs both *in vitro* and *in vivo*.

In summary, we showed in this study that PGE_2 exerted biphasic effects on hTSCs: at low concentrations, PGE_2 enhanced their proliferation and expression of stem cell markers, whereas high concentrations of PGE_2 were detrimental to hTSCs, because they reduced their proliferation and induced non-tenocyte differentiation. These results suggest that, on one hand, low levels of PGE_2 promote tendon homeostasis by maintaining hTSCs and tenogenesis; on the other hand, high levels of PGE_2 in tendons may induce differentiation of hTSCs into non-tenocytes and thus lead to the development of the degenerative tendinopathy often seen in clinical settings.

Acknowledgments

We thank Dr. Nirmala Xavier for assistance in the preparation of this manuscript.

Author Contributions

Conceived and designed the experiments: JHW. Performed the experiments: JZ. Analyzed the data: JHW. Contributed reagents/materials/analysis tools: JHW. Wrote the paper: JZ JHW. Supervised the experiments, analyzed and interpreted results: JHW.

References

1. Meknas K, Johansen O, Steigen SE, Olsen R, Jorgensen L, et al. (2012) Could tendinosis be involved in osteoarthritis? Scand J Med Sci Sports 22: 627–634.
2. Wang JH, Iosifidis MI, Fu FH (2006) Biomechanical basis for tendinopathy. Clin Orthop Relat Res 443: 320–332.
3. Wang JH, Jia F, Yang G, Yang S, Campbell BH, et al. (2003) Cyclic mechanical stretching of human tendon fibroblasts increases the production of prostaglandin E2 and levels of cyclooxygenase expression: a novel in vitro model study. Connect Tissue Res 44: 128–133.
4. Almekinders LC, Banes AJ, Ballenger CA (1993) Effects of repetitive motion on human fibroblasts. Med Sci Sports Exerc 25: 603–607.
5. Langberg H, Skovgaard D, Karamouzis M, Bulow J, Kjaer M (1999) Metabolism and inflammatory mediators in the peritendinous space measured by microdialysis during intermittent isometric exercise in humans. J Physiol 515 (Pt 3): 919–927.
6. Zhang J, Wang JH (2010) Production of PGE(2) increases in tendons subjected to repetitive mechanical loading and induces differentiation of tendon stem cells into non-tenocytes. J Orthop Res 28: 198–203.
7. Bi Y, Ehirchiou D, Kilts TM, Inkson CA, Embree MC, et al. (2007) Identification of tendon stem/progenitor cells and the role of the extracellular matrix in their niche. Nat Med 13: 1219–1227.
8. Zhang J, Wang JH (2010) Characterization of differential properties of rabbit tendon stem cells and tenocytes. BMC Musculoskelet Disord 11: 10.
9. Zhang J, Wang JH (2010) Mechanobiological response of tendon stem cells: implications of tendon homeostasis and pathogenesis of tendinopathy. J Orthop Res 28: 639–643.
10. Langberg H, Boushel R, Skovgaard D, Risum N, Kjaer M (2003) Cyclo-oxygenase-2 mediated prostaglandin release regulates blood flow in connective tissue during mechanical loading in humans. J Physiol 551: 683–689.

11. Alfredson H, Thorsen K, Lorentzon R (1999) In situ microdialysis in tendon tissue: high levels of glutamate, but not prostaglandin E2 in chronic Achilles tendon pain. Knee Surg Sports Traumatol Arthrosc 7: 378–381.
12. Tang L, Loutzenhiser K, Loutzenhiser R (2000) Biphasic actions of prostaglandin E(2) on the renal afferent arteriole : role of EP(3) and EP(4) receptors. Circ Res 86: 663–670.
13. Kleiveland CR, Kassem M, Lea T (2008) Human mesenchymal stem cell proliferation is regulated by PGE2 through differential activation of cAMP-dependent protein kinase isoforms. Exp Cell Res 314: 1831–1838.
14. Yamato M, Nagahama K, Kotani T, Kato S, Takeuchi K (2005) Biphasic effect of prostaglandin E2 in a rat model of esophagitis mediated by EP1 receptors: relation to pepsin secretion. Digestion 72: 109–118.
15. Yau L, Zahradka P (2003) PGE2 stimulates vascular smooth muscle cell proliferation via the EP2 receptor. Molecular and Cellular Endocrinology 203: 77–90.
16. Draper J, Pigott C, Thomson J, Andrews P (2002) Surface antigens of human embryonic stem cells: changes upon differ- entiation in culture. J Anat 200: 249–258.
17. Ying QL, Nichols J, Chambers I, Smith A (2003) BMP induction of Id proteins suppresses differentiation and sustains embryonic stem cell self-renewal in collaboration with STAT3. Cell 115: 281–292.
18. Loh Y, Wu Q, Chew J, Vega V, Zhang W, et al. (2006) The Oct4 and Nanog transcription network regulates pluripotency in mouse embryonic stem cells. Nat Genet 38: 431–440.
19. Chambers I, Silva J, Colby D, Nichols J, Nijmeijer B, et al. (2007) Nanog safeguards pluripotency and mediates germline development. Nature 450: 1230–1234.

20. Mitsui K, Tokuzawa Y, Itoh H, Segawa K, Murakami M, et al. (2003) The homeoprotein Nanog is required for maintenance of pluripotency in mouse epiblast and ES cells. Cell 113.

21. Niwa H, Toyooka Y, Shimosato D, Strumpf D, Takahashi K, et al. (2005) Interaction between Oct3/4 and Cdx2 determines trophectodermdifferentiation. Cell 123: 917–929.

22. Zhang J, Wang JH (2012) BMP-2 mediates PGE(2) -induced reduction of proliferation and osteogenic differentiation of human tendon stem cells. J Orthop Res 30: 47–52.

23. Khan MH, Li Z, Wang JH (2005) Repeated exposure of tendon to prostaglandin-e2 leads to localized tendon degeneration. Clin J Sport Med 15: 27–33.

24. Riley G (2004) The pathogenesis of tendinopathy. A molecular perspective. Rheumatology (Oxford) 43: 131–142.

25. Raisz L (1999) Prostaglandins and bone: physiology and pathophysiology. Osteoarthritis Cartilage 7: 419–421.

26. Arikawa T, Omura K, Morita I (2004) Regulation of bone morphogenetic protein-2 expression by endogenous prostaglandin E2 in human mesenchymal stem cells. J Cell Physiol 200: 400–406.

27. Liou J, Ellent D, Lee S, Goldsby J, Ko B, et al. (2007) Cyclooxygenase-2-derived prostaglandin e2 protects mouse embryonic stem cells from apoptosis. Stem Cells 25: 1096–1103.

28. Hoggatt J, Singh P, Sampath J, Pelus L (2009) Prostaglandin E2 enhances hematopoietic stem cell homing, survival, and proliferation. Blood 113: 5444–5455.

29. Ferry S, Afshari H, Lee J, Dahners L, Weinhold P (2012) Effect of prostaglandin E2 injection on the structural properties of the rat patellar tendon. Sports Med Arthrosc Rehabil Ther Technol 4: 2.

30. Breyer RM, Bagdassarian CK, Myers SA, Breyer MD (2001) Prostanoid receptors: subtypes and signaling. Annu Rev Pharmacol Toxicol 41: 661–690.

31. Markworth JF, Vella LD, Lingard BS, Tull DL, Rupasinghe TW, et al. (2013) Human inflammatory and resolving lipid mediator responses to resistance exercise and ibuprofen treatment. Am J Physiol Regul Integr Comp Physiol.

32. Chen YC, Hsu HS, Chen YW, Tsai TH, How CK, et al. (2008) Oct-4 expression maintained cancer stem-like properties in lung cancer-derived CD133-positive cells. PLoS One 3: e2637.

33. Ling GQ, Chen DB, Wang BQ, Zhang LS (2012) Expression of the pluripotency markers Oct3/4, Nanog and Sox2 in human breast cancer cell lines. Oncol Lett 4: 1264–1268.

34. Minear MA, Crosslin DR, Sutton BS, Connelly JJ, Nelson SC, et al. (2011) Polymorphic variants in tenascin-C (TNC) associated with atherosclerosis and coronary artery disease. Hum Genet 129: 641–654.

35. Kohjima M, Enjoji M, Higuchi N, Kotoh K, Kato M, et al. (2007) NIM811, a nonimmunosuppressive cyclosporine analogue, suppresses collagen production and enhances collagenase activity in hepatic stellate cells. Liver International 27: 1273–1281.

36. Zhang Y, Ba Y, Liu C, Sun G, Ding L, et al. (2007) PGC-1alpha induces apoptosis in human epithelial ovarian cancer cells through a PPARgamma-dependent pathway. Cell Res 17: 363–373.

37. Tew SR, Hardingham TE (2006) Regulation of SOX9 mRNA in human articular chondrocytes involving p38 MAPK activation and mRNA stabilization. J Biol Chem 281: 39471–39479.

38. Shui C, Spelsberg TC, Riggs BL, Khosla S (2003) Changes in Runx2/Cbfa1 expression and activity during osteoblastic differentiation of human bone marrow stromal cells. J Bone Miner Res 18: 213–221.

8

Comparative Evaluation of Differentiation Potential of Menstrual Blood- *Versus* Bone Marrow- Derived Stem Cells into Hepatocyte-Like Cells

Sayeh Khanjani[1]♥, Manijeh Khanmohammadi[1]♥, Amir-Hassan Zarnani[2,3], Mohammad-Mehdi Akhondi[1], Ali Ahani[1], Zahra Ghaempanah[1], Mohammad Mehdi Naderi[1], Saman Eghtesad[4], Somaieh Kazemnejad[1]*

1 Reproductive Biotechnology Research Center, Avicenna Research Institute, ACECR, Tehran, Iran, **2** Nanobiotechnology Research Center, Avicenna Research Institute, ACECR, Tehran, Iran, **3** Immunology Research Center, Iran University of Medical Sciences, Tehran, Iran, **4** Department of Biochemistry and Molecular Biology, University of Maryland School of Medicine, Baltimore, Maryland, United States of America

Abstract

Menstrual blood has been introduced as an easily accessible and refreshing stem cell source with no ethical consideration. Although recent works have shown that menstrual blood stem cells (MenSCs) possess multi lineage differentiation capacity, their efficiency of hepatic differentiation in comparison to other stem cell resources has not been addressed so far. The aim of this study was to investigate hepatic differentiation capacity of MenSCs compared to bone marrow-derived stem cells (BMSCs) under protocols developed by different concentrations of hepatocyte growth factor (HGF) and oncostatin M (OSM) in combination with other components in serum supplemented or serum-free culture media. Such comparison was made after assessment of immunophenotye, trans-differentiation potential, immunogenicity and tumorigeicity of these cell types. The differential expression of mature hepatocyte markers such as albumin (ALB), cytokeratin 18 (CK-18), tyrosine aminotransferase and cholesterol 7 alpha-hydroxylase activities *(CYP7A1)* at both mRNA and protein levels in differentiating MenSCs was significantly higher in upper concentration of HGF and OSM (P1) compared to lower concentration of these factors (P2). Moreover, omission of serum during differentiation process (P3) caused typical improvement in functions assigned to hepatocytes in differentiated MenSCs. While up-regulation level of *ALB* and *CYP7A1* was higher in differentiated MenSCs compared to driven BMSCs, expression level of *CK-18*, detected level of produced ALB and glycogen accumulation were lower or not significantly different. Therefore, based on the overall comparable hepatic differentiation ability of MenSCs with BMSCs, and also accessibility, refreshing nature and lack of ethical issues of MenSCs, these cells could be suggested as an apt and safe alternative to BMSCs for future stem cell therapy of chronic liver diseases.

Editor: Maria Cristina Vinci, Cardiological Center Monzino, Italy

Funding: This research was supported by a research grant from the Iranian Stem Cell Network, a non-commercial organization. The funders had no role in study design, data collection and analysis, decision to publish, or preparation of the manuscript.

Competing Interests: The authors have declared that no competing interests exist.

* E-mail: s.kazemnejad@avicenna.ac.ir

♥ These authors contributed equally to this work.

Introduction

Cell therapy, using human hepatocytes, is being regarded worldwide as an alternative approach to organ transplantation for liver failure. Major obstacles, including donor organ shortage have encouraged scientists and physicians to take advantage of stem cells for cell therapy of liver disorders [1]. Adult bone marrow has commonly been known as the most conventional stem cell source in the field of regenerative medicine and tissue engineering, including liver tissue engineering, because bone marrow-derived mesenchymal stem cells (BMSCs) do not present the ethical issues of embryonic stem cells (ESCs), have hepatogenic differentiation ability *in vitro* and *in vivo* [2,3] and exhibit immunosuppressive capabilities [4,5]. However, problems such as less availability, invasive methods for sample collection and lower proliferation capacity in comparison with ESCs limit applicability of BMSCs for clinical therapy of liver diseases. Pertaining to other sources of stem cells and regardless of great achievements in generating terminally-differentiated hepatocyte-like cells from human induced pluripotent stem (iPS) cells, limitations such as the risk of tumor formation are yet to be addressed in this stem cell type [6,7].

Several studies have reported that menstrual blood (MB) contains a unique population of cells with properties similar to adult stem cells [8–11]. It has been proposed that MB contains circulating BMSCs, which contribute to endometrial regeneration [12]. Menstrual blood-derived stem cells (MenSCs) exhibit a long term self-renewal ability, greater proliferation capacity compared to BMSCs and have minimal risk of karyotypic abnormalities [8–11,13]. In addition, recent studies have showed that reprogramming efficiency for generation of iPS cells could be increased using MenSCs as a cell source even in the absence of ectopic expression of c-Myc [14,15]. These characteristics, as well as the ease of access and the possibility of cyclic sample collection, make MB an appropriate stem cell supply for tissue engineering and regenerative medicine.

We have previously presented evidence on the potential of MenSCs to generate hepatocyte-like cells [16]. In the present study, we sought to compare for the first time the hepatogenic differentiation potential of MenSCs to that of BMSCs.

To date, different concentrations of hepatocyte growth factor (HGF) and oncostatin M (OSM), as main factors involved in hepatocyte development and maturation, in combination with other cytokines and growth factors, such as epidermal growth factor (EGF), basic fibroblast growth factor (b-FGF) and insulin-like growth factor (IGF), have been used to generate both biochemically- and metabolically-active hepatocyte-like cells from BMSCs. Indeed, chemical compounds such as dexamethasone (Dexa), retinoic acid (RA), sodium butyrate, nicotinamide (NTA), norepinephrine, and dimethylsulfoxide (DMSO) are known to promote hepatic differentiation of BMSCs [17–20,21]. Moreover, much effort is underway to induce hepatic differentiation of BMSCs under serum-free conditions for human trial applications [22–25,26]. In this study, we have presented some data about: 1) unique immunophenotypic characteristics and trans-differentiation capability of MenSCs compared to BMSCs, and 2) chromosomal stability, and non-immunogenic and non-tumorigenic nature of MenSC. As a main goal, differentiation potential of MenSCs into hepatocyte-like cells has been compared to BMSCs under three protocols developed using combinations of HGF and OSM with other components in both serum-supplemented and serum-free culture media.

Materials and Methods

Isolation of MenSCs and BMSCs

Menstrual blood was collected from healthy females aged 25–35 years using a Diva cup (Diva International Co., Lunette, Finland) on day 2 of menstrual cycle. All donors were given a complete description of the study and provided written informed consent before sampling. This study was approved by the Medical Ethics Committee of Avicenna Research Institute.

The contents of Diva cup were transferred into tubes containing 2.5 µg/ml fungizone (Gibco, Scotland, UK), 100 µg/mL strepto-mycin, 100 U/mL penicillin (Sigma-Aldrich, MO, USA) and 0.5 mM EDTA in phosphate buffered saline (PBS) without Ca^{2+} or Mg^{2+}. Isolation of stem cells from menstrual blood was performed as described previously [27,28].

Bone marrow aspirates (5–10 mL) were obtained from iliac crests of human donors at the Bone Marrow Transplantation Center, Shariati Hospital, Tehran, Iran. Samples were harvested after getting signed informed consent. This work was approved by the Medical Ethics Committee, Ministry of Health, Iran. Bone marrow stem cells (BMSCs) were isolated using a combination of density gradient centrifugation and plastic adherence as described in our previous studies [29].

Characterization of immunophenotypic properties of MenSCs versus BMSCs

The expression of CD106, CD166, CD105 and CD146 as mesenchymal and OCT-4 as embryonic stem cell markers and CD45, CD133 and CD14 as hematopoietic cell markers were evaluated by flow cytometric analysis. Briefly, aliquots of 10^5 cells/100 µl were incubated separately with PE-conjugated mouse anti-human CD133 (clone TMP4; eBioscience, CA, USA), CD14 (clone M5E2; BD Pharmingen, CA, USA), CD106 (clone STA; eBioscience), CD105 (clone 43A3; BioLegend, CA, USA), CD146 (clone P1H12; BD Pharmingen), CD45 (clone HI30; BD Pharmingen) or CD166 (clone 3A6; MBL International, Woburn, MA) for 40 minutes (min). To assess OCT-4 expression, the 0.1%

saponin-permeabilized cells with were treated with rabbit anti-human OCT-4 antibody (Abcam) for 40 min followed by 30 min incubation with FITC-conjugated goat anti-rabbit Ig (Sigma). Next, all cell suspensions were fixed in 1% formaldehyde solution and examined using a flow cytometer (Partec PAS, Münster, Germany) in reference to appropriate isotype controls (IgG2a κ for CD14 and IgG1 for CD105, CD146, CD45, CD106 and CD166).

Indeed, cells were fixed in acetone at $-20°C$ for 5 min and then subjected to immunofluorescent staining for OCT-4, vimentin and GFAP. In brief, the fixed cells were permeabilized with 0.4% triton X-100 for 20 min. After washing step, cells were incubated for 1 h at room temperature (RT) with rabbit anti-human OCT-4 polyclonal antibody (Abcam), mouse anti-human vimentin mono-clonal antibody (clone V9, 1:200; Sigma) or rabbit anti-human GFAP monoclonal antibody (clone name:EP672Y, 1:250). As reagent negative control, the cells were treated in parallel with the same concentrations of normal rabbit irrelevant IgG for OCT-4 and GFAP and mouse irrelevant IgG1 for vimentin.

Subsequently, the cells were washed three times with PBS and incubated with FITC-labeled goat anti-rabbit IgG (Sigma) or FITC-labeled sheep anti-mouse IgG (Avicenna Research Institute) at RT for 45 minutes in the dark. Thereafter, cells were incubated with 4', 6 diamidino-2-phenylindole (DAPI; 1:1000) (Sigma-Aldrich) for nuclear staining. The cells were visualized and photomicrographed using an epifluorescence microscope (Olympus BX51 microscope, Tokyo, Japan) connected to digital camera (Olympus DP71, Tokyo, Japan).

Multi-lineage differentiation potential of MenSCs and BMSCs

To further characterization of isolated MenSCs in comparison with BMSCs, we assessed differentiation ability of these cells into osteoblasts, chondrocytes and adipocytes as described previously (27, 28). The differentiated cells into osteoblasts were identified by specific histochemical staining for calcium with Alizarin red staining (Sigma-Aldrich). Chondrogenesis was assessed by immunofluorescence staining using primary monoclonal mouse anti-human Collagen type II (clone 5B2.5, 1:500; Abcam) and FITC-labeled goat anti-mouse IgG (Abcam). Adipogenic-induced cells were stained for fat vacuoles using the Oil red O staining. Control cultures without the differentiation stimuli were maintained in parallel to the differentiation experiments and stained in the same manner.

Multiplex Ligation-dependent Probe Amplification (MLPA)

To investigate chromosomal stability of MenSCs during passages, MLPA analysis was performed on genomic DNA of cells at passages 2 and 12 using the SALSA MLPA kit P036-E1 Human telomer3 (MRC-Holland, Netherlands) according to the manufacturer's protocol. Briefly, a total of 100 ng of genomic DNA in a final volume of 5 µl was denatured and hybridized to SALSA probe mix, followed by incubation at 60°C for 18 hr. Subsequently, the annealed probes were ligated using provided Ligase-65 mix at 54°C for 15 min. In the next step, 10 µl of ligated products, as template, were used for DNA amplification. The PCR amplicons were run on a Genetic Analyzer 3130 (Applied Biosystems, USA), and the results were analyzed by GeneMarker software version 2.4 (SoftGenetics, USA). The normal pattern was expected to produce a normalized signal value ratio of 1:1; any value out of the ranges <0.75 or >1.35 was considered as abnormal and corresponded to a deletion or duplication, respectively. In each MLPA reaction, whole blood of adult people with no evidence of genetic anomalies, cancerous tissue of colorectal cancer with chromosomal abnormality and aborted fetus with monosomy 21 were simultaneously used as controls. In addition, all control samples were screened for chromosomal abnormality through karyotyping and a normal 550 GTG banding protocol [30].

Examination of MenSCs immunogenicity and tumorigenicity

Testing of *in vivo* immune response to MenSCs was carried out in five immunocompetent male C57BL/6 mice aged 6 to 8 weeks. Moreover, five nude mice were used to evaluate MenSCs tumorigenicity. To this end, MenSCs at passages 2–4 were harvested using trypsin-EDTA and cell density was adjusted to 1×10^6 cells/ml in serum- free medium. After that, 0.2 ml of cell suspension was subcutaneously injected in the dorsolateral flank of C57BL/6 mice and nude mice. After two weeks, 20–30 μl of sera was obtained from C57BL/6 mice and sera immunoreactivity to cultured MenSCs was evaluated using immunofluorescence staining.

In order to test tumorigenicity potential, the treated nude mice were scarified and autopsied 12 weeks after cells injection. The inoculation site from the deep aspect was inspected and excised. Moreover, to evaluate the metastatic potential of the MenSCs, the ipsilateral axillary lymph node, spleen, liver, lung, kidneys and heart were excised for histological examination using hematoxylin and eosin (H & E) staining. All procedures were approved by the animal care and ethics committee of Avicenna Research Institute.

Hepatogenic differentiation

MenSCs and BMSCs were plated at a density of 1.5×10^4 cells/cm^2 in 1% ECM gel (derived from Engelbreth-Holm- Swarm mouse sarcoma, Sigma-Aldrich)-coated T-25 flasks. In order to induce hepatogenic differentiation, sequential exposures to three different combinations of cytokines, growth factors and hormones were examined. In the commitment step of protocol 1 (P1), 60–70% confluent MenSCs cultured in DMEM-F12 medium were supplemented with 10% fetal bovine serum (FBS) and fortified with 20 ng/ml EGF and 10 ng/ml b-FGF (Sigma-Aldrich) for 2 days prior to differentiation step. Differentiation was induced by treating cells with 5% FBS-supplemented media containing 10^{-7} M Dexa, 1% (insulin, transferrin, selenium pre-mix) ITS+1, 50 μg/ml NTA and 40 ng/ml HGF (all from Sigma-Aldrich). After 14 days, differentiation step was followed by maturation step using a substituted media containing 5% FBS, 10^{-7} M Dexa, 1% ITS +1 and 20 ng/ml OSM (Sigma- Aldrich) for an additional 14 days. Protocol 2 (P2) was done in a similar manner as P1, but with half concentration of HGF (20 ng/ml) and OSM (10 ng/ml). In protocol 3 (P3), cells at 100% confluence were serum-deprived during the treatment period. In this protocol, cells were committed in serum-free DMEM medium supplemented with 20 ng/ml EGF and 10 ng/ml b-FGF for 2 days. Differentiation and maturation steps were conducted as in P1, but in serum-free media. Media in all culture conditions were changed every two days and each protocol lasted up to 30 days. Control cultures without the differentiation stimuli were maintained in parallel to the differentiation experiments in the same manner. The isolated hepatocytes from aborted fetuses (8–12 weeks) or HepG2 cell line were used as positive control. The summary of the differentiation protocols is shown in Figure 1.

Immunofluorescence staining of albumin (ALB), cytokeratin 18 (CK-18) and Alpha-fetoprotein (AFP)

On the day 30 of differentiation, cells were harvested enzymatically, using 0.25% trypsin-EDTA solution. Cytospins were prepared by centrifugation of the cell suspension onto poly-L-lysine coated glass slides. The cells (10^4 cells per slide) were fixed in acetone at $-20°$C for 5 min and incubated overnight at $4°$C with primary monoclonal mouse anti-human ALB (clone HAS-11, 1:100; Sigma-Aldrich), AFP (clone C3, 1:200; Sigma-Aldrich) or CK-18 (clone RGE53, 1:100; Chemicon International, Temecula, CA) antibodies. As negative control, cells were treated in parallel with the mouse irrelevant IgG2a for ALB and AFP and IgG1 for CK-18 at the same dilutions. Subsequently, the cells were washed three times with PBS and incubated with FITC-labeled sheep anti-mouse IgG (Avicenna Research Institute) at RT for 45 min in the dark. After washing with PBS, the cells were processed as for OCT-4 immunostaining. Human HepG2 hepatoma cells (National Cell Bank, Pasteur Institute, Tehran, Iran) were simultaneously stained for ALB and CK-18 and considered as positive control. The green fluorescence intensity of 30 protein marker/DAPI positive cells was determined using ImageJ software (ImageJ, NIH, Bethesda, Maryland, USA, http://imagej.nih.gov/ij/, 1997–2012).

Quantitative real-time reverse transcription–polymerase chain reaction (QRT-PCR)

Expression patterns of mature hepatic markers, *ALB*, tyrosine aminotransferase *(TAT)*, cytokeratin-18 *(CK-18)* [2,18,23,31,32] and cytokeratin-19 *(CK-19)* as oval cell marker [33,34,35] were investigated in MenSCs and BMSCs, differentiated under different protocols in reference to untreated cells, using qRT-PCR. Isolated hepatocytes from human aborted fetuses (8–12 weeks) were used as positive control. Total RNA was extracted from $7 \pm 2 \times 10^5$ undifferentiated and differentiated cells using RNEasy mini kit (Qiagen, CA, USA) per the manufacturer's instructions. Reverse transcription was performed using 2 μg DNAse (Fermentas Inc, MD, USA)-treated RNA, 1 μl SuperScriptTM II Reverse Transcriptase (200 U) (Life Technologies, CA, USA), 20 pM N6 Random-Hexamer, 20 pM dNTP Mix, 4 μl 5× First Strand buffer, 2 μl Dithiothreitol (0.1 M) and 1 μl RiboLockTM RNase inhibitor (all from Fermentas Inc) in a thermocycler (Eppendorf, Germany) at $25°$C for 10 min, $42°$C for 50 min and $70°$C for 15 min. Next, 2 μl of cDNA was mixed with 1× SYBR Premix EX TaqTM (Takara Bio Inc, Japan), 0.2 μM of each primer [16] and 0.4 μl ROXTM Reference Dye 50×, and amplified using a 7500 Real-Time PCR System (Applied Biosystems, MA, USA) as follows: initial denaturation at $95°$C for 10 seconds (sec), 40 cycles of a two-step PCR ($95°$C for 5 sec, $60°$C for 30 sec), dissociation stage at $95°$C for 15 sec, $60°$C for 1 min and $95°$C for 15 sec. The amplified genes were sequenced by 3130 Genetic Analyzer (Applied Biosystems, CA, USA). Mean efficiencies and crossing point values for each gene was determined with LinRegPCR (version 11.0) [36] and normalized to values for GAPDH in differentiated cells in reference to undifferentiated cells using relative expression software tool-2009 (REST-2009) (available at http://www.gene-quantification.info).

To assess expression of *CYP7A1*, 1 μL of cDNA was admixed with 12.5 μL reaction master mix (Amplicon) and 1 μL of each primer (Table 1). After initial denaturation at $94°$C for 3 min, PCR amplification was continued at $94°$C for 30 s, $60°$C for 30 s, and $72°$C for 30 s for total cycles of 35, and a final extension was performed at $72°$C for 7 min. GAPDH amplification was used as an internal standard. The amplified DNA fragments were

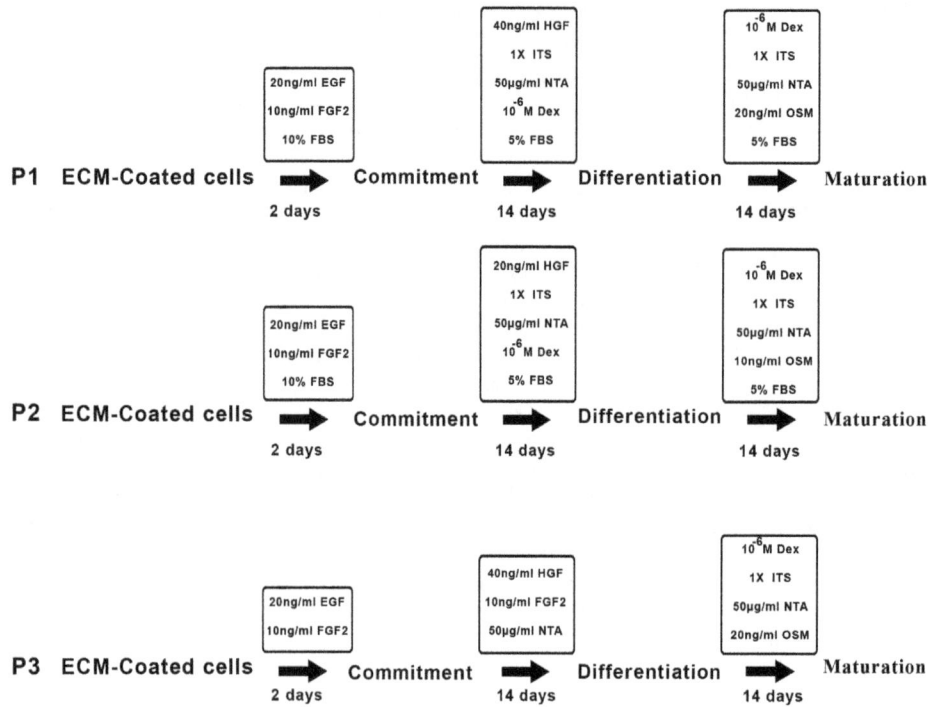

Figure 1. Schematic diagram of three-stage differentiation protocols. Both MenSCs and BMSCs were sequentially treated by three combinations of cytokines, growth factors, and hormones under commitment, differentiation and maturation steps.

electrophoresed on a 2% agarose gel and visualized by an ultraviolet transilluminator (Uvitec- USA). For semi-quantitative determination, specific band density was normalized to that of the corresponding GAPDH using AlphaEase software (Genetic Technologies, Inc.).

Albumin secretion

We quantified ALB in the media to assess ALB secretion ability of differentiated cells, a reliable assay to evaluate hepatic metabolic function [2,32]. Sample culture media was collected on days 0, 15,

20 and 25 of differentiation and measured by human ALB ELISA quantitative Kit (Immunology Consultant Lab, OR, USA) according to the manufacturer's instructions. The measurements were performed in triplicate.

Glycogen storage

Intracellular glycogen was analyzed by Periodic Acid-Schiff (PAS) staining. The cells were seeded on poly L-lysine mounted slides and fixed in 4% paraformaldehyde. The slides were oxidized in 1% periodic acid for 5 min and rinsed three times in deionized

Table 1. Sequences of the primers used for analysis of cells differentiation into hepatocyte-like cells.

Name of Gene	Sequence	Product size (bp)	NCBI Accession number
ALB	F 5'-GAGACCAGAGGTTGATGTGATG-3'	186	NM_00047
	R 5'- AGGCAGGCAGCTTTATCAGCA-3'		
CK-18	F 5'-TTGATGACACCAATATCACACGA-3'	202	NM_000224
	R 5'-TATTGGGCCCGGATGTCTG-3'		
CK-19	F 5'-GCGGCCAACGGCGAGCTA-3'	154	NM_002276
	R 5'-GCAGGACAATCCTGGAGTTCTC-3'		
CYP7A1	F 5'-CAAGCAAACACCATTCCAGCGAC-3'	388	NM_000780
	R 5'-ATAGGATTGCCTTCCAAGCTGAC-3'		
TAT	F 5'- 5CTTCTGGGGCTATGTACCTCA-3'	165	NM_000353
	R 5'- GGACTGTGATGACCACTCGGAT-3'		
GAPDH	F 5'-CTCTCTGCTCCTCCTGTTCG-3'	114	NM_002046
	R 5'-ACGACCAAATCCGTTGACTC-3'		

ALB: Albumin, CK-18: Cytokeratin-18, CK-19: Cytokeratin-19, CYP7A1: Cytochrome P450 7A1, TAT: Tyrosine aminotransferase, GAPDH: Glyceraldehyde 3-phosphate dehydrogenase.

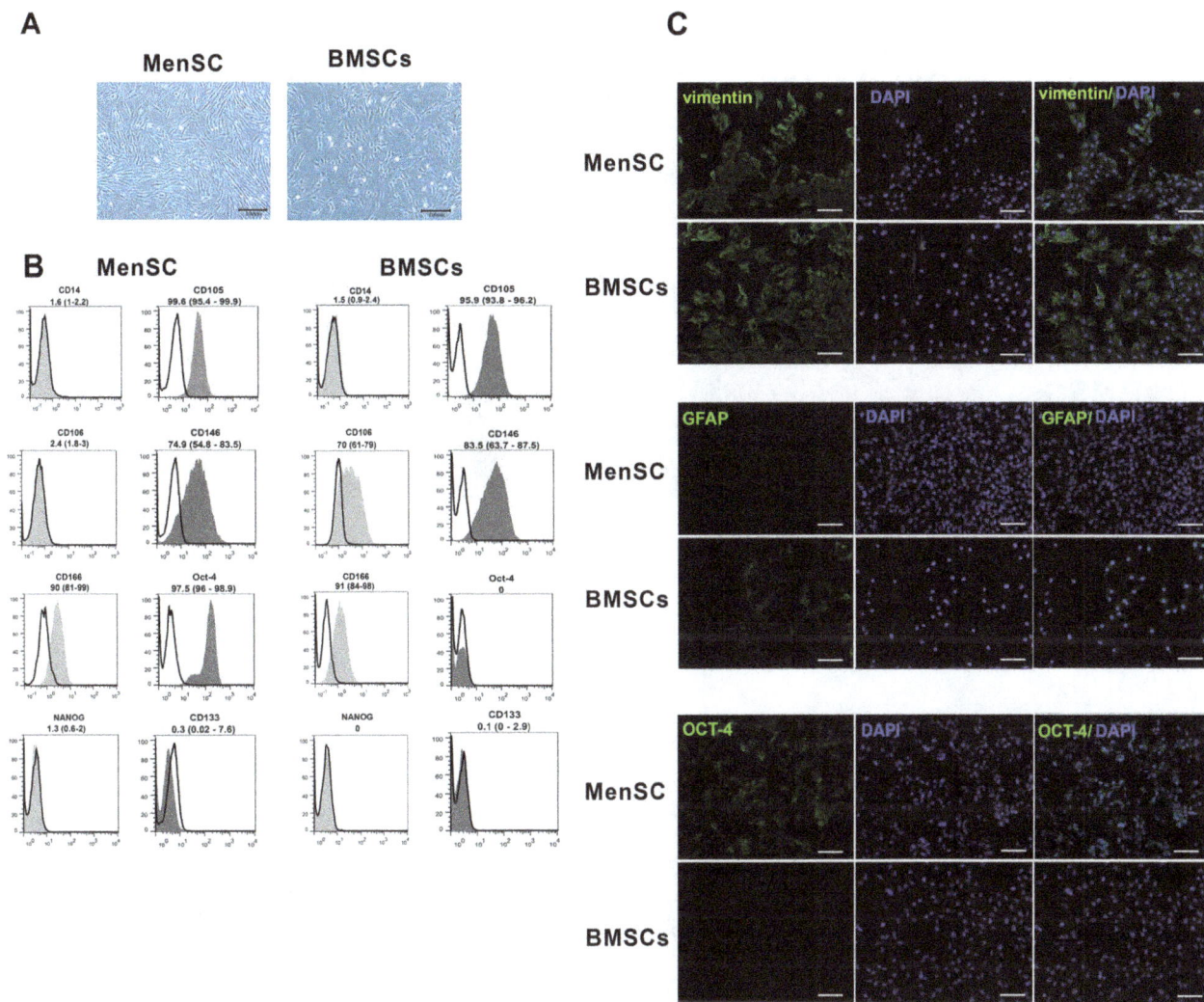

Figure 2. Phenotypic characterization of the MenSCs and BMSCs. (A) Morphological appearance of cultured MenSCs and BMSCs at passage 1; Scale bar: 100 μm. (B) Representative histograms of MenSC and BMSCs immunophenotyping by flow cytometry. CD markers are demonstrated in gray and the respective isotype control is shown as colorless. The results are presented as median (range) of 3–5 independent experiments. (C) Immunofluorescence staining of OCT-4, vimentin and GFAP in cultured MenSCs and BMSCs. Scale bar: 100 μm.

water. Slides were then treated with Schiff's reagent (Sigma-Aldrich) for 15 min and subsequently counterstained with Mayer's hematoxylin for 1 min. The slides were inspected carefully by three independent persons under a light microscope (Olympus BX51) and scored according to intensity level of PAS stain from 1+ to 4+.

Statistical analysis

All experiments were performed using cells at passages 2–4 from 3–6 donors. All measurements were performed in triplicate. The results of flow cytometry were presented as median ± range. ALB quantification using ELISA was analyzed using Mann-Whitney U-test and the data were expressed as mean ± standard deviation (SD). For the statistical analysis, the SPSS 13 software was used and P value<0.05 was considered significant.

Statistical analysis of relative gene expression results in real time PCR was performed using REST© freeware according to formula presented by Pfaffl et al [37].

Results

In vitro and *in vivo* characterization of cultured cells

The MenSCs like BMSCs had a same spindle-shaped and fibroblast-like morphology (Fig. 2A). The immunophenotyping analysis of the cells by flow cytometry and immunofluorescence staining revealed that MenSCs in a similar manner with BMSCs typically expressed CD105, CD166, CD146 and vimentin, while unlike BMSCs failed to express CD106 and GFAP. In contrast, expression of OCT-4 was only observed in MenSCs. Lack of CD133 and CD45 in cultured cells reflected a non- hematopoietic stem cell phenotype (Fig. 2B and 2C).

MenSCs differentiated into chondrocytes had strong immuno-reactivity with monoclonal antibody against Collagen type II in a similar manner with differentiated BMSCs. Mineralization was also pronounced in both osteoblast-differentiated MenSCs and BMSCs as showed by Alizarin red staining. However, the mineralization potential of MenSCs was much lower than that of BMSCs. Meanwhile, unlike formation of oil droplets in

A

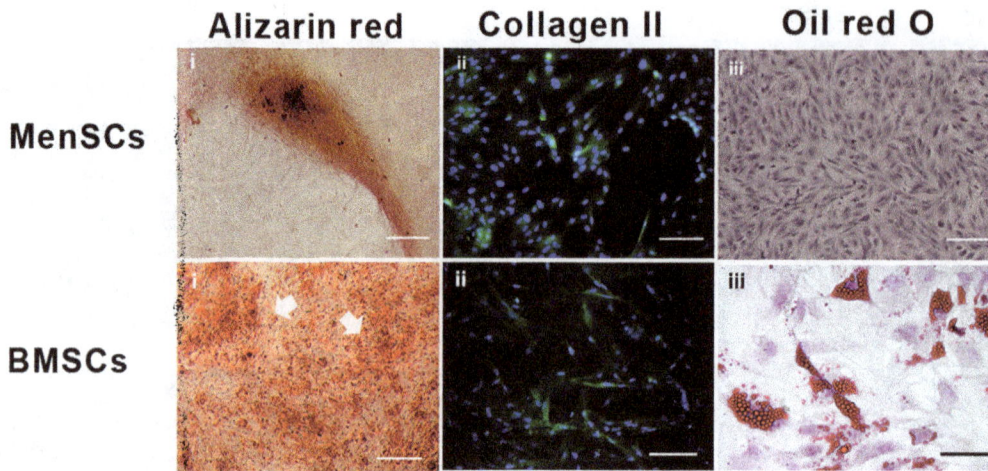

B

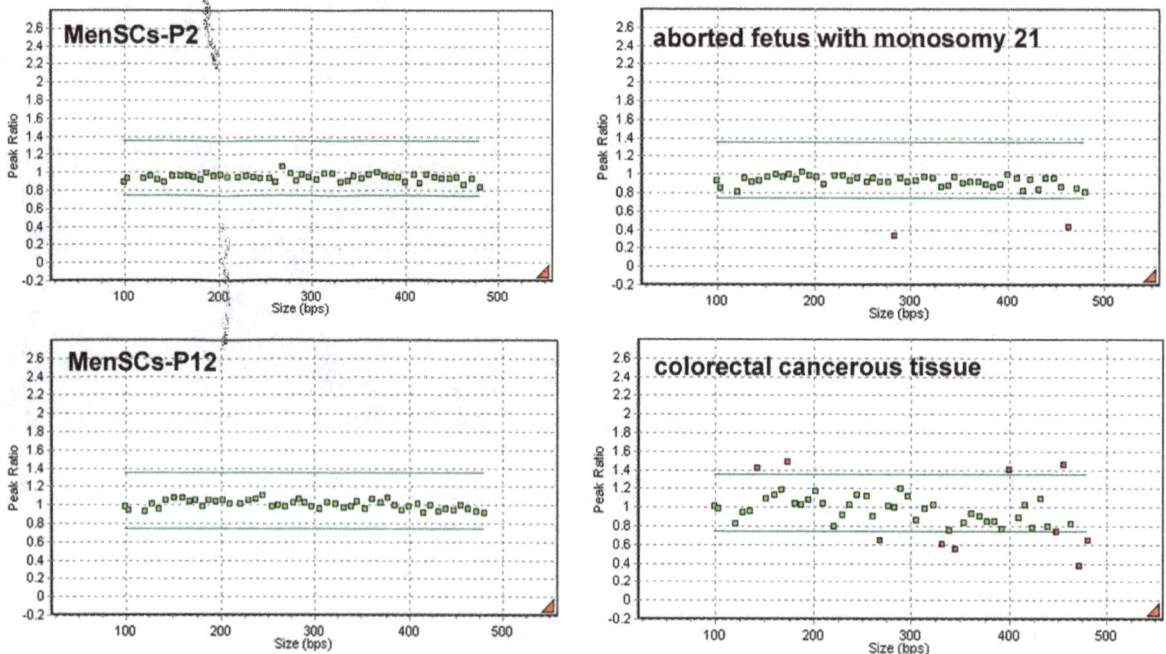

Figure 3. Evaluation of multipotency and chromosomal stability of isolated MenSCs *versus* BMSCs. (A) MenSC and BMSCs differentiation into osteoblasts (ii), chondrocytes (iii) and adipocytes (iiii) judged by Alizarin red staining, immuostaining of Collagen type II and Oil red O staining, respectively; Scale bar: 100 μm. (B) Chromatograms illustrating no chromosomal aberrations in MenSCs at passage 12 compared to cells at passage 2. GeneMarker plots showing results of MLPA analysis. Green lines illustrated the upper and lower limits of acceptable ranges of variations in MLPA analysis. Green dots show the chromosomal locations which are balanced and the red dots in the upper side of the plots show chromosomal gain and red dots in lower side of the plots show chromosomal loss.

differentiated BMSCs, Oil red O staining of the differentiated MenSCs into adipocytes was negative (Fig. 3A).

MLPA analysis showed that MenSCs, in contrast to positive controls, maintained diploid phenotype without chromosomal aberrations during passages (Fig. 3C).

According to histological examination of nude mice injected with MenSCs (Fig. 4A & B), no evidence of tumor growth was found in inoculation site and the examined tissues had no morphological characteristics of tumor as judged by H & E staining. Moreover, analysis of immunoreactivity of mice sera to

cultured MenSCs revealed no positive reaction in immunofluorescence staining (Fig. 4C).

Phenotypic characterization of differentiated cells into hepatocyte-like cells

Morphological assessment of the cells using phase contrast microscopy (Fig. 5) showed that under all differentiation conditions, the fibroblast-like morphology of MenSCs and BMSCs did not change significantly during the commitment step. The

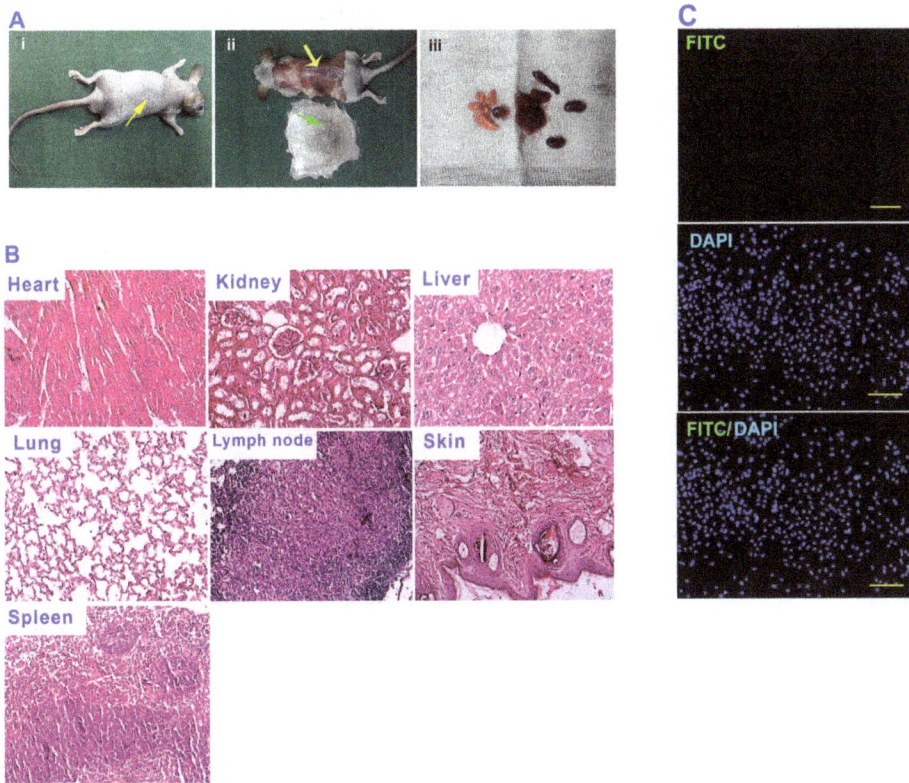

Figure 4. *In vivo* assay of tumorigenicity and immunological reaction of MenSCs. (A) 2×10^6 cells were subcutaneously injected in the dorsolateral part of the flank of nude mice (i), No tumor formation was observed in treated mice (ii), The excised tissues were fixed in buffered formalin, embedded in paraffin, and sectioned in 5-μm sections (iii). (B) The sectioned tissues were evaluated using H & E staining. (C) Immunoreactivity of mice sera to cultured MenSCs was evaluated using immunofluorescence staining. The cells (2×10^4 cells per slide) were fixed in acetone at $-20°C$ for 5 min and were incubated for 1 hour at $4°C$ with mice sera. Subsequently, the cells were washed three times with PBS and incubated with FITC-labeled sheep anti-mouse IgG at RT for 45 min in the dark. DAPI was used for nuclear staining.

hepatocyte-like morphology, evidenced by binucleation and cytoplasmic granulation, was observed when both cell types were exposed to differentiation media under protocols 1 and 3 (P1 and P3), with more polyhedral contours and binuclation under P3 compared to P1. These morphological changes were more obvious in differentiated MenSCs compared to differentiated BMSCs. The MenSCs and BMSCs cultured under P2 did not exhibit significant changes in cell morphology during differentiation.

Immunofluorescent staining of Albumin, CK18 and AFP

Albumin (ALB) was not detectable in undifferentiated MenSCs or BMSCs, whereas there was a significant level of ALB expression at the end of differentiation process in both MenSCs and BMSCs under all conditions. The intensity of ALB expression in MenSCs was dependent on differentiation protocol, with the highest level under P3 (44.86 ± 18.34) and lowest under P2 (18.85 ± 4.15). The mean intensity of ALB expression was 27.45 ± 10 in differentiated cells under P1 (Fig. 6A). Nonetheless, the differentiated MenSCs showed a considerably higher level of ALB expression compared to differentiated BMSCs under P1 and P3. The mean intensity of ALB expression was highest when BMSCs differentiated under P3 (25.26 ± 15.2) and lowest (12.85 ± 3.2) in differentiated BMSCs under P2.

Expression of CK-18 in differentiated MenSCs under P1 and P2 was not different with undifferentiated cells, but, the treated MenSCs using P3 showed significantly higher expression level of this protein compared with other groups (Fig. 6B). In contrast,

relatively the same levels of AFP expression were observed in MenSCs differentiated by either of protocols and undifferentiated cells (Fig. 6C).

Hepatic gene expression

As shown in figure 7A, the MenSCs showed a significant up-regulation of *ALB* gene following differentiation under P1 (4.37 ± 0.50 fold, $P = 0.001$) and P3 (3.7 ± 0.63 fold, $P = 0.001$), whereas no significant change was detected in *ALB* expression induced by P2 compared to undifferentiated MenSCs (2.54 ± 0.6, $P = 0.17$). While the up-regulation level of *ALB* gene in differentiated MenSCs by P1 was ~2 fold less than that in differentiated BMSCs, up-regulation level of this gene was approximately 3 fold higher in differentiated MenSCs under P3 compared to differentiated BMSCs ($P = 0.001$). In addition, a significant up-regulation of *CK-18* mRNA was observed in both MenSCs and BMSCs differentiated under all protocols ($P = 0.001$). The up-regulation level of this gene in differentiated MenSCs under P1 and P2 was respectively ~10 and ~8.6 fold higher compared to differentiated BMSCs ($P = 0.005$). Although, the up-regulation level of *CK-18* mRNA in both differentiated MenSCs and BMSCs by P3 was significantly greater than that of cells differentiated under P1 or P2 ($P = 0.001$), differentiated MenSCs under P3 exhibited significantly lower level of CK-18 in comparison with treated BMSCs under the same protocol ($P = 0.001$). Moreover, a slight up-regulation of CK-19 was observed in differentiated MenSCs under all protocols (P1: 0.23 ± 0.18; $P = 0.001$, P2: 0.35 ± 0.2; $P = 0.001$ and P3:

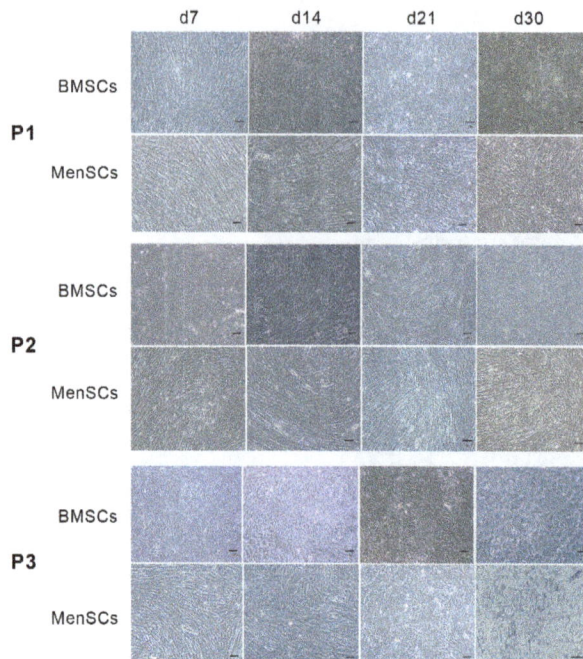

Figure 5. Morphology of MenSCs compared to BMSCs during differentiation by three protocols. The gradual change of MenSC morphology compared with BMSCs under each differentiation protocol (P1–P3) has been shown by phase contrast photographs. Scale bar: 100 µm.

1.6 ± 0.69; $P=0.001$). The up-regulation level of CK-19 in treated MenSCs under P1 and P2 was further than differentiated BMSCs ($P=0.001$). Furthermore, a typical up-regulation of TAT gene was beheld in the all diverted MenSCs in ratio to undifferentiated cells. The TAT over-expression was higher in cells differentiated under P2 (1251 ± 170, $P=0.001$) and P3 (2331 ± 777, $P=0.005$) compared to the cells induced by P1 (300 ± 108, $P=0.001$).

Although the expression level of *ALB* gene in both differentiated BMSCs and MenSCs under all protocols was lower than those observed in isolated fetal hepatocytes, the expression levels of *TAT*, *CK-18* and *CK-19* genes in both differentiated cells were significantly higher than those of isolated fetal hepatocytes in a protocol-dependent manner (Fig. 7B).

Functionality assay

Glycogen storage and ALB secretion were evaluated to determine whether the hepatocyte-like cells also had functional features related to hepatocytes. Our analysis for ALB secretion indicated that undifferentiated MenSCs and BMSCs did not secrete ALB, whereas after hepatic induction under all protocols, a time-dependent increase in the level of secreted ALB was observed (Fig. 8A). Although ALB level had no significant difference between differentiated MenSCs and BMSCs, there was a significant difference in ALB level of differentiated cells depending on the differentiation protocol. In both cell types, the secreted ALB by differentiated cells under P1 (MenSCs: 467.3 ± 16, BMSCs: 431 ± 13.86) and P3 (MenSCs: 443.95 ± 6.15, BMSCs: 446.25 ± 5.47) was in a higher level compared to cells induced with P2 (MenSCs: 350 ± 28.7, BMSCs: 398 ± 5.23)

Based on the findings of PAS staining (Fig. 8B), all differentiation protocols resulted in glycogen accumulation in derived MenSCs or BMSCs, however the intensity of PAS staining varied

dependent on the differentiation protocol and cell type. The highest PAS intensity was in both cell types was observed in P3 (4+). While the intensity of PAS staining had no gross difference between BMSCs induced by P1 and P2 (both 2+), the differentiated MenSCs under P1 exhibited higher level of glycogen deposits compared to those induced by P2 (3+ versus 1+). Furthermore, mRNA expression of Cytochrome P450 7A1 (*CYP7A1*) was examined for further functional characterization of differentiated hepatocyte-like cells. As shown in figure 8C, MenSCs in undifferentiated state did not express the *CYP7A1* gene, however, expression of this marker was detectable on the last day of differentiation process under all protocols. Unlike differentiated BMSCs that expressed high levels of *CYP7A1* under all protocols, differentiated MenSCs exhibited a protocol-dependent variation in *CYP7A1* expression. Interestingly, in consistence with data obtained by other functionality tests, differentiated MenSCs under P3 expressed the highest level of *CYP7A1* gene.

Discussion

Menstrual blood is an interesting stem cell source in the field of regenerative medicine and tissue engineering, because it is abundant and easy to access. Furthermore, MenSCs isolated from MB can be expanded in quantities relevant to clinical applications without chromosomal abnormalities. Non-immunogenic nature and lack of tumor formation capability, as shown in our study, are among other outstanding advantages that make MenSCs a suitable candidate for cell therapy. While current reports of animal studies on MenSCs-based therapy for some diseases such as cardiac [38], muscular [39] and nervous system disorders [40,41] hold promise for cell-based therapy using these stem cells, there is very limited information about MenSCs potential to generate hepatocytes [8,16], particularly compared with other known stem cells such as BMSCs. In this study, some protocol-dependent differences were beheld between hepatogenic differentiation potential of MenSCs and BMSCs.

We showed recently that MenSCs possess some markers of mesenchymal stem cells such as CD44, CD73, CD29 and CD90 but fail to express STRO-1 [27,28]. To further characterization of these cells, in this study we evaluated expression of other mesenchymal and embryonic stem cell markers. The high expression level of BMSCs markers such as CD146, CD166, CD105 and vimentin in parallel to OCT-4, the later being an embryonic stem cell marker, suggests a dual characteristic for MenSCs. On the other hand, lack of some other mesenchymal (STRO-1, CD106, and GFAP) and embryonic (NANOG and SSEA-4) stem cell markers distinguishes MenSCs from these stem cell types [27,28].

Our preliminary evaluation indicated a trans-differentiation ability of MenSC population into chondrogenic and osteogenic lineages in a similar manner with BMSCs. However, differentiation capability of MenSCs to adipocytes was significantly lower than that of BMSCs. Considering aforesaid differences, comparative evaluation of MenSCs and BMSCs in terms of differentiation into other lineages including hepatic lineage would help us to understand other potential distinctive characteristics that they may present. To our best knowledge, such data has not been reported so far. In the present study, to determine hepatogenic differentiation potential of MenSCs especially compared with BMSCs, we developed three protocols (P1-P3) using HGF and OSM, in two combinations with other components in both serum-supplemented and serum-free culture media. All differentiation processes were entirely examined on ECM-seeded cells, considering the well-

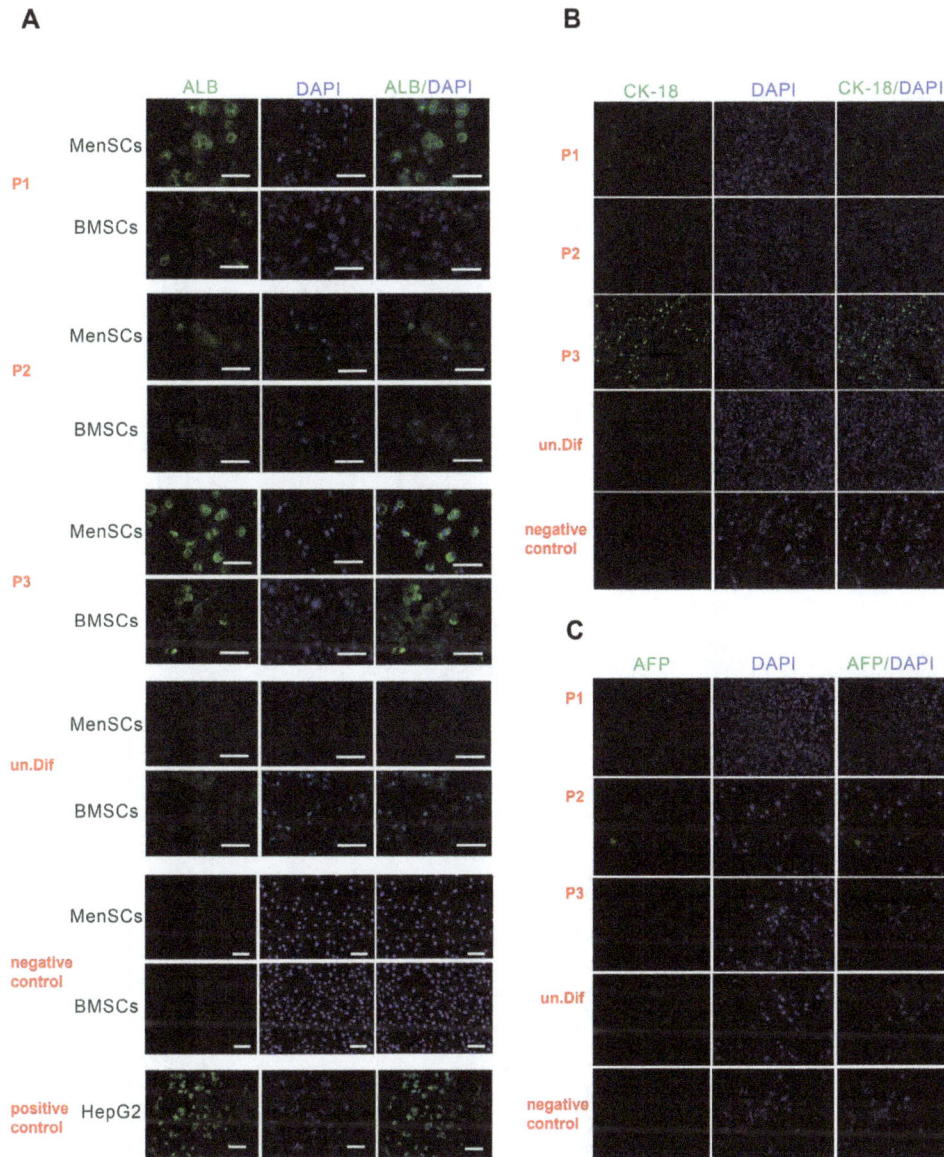

Figure 6. Immunofluorescent staining of differentiated MenSCs and BMSCs. Expression of ALB in differentiated MenSCs and BMSCs (A) and expression of CK-18 (B) and AFP (C) in differentiated MenSCs was examined by immunofluorescence staining with DAPI nuclear staining; scale bar: 100 μm. As negative control, cells were treated in parallel with the mouse irrelevant IgG2a for ALB and AFP and IgG1 for CK-18. Human HepG2 hepatoma cells were considered as positive control.

known growth and hepatogenic differentiation promoting effect of ECM on stem cells [42,43].

Here we showed that all induced MenSCs using three protocols acquired some features of functional mature hepatocytes such as mRNA and/or protein expression of ALB, CK-18, TAT, CYP7A1 and also ALB secretion and glycogen accumulation (table 2). Moreover, the differentiated MenSCs showed a slight expression of oval cell (mature hepatocyte precursor) markers, CK-19 and AFP [33,34,35], suggesting presence of a small portion of these cells in mature hepatocyte population.

Based on our results, degree of differentiation was protocol-dependent. In comparison with differentiated MenSCs in inducing media containing 20 ng/ml HGF and 10 ng/ml OSM (P2), cells differentiated under higher concentration of these factors (P1) displayed upper level of ALB at the protein level. In addition, the

mean intensity of ALB signal in differentiated MenSCs was higher than differentiated BMSCs under P1 and P3. These differences were also reflected in the results of secreted ALB levels. In consistent with these results, expression level of CYP7A1 and ALB mRNA was significantly higher in differentiated MenSCs under P1 in reference to those developed by P2. Moreover, differentiated hepatocyte-like cells were able to accumulate more glycogen in differentiated cells induced by P1 compared to those by P2.

Therefore, although expression level of CK-18 mRNA had no significant difference between induced MenSCs by P1 and P2, accumulative data showed P1 was more efficient to differentiate MenSCs into hepatocyte-like cells compared to P2.

In a similar manner with MenSCs, differentiated BMSCs under P1 showed a higher level of mature hepatocyte markers including ALB mRNA, CYP7A1 mRNA and produced ALB protein

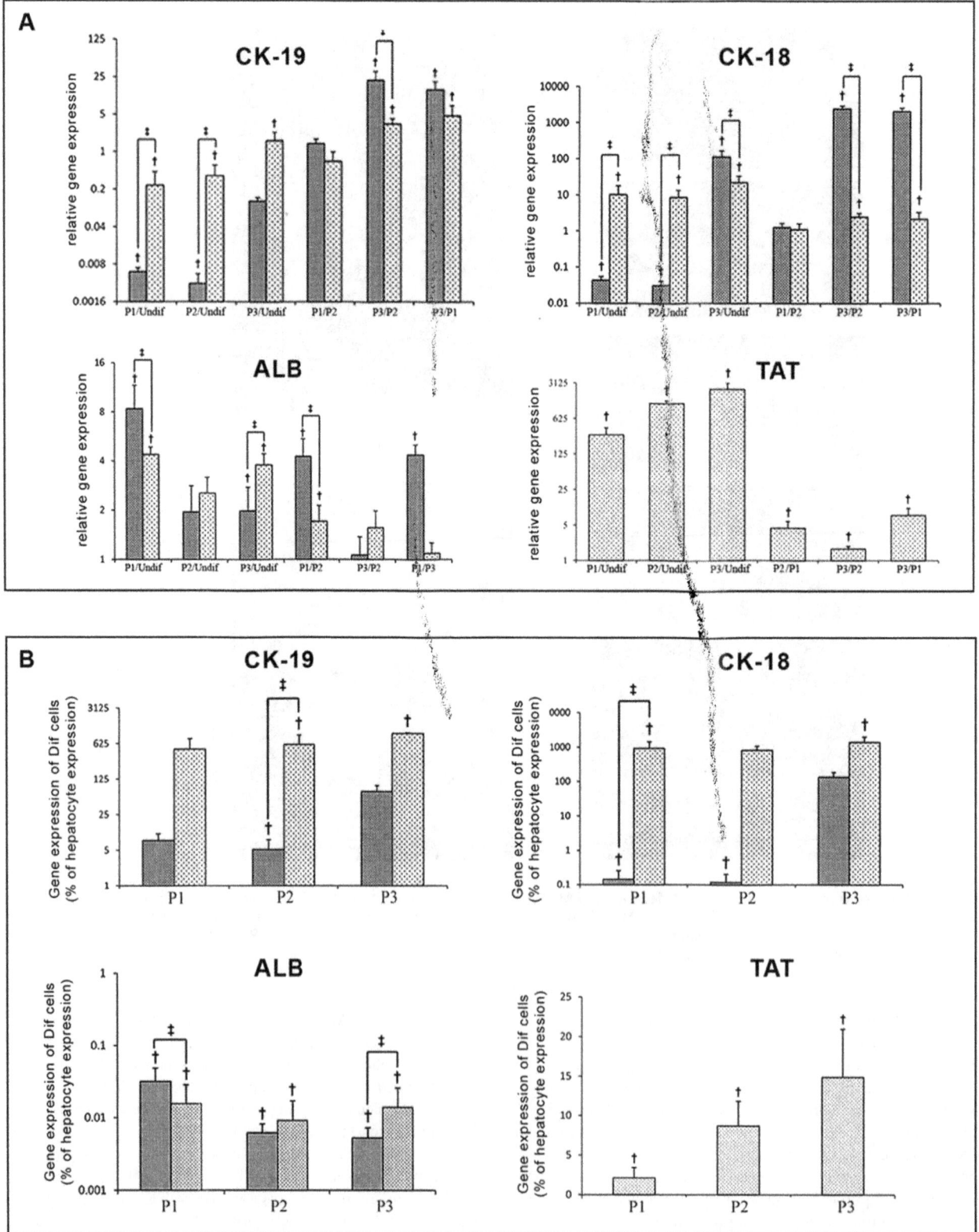

Figure 7. Quantitative RT-PCR results of differentiated cells using three protocols. (A) Data of differentiated MenSCs and BMSCs were normalized to corresponding GAPDH and calculated in reference to undifferentiated cells. Each bar in each differentiation protocol represents the gene expression in ratio to undifferentiated cells (the first three pairs of bars). The second three pairs of bars represent comparisons of a given marker expression between two protocols in each stem cell. † indicates significant difference between differentiated and undifferentiated status of the same

stem cell (P<0.05), ‡ indicates significant difference (P<0.05) between MenSCs and BMSCs. (B) Relative gene expression of differentiated cells compared to isolated adult hepatocytes. Results are shown as % of hepatocyte expression level.

compared to those induced by P2. While some parameters such as CYP7A1 and CK-18 and secreted albumin showed no significant differences between differentiated MenSCs and BMSCs, the expression level of ALB and CK-18 at mRNA and/or protein level was significantly different between these two cell types. The greatest expression of hepatic markers in both MenSCs and BMSCs induced by P1 compared to those differentiated by P2 presents an additional evidence of critical role of HGF and OSM in stem cell differentiation which has also been reported by others [29,26,44]. On the other hand, it sounds that the source of stem cells serves a fundamental role on expression pattern of mature hepatic markers. In consistent with the this conclusion, significant differences in hepatic differentiation potential were reported between stem cells derived from various origins including placental-derived stem cells and BMSCs [45].

In sharp contrast to the main role of serum in stem cell expansion, serum-free conditions have been applied on a routine basis for hepatic differentiation of embryonic or adult stem cells [46,47,26]. In our knowledge, there are no reports available on hepatogenic fate of stem cells in serum-free media compared with that in serum-fortified media. Surprisingly, we found that omission of FBS during three steps of differentiation process (P3) causes typical improvement in functions assigned to hepatocytes. Some morphological characteristics of the differentiated MenSCs under P3, such as binucleation, granulation of the cytoplasm, and formation of polyhedral contours, resembled native hepatocytes. In addition to the morphological appearance, the expression of ALB protein, TAT, CK-18 and CYP7A1 in mRNA and/or protein levels under P3 was typically superior compared to those of MenSCs differentiated under both P1 and P2. Although ALB mRNA and secreted level of ALB showed no significant differences between MenSCs treated under P3 and P1, the further up-regulation of other mentioned parameters suggested more efficacy of P3 than P1 and P2 to develop hepatocyte-like cells from

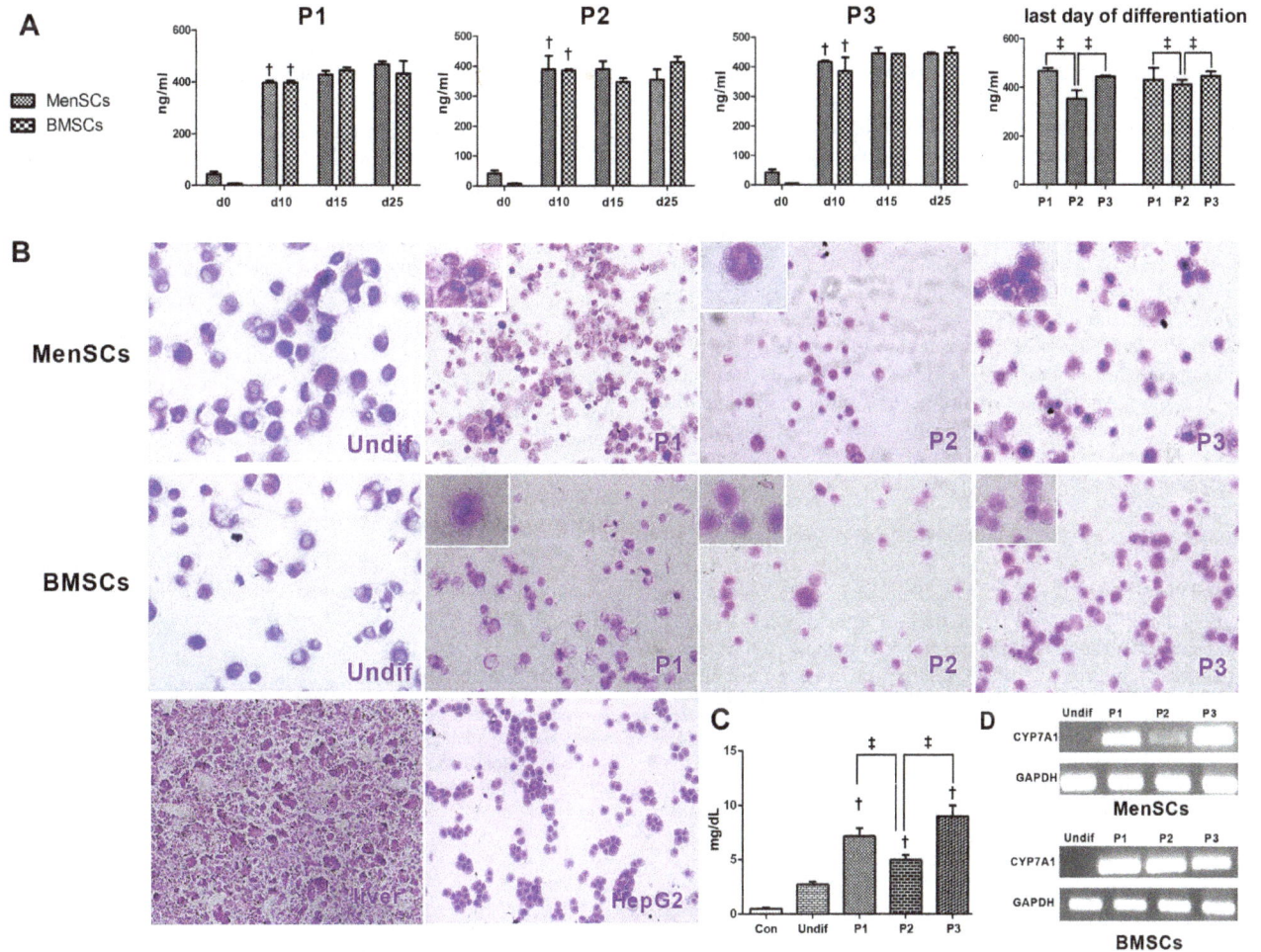

Figure 8. Functionality characteristics of differentiated MenSCs compared to BMSCs. (A) ALB levels (ng/ml/48 h) in cell supernatants at days 0, 10, 15, and 25 of differentiation. † indicates significant difference between specified day and the previous time period of differentiation in the same stem cell (P<0.05), ‡ indicates significant difference (P<0.05) between differentiated MenSCs and BMSCs at last day of differentiation. (B) PAS staining of glycogen storage in fetus liver and HepG2 as positive control, undifferentiated and differentiated MenSCs and BMSCs by various protocols (P1–P3). (C) Expression pattern of Cytochrome P450 7A1 (CYP7A1) in reference to GAPDH in differentiated MenSCs and BMSCs by various protocols. Undif: undifferentiated cells, W: water.

Table 2. Comparative analysis of hepatic markers expression in MenSCs and BMSCs using three protocols.

Cell type		Differentiated MenSCs			Differentiated BMSCs		
Differentiation protocol		P1	P2	P3	P1	P2	P3
Immunofluorescent staining	ALB	+++	+	++++	++	+	++
	CK-18	+	+	+++	ND.	ND.	ND.
	AFP	+	+	+	ND.	ND.	ND.
QRT-PCR	ALB	++	+	++	++++	+	+
	CK-18	++	+	+++	+	+	++++
	CK-19	+	+	+	+	+	+++
	TAT	+	++	+++	ND.	ND.	ND.
	CYP7A1	+++	+	++++	+++	++	++
PAS	Glycogen	+++	+	++++	++	++	++++
ELISA	ALB	++	+	++	++	+	++

The up-regulation levels of hepatic markers compared to undifferentiated cells is scored between 1+ to 4+. ND: not determined.

MenSCs. In line with this conclusion, P3-induced MenSCs accumulated more glycogen than those primed by P1 or P2. Although molecular mechanisms governing the higher efficiency of supplemented serum-free in comparison with serum-fortified media for MenSC differentiation remain to be determined, data suggest that FBS deprivation from differentiation media may serve as a triggering factor for MenSCs differentiation.

Compared to differentiated MenSCs, driven BMSCs exhibited more variable expression pattern during differentiation under P1 and P3. While up-regulation level of ALB and CYP7A1 was higher in BMSCs induced by P1 compared to P3, the opposite pattern was found in the case of CK-18 expression and glycogen accumulation. Indeed, BMSCs differentiated under P1 and P3 showed the same levels of ALB secretion. Thus, based on these results superiority of P3 over P1 in BMSCs is not as certain as MenSCs and further studies are needed to counsel a robust notion. In addition, with respect to high costs and safety problems with FBS, serum-free composition of P3 would be advantageous over P1 that contains FBS during differentiation of both MenSCs and BMSCs.

Differentiated MenSCs and BMSCs showed upper level of CK-18, CK-19 and TAT and lower level of ALB compared with isolated fetal hepatocytes. Such inconsistency has also been reported for hepatocyte-like cells derived from iPS cells [48,49]. It indicates that the other supplementary strategies, such as three-dimensional cultures or co-culture may be required to further induce the acquisition of native mature hepatocyte nature.

Taken together, the evidence presented in this study indicates MenSCs are a stem cell population with unique immunopheno-type and differentiation potential and no risk of tumor formation and immunological reaction. We proved MenSCs have capability

to generate functional hepatocyte-like cells, but with different pattern compared to BMSCs depending on critical growth factor concentration and culture media condition. P3 enriched in vitro production of functional MenSC-derived hepatocyte-like cells more than P1 and especially P2. However, in the case of BMSCs, P1 and P3 exhibited rather same efficacy. Differences in protocol-dependent expression pattern of hepatogenic markers between these cells may be attributed to different immunophenotypic features and signaling machinery.

On the other hand, regarding overall expression pattern of hepatic markers in differentiated MenSCs with BMSCs, it sounds that hepatic differentiation potential of these two stem cells had no significant predominance to each other. Here again, considering the refreshing nature, accessibility and lack of ethical issues, MenSCs are superior over BMSCs. Further studies using *in vivo* models and more functional tests are needed before a firm conclusion on applicability of MenSCs-derived hepatocyte-like cells in the clinic or could be made.

Acknowledgments

The authors would like to thank Dr. Saeed Talebi, Miss. Haleh Edalatkhah, Dr. Bozorgmehr and Mr. Ebrahim Mirzadegan for their technical assistance in QRT-PCR and flow cytometry analysis.

Author Contributions

Conceived and designed the experiments: S. Kazemnejad. Performed the experiments: S. Khanjani AA MMN. Analyzed the data: MK AHZ. Contributed reagents/materials/analysis tools: ZG MMA. Wrote the paper: S. Kazemnejad. Grammatically corrected paper: SE.

References

1. Allameh A, Kazemnejad S (2012) Safety evaluation of stem cells used for clinical cell therapy in chronic liver diseases; with emphasize on biochemical markers. Clin Biochem 45: 385–396.
2. Snykers S, De Kock J, Rogiers V, Vanhaecke T (2009) In vitro differentiation of embryonic and adult stem cells into hepatocytes: state of the art. Stem Cells 27: 577–605.
3. Henningson CT Jr, Stanislaus MA, Gewirtz AM (2003) Embryonic and adult stem cell therapy. J Allergy Clin Immunol 111(S2): 5745–5753.
4. Zhao S, Wehner R, Bornhäuser M, Wassmuth R, Bachmann M, et al. (2010) Immunomodulatory properties of mesenchymal stromal cells and their therapeutic consequences for immune-mediated disorders. Stem Cells 19: 607–614.

5. Nauta AJ, Fibbe WE (2007) Immunomodulatory properties of mesenchymal stromal cells. Blood 110: 3499–3506.
6. Takahashi K, Tanabe K, Ohnuki M, Narita M, Ichisaka T, et al. (2007) Induction of pluripotent stem cells from adult human fibroblasts by defined factors. Cell 131: 861–872.
7. Gutierrez-Aranda I, Ramos-Mejia V, Bueno C, Munoz-Lopez M, Real PJ, et al. (2010) Human Induced Pluripotent Stem Cells Develop Teratoma More Efficiently and Faster than Human Embryonic Stem Cells Regardless of the Site of Injection. Stem Cells 28: 1568–1570.
8. Meng X, Ichim TE, Zhong J, Rogers A, Yin Z, et al. (2007) Endometrial regenerative cells: a novel stem cell population. J Transl Med 5: 57–66.

9. Patel AN, Park E, Kuzman M, Benetti F, Silva FJ, et al. (2008) Multipotent menstrual blood stromal stem cells: isolation, characterization, and differentiation. Cell Transplant 17: 303–311.

10. Musina RA, Belyavski AV, Tarusova OV, Solovyova EV, Sukhikh GT (2008) Endometrial mesenchymal stem cells isolated from the menstrual blood. Bull Exp Biol Med 145: 539–543.

11. Masuda H, Matsuzaki Y, Hiratsu E, Ono M, Nagashima T, et al. (2010) Stem cell-like properties of the endometrial side population: implication in endometrial regeneration. PLoS One 5(4): e10387.

12. Du H, Taylor HS (2007) Contribution of bone marrow-derived stem cells to endometrium and endometriosis. Stem Cells 25: 2082–2086.

13. Allickson JG, Sanchez A, Yefimenko N, Borlongan CV, Sanberg PR (2011) Recent Studies Assessing the Proliferative Capability of a Novel Adult Stem Cell Identified in Menstrual Blood. Open Stem Cell J 3: 4–10.

14. de Carvalho Rodrigues D, Asensi KD, Vairo L, Azevedo-Pereira RL, Silva R, et al. (2012) Human menstrual blood-derived mesenchymal cells as a cell source of rapid and efficient nuclear reprogramming. Cell Transplant 21: 2215–2224.

15. Li Y, Li X, Zhao H, Feng R, Zhang X, et al. (2013) Efficient Induction of Pluripotent Stem Cells from Menstrual Blood. Stem Cells Dev 22: 1147–1158.

16. Khanjani S, Khanmohammadi M, Zarnani AH, Talebi S, Edalatkhah H, et al. (2013) Efficient generation of functional hepatocyte-like cells from menstrual blood derived stem cells. J Tis Eng Reg Med In press.

17. Chivu M, Dima SO, Stancu CI, Dobrea C, Uscatescu V, et al. (2009) In vitro hepatic differentiation of human bone marrow mesenchymal stem cells under differential exposure to liver-specific factors. Transl Res 154: 122–132.

18. Stock P, Staege MS, Müller LP, Sgodda M, Völker A, et al. (2008) Hepatocytes derived from adult stem cells. Transplant Proc 40: 620–623.

19. Banas A, Yamamoto Y, Teratani T, Ochiya T (2007) Stem cell plasticity: learning from hepatogenic differentiation strategies. Dev Dyn 236: 3228–3241.

20. Ji R, Zhang N, You N, Li Q, Liu W, et al. (2012) The differentiation of MSCs into functional hepatocyte-like cells in a liver biomatrix scaffold and their transplantation into liver-fibrotic mice. Biomaterials 33: 8995–9008.

21. Forte G, Minieri M, Cossa P, Antenucci D, Sala M, et al. (2006) Hepatocyte growth factor effects on mesenchymal stem cells: proliferation, migration, and differentiation. Stem Cells 24: 23–33.

22. Kazemnejad S, Allameh A, Gharehbaghian A, Soleimani M, Amirizadeh N, et al. (2008) Efficient replacing of fetal bovine serum with human platelet releasate during propagation and differentiation of human bone-marrow-derived mesenchymal stem cells to functional hepatocyte-like cells. Vox Sang 95: 149–158.

23. Lee KD, Kuo TK, Whang-Peng J, Chung YF, Lin CT, et al. (2004) In vitro hepatic differentiation of human mesenchymal stem cells. Hepatology 40: 1275–1284.

24. Snykers S, Vanhaecke T, Papeleu P, Luttun A, Jiang Y, et al. (2006) Sequential exposure to cytokines reflecting embryogenesis: the key for in vitro differentiation of adult bone marrow stem cells into functional hepatocyte-like cells. Toxicol Sci 94: 330–341.

25. Kinoshita T, Miyajima A (2002) Cytokine regulation of liver development. Biochim Biophys Acta 1592: 303–312.

26. Seo MJ, Suh SY, Bae YC, Jung JS (2005) Differentiation of human adipose stromal cells into hepatic lineage in vitro and in vivo. Biochem Biophys Res Commun 328: 258–264.

27. Darzi S, Zarnani AH, Jeddi-Tehrani M, Entezami K, Mirzadegan E, et al. (2012) Osteogenic differentiation of stem cells derived from menstrual blood versus bone marrow in the presence of human platelet releasate. Tissue Eng Part A 18: 1720–1728.

28. Kazemnejad S, Zarnani AH, Khanmohammadi M, Mobini S (2013) Chondrogenic differentiation of Menstrual Blood-Derived Stem Cells on Nanofibrous Scaffolds. Methods Mol Biol 1058: 149–169

29. Kazemnejad S, Allameh A, Soleimani M, Gharehbaghian A, Mohammadi Y, et al. (2009) Biochemical and molecular characterization of hepatocyte-like cells derived from human bone marrow mesenchymal stem cells on a novel three-dimensional biocompatible nanofibrous scaffold. J Gastroenterol Hepatol 24: 278–287.

30. Schouten JP, McElgunn CJ, Waaijer R, Zwijnenburg D, Diepvens F, et al. (2002) Relative quantification of 40 nucleic acid sequences by multiplex ligation-dependent probe amplification. Nucleic Acids Res 30(12):e57.

31. Feldstein AE, Wieckowska A, Lopez AR, Liu YC, Zein NN, et al. (2009) Cytokeratin-18 fragment levels as noninvasive biomarkers for nonalcoholic steatohepatitis: a multicenter validation study. Hepatology 50: 1072–1078.

32. Chen Y, Wong PP, Sjeklocha L, Steer CJ, Sahin MB (2012) Mature hepatocytes exhibit unexpected plasticity by direct dedifferentiation into liver progenitor cells in culture. Hepatology 55: 563–574

33. Grisham JW, Thorgeirsson SS (1997) Liver Stem Cells. In: Potten CS, editor. Stem Cells. London: Academic Press. pp. 233–282.

34. Sell S (2001) Heterogeneity and plasticity of hepatocyte lineage cells. Hepatology 33: 738–750.

35. Oertel MA, Shafritz D (2008) Stem cells, cell transplantation and liver repopulation. Biochimica et Biophysica Acta 1782: 61–74

36. Ruijter JM, Ramakers C, Hoogaars WM, Karlen Y, Bakker O, et al. (2009) Amplification efficiency: linking baseline and bias in the analysis of quantitative PCR data. Nucleic Acids Res 37: e45.

37. Pfaffl MW, Horgan GW, Dempfle L (2002) Relative expression software tool (REST) for group-wise comparison and statistical analysis of relative expression results in real-time PCR. Nucleic Acids Res 30: e36.

38. Hida N, Nishiyama N, Miyoshi S, Kira S, Segawa K, et al. (2008) Novel cardiac precursor-like cells from human menstrual blood-derived mesenchymal cells. Stem Cells 26: 1695–1704.

39. Cui CH, Uyama T, Miyado K, Terai M, Kyo S, et al. (2007) Menstrual blood-derived cells confer human dystrophin expression in the murine model of Duchenne muscular dystrophy via cell fusion and myogenic transdifferentiation. Mol Biol Cell 18: 1586–1594.

40. Sanberg PR, Eve DJ, Willing AE, Garbuzova-Davis S, Tan J, et al. (2011) The treatment of neurodegenerative disorders using umbilical cord blood and menstrual blood-derived stem cells. Cell Transplant 20: 85–94.

41. Rodrigues MC, Voltarelli J, Sanberg PR, Allickson JG, Kuzmin-Nichols N, et al. (2012) Recent progress in cell therapy for basal ganglia disorders with emphasis on menstrual blood transplantation in stroke. Neurosci Biobehav Rev 36(1): 177–190.

42. Schwartz RE, Reyes M, Koodie L, Jiang Y, Blackstad M, et al. (2002) Multipotent adult progenitor cells from bone marrow differentiate into functional hepatocyte-like cells. J Clin Invest 109: 1291–1302.

43. Mooney DJ, Langer R, Ingber DE (1995) Cytoskeletal filament assembly and the control of cell spreading and function by extracellular matrix. J Cell Sci 108: 2311–2320.

44. Taléns-Visconti R, Bonora A, Jover R, Mirabet V, Carbonell F, et al. (2006) Hepatogenic differentiation of human mesenchy- mal stem cells from adipose tissue in comparison with bone marrow mesenchymal stem cells. World J Gastroenterol 12: 5834–5845.

45. Lee HJ, Jung J, Jin Cho K, Kyou Lee C, Hwang SG, et al. (2012) Comparisonofinvitrohepatogenicdifferentiationpotentialbetweenvarious placenta-derived stem cells and other adult stem cells as an alternative source of functional hepatocytes. Differentiation 84: 223–231

46. Agarwal S, Holton KL, Lanza R (2008) Efficient differentiation of functional hepatocytes from human embryonic stem cells. Stem Cells 26: 1117–1127.

47. Hay DC, Zhao D, Fletcher J, Hewitt ZA, McLean D, et al. (2008) Efficient differentiation of hepatocytes from human embryonic stem cells exhibiting markers recapitulating liver development in vivo. Stem Cells 26: 894–902.

48. Si-Tayeb K, Noto FK, Nagaoka M, Li J, Battle MA, et al. (2010) Highly efficient generation of human hepatocyte-like cells from induced pluripotent stem cells. Hepatology 51: 297–305.

49. Sancho-Bru P, Roelandt P, Narain N, Pauwelyn K, Notelaers T, et al. (2011) Directed differentiation of murine-induced pluripotent stem cells to functional hepatocyte-like cells. J Hepatol 54: 98–107.

9

PTEN Regulates BCRP/ABCG2 and the Side Population through the PI3K/Akt Pathway in Chronic Myeloid Leukemia

Fang-Fang Huang[1❸], **Li Zhang**[2❸], **Deng-Shu Wu**[1], **Xiao-Yu Yuan**[1], **Fang-Ping Chen**[1], **Hui Zeng**[1*], **Yan-Hui Yu**[1], **Xie-Lan Zhao**[1]

1 Department of Hematology, Xiang Ya Hospital, Central South University, Changsha, Hunan, China, 2 Department of Hematology, West China Hospital, Si Chuan University, Chengdu, Sichuan, China

Abstract

A small population of cancer stem cells named the "side population" (SP) has been demonstrated to be responsible for the persistence of many solid tumors. However, the role of the SP in leukemic pathogenesis remains controversial. The resistance of leukemic stem cells to targeted therapies, such as tyrosine kinase inhibitors (TKIs), results in therapeutic failure or refractory/relapsed disease in chronic myeloid leukemia (CML). The drug pump, ATP-binding cassette sub-family G member 2 (ABCG2), is well known as a specific marker of the SP and could be controlled by several pathways, including the PI3K/Akt pathway. Our data demonstrated that compared with wild-type K562 cells, the higher percentage of ABCG2+ cells corresponded to the higher SP fraction in K562/ABCG2 (ABCG2 overexpressing) and K562/IMR (resistance to imatinib) cells, which exhibited enhanced drug resistance along with downregulated phosphatase and tensin homologue deleted on chromosome -10 (PTEN) and activated phosphorylated-Akt (p-Akt). PTEN and p-Akt downregulation could be abrogated by both the PI3K inhibitor LY294002 and the mTOR inhibitor rapamycin. Moreover, in CML patients in the accelerated phase/ blastic phase (AP/BP), increased SP phenotype rather than ABCG2 expression was accompanied by the loss of PTEN protein and the up-regulation of p-Akt expression. These results suggested that the expression of ABCG2 and the SP may be regulated by PTEN through the PI3K/Akt pathway, which would be a potentially effective strategy for targeting CML stem cells.

Editor: Giuseppe Viglietto, UNIVERSITY MAGNA GRAECIA, Italy

Funding: This work was supported by the National Natural Science Foundation of China (Grant No. 81100328), the Doctoral Fund of Ministry of Education of China (Grant No. 20110162120010), and the Fundamental Research Funds for the Central Universities (Grant No. 2011QNZT151) to HZ and the National Natural Science Foundation of China (Grant No. 81170477) to F-PC. The funders had no role in study design, data collection and analysis, decision to publish, or preparation of the manuscript.

Competing Interests: The authors have declared that no competing interests exist.

* E-mail: androps2011@hotmail.com

❸ These authors contributed equally to this work.

Introduction

Chronic myeloid leukemia (CML) is a clonal bone marrow stem cell disorder that accounts for 7–20% of all leukemia cases and has an estimated incidence of 1–2 per 100,000 worldwide [1]. CML arises by a reciprocal translocation between the long arms of chromosome 9 and chromosome 22 in an early hematopoietic stem cell (HSC) to produce the Philadelphia chromosome [2,3,4]. Although tyrosine kinase inhibitors (TKI) such as imatinib mesylate, nilotinib and dasatinib have been proven to be highly effective in the treatment of CML [5,6,7], a considerable number of the patients unfortunately face relapse or are unable to obtain complete remission during TKIs therapy [8,9,10]. The relative quiescence of CML stem cells or the overexpression of drug transporters are currently considered the main factors contributing to impaired effectiveness for CML treatments [11,12,13].

The side population (SP), which can be identified and sorted by the efflux of Hoechst 33342, expresses stem cell properties, such as pluripotency and differentiation ability. ATP-binding cassette sub-family G member 2 (ABCG2), which is also known as breast

cancer resistance protein (BCRP), is defined as a specific marker of the SP in a variety types of stem cells based on its ability to efflux Hoechst 33342 [14,15,16]. Previous results from adult acute myeloid leukemia demonstrated that SP cells may represent candidate leukemia stem cells. However, the role of ABCG2 expression and the SP phenotype in the mechanism of resistance to TKI in CML stem cells remains unclear [17]. Interestingly, the tumor suppressor gene phosphatase and tensin homologue deleted on chromosome-10 (PTEN), which is often deleted or inactivated in many solid tumor types [18,19,20], has also been shown to be down-regulated by BCR-ABL in CML stem cells, and its deletion can accelerate CML development through the regulation of its downstream target, Akt1 [21]. Moreover, PTEN was described as regulating the SP but not the expression of ABCG2 in glioma tumor stem-like cells through the PI3K/Akt pathway [22]. We speculate that the crosstalk between ABCG2 and PTEN in CML mediates therapeutic resistance and disease progression in CML cells, particularly within the SP compartment. As such, we analyzed data from both CML cell lines and clinical samples from CML patients (Tab. 1).

Materials and Methods

Cell lines and culture condition

K562 cells were purchased from a cell resource center (Xiang-Ya Medical College, Central South University, Hunan, China). K562/IMR and K562/AO2 cells were kindly obtained from the Institute of Hematology and Blood Diseases Hospital (Chinese Academy of Medical Sciences and Peking Union Medical College, Tianjin, China) and the First Affiliated Hospital of Zhengzhou University (Zhengzhou, China), respectively. Cell lines were routinely maintained in RPMI-1640 medium (GIBCO, NY, USA) supplemented with 10% fetal bovine serum (FBS; HyClone, MA, USA) and 1% penicillin/streptomycin (Sigma, MO, USA) in the humidified atmosphere of a 5% CO_2 incubator at $37^\circ C$. The PI3K inhibitor LY294002 (Invitrogen, Carlsbad, CA, USA) and the mTOR inhibitor rapamycin (Invitrogen, Carlsbad, CA, USA) were added to leukemia cells for 72 hours prior to mitoxantrone in some experiments.

Patient characteristics

From 2010 to 2012, bone marrow samples were obtained from 96 CML patients and 10 healthy candidate donors for hematopoietic stem cell transplantation as controls enrolled at the Xiang-Ya Hospital of Central South University, Hunan, China (Table 1). All patients and donors gave informed consent. The protocol was approved by the Medical Ethic committee of Xiangya Hospital, Central South University. Participants provided their written informed consent to participate in this study. The diagnosis and classification of the leukemia were based on 2008 World Health Organization's criteria. Mononuclear cells (MNCs) were obtained by density centrifugation over Ficoll-Paque (Sigma, St Louis, MO, USA) and stored at $-80^\circ C$.

Cytotoxicity assay

Cells were cultured with various concentrations of the indicated agents. Cell viability was determined by a CCK8 assay (Nan Jing Key Gen Biotechnology, Nan Jing, China). Briefly, cells were seeded in 96-well culture plates (8×10^3 per well) in 100 μL media for 12 h. Subsequently, different concentrations of mitoxantrone (0.01–1.0 μM) were added to the wells and incubated for 72 h. At the end of the treatment, 10 μM of CCK8 solution was added to each well for 1 h culture at $37^\circ C$. Absorbance was measured with a spectrophotometer (Thermo Scientific Evolution 600, China) at a wavelength of 450 nm and compared with 630 nm.

Apoptosis assessment

After a 48-h culture, at least 1×10^5 untreated and mitoxantrone-treated cells were collected and washed twice with cold PBS, stained with 5 μl of Annexin V-FITC and 5 μl of propidium

iodide (PI) for 15 min, and subjected to flow cytometry (Becton Dickinson, CA, USA) to analyze apoptosis.

Lentiviral infection of cell lines

The lentiviral constructs PsPAX2, VSVG, pSIN4-EF2-ABCG2-IRES-Neo (from Dr. Ren-He Xu's laboratory) and pSIN4-EF2-EGFP-IRES-Neo plasmids (from Dr. James Thomson's laboratory) were used to make viral stocks by transfection of 293FT cells using Lipofectamine 2000 (Invitrogen, Carlsbad, CA, USA), as previously described. The lentiviral supernatants were harvested 72 h post-transfection and were filtered (0.45 μM) prior to infection of the cell lines. After a 48-h infection (MOI: 5–50), the K562 cell lines were allowed to recover for 24 h in fresh media and were thereafter referred to K562/ABCG2 cells. The cells were then allowed to grow for 72 h before being subjected to additional assays.

Western blot analysis

Protein (50 μg/sample) was separated in 8% sodium dodecyl sulfate-polyacrylamide gel electrophoresis (SDS-PAGE) and then transferred electrophoretically to polyvinylidene fluoride (PVDF) membranes. The membranes were saturated in PBS-T containing 5% nonfat milk (blocking buffer) overnight at $4^\circ C$ with the primary antibodies. The dilutions of the antibodies used for western blotting were as follows: PI3K p85 (Tyr458)/p55 (Tyr199) 1:1000 (Cell Signaling Technology, Beverly, MA, USA), PI3K 1:1000 (Abcam, MO, USA), phosphorylated-Akt (p-Akt) (S473) 1:400 (Cell Signaling Technology, Beverly, MA, USA), PTEN 1:200 (Abcam, MO, USA) and β-actin 1:1000 (Abcam, MO, USA). The membranes were then incubated for 60 min at room temperature with an HRP-linked secondary antibody (Santa Cruz Biotechnology, Santa Cruz, CA, USA). Protein bands were visualized using the Chei Doc^{TM}MP imaging system (BioRad). The blots shown are representative of three different experiments.

RNA isolation, RT-PCR and real-time RT-PCR

Total RNA (2 μg) was reverse transcribed using the M-MLV First Strand Kit (Invitrogen, Carlsbad, CA, USA). The transcript levels for the genes of interest were normalized to the GAPDH transcripts. The gene-specific primers and RT-PCR conditions are summarized in Table 2. Real-time PCR was performed using a SYBR qPCR Mix (Toyobo, TOYOBO CO., LTD, JAPAN) on an ABI StepOnePlus (Applied Biosystems, Foster City, CA, USA) with the specific primers. The thermal cycler conditions were as follows: $95^\circ C$ for 1 min, followed by 40 cycles of $95^\circ C$ for 10 s, $61^\circ C$ for 20 s, and $72^\circ C$ for 50 s.

Cellular surface expression of ABCG2

For the phenotypic analysis of the cell lines, the cells were washed in PBS and then stained for 30 min at room temperature with a PE anti-human CD338 antibody (ABCG2, 1:100, Biolegend, San Diego, CA, USA), and a PE-IgG2b isotype control (1:100, Biolegend, San Diego, CA, USA) used as a negative control. Finally, the cells were washed twice with ice-cold PBS and then analyzed by flow cytometry (Becton Dickinson, Mountain View, CA, USA).

Side population analysis

The cell suspensions were labeled with Hoechst 33342 (Invitrogen, Carlsbad, CA, USA) dye for side population analysis according to standard protocol [23]. Cells were briefly resuspended in pre-warmed Hepes buffer containing 2% FBS at a density of 10^6/ml. Hoechst 33342 dye (Invitrogen, Carlsbad, CA,

Table 1. Characteristics of patients with CML (n = 96).

Stage of CML	Total no. of Pts	M/F	Age, median (range) [years]
CML-CP	61	32/29	40/(21–61)
CML-AP/BP	35	15/22	42/(24–65)

Table 2. PCR primers and conditions.

Gene	Strand	Primer sequences	Annealing temperature
PTEN	sense	5′-ACCAGGACCAGAGGAAACCT-3′	61.5°C
	anti-sense	5′-GCT AGCCTCTGGATTTGACG-3′	
ABCG2	sense	5′-ATTGAAGGCAAAGGCAGATG-3′	61.0°C
	anti-sense	5′-TGAGTCCTGGGCAGAAGTTT-3′	
GAPDH	sense	5′- AGGTGACACTATAGAATAAA GGTGAAGGTCGGAGT CAA -3′	68.0°C
	anti-sense	5′- GTACGACTCACTATAGGGAGA TCTCGCTCCTGGAAGATG -3′	
BCR/ABL (b3a2)	sense	5′- GCATTCCGCTGACCATCAATA -3′	58.5°C
	anti-sense	5′- TCCAACGAGCGGCTTCAC -3′	

USA) was then added at a final concentration of 5 µg/ml (37°C for 90 min) in the presence or absence of verapamil (50 µM; Sigma) with intermittent shaking. The cells were counterstained with 2 µg/ml propidium iodide (PI) (Sigma–Aldrich) to label dead cells and were then analyzed using a FACS Vantage SE cell sorter (Becton Dickinson, Mountain View, CA, USA) by a dual-wavelength analysis (450/20 and 675/20 nm) after excitation at 350 nm.

Small interfering RNA transfection

Synthetic control small interfering RNA (siRNA) and siRNA against PTEN were purchased from Santa Cruz Biotechnology. Approximately 2×10^5 cells were seeded in 6-well culture plates and incubated for 72 h prior to transfected with 80 nM siRNA using Lipofectamine 2000 (Invitrogen Corp., Carlsbad, CA, USA) as directed by the manufacturer.

Statistical analysis

The comparisons of the chronic myeloid leukemia patients and normal donors or among the cell lines were made using the Statistical Package for the Social Sciences (SPSS) version 17.0. Differences were considered significant for $P<0.05$. The differences between the populations were calculated using Student's t test or one-way ANOVA, as appropriate. The diagrams were created using GraphPad Prism 5 software.

Results

ABCG2 overexpression decreased drug sensitivity and drug-induced inhibition of DNA synthesis in leukemia cell lines

Cellular resistance to imatinib can arise through a variety of mechanisms, including mutations in the BCR-ABL kinase and increased efflux by multidrug resistance transporters such as P-glycoprotein (P-gp) and ABCG2 [24,25,26]. Imatinib stimulates ABCG2-specific ATPase activity and is both a substrate and an inhibitor of ABCG2 [13,27]. To determine whether ABCG2 is related to CML resistance, the K562, K562/ABCG2 (ABCG2 overexpressing), K562/AO2 (resistance to adriamycin) and K562/IMR (resistance to imatinib) cell lines were exposed to 10 nM, 100 nM or 1 µM of mitoxantone. Compared with the wild-type K562 cells, increased mitoxantrone resistance was observed in K562/ABCG2, K562/AO2 and K562/IMR cells after 72-h treatment with 100 nM or 1 µM of mitoxantone ($P<0.05$) (Fig. 1a and Table 3), with comparable mitoxantrone-induced apoptosis ($P>0.05$) (Fig. 1b). The results indicated that ABCG2 might attenuate the cytotoxic effects of mitoxantrone in CML.

Next, we evaluated the effect of ABCG2 overexpression on the mitoxantrone-induced cell cycle changes in leukemia cell lines. As shown in Fig. 1c and Fig. S4, after 72-h treatment with mitoxantrone, the proportion of K562, K562/ABCG2, and K562/AO2 cells in S phase was significantly decreased in concentration-dependent manner ($P<0.05$), indicating an inhibition in DNA synthesis by mitoxantrone. However, compared with wild-type K562 cells, the subpopulation in the S phase was remarkably higher in K562/ABCG2, K562/AO2 and K562/IMR cells when treated with 100 nM or 1 µM of mitoxantone ($P<0.05$). These findings suggest that S-phase blockage might be an underlying mechanism contributing to drug resistance and the pivotal cross between drug resistance and tumor growth.

ABCG2 expression regulated the SP phenotype, and PTEN protein deletion increased p-Akt in resistant leukemia cells

As reported, PTEN deletion influences the disease progression of both CML and B-ALL by regulating its downstream target Akt1 [21], through which the BCR-ABL fusion gene regulates ABCG2 expression [28]. Thereby, we evaluated the endogenous levels of PTEN, PI3K, p-PI3K, p-Akt and ABCG2 in four leukemia cell lines. Compared with K562 cells, the expression level of PTEN protein was decreased significantly in all drug-resistant cells, with upregulated levels of BCR-ABL transcript detected by RT-PCR (Fig. 2a) (Tab. 2) and upregulated p-Akt protein detected by western blot (Fig. 2b). Additionally, ABCG2 mRNA levels (Fig. 2a) were significantly higher and the ABCG2+ population (Fig. 2c and d) was significantly larger in K562/ABCG2 and K562/IMR cells than in K562 and K562/AO2 cells. The siRNA-mediated knockdown of PTEN significantly increased ABCG2 transcript level in K562/ABCG2 cell (Fig. S3). These results suggested that the down-regulation of PTEN along with both p-Akt and ABCG2 upregulation might contribute to drug resistance in K562/ABCG2 and K562/IMR cells but not in K562/AO2 cells.

Because functional ABCG2 has been reported to be overexpressed on the surface of primary CML stem cells [29], we combined surface marker characteristics with Hoechst dye efflux to explore the interaction of the PI3K/Akt pathway, ABCG2 expression and the SP fraction specifically in the primitive stem cell subset. Flow cytometry confirmed higher fractions of SP cells in the K562/ABCG2 and K562/IMR cell lines compared with K562 and K562/AO2 cells ($P<0.05$) (Fig. 2e and Fig. S1). In addition, ABCG2 transcript levels were quantified by RT-PCR in Hoechst 33342-labeled sorted SP and non-SP fractions. Higher expression level of ABCG2 was present in the SP fraction compared to the non-SP fraction in K562/ABCG2 cell, whereas

a

Apoptosis assay

K562 cell
K562/ABCG2 cell
K562/AO2 cell
K562/IMR cell

b

Mitoxantrone 1 μM

K562 cell — 34.46% / 30.13%
K562/ABCG2 cell — 11.46% / 11.38%
K562/AO2 cell — 12.63% / 8.33%
K562/IMR cell — 10.36% / 5.78%

Propidium Iodide (PI)

Annexin V-FITC

c

Cell cycle assay

K562 cell
K562/ABCG2 cell
K562/AO2 cell
K562/IMR cell

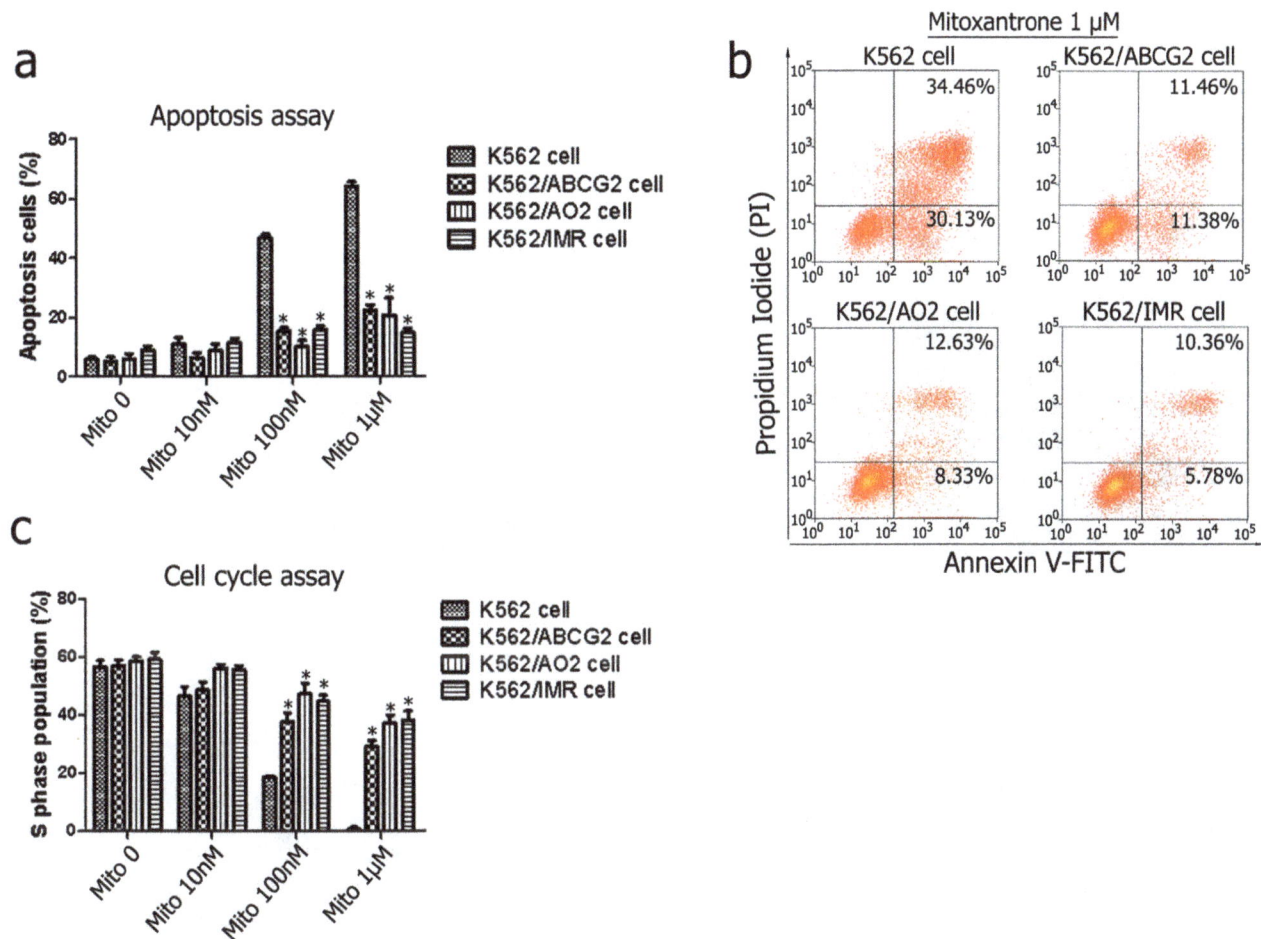

Figure 1. K562 cells with ABCG2 overexpression exhibited attenuated apoptosis and were no longer arrested in S phase by mitoxantone treatment. (**a**, **b**) After exposure to 10 nM, 100 nM or 1 μM of mitoxantone for 72 h, the cell lines were subjected to flow cytometry to quantify the number of apoptotic cells, and an increased drug resistance was observed in K562/ABCG2, K562/AO2 and K562/IMR cells compared with wild-type K562 cells. (**c**) The cell cycle phase was determined by flow cytometry after treatment with different concentrations of mitoxantone, which decreased the inhibition of DNA synthesis at S phase in K562/ABCG2, K562/AO2 and K562/IMR cells compared with wild-type K562 cells. The histogram represented the means ± s.d. for three replicates. *$P < 0.05$.

Table 3. The ratio of viable cells after 72-h treatment with mitoxantone at a range of concentrations in the presence of LY294002 or rapamycin by CKK8 assay.

Mitoxantrone	K562 −	K562 +LY	K562 +Rapa	K562/ABCG2 −	K562/ABCG2 +LY	K562/ABCG2 +Rapa	K562/AO2 −	K562/AO2 +LY	K562/AO2 +Rapa	K562/IMR −	K562/IMR +LY	K562/IMR +Rapa
10 nM	0.90±	0.80±	0.80±	0.91±	0.81±	0.79±	0.91±	0.85±	0.82±	0.91±	0.83±	0.79±
	0.04	0.07	0.07	0.04	0.08	0.08	0.05	0.04	0.06	0.06	0.07	0.09
100 nM	0.60±	0.48±	0.53±	0.82±	0.58±	0.52±	0.85±	0.76±	0.69±	0.53±	0.50±	0.50±
	0.08	0.07‡	0.07§	0.06†	0.07*‡	0.04*§	0.05†	0.07	0.07	0.05†	0.08*‡	0.09*§
1 μM	0.39±	0.17±	0.23±	0.72±	0.24±	0.23±	0.80±	0.43±	0.38±	0.83±	0.22±	0.19±
	0.05	0.05*‡	0.06*§	0.08†	0.07*‡	0.04*§	0.07†	0.06*	0.06*	0.08†	0.08*‡	0.06*§

Each value represents the mean value ± s.d. from three individual experiments.
* Versus cells without LY294002 or rapamycin treatment within each cell group, $P < 0.05$.
†Versus K562 wild-type cells without LY294002 or rapamycin when treated with the same concentration of mitoxantone, $P < 0.05$.
‡Versus K562/AO2 cells among all of the experimental cells treated with LY294002, $P < 0.05$.
§Versus K562/AO2 cells among all the experimented cells treated with rapamycin, $P < 0.05$.

a

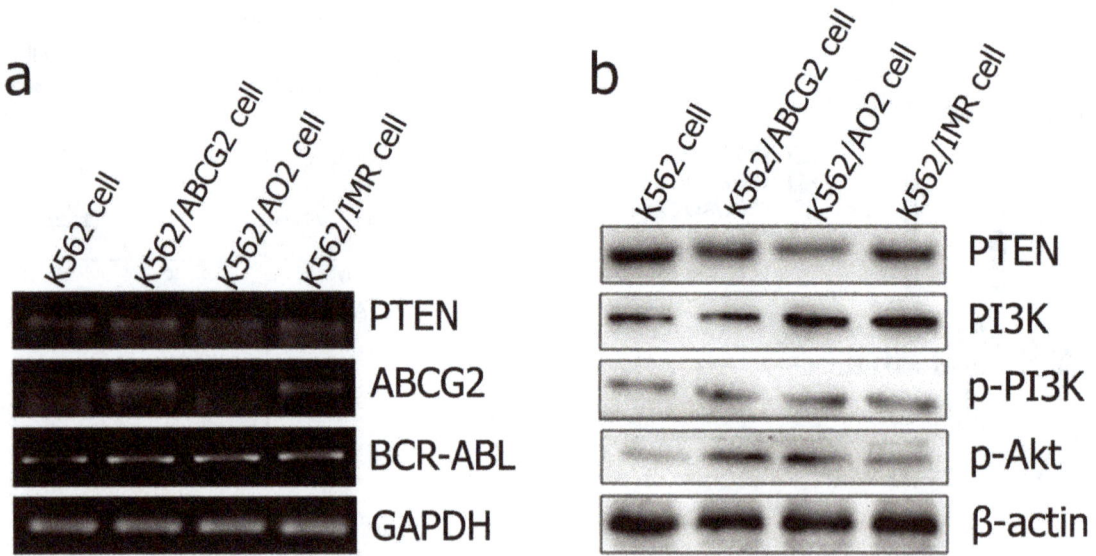

b

c

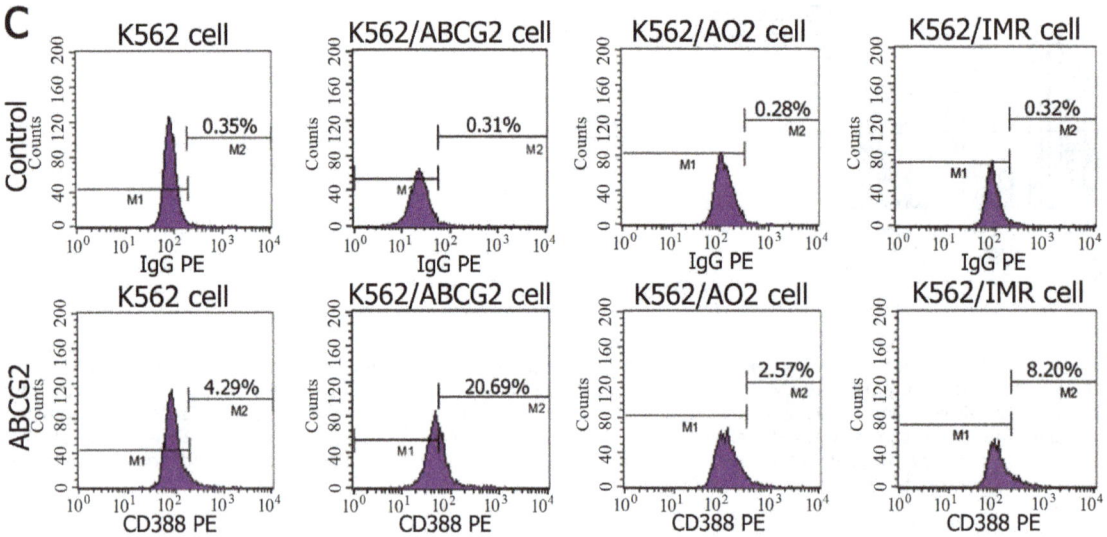

d

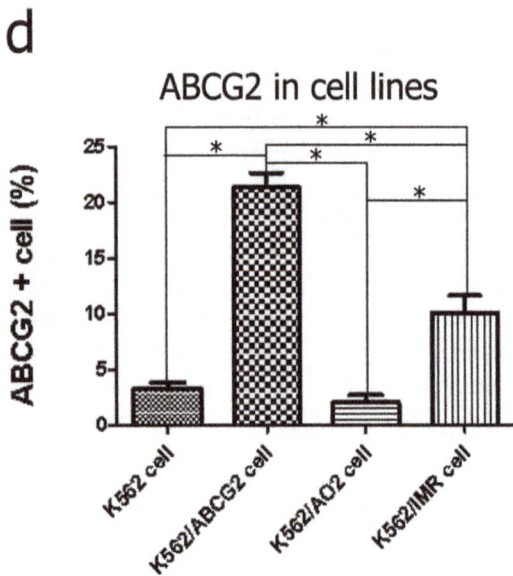

e

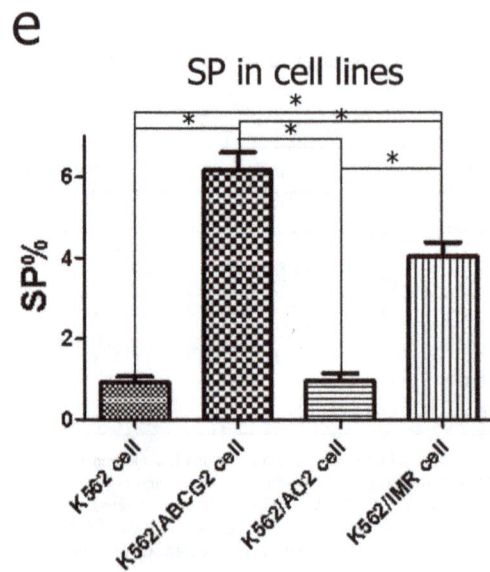

Figure 2. Detection of the PTEN/PI3K/Akt signal pathway, ABCG2 and SP in leukemia cells. (a) BCR-ABL transcript expression was upregulated in all drug-resistant cells, whereas higher ABCG2 transcript was only detected in K562/ABCG2 and K562/IMR cells. (b) Western blot analysis revealed decreased PTEN protein but increased p-Akt expression in K562/ABCG2, K562/AO2 and K562/IMR cells. (c) Expression of CD338 (anti-ABCG2) on the cell surface were detected by FCM. IgG2b-PE was used as an isotype control. (d) The distribution of ABCG2+ cells was detected by flow cytometry (FCM). The histogram demonstrates a higher ratio of ABCG2+ cells in K562/ABCG2 and K562/IMR cell lines. The results were represented as the mean ± s.d. of three experiments. *P<0.05. (e) The increased SP fraction was observed in K562/ABCG2 and K562/IMR cells. Each sample was incubated with 50 μM verapamil as a control, and only PI-negative (live) cells were gated to be analyzed.

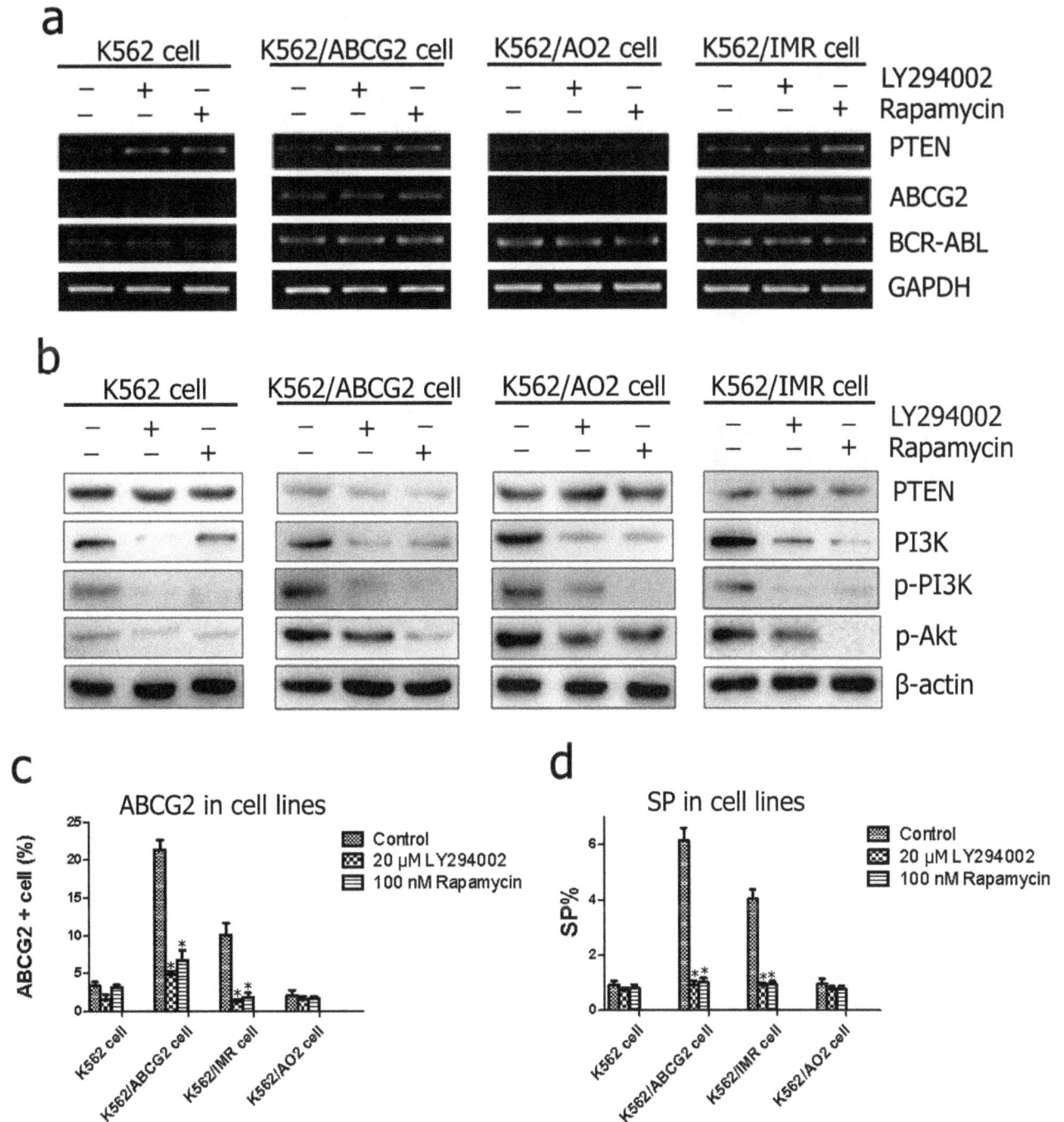

Figure 3. PI3K/Akt pathway activation and the fraction of ABCG2+ and SP cells was detected by RT-PCR, western blotting and FACS analysis in leukemia cells before or after incubation with 20 μM LY294002 or 100 nM rapamycin for 72 h. (a, b) After LY294002 or rapamycin treatment, activation of the PTEN transcript and down-regulation of p-PI3K and p-Akt were detected in four cell lines. (c) A reduced fraction of ABCG2 cells in K562/ABCG2 and K562/IMR cells was observed after treatment. (d) The increased SP fraction was abolished by LY294002 or rapamycin treatment in K562/ABCG2 and K562/IMR cells. All results are presented as the means ± s.d. from three independent experiments. *P<0.05.

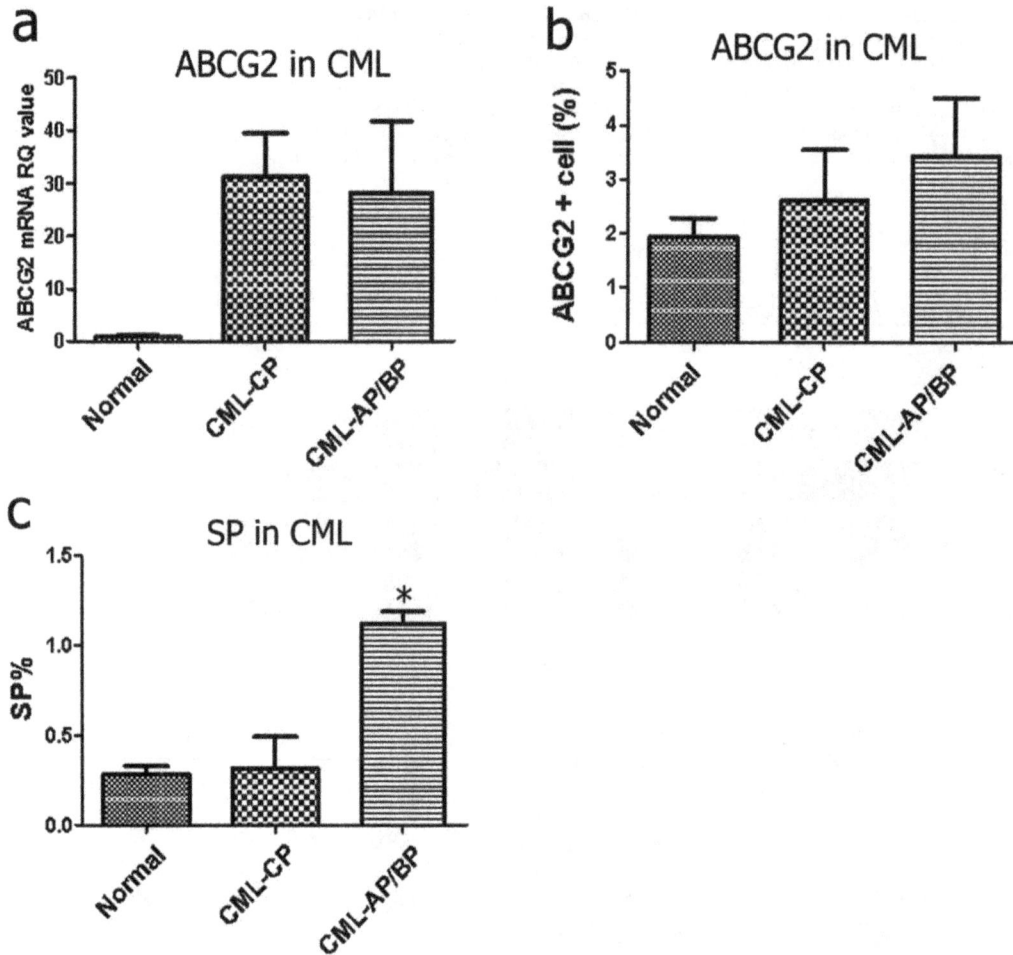

Figure 4. Endogenous expression of ABCG2 and SP fractions in CML patients. (a) The ABCG2 transcript in CML patients was determined by real-time RT-PCR using the RQ values with *GAPDH* mRNA as the endogenous control. In contrast to normal donors, no substantial difference in the ABCG2 transcript levels was detected in CML patients. *$P<0.05$. (b) ABCG2+ cells were labeled by CD338-PE and detected by FACS, using IgG2b-PE as the isotype control. There were no significant differences among the groups. *$P<0.05$. (c) Significantly higher percentages of SP cells were observed in CML blasts at AP/BP using flow cytometry. *$P<0.05$.

no similar response was yielded in K562 cell (Fig. S2a–b) These findings further demonstrated that ABCG2 might contribute essentially to the SP phenotype.

PTEN regulated the ABCG2+ fraction, SP phenotype and drug sensitivity through the PI3K/Akt pathway in leukemia cells

To determine whether the PI3K/Akt pathway is involved in drug resistance through the regulation of ABCG2, leukemia cells were treated with the PI3K inhibitor LY294002 (20 µM) or the mTOR inhibitor rapamycin (100 nM) for 72 h. Although the mRNA levels of both ABCG2 and BCR-ABL did not significantly change (Fig. 3a), activation of PTEN transcript and the down-regulation of both p-PI3K and p-Akt (Fig. 3b), as well as a reduced proportion of ABCG2+ cells (Fig. 3c), were observed in treated K562/ABCG2 and K562/IMR cells, suggesting that PI3K and Akt activity might regulate ABCG2 expression. Furthermore, we observed that incubation with LY294002 or rapamycin decreased the fractions of SP cells in K562/ABCG2 and K562/IMR cells ($P<0.05$) (Fig. 3d and Fig. S1). Taken together, these results

indicated that the PI3K/Akt pathway participated in regulating the SP fraction through ABCG2.

Furthermore, as shown in Table 3, CCK8 analysis demonstrated that the incubation of leukemia cells with LY294002 or rapamycin strongly increased the sensitivity of leukemia cells to 100 nM or 1 µM of mitoxantrone ($P<0.05$), suggesting that chemosensitivity to mitoxantrone is tightly correlated with decreased p-PI3K and p-Akt expression in the resistant K562 cells. When pretreated with either LY294002 or rapamycin, K562/ABCG2 and K562/IMR cells exhibited a more significant decrease in the proportion of viable cells after treatment with 100 nM or 1 µM mitoxantrone than did K562/AO2 cells ($P<0.05$). These data suggest that p-PI3K, p-Akt or ABCG2 activity may fractionally participate in chemoresistance in K562/AO2 cells.

No substantial difference in the ABCG2+ fraction but an increased SP fraction in CML patients in AP/BP compared with the control group

Based on the results of CML cell lines, the endogenous mRNA and protein levels of *ABCG2* in 61 consecutive CML patients with

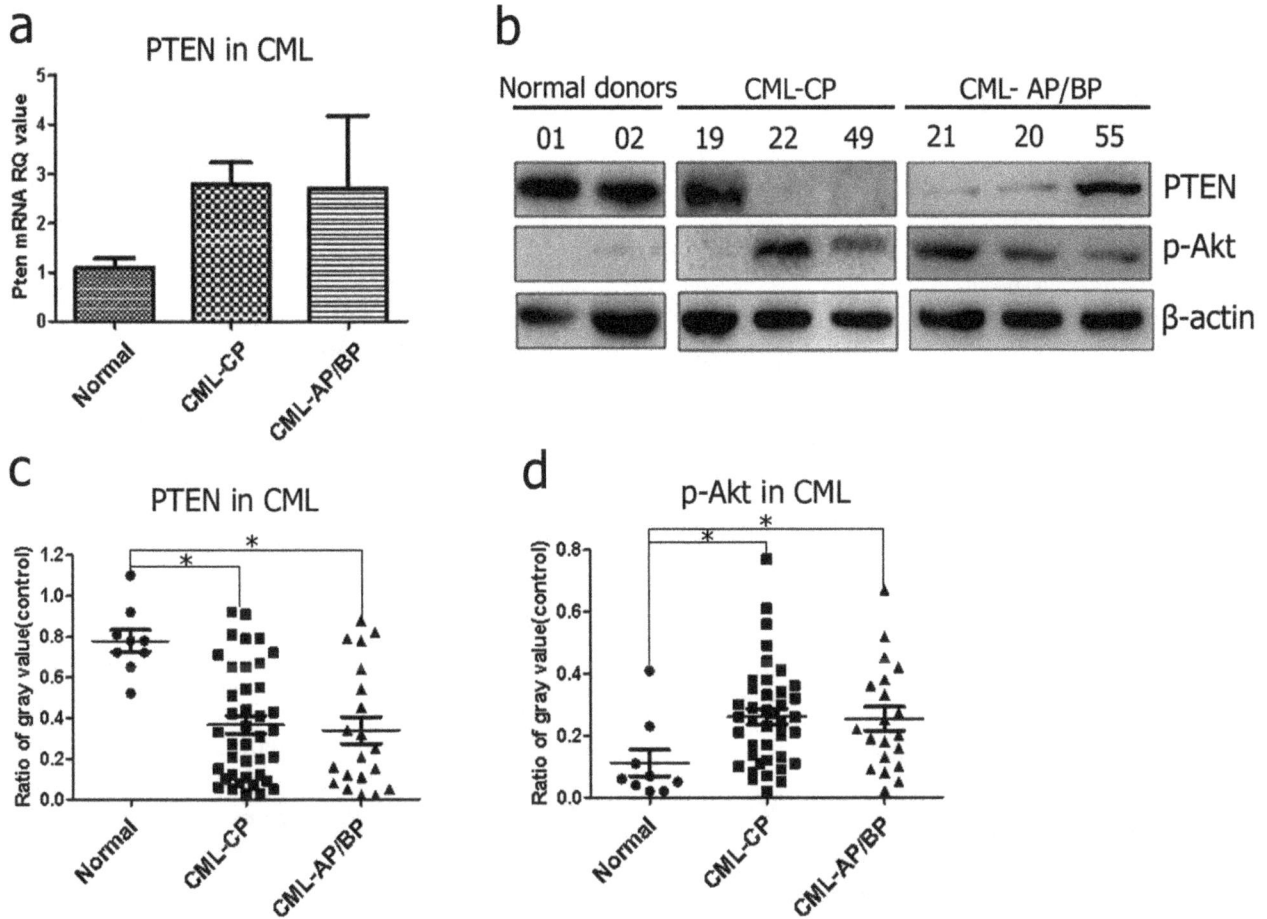

Figure 5. Endogenous expression of the PTEN/PI3K/Akt pathway in CML patients. (a) Endogenous levels of *PTEN* mRNA in CML patients and normal donors were examined by real-time RT-PCR. *PTEN* transcript was not notably decreased in leukemia blasts compared with the normal donors. *$P<0.05$. **(b–d)** These results suggested that p-Akt might be negatively regulated by PTEN in CML blasts by western blot analysis. The ratio of the gray value was calculated by band intensities using ImageJ software (Wayne Rasband, NIH). β-actin was used as a control for protein loading. All histograms represent the mean of one experiment performed in triplicate ± s.d.

Ph-positive metaphases and/or BCR-ABL–positive transcripts in the CP (chronic phase) and in 35 patients in AP/BP (accelerated phase/blastic phase) were detected by real-time RT-PCR and FACS analysis, respectively, at our hospital (Tab. 1). ABCG2+ cells and the mRNA levels in CML patients in different phases exhibited no substantial differences ($P>0.05$) but were obviously higher than in normal donors (Fig. 4a–b).

Stem cells are frequently identified as the "side population" by flow cytometry based on ABCG2-mediated efflux of Hoechst 33342 dye. Incubation of the leukemia cells with 50 μM verapamil, known to block ABC transporter activity, abolished part of the SP as a negative control. ABCG2 transcript differed significantly between SP and non-SP cells in some CML patient (Fig. S2c–d). Flow cytometry revealed a very small SP fraction, ranging from 0.11% to 1.14% in the cases, and a significantly higher percentage of the SP cells in CML patients at AP/BP compared with the healthy donors ($P<0.05$) (Fig. 4c). Our data suggest that an increase in the SP fraction might confer a survival advantage to CML cells.

The SP phenotype was correlated with low PTEN and high p-Akt levels in CML patients. PTEN maintains normal hematopoietic stem cells and prevents leukemia development from leukemia stem cells [30,31]. To determine whether increased ratios of the

SP would be related to the down-regulation of PTEN, both transcript and protein levels of PTEN were detected in CML patients. In leukemia blasts, the levels of PTEN transcript were surprisingly increased compared to the normal donors ($P>0.05$) (Fig. 5a). However, PTEN protein was remarkably decreased in CML patients in AP/BP compared with other patients ($P<0.05$), suggesting that low PTEN protein accompanying the SP phenotype was limited to the status of the disease. Furthermore, p-Akt was activated in some CML samples with low expression of PTEN protein ($P<0.05$) (Fig. 5b–d). Our study further demonstrated that the SP fraction may contribute to the progression of CML, as indicated by decreased PTEN protein expression and Akt activation.

Discussion

In light of the crucial role of the BCR-ABL tyrosine kinase in chronic myelogenous leukemia, TKIs have become the first-line therapy for most patients with chronic myelogenous leukemia [4,32,33]. However, TKIs do not kill CML stem cells [34]. Although more than 80% of CML patients in chronic phase can achieve an ongoing complete hematologic response after treatment with imatinib, a considerable number of cases eventually progress to the accelerated phase and even blast crisis [35,36]. Additionally, BCR-ABL-positive malignant cell clones have been shown to persist within the CD34+ stem cell fractions, even in CML patients for whom imatinib had induced a complete cytogenetic remission [37]. Given that CML patients harbor quiescent CML stem cells that may serve as reservoir for disease progression to blast crisis, there is a strong possibility of the existence of imatinib-refractory CML stem cells.

The drug pump ABCG2 is a multidrug resistance protein and is well known as a specific phenotype of the SP cells with stem-like properties [15]. Consistent with previous reports that imatinib is an inhibitor and substrate for ABCG2 [13,27,28], we observed upregulation of ABCG2 in K562/IMR cells and K562/ABCG2 cells, both of which accordingly exhibited higher ratios of SP cells and decreased susceptibility to mitoxantrone due to higher levels of ABCG2. Furthermore, PTEN was revealed to be involved in ABCG2-mediated multi-drug resistance for CML through the PI3K/Akt signaling pathway in our study. Lower PTEN expression was observed in drug-resistant CML cells; consistent with our reports in human embryonic stem cells, the over-expression of ABCG2 in H9 cells leads to p-Akt activation [38]. Higher p-Akt expression levels were also detected in both K562/ABCG2 and K562/IMR cells. Second, consistent with previous studies in human glioma, primary esophageal carcinoma and epithelial carcinoma cells [22,39,40], our data demonstrated increased drug sensitivity, down-regulation of p-PI3K and p-Akt, suppression of the ABCG2 and a decrease in the SP fraction in K562/ABCG2 and K562/IMR cells after LY294002 or rapamycin treatment. These findings suggest that the chemotherapeutic sensitivity and the fractions of ABCG2 and SP cells in drug-resistant CML might be mediated by the PI3K/Akt signaling pathway, which is consistent with our previous reports in acute leukemia [41]. Therefore, accumulating data provide underlying connections among the PTEN, PI3K/Akt pathway, multidrug resistance transporters, stem-like character, and therapeutic resistance, suggesting that activation of this pathway also enhances the ability of CML cancer stem-like cells to expel drugs.

This study further focused on the complex interaction between PTEN and ABCG2 in CML cases to explore whether such regulation existed in the clinic process. Inconsistent with data from CML cell lines, no substantial differences in the ABCG2+ phenotype and ABCG2 mRNA levels were detected among different stages of CML blasts and normal cells. Additionally, the research in glioma tumor reported that PI3K inhibitor treatment changed the activity of ABCG2 in neurospheres, but the expression levels of the mRNA and protein were unaffected [22]. These data indicate that ABCG2 function, rather than its mRNA or protein expression, might play a more important role in initiation and progress of CML. SP cells are also found in a variety of mammalian species, including humans, where their frequency is low [42]. Inconsistent with ABCG2 expression, a higher percentage of SP compared with the donors was observed in the AP/BP group, which partly suggested that monitoring the SP ratio could predict disease progression and might be an optimal indicator to represent in vivo ABCG2 function. Meanwhile, similar to the results from CML cell lines, this study also demonstrated that absent/low expression of PTEN at the protein level and subsequent p-Akt activation in the CML groups might promote the acceleration of CML development and an increased SP ratio. Nevertheless, more efforts are needed to reveal the precise mechanism of how loss of the tumor suppressor PTEN regulates ABCG2 function and further enhances the SP phenotype through the PI3K/Akt pathway in CML.

In summary, to investigate the perplexing relationships among PTEN, ABCG2 and the SP in CML, our studies demonstrated that PTEN played an essential role in regulating the SP in CML through the PI3K/Akt signaling pathway in vitro. Then, our studied revealed that the SP phenotype and ABCG2 function rather than ABCG2 expression was correlated with drug resistance and disease progression in CML patients, which was mediated at least partially by p-Akt activation. Therefore, intervention in the functional enhancement of p-Akt mediated by the loss of PTEN inhibition would provide a potential therapeutic strategy for targeting CML stem cells.

Supporting Information

Figure S1 The distribution of the SP phenotype was assessed by flow cytometry in cell lines before and after treatment with LY294002 or rapamycin. Each sample was incubated with 50 µM verapamil as a control, and only PI-negative (live) cells were gated to be analyzed. The ABCG2+ population was significantly larger in K562/ABCG2 and K562/IMR cells than in the other cell types.

Figure S2 ABCG2 transcript in the SP fraction. ABCG2 mRNA was analyzed by RT-PCR in the flow cytometry-selected SP fraction and compared with the non-SP fraction in K562 cells (**a**), K562/ABCG2 cells (**b**), CML-CP patient No. 23 (**c**) and CML-AP/BP patient No. 9 (**d**). GAPDH was used as a control.

Figure S3 siRNA directed against PTEN specifically inhibited PTEN expression in leukemia cell lines. One hundred nanomolar siRNA directed against PTEN specifically inhibited PTEN expression in the K562/ABCG2 cell line. RT-PCR was performed 48 h after the leukemia cells were treated with PTEN siRNA or control siRNA to evaluate PTEN and ABCG2 expression.

Figure S4 K562 cells overexpressing ABCG2 overcame mitoxantrone-induced S-phase arrest. (**a, b**) After exposed to 10 nM, 100 nM or 1 µM of mitoxantone for 72 h, the cell lines were subjected to flow cytometry to determine the cell cycle distribution, and a decreased the inhibition of DNA synthesis at S phase was observed in the K562/ABCG2, K562/AO2 and K562/IMR cells compared with wild-type K562 cells. The histogram represented the means ± s.d. for three replicate determinations. *$P<0.05$.

Acknowledgments

We thank Dr. James Thomson for the pSIN4-EF2-EGFP-IRES-Neo lentiviral plasmid, Dr. Ren-He Xu for the pSIN4-EF2-ABCG2-IRES-Neo lentiviral plasmid, Hui-En Zhan and Yan-Hong Zhou for technical assistance, and Dr. Bei Liu, Zheng Zhang and Xiao-Ping Chen for their critical reading of the manuscript.

Author Contributions

Conceived and designed the experiments: HZ FPC. Performed the experiments: FFH LZ. Analyzed the data: FFH LZ HZ. Contributed reagents/materials/analysis tools: DSW YHY XYY XLZ. Wrote the paper: FFH LZ HZ. Obtained permission for use of cell line: DSW. Provided the experiment station: FPC HZ.

References

1. Jemal A, Siegel R, Ward E, Hao Y, Xu J, et al. (2009) Cancer statistics, 2009. CA Cancer J Clin 59: 225–249.
2. Nowell PC, Hungerford DA (1960) Chromosome studies on normal and leukemic human leukocytes. J Natl Cancer Inst 25: 85–109.
3. Rowley JD (1973) Letter: A new consistent chromosomal abnormality in chronic myelogenous leukaemia identified by quinacrine fluorescence and Giemsa staining. Nature 243: 290–293.
4. Lugo TG, Pendergast AM, Muller AJ, Witte ON (1990) Tyrosine kinase activity and transformation potency of bcr-abl oncogene products. Science 247: 1079–1082.
5. Deininger M, Buchdunger E, Druker BJ (2005) The development of imatinib as a therapeutic agent for chronic myeloid leukemia. Blood 105: 2640–2653.
6. Kantarjian HM, Giles F, Gattermann N, Bhalla K, Alimena G, et al. (2007) Nilotinib (formerly AMN107), a highly selective BCR-ABL tyrosine kinase inhibitor, is effective in patients with Philadelphia chromosome-positive chronic myelogenous leukemia in chronic phase following imatinib resistance and intolerance. Blood 110: 3540–3546.
7. Ottmann O, Dombret H, Martinelli G, Simonsson B, Guilhot F, et al. (2007) Dasatinib induces rapid hematologic and cytogenetic responses in adult patients with Philadelphia chromosome positive acute lymphoblastic leukemia with resistance or intolerance to imatinib: interim results of a phase 2 study. Blood 110: 2309–2315.
8. Cortes J, Talpaz M, O'Brien S, Jones D, Luthra R, et al. (2005) Molecular responses in patients with chronic myelogenous leukemia in chronic phase treated with imatinib mesylate. Clin Cancer Res 11: 3425–3432.
9. Albano F, Anelli L, Zagaria A, Coccaro N, D'Addabbo P, et al. (2010) Genomic segmental duplications on the basis of the t(9;22) rearrangement in chronic myeloid leukemia. Oncogene 29: 2509–2516.
10. Barnes DJ, Schultheis B, Adedeji S, Melo JV (2005) Dose-dependent effects of Bcr-Abl in cell line models of different stages of chronic myeloid leukemia. Oncogene 24: 6432–6440.
11. Holyoake TL, Jiang X, Jorgensen HG, Graham S, Alcorn MJ, et al. (2001) Primitive quiescent leukemic cells from patients with chronic myeloid leukemia spontaneously initiate factor-independent growth in vitro in association with up-regulation of expression of interleukin-3. Blood 97: 720–728.
12. Thomas J, Wang L, Clark RE, Pirmohamed M (2004) Active transport of imatinib into and out of cells: implications for drug resistance. Blood 104: 3739–3745.
13. Burger H, van Tol H, Boersma AW, Brok M, Wiemer EA, et al. (2004) Imatinib mesylate (STI571) is a substrate for the breast cancer resistance protein (BCRP)/ABCG2 drug pump. Blood 104: 2940–2942.
14. Doyle LA, Yang W, Abruzzo LV, Krogmann T, Gao Y, et al. (1998) A multidrug resistance transporter from human MCF-7 breast cancer cells. Proc Natl Acad Sci U S A 95: 15665–15670.
15. Zhou S, Schuetz JD, Bunting KD, Colapietro AM, Sampath J, et al. (2001) The ABC transporter Bcrp1/ABCG2 is expressed in a wide variety of stem cells and is a molecular determinant of the side-population phenotype. Nat Med 7: 1028–1034.
16. Ross DD, Nakanishi T (2010) Impact of breast cancer resistance protein on cancer treatment outcomes. Methods Mol Biol 596: 251–290.
17. Doyle L, Ross DD (2003) Multidrug resistance mediated by the breast cancer resistance protein BCRP (ABCG2). Oncogene 22: 7340–7358.
18. Li J, Yen C, Liaw D, Podsypanina K, Bose S, et al. (1997) PTEN, a putative protein tyrosine phosphatase gene mutated in human brain, breast, and prostate cancer. Science 275: 1943–1947.
19. Peiffer SL, Herzog TJ, Tribune DJ, Mutch DG, Gersell DJ, et al. (1995) Allelic loss of sequences from the long arm of chromosome 10 and replication errors in endometrial cancers. Cancer Res 55: 1922–1926.
20. Gronbaek K, Zeuthen J, Guldberg P, Ralfkiaer E, Hou-Jensen K (1998) Alterations of the MMAC1/PTEN gene in lymphoid malignancies. Blood 91: 4388–4390.
21. Peng C, Chen Y, Yang Z, Zhang H, Osterby L, et al. (2010) PTEN is a tumor suppressor in CML stem cells and BCR-ABL-induced leukemias in mice. Blood 115: 626–635.
22. Bleau AM, Hambardzumyan D, Ozawa T, Fomchenko EI, Huse JT, et al. (2009) PTEN/PI3K/Akt pathway regulates the side population phenotype and ABCG2 activity in glioma tumor stem-like cells. Cell Stem Cell 4: 226–235.
23. Goodell MA (2005) Stem cell identification and sorting using the Hoechst 33342 side population (SP). Curr Protoc Cytom Chapter 9: Unit9 18.
24. Weisberg E, Griffin JD (2003) Resistance to imatinib (Glivec): update on clinical mechanisms. Drug Resist Updat 6: 231–238.
25. Illmer T, Schaich M, Platzbecker U, Freiberg-Richter J, Oelschlagel U, et al. (2004) P-glycoprotein-mediated drug efflux is a resistance mechanism of chronic myelogenous leukemia cells to treatment with imatinib mesylate. Leukemia 18: 401–408.
26. Ozvegy-Laczka C, Hegedus T, Varady G, Ujhelly O, Schuetz JD, et al. (2004) High-affinity interaction of tyrosine kinase inhibitors with the ABCG2 multidrug transporter. Mol Pharmacol 65: 1485–1495.
27. Houghton PJ, Germain GS, Harwood FC, Schuetz JD, Stewart CF, et al. (2004) Imatinib mesylate is a potent inhibitor of the ABCG2 (BCRP) transporter and reverses resistance to topotecan and SN-38 in vitro. Cancer Res 64: 2333–2337.
28. Nakanishi T, Shiozawa K, Hassel BA, Ross DD (2006) Complex interaction of BCRP/ABCG2 and imatinib in BCR-ABL-expressing cells: BCRP-mediated resistance to imatinib is attenuated by imatinib-induced reduction of BCRP expression. Blood 108: 678–684.
29. Jordanides NE, Jorgensen HG, Holyoake TL, Mountford JC (2006) Functional ABCG2 is overexpressed on primary CML CD34+ cells and is inhibited by imatinib mesylate. Blood 108: 1370–1373.
30. Zhang J, Grindley JC, Yin T, Jayasinghe S, He XC, et al. (2006) PTEN maintains haematopoietic stem cells and acts in lineage choice and leukaemia prevention. Nature 441: 518–522.
31. Yilmaz OH, Valdez R, Theisen BK, Guo W, Ferguson DO, et al. (2006) Pten dependence distinguishes haematopoietic stem cells from leukaemia-initiating cells. Nature 441: 475–482.
32. Daley GQ, Van Etten RA, Baltimore D (1990) Induction of chronic myelogenous leukemia in mice by the P210bcr/abl gene of the Philadelphia chromosome. Science 247: 824–830.
33. Kantarjian H, Giles F, Wunderle L, Bhalla K, O'Brien S, et al. (2006) Nilotinib in imatinib-resistant CML and Philadelphia chromosome-positive ALL. N Engl J Med 354: 2542–2551.
34. Davies A, Jordanides NE, Giannoudis A, Lucas CM, Hatziieremia S, et al. (2009) Nilotinib concentration in cell lines and primary CD34(+) chronic myeloid leukemia cells is not mediated by active uptake or efflux by major drug transporters. Leukemia 23: 1999–2006.
35. Druker BJ, Sawyers CL, Kantarjian H, Resta DJ, Reese SF, et al. (2001) Activity of a specific inhibitor of the BCR-ABL tyrosine kinase in the blast crisis of chronic myeloid leukemia and acute lymphoblastic leukemia with the Philadelphia chromosome. N Engl J Med 344: 1038–1042.
36. Cortes J, O'Brien S, Kantarjian H (2004) Discontinuation of imatinib therapy after achieving a molecular response. Blood 104: 2204–2205.
37. Bhatia R, Holtz M, Niu N, Gray R, Snyder DS, et al. (2003) Persistence of malignant hematopoietic progenitors in chronic myelogenous leukemia patients in complete cytogenetic remission following imatinib mesylate treatment. Blood 101: 4701–4707.
38. Zeng H, Park JW, Guo M, Lin G, Crandall L, et al. (2009) Lack of ABCG2 expression and side population properties in human pluripotent stem cells. Stem Cells 27: 2435–2445.
39. Li H, Gao Q, Guo L, Lu SH (2011) The PTEN/PI3K/Akt pathway regulates stem-like cells in primary esophageal carcinoma cells. Cancer Biol Ther 11: 950–958.
40. Hegedus C, Truta-Feles K, Antalffy G, Brozik A, Kasza I, et al. (2012) PI3-kinase and mTOR inhibitors differently modulate the function of the ABCG2 multidrug transporter. Biochem Biophys Res Commun 420: 869–874.
41. Huang FF, Wu DS, Zhang L, Yu YH, Yuan XY, et al. (2013) Inactivation of PTEN increases ABCG2 expression and the side population through the PI3K/Akt pathway in adult acute leukemia. Cancer Lett 336: 96–105.
42. Goodell MA, Rosenzweig M, Kim H, Marks DF, DeMaria M, et al. (1997) Dye efflux studies suggest that hematopoietic stem cells expressing low or undetectable levels of CD34 antigen exist in multiple species. Nat Med 3: 1337–1345.

10

Human Tendon Stem Cells Better Maintain Their Stemness in Hypoxic Culture Conditions

Jianying Zhang, James H.-C. Wang*

MechanoBiology Laboratory, Departments of Orthopaedic Surgery, Bioengineering, Mechanical Engineering and Materials Science, and Physical Medicine and Rehabilitation, University of Pittsburgh, Pittsburgh, Pennsylvania, United States of America

Abstract

Tissues and organs *in vivo* are under a hypoxic condition; that is, the oxygen tension is typically much lower than in ambient air. However, the effects of such a hypoxic condition on tendon stem cells, a recently identified tendon cell, remain incompletely defined. In cell culture experiments, we subjected human tendon stem cells (hTSCs) to a hypoxic condition with 5% O_2, while subjecting control cells to a normaxic condition with 20% O_2. We found that hTSCs at 5% O_2 had significantly greater cell proliferation than those at 20% O_2. Moreover, the expression of two stem cell marker genes, Nanog and Oct-4, was upregulated in the cells cultured in 5% O_2. Finally, in cultures under 5% O_2, more hTSCs expressed the stem cell markers nucleostemin, Oct-4, Nanog and SSEA-4. In an *in vivo* experiment, we found that when both cell groups were implanted with tendon-derived matrix, more tendon-like structures formed in the 5% O_2 treated hTSCs than in 20% O_2 treated hTSCs. Additionally, when both cell groups were implanted with Matrigel, the 5% O_2 treated hTSCs showed more extensive formation of fatty, cartilage-like and bone-like tissues than the 20% O_2 treated cells. Together, the findings of this study show that oxygen tension is a niche factor that regulates the stemness of hTSCs, and that less oxygen is better for maintaining hTSCs in culture and expanding them for cell therapy of tendon injuries.

Editor: Sudha Agarwal, Ohio State University, United States of America

Funding: This project was funded by National Institutes of Health grants AR049921 and AR06139. The funders had no role in study design, data collection and analysis, decision to publish, or preparation of the manuscript.

Competing Interests: The authors have declared that no competing interests exist.

* E-mail: wanghc@pitt.edu

Introduction

Tendons connect muscles to bones to enable joint movement. As a result, they are subjected to large mechanical loads and hence are frequently injured. Full recovery of injured tendons requires a long, complex healing process, particularly in the case of complete tendon rupture when tendon retraction occurs. Moreover, healed tendons consist of scar tissue that has lower mechanical strength than normal tendon tissue. This mechanical weakness not only impairs normal tendon function and joint kinematics, but also predisposes patients to further tendon injury [1].

Restoring normal structure and function to injured tendons is challenging and a number of ways are being discovered to promote tendon regeneration after injury. Tissue engineering is one such approach that uses cells, scaffolds and growth factors to effectively repair or regenerate injured tendons more effectively. Cell therapy in particular, is important in tissue engineering to repair injured tendons or other tissues. For example, bone marrow mesenchymal stem cells (BMSCs) in conjugation with collagen gels, have been used to repair injured tendons [2] although these have resulted in ectopic bone formation in rabbit tendon injury models [3]. In addition, embryonic stem cells (ESCs) have also been used to repair injured tendons. However, ESCs implantation could result in teratoma formation, which occurs due to difficulty in controlling ESCs differentiation *in vivo* when compared to adult stem cells such as BMSCs. These and other studies clearly indicate that stem cells from non-tendinous tissues may not be optimal to restore the normal structure and function of injured tendons using cell therapy.

Implantation of autologous tenocytes, which are resident tendon cells responsible for the maintenance and repair of tendons has resulted only in a slight improvement in tendon quality [4]. A new type of recently discovered tendon cells called tendon stem cells (TSCs) have a great potential to repair injured tendons and have been identified in humans, rabbits, rats and mice [5–7]. Like adult stem cells, TSCs have the capacity for self-renewal, which enables them to make more stem cells by cell division and also possess multi-differentiation potential, which enables them to become specialized cell types. Under normal conditions, TSCs differentiate into tenocytes [6]. However, when implanted with engineered tendon matrix (ETM), TSCs form tendon-like tissues in nude rats [8]. Therefore, TSCs may be an ideal cell source for tissue engineering approaches that could effectively repair injured tendons.

To obtain sufficient numbers of cells for cell therapy of injured tendons, TSCs must be expanded in culture. However, under regular culture conditions that use 95% air and 5% CO_2, TSCs tend to differentiate and consequently lose their stemness quickly. *In vivo*, tendons, which are collagen-rich structures with only a few blood vessels, have low oxygen levels when compared to vascular-rich organs and tissues such as the lungs, heart, liver and kidneys where the oxygen levels range from 10 to 13% [9]. However, the effects of low oxygen concentrations on TSCs have not been completely defined yet. In this study, we tested the hypothesis that under hypoxic conditions TSCs better maintain their stemness. Indeed, our findings show that low oxygen tension enhances the stemness of TSCs, which was characterized by quicker cell proliferation, higher expression of stem cell markers *in vitro* and more extensive formation of tendon-like and non-tendon-like tissues *in vivo*.

Materials and Methods

Ethics Statement

Normal human knee tissues were obtained within 24 hours of death of donors from the Gift of Hope Organ and Tissue Donor Network (Elmhurst, IL) with approval from the local ethics committee (Gift of Hope Organ and Tissue Donor Network). Written consent from the families was obtained and approved by the Gift of Hope Organ and Tissue donor Network. Tissue specimens were obtained for investigation only. The protocol to use human tendon tissues for subsequent cell culture and animal studies was approved by the University of Pittsburgh IRB.

This project did not involve human subjects, and the authors conducting research did not obtain data through intervention or interaction with individuals or obtain identifiable private information.

In addition, protocol for the use of rats for *in vivo* experimentation was approved by the University of Pittsburgh IACUC. All animal surgery was performed under general anesthesia and efforts were made to minimize suffering.

Control of Hypoxic and Normoxic Culture Conditions

We used a dedicated tri-gas incubator (Thermo Scientific Heracell 150i, Thermo Scientific, and Pittsburgh, PA) to achieve hypoxic conditions (5% O_2) in cell culture experiments. Concentration of oxygen in the incubator was precisely controlled by two gas controllers and an oxygen sensor. Nitrogen and carbon dioxide gases were supplied using a nitrogen gas controller (Thermo Scientific) connected to two nitrogen tanks and a carbon dioxide gas controller connected to two carbon dioxide tanks. The set up was such that the supply of gas automatically switched from the first to the second tank when the first tank was empty. To avoid air flow into the incubator during brief openings of the door the incubator was separated into three isolated chambers with each chamber closed by double doors. With these control devices in place, oxygen concentration in the incubator was maintained at a constant level of 5% during all cell culture experiments.

To maintain normoxic culture conditions (20% O_2) a regular tissue culture incubator (Thermo Scientific) was used. About 20% O_2 concentration was achieved inside the incubator by feeding 95% air and 5% carbon dioxide from tanks.

Human TSC Culture

Human TSC (hTSCs) were obtained from the patellar tendons of six young adult donors aged 26 to 49 years following our previously published method [6].

Cell Proliferation Experiment

hTSCs were seeded into 6-well culture plates at a density of 40,000 cells/well in 3 ml DMEM growth medium with 20% FBS and maintained in the tri-gas incubator to achieve a 5% O_2 culture condition or the regular incubator to provide a 20% O_2 culture condition. Replacement medium for the cells cultured in the tri-gas incubator was prepared by pre-conditioning the medium in the tri-gas incubator for at least 30 min before use. The medium was changed every two days under both hypoxic and normoxic culture conditions. Colony

formation by hTSCs cultured in the two oxygen conditions was tested by staining with methyl violet. Cell proliferation was determined by counting cells on days 1, 2, 6 and 12 after seeding, as previously described [10].

Stem Cell Marker Expression

To characterize the stemness of hTSCs in hypoxic and normaxic culture conditions, we determined differential expression of stem cell markers in hTSCs in both culture conditions. Cells were seeded into 12-well plates at a density of 20,000 cells/well with 1.5 ml medium and cultured either with 5 or 20% O_2 for 3–5 days. Expression of four stem cell markers including nucleostemin (NS), octamer-binding transcription factor 4 (Oct-4), Nanog and stage-specific embryonic antigen-4 (SSEA-4) was measured using immunocytochemistry. Briefly, hTSCs were fixed in 4% paraformaldehyde in PBS for 20 min at room temperature. For Oct-4, Nanog and nucleostemin staining fixed cells were treated with 0.5% Triton-X-100 in PBS for 15 min and washed with 2% mouse or goat serum-PBS for 30 min. The cells were then incubated with either mouse anti-human Oct-4 (1:500), rabbit anti-human Nanog (1:500) or goat anti-human nucleostemin (1:500) overnight at 4C. After washing in PBS three times, the cells were again incubated with either Cy-3-conjugated goat anti-mouse IgG antibodies (1:1000), Cy3-conjugated goat anti-rabbit IgG (1:500) or Cy-3-conjugated donkey anti-goat IgG antibodies (1:500) for 2 hrs at room temperature to detect Oct-4, Nanog and nucleostemin respectively. To stain for SSEA-4, fixed cells were blocked with 2% mouse serum for 1 hr and incubated with mouse anti-human SSEA-4 antibody (1:500) for 2 hrs at room temperature. After subsequent washing with PBS, TSCs were treated with Cy3-conjugated goat anti-mouse IgG antibody (1:1000) for 1 hr at room temperature. Stained cells were then examined using fluorescence microscopy. All antibodies were obtained from Chemicon International (Temecula, CA), BD Biosciences (Franklin Lakes, NJ), Neuromics (Edina, MN) or Santa Cruz Biotechnology Inc. (Santa Cruz, CA).

Multi-differentiation Potentials

The differentiation capacity of hTSCs in hypoxic and normaxic culture conditions was examined *in vitro* by testing their abilities to undergo adipogenesis, chondrogenesis and osteogenesis. Cells at passage 1 were seeded into 6-well plates at a density of 24×10^4 cells/well in basic growth medium (DMEM plus 10% FBS) and cultured in either 5 or 20% O_2. To measure adipogenic potential, hTSCs were cultured in adipogenic induction medium (Millipore, Billerica, MA) that consists of basic growth medium supplemented with dexamethasone (1 µM), insulin (10 µg/ml), indomethacin (100 µM) and isobutylmethylxanthine (0.5 mM). To determine chondrogenic potential, hTSCs were cultured in basic growth medium supplemented with proline (40 µg/ml), dexamethasone (39 ng/ml), TGF-β3 (10 ng/ml), ascorbic 2-phosphate (50 µg/ml), sodium pyruvate (100 µg/ml) and insulin-transferrin-selenious acid mix (50 mg/ml) (BD Bioscience, Bedford, MA). Finally, osteogenic potential of hTSCs in both hypoxic and normaxic culture conditions was studied by culturing cells in osteogenic induction medium (Millipore, Billerica, MA) consisting of basic growth medium supplemented with dexamethasone (0.1 µM), ascorbic 2-phosphate (0.2 mM) and glycerol 2-phosphate (10 mM). hTSCs were grown in above three media for 21 days followed by Oil red O assay for adipogenesis, Safranin O assay for chondrogenesis and Alizarin red S assay for osteogenesis as described previously [6].

Semi-quantification of the Extent of hTSC Differentiation

For the semi-quantification of cell differentiation, twelve images of each well were randomly taken under a microscope (Nikon eclipse, TE2000-U). Then areas with positive staining were manually identified from each picture and computed by SPOT^{TM} imaging software (Diagnostic Instruments, Inc., Sterling Heights, MI). Proportion of positive staining was calculated by dividing the positively stained area by the total area viewed under the microscope. These values were obtained for all twelve images of a well and their average was used to represent the percentage of positive staining, which is the extent of cell differentiation in the respective induction media described under multi-differentiation potentials.

Quantitative Real-time PCR (qRT-PCR)

To measure the stemness of hTSCs under hypoxic (5% O_2) and normaxic culture conditions (20% O_2) we performed qRT-PCR analysis. Total RNA was extracted from hTSCs using an RNeasy Mini Kit with an on-column DNase I digest (Qiagen). First-strand cDNA was synthesized by reverse transcribing 1 μg total RNA with SuperScript II (Invitrogen) in a 20 μl reaction volume. The conditions for cDNA synthesis included: 65°C for 5 min followed by cooling at 4°C for 1 min, then 42°C for 50 min and finally 72°C for 15 min. qRT-PCR was carried out using 2 μl cDNA (approximately 100 ng RNA) in a 25 μl PCR reaction volume using QIAGEN QuantiTect SYBR Green PCR Kit (Qiagen) in a Chromo 4 Detector (MJ Research). To determine stemness of TSCs gene-specific primers of human Oct-4, Nanog, and tenocyte-related genes, including collagen type I and tenascin C were used. Glyceraldehyde-3-phosphate dehydrogenase (GAPDH) was used as an internal control. Forward and reverse primers for all genes were designed based on previously published sequences [11–13] and were synthesized by Invitrogen (Carlsbad, CA). Relative gene expression levels in hTSCs under hypoxic and normaxic culture conditions were determined using the formula $\Delta\Delta CT = (CT_{target} - CT_{GAPDH})_{Hypoxia} - (CT_{target} - CT_{GAPDH})_{Normoxia}$, where CT represents cycle threshold of each RNA sample. At least three replicates were performed for each gene and each experimental condition.

Preparation of hTSCs for in vivo Implantation

hTSCs for implantation were prepared by plating cells from passage 2 into two 24-well plates at a seeding density of 6×10^4/ well and were allowed to grow in 5 or 20% O_2 culture conditions. After one week, cells under both culture conditions were collected and each was mixed separately with 0.5 ml 5% ETM made from rabbit patellar tendon samples according to our previously published method [8] or 0.5 ml Matrigel (Cat. # 354234, BD Biosciences, Bedford, MA). The cell-ETM or cell-Matrigel composites were then reseeded into a 24-well plate and cultured overnight with 5 or 20% O_2 to maintain hypoxic or normaxic conditions respectively.

In vivo Implantation Experiment

Four 10 weeks old female nude rats weighing between 200–250 g were used for hTSC implantation experiments. Protocol for the use of rats for in vivo experimentation was approved by the University of Pittsburgh IACUC. Before implantation, all rats were given general anesthesia by intramuscular injection of a mixture of ketamine hydrochloride (75 mg/kg body weight) and xylazine hydrochloride (5 mg/kg body weight). Two rats each were implanted with hTSCs cultured in 5 or 20% O_2 conditions. A total of six distinct wounds were made on the back of each rat

and each wound was filled with a piece of cell-ETM or cell-Matrigel composite. Three weeks after implantation, the wound sites were opened and tissues in the area were harvested. The tissue samples were then immersed in frozen section medium (Neg 50; Richard-Allan Scientific; Kalamazoo, MI) in pre-labeled base molds and were quickly frozen in 2-methylbutane chilled with liquid nitrogen. Frozen tissue blocks were then placed on dry ice and stored in -80°C until further use for histological and immunohistochemical analyses. At least three replicates were performed for each experimental condition.

Detection of hTSC Differentiation in vivo

Frozen tissue blocks were cut into 8 μm thick sections, fixed in 4% paraformaldehyde for 15 min and stained with mouse anti-human collagen type I (1:100, Millipore, Cat. #MAB1340; Temecula, CA), mouse anti-human adiponectin (1:300, Millipore; Cat. #MAB3604; Temecula, CA), mouse anti-human collagen type II (1:100, Millipore, Cat. #MAB1330, Temecula, CA) and mouse anti-human osteocalcin (1:200, Abcam, Cat #13418, Cambridge, MA) at room temperature for 2 hrs. Cy-3 conjugated goat anti-mouse IgG (1:500, Jackson ImmunoResearch Laboratories, Inc., Cat. #115-165-146, West Grove, PA) was used as the secondary antibody to detect collagen type I, collagen type II and osteocalcin at room temperature for 2 hrs. FITC-conjugated goat anti-mouse IgM (1:500, Santa Cruz Biotechnology, Cat. #sc-2082, Santa Cruz, CA) was used as the secondary antibody to detect adiponectin. The tissue sections were also treated with Hoechst 33342 (Sigma, Cat. #B2261, St. Louis, MO) to stain nuclei.

Statistical Analysis

One-way analysis of variance (ANOVA) followed by either Fisher's predicted least-square difference (PLSD) for multiple comparisons or two tailed, paired or unpaired student t-test were performed wherever applicable. Differences between two groups (hypoxic vs. normaxic conditions) were considered significant when P-value was below 0.05.

Results

To determine the effects of hypoxia, similar numbers of hTSCs were seeded and cultured in both hypoxic and normaxic conditions. We found that hTSCs cultured in 5% O_2 formed more colonies that were also larger than cells cultured in 20% O_2 (Figure 1A, 1B). Colonies formed in 5% O_2 were twice as many as those in 20% O_2 (Figure 1B) and the colony size in 5% O_2 was on an average about 2.8 times larger than those in 20% O_2 (Figure 1C). In addition, proliferation of hTSCs in both culture conditions increased in the days following seeding and the number of cells in 5% O_2 was higher than in 20% O_2 from day 1 through days 2, 6 and 12 (Figure 2).

Expression of stem cell markers NS, Oct-4, Nanog and SSEA-4 determined by immunocytochemistry was also higher in colonies cultured in 5% O_2 compared to hTSCs grown in 20% O_2 (Figure 3A). Semi-quantification of the immuno-stained cells further showed that more than 90% of hTSCs cultured in 5% O_2 were NS positive compared to 66% in hTSCs cultured in 20% O_2. Oct-4 expression was also higher (98%) in hTSCs cultured at 5% O_2 when compare to cells cultured in 20% O_2 (44%). Similarly, the expression of Nanog and SSEA-4 in hTSCs cultured in 5% O_2 was 95 and 88% respectively when compared to the lower percentages (55 and 49%) observed in hTSCs cultured in 20% O_2 (Figure 3B). Consistent with these results RT-PCR analysis also showed higher expression levels of both Oct-4 and Nanog genes in

A

B

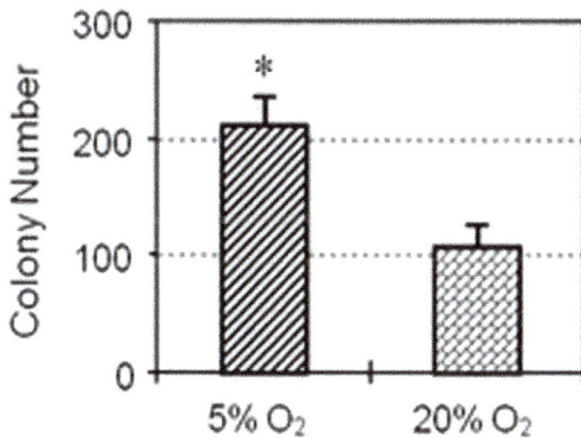

C

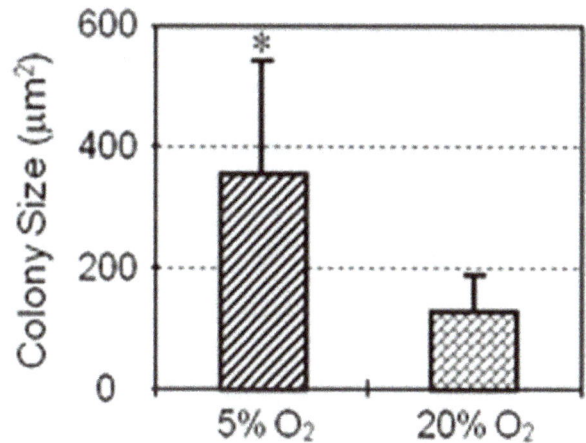

Figure 1. A. Colony formation by hTSCs under hypoxic and normaxic culture conditions. T-25 flask maintained in 5% O_2 (left) shows a number of large methyl violet stained hTSC colonies while the flask in 20% O_2 (right) has fewer and smaller colonies. B. Quantification of hTSC colony numbers under hypoxic and normaxic conditions. Colony numbers in 5% O_2 were twice that in 20% O_2. Two flasks of hTSCs each were used to calculate colony numbers in hypoxic and normaxic conditions. C. Quantification of hTSC colony sizes under hypoxic and normaxic conditions. Colony size of hTSCs in hypoxic condition (5% O_2) was about 2.8 times larger than in normaxic condition (20% O_2). Asterisks represent P<0.05.

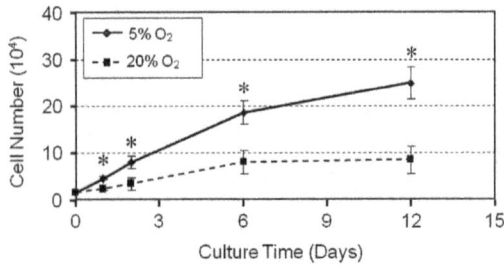

Figure 2. Proliferation of hTSCs cultured under hypoxic and normaxic culture conditions. hTSCs were grown in DMEM growth medium with FBS under hypoxic or normaxic conditions and colony formation was determined by counting cells stained with methyl violet. While the cells grew at both culture conditions, at all time points (days 1, 2, 6, and 12), hTSCs at 5% O_2 grew significantly quicker than at 20% O_2.

5% O_2 compared to 20% O_2 (Figure 4). Cells cultured in both conditions showed no significant difference in the expression of the tenocyte-related gene, collagen type I, but the level of tenascin C expression was more than 2-fold higher in 5% O_2 compared to 20% O_2 (Figure 5A). Moreover, the expressions of non-tenocyte-related genes Sox-9 and Runx-2 were significantly lower in 5% O_2 than in 20% O_2 culture conditions while expression of PPARγ was marginally lower (Figure 5B).

We next examined the multi-differentiation capacity of TSCs under hypoxic and normaxic culture conditions. After 21 days in culture, the degree of adipogenesis, chondrogenesis and osteogenesis of hTSCs was more extensive in 5% O_2 condition compared to 20% O_2 (Figure 6A). Semi-quantitative analysis was also consistent with these results with the percentages of hTSCs that differentiated into adipocytes, osteocytes and chondrocytes about 51, 90 and 54% respectively in 5% O_2 conditions when compared to lesser percentages of 31, 46 and 31% respectively in 20% O_2 (Figure 6B).

To further characterize hTSCs after exposure to hypoxic and normaxic conditions, we implanted the cells grown under the two conditions into nude rats subcutaneously. Three weeks after implantation hTSCs cultured in 5% O_2 and embedded in ETM resulted in extensive formation of bands that corresponded to tendon-like structures, as evidenced by strong staining for human collagen type I (hCT-I). In contrast, hTSCs cultured in 20% O_2 before embedding in ETM and transplantation, showed only discreet areas which were stained positive for hCT-I. Staining for adiponectin (a marker for adipogenesis), collagen type II (a marker for chondrogenesis) and osteocalcin (a marker for osteogenesis) showed minimal formation of fatty, cartilage, and bony tissues in 5% O_2 treated hTSCs and embedded in ETM (ETM-5% O_2) whereas the cells treated with 20% O_2 and embedded in ETM (ETM-20% O_2) formed well- developed cartilage and bony tissues. Similarly, when 5 and 20% O_2 treated hTSCs were embedded in Matrigel and implanted, we observed formation of all four types of tissues (tendinous, fatty, cartilage-like and bony tissues), which was more extensive in 5% O_2 treated hTSCs compared to 20% O_2 treated hTSCs (Figure 7). Semi-quantitative analysis further showed that compared to normaxic condition, hypoxic condition resulted in higher amounts of both tendinous protein (collagen type I) and non-tendinous proteins (adiponectin, collagen type II and osteocalcin) (data not shown).

Discussion

Tendon injury is common in both occupational and athletic settings. Currently, there are no effective means to restore normal structure and function to injured tendons. TSCs, which were only recently identified, are tendon-specific adult stem cells that are thought to play a critical role in the repair of injured tendons. Therefore, TSCs may be an optimal cell source for effective tissue engineering of injured tendons. However, the major obstacle using such cell therapy is that once TSCs are isolated from tendons and grown in a conventional *in vitro* environment, they tend to differentiate quickly. Considering that TSCs *in vivo* are under hypoxic conditions due to poor vascularity in tendon substances, we designed this study to investigate the effects of hypoxic conditions on hTSCs. By performing cell culture experiments, we have shown that the stemness of hTSCs cultured in 5% O_2 was better than hTSCs at 20% oxygen levels. Specifically, under the hypoxic culture condition (5% O_2), growth of hTSCs was faster, higher number of cells expressed stem cell markers (NS, Oct-4, Nanog and SSEA-4) and the expression level of stem cell marker genes (Oct-4 and Nanog) was also significantly higher. In addition, while tenocyte-related gene expression levels were similar under both hypoxic and normaxic conditions (5% O_2 vs. 20% O_2), non-tenocyte related gene expression at hypoxic condition was significantly lower than in normaxic condition. Moreover, hTSCs cultured under the hypoxic condition (5% O_2) exhibited more potent multi-differentiation capacity in terms of adipogenesis, chondrogenesis and osteogenesis. By performing an *in vivo* implantation experiment, we were also able to show that when implanted together with ETM, hTSCs in hypoxic condition produced more extensive tendon-like tissues than in normaxic condition. The hypoxic condition also resulted in more tendinous and non-tendinous tissues than normaxic condition when both were implanted with Matrigel. The *in vivo* results further showed that hypoxic condition enhanced multi-differentiation potential of hTSCs.

The findings of this study show that oxygen is an important niche factor for the maintenance of stemness by hTSCs. These findings also indicate that for effective tissue engineering of injured tendons, TSCs should be cultured in a hypoxic environment. This hypoxic condition can promote TSCs' self-renewal, thus allowing sufficient numbers of TSCs to be obtained for tissue engineering, which may repair injured tendons more effectively. The high self-renewal rate of hTSCs under a hypoxic condition is consistent with the concept that adult stem cells like hTSCs ensure maintenance of their pool for tissue repair or regeneration when the tissue is injured [14]. Finally, this study also indicates that caution should be exercised before using the so-called hyperbaric oxygen therapy to treat injured tendons, at levels as high as >20% O_2 that could potentially deplete the TSCs pool quickly by promoting their differentiation into specialized cell types and consequently could hinder the repair of tendons after re-injury.

Many studies have investigated the effects of various hypoxic conditions on cells. For example, compared to normaxic condition (20% O_2), hypoxic condition (1.5% to 5% O_2) increases proliferation of human mesenchymal stem cells (hMSCs) [15]. In addition, hMSCs grown in 2% O_2 exhibited enhanced colony-forming capabilities and had a higher expression of Oct-4 [16,17]. Hypoxic conditions also produced greater numbers of stem cell colonies that proliferated more rapidly in culture. Rat MSCs cultured in 5% O_2 produced more bone than cells cultured in 20% O_2 when the cells were loaded into porous ceramic cubes and implanted into animals [18]. In addition, hESCs in 20% O_2

A

B

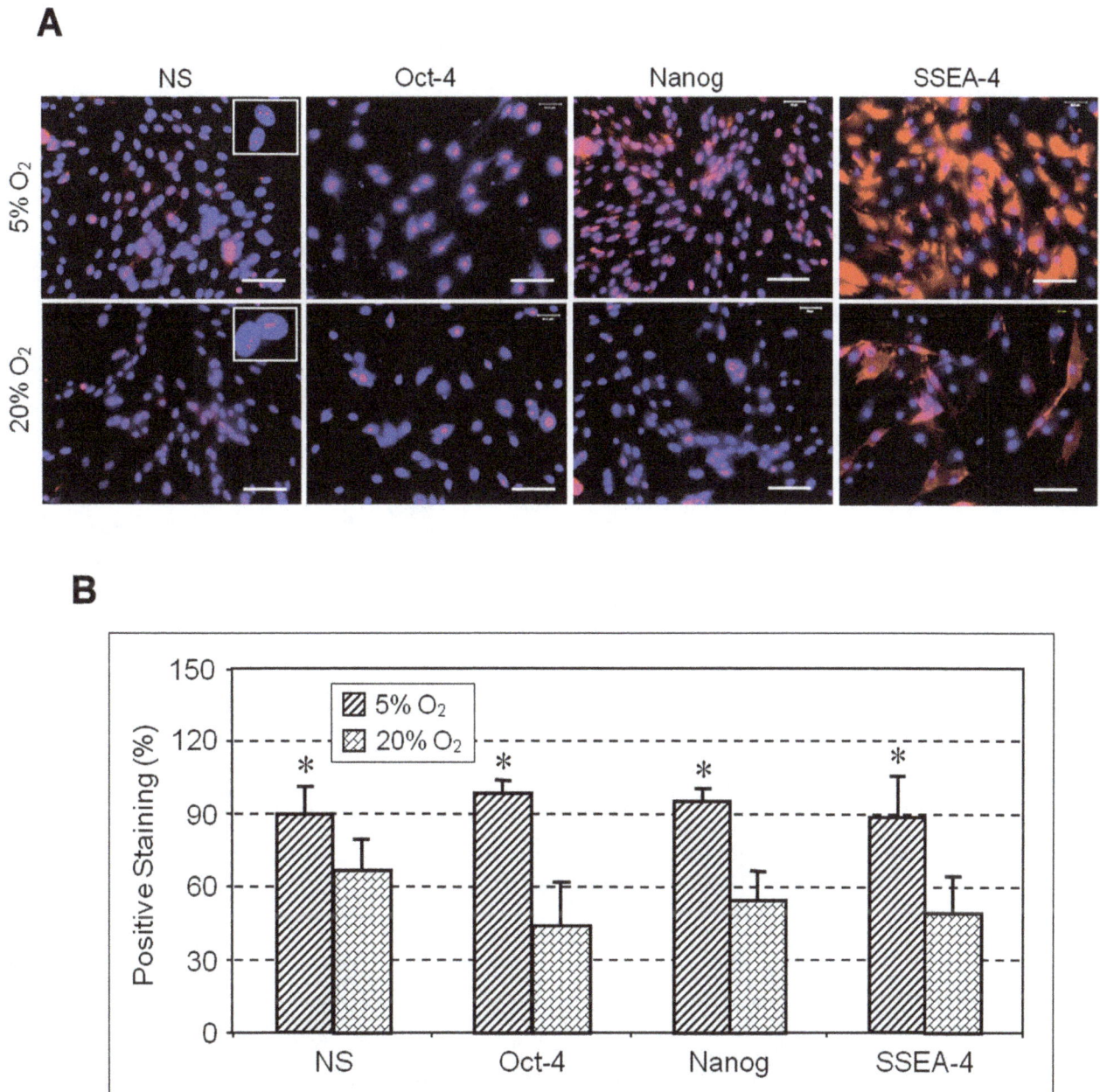

Figure 3. A. The expression of stem cell markers by hTSCs under hypoxic and normaxic culture conditions. hTSCs grown under hypoxic or normaxic conditions were analyzed by immunocytochemistry using specific antibodies to determine stem cell marker expression (See materials and methods for details). Compared to normaxic condition (20% O_2), more hTSCs at hypoxic condition (5% O_2) expressed nucleostemin (NS), Oct-4, Nanog, and SSEA-4, all of which are known stem cell markers. Insets indicate NC proteins in the nuclei of hTSCs. Nuclei were stained with Hoechst 33342. Scale bars: 100 μm. B. Semi-quantification of stem cell markers by staining. hTSCs specifically stained for NS, Oct-4, Nanog and SSEA-4 by immunocytochemical staining were counted to calculate percentage staining. As indicated, significantly higher percentages of hTSCs cultured under 5% O_2 conditions expressed the stem cell markers (NS, Oct-4, Nanog, and SSEA-4) compared to those cultured under 20% O_2 conditions (*$P<0.05$, with respect to hTSCs under 20% O_2 culture conditions). Scale bars: 100 μm.

culture condition showed decreased cell proliferation and reduced expression of Nanog and Oct-4 genes, and Oct-4 protein, compared to 5% O_2 culture condition [19]. Our results were consistent with the findings of these previous studies.

However, while our study found that the hypoxic condition at 5% O_2 enhanced differentiation potential of hTSCs, a previous study showed a decrease in the multi-differentiation potential of hTSCs under hypoxic condition (2% O_2) [20]. There are several possible reasons for this discrepancy. First, our study

used a tri-gas incubator, whereas their study used a hypoxic chamber that controlled oxygen levels in a regular incubator with 20% O_2. The two different means of controlling oxygen concentrations could result in huge differences in the conditions under which hTSCs were cultured; i.e., nearly constant oxygen levels vs. fluctuating oxygen levels during culture experiments. Second, the initial states of hTSCs in both studies could have been different. For example, Oct-4 expressing hTSCs vs. tendon progenitor cells that do not express Oct-4 that consequently

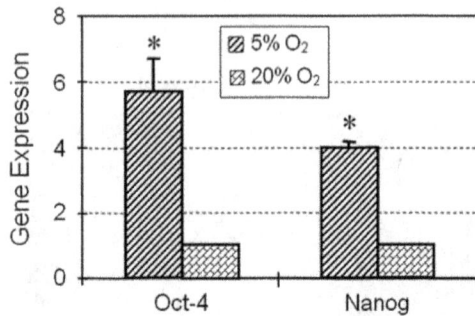

Figure 4. Stem cell gene analysis by qRT-PCR. Total RNA extracted from hTSCs grown under hypoxic or normaxic conditions was used to synthesize cDNA, which was used as a template in qRT-PCR using primers specific to Oct-4 and Nanog. GAPDH was used as an internal control. Y- axis represents relative gene expression when compared to GAPDH expression levels. Ct values were normalized against hTSCs cultured under 20% O_2. Both stem cell marker genes (Oct-4 and Nanog) cultured at 5% O_2 culture conditions were expressed at significantly higher g005levels than those cultured at 20% O_2 culture conditions.

resulted in differences in cellular responses to similar hypoxic conditions. Finally, there could be differences in experimental conditions used in hTSCs culture including the density of cells, depth of medium, pre-conditioning of medium and cellular respiration, all of which could alter the oxygen tension at the surface of cultured cells, consequently leading to differential responses of hTSCs to hypoxic conditions.

Because there are no specific stem cell markers for hTSCs, we used general stem cell markers (NS, Oct-4, Nanog, and SSEA-4) to characterize their stemness under both hypoxic and normaxic conditions. Nucleostemin (NS), that controls cell cycle progression, is exclusively expressed in stem cells, and is therefore not expressed in committed and terminally differentiated cells [21]. Nanog, a unique homeobox transcription factor, was reported to be expressed in pluripotent stem cells, and its expression was associated with stem cell differentiation [22]. Typically expressed in embryonic stem cells (ESCs) during development, Oct-4 is a transcription factor that is known to mediate pluripotency in ESCs [23]. Oct-4 is also essential for maintaining pluripotent stem cells, and is not expressed in differentiated cells [24]. Finally, SSEA-4 is a transcription factor specific to undifferentiated pluripotent human or mouse stem cells [25–27]. Thus, the higher expression levels of these stem cell markers in hypoxic condition (5% O_2) observed in this study indicate that more hTSCs were kept in an undifferentiated state and self-renewed when they were cultured at hypoxic condition (5% O_2) than at normaxic condition (20% O_2).

It is generally accepted that 3 to 5% oxygen levels are present in tissues, although the actual O_2 concentration *in situ* depends on vascularization of the tissue and its metabolic activity [15]. To our best knowledge, the physiological oxygen tension of the human patellar tendon remains unknown. In the articular cartilage, however, oxygen tension is known to be less than 10% at the surface and less than 1% in the deepest layer [28]. Considering that tendons are largely avascular, it is likely that their oxygen tension is higher than 1% but lower than 10%. This is the reason we chose a 5% O_2 level in this study. Use of 5% O_2 level also makes it possible to control oxygen levels in an incubator more precisely, as too low levels of oxygen, which creates a high gradient of oxygen against the environment, is technically demanding in terms of precisely controlling constant oxygen levels to culture cells.

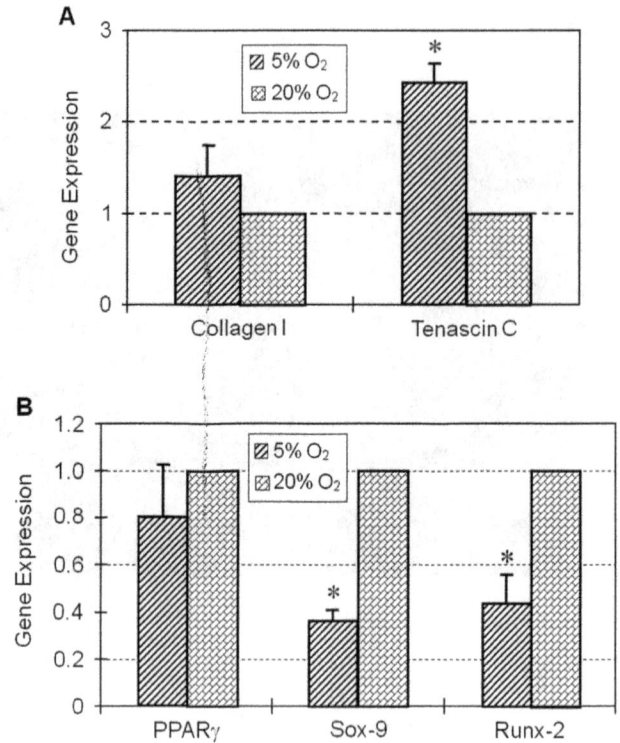

Figure 5. A. Tenocyte related gene expression by hTSCs under hypoxic and normaxic culture conditions. Total RNA extracted from hTSCs grown under hypoxic or normaxic conditions was used to synthesize cDNA, which was used as a template in qRT-PCR using primers specific to Collagen-1 and Tenascin C. GAPDH was used as an internal control. Y- axis represents relative gene expression when compared to GAPDH expression levels. Ct values were normalized against hTSCs cultured under 20% O_2. At both oxygen conditions (5% and 20% O_2), there was no significant difference in the expression of collagen type I, but the expression of tenascin C in the hypoxic group was significantly higher than in the normaxic group (*P<0.05). B. Non-tenocyte related gene expression by hTSCs under the above two oxygen conditions. Total RNA extracted from hTSCs grown under hypoxic or normaxic conditions to synthesize cDNA, which was used as a template in qRT-PCR using primers specific to PPARγ, Sox-9 and Runx-2. GAPDH was used as an internal control. Y- axis represents relative gene expression when compared to GAPDH expression. Ct values were normalized against hTSCs cultured under 20% O_2. The cellular expression of PPARγ, a marker for adipogenesis, was not significantly different in 5 and 20% O_2 conditions. However, Sox-9 and Runx-2 (markers for chondrogenesis and osteogenesis, respectively) were expressed at significantly lower levels when hTSCs were cultured at 5% O_2 condition in comparison to 20% O_2 (*P<0.05, respective to hTSCs that were under normaxic conditions).

There are a few limitations associated with this study. First, we grew hTSCs in plastic dishes, which itself is "foreign" to hTSCs and therefore may cause cell differentiation in culture. Our previous study showed that TSCs grown on tendon matrix coated plastic surfaces can encourage self-renewal of TSCs. Therefore, it seems reasonable to speculate that culture of hTSCs in tendon matrix under a hypoxic condition will result in an even higher stemness of hTSCs, especially in long term cell culture. Second, tendons *in vivo* are constantly subjected to mechanical loading because of their role in the transmission of muscular forces to bones. Mechanical loading, however, was not included in our cell culture experiment although, our previous study showed that mechanical loading itself can regulate TSC functions including proliferation and differentiation [29]. Therefore, future studies

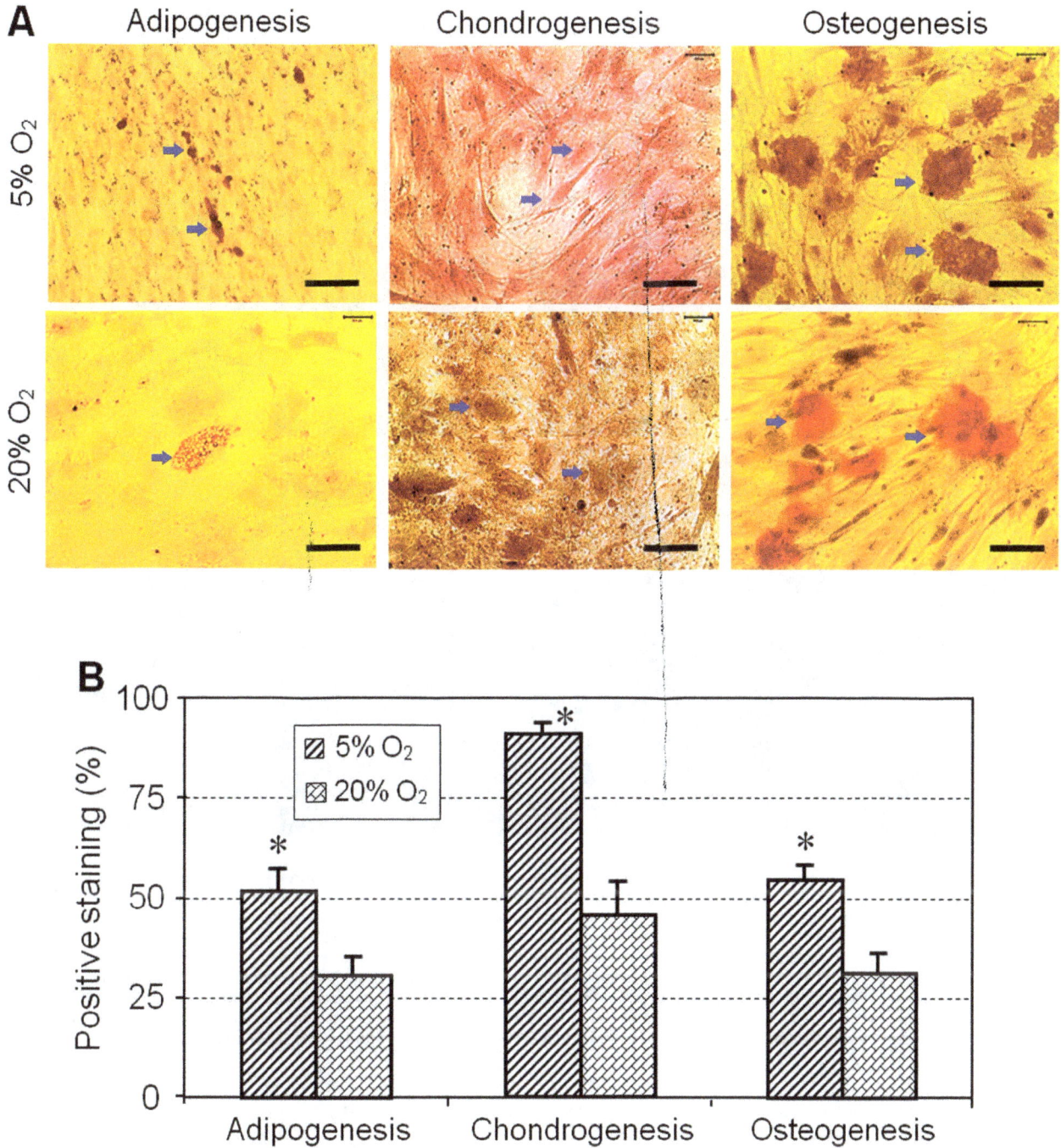

Figure 6. A. Multi-differentiation capacity of hTSCs under hypoxic and normaxic culture conditions. hTSCs were separately grown under both hypoxic or normaxic conditions in adipogenic, chondrogenic and osteogenic induction media for 21 days followed by staining with Oil red O for adipogenesis, Safranin O for chondrogenesis and Alizarin red S for osteogenesis. It is apparent that compared to hTSCs at 20% O_2 condition, cells grown at 5% O_2 culture condition formed more extensive lipids, proteoglycan accumulation, and calcium deposition, as revealed by Oil red O assay, Safranin O assay, and Alizarin red S assay, respectively. Positively stained cells are indicated by arrows. Scale bars: 100 µm. B. Semi-quantification of the staining results by three assays. Positively stained cells were counted to calculate percentage staining. More hTSCs at 5% O_2 condition were found to differentiate into adipocytes, chondrocytes and osteocytes than hTSCs at 20% O_2 condition (*$P < 0.05$, with respect to hTSCs that were under normaxic condition).

should investigate the combined effects of hypoxic conditions and mechanical loading on TSCs. Finally, the molecular mechanisms that are responsible for enhanced stemness in hTSCs as shown in this study are yet to be determined. Nevertheless, it is known that when cells sense changes in oxygen availability, they initiate survival responses by inducing and increasing the expression of hypoxic inducible factor (HIF) in hMSCs [15]. HIF-2α, an isoform of HIF, was also reported to regulate hESC pluripotency and proliferation under hypoxic conditions [19]. Therefore, it is possible that HIF also regulates hypoxic responses of hTSCs. This

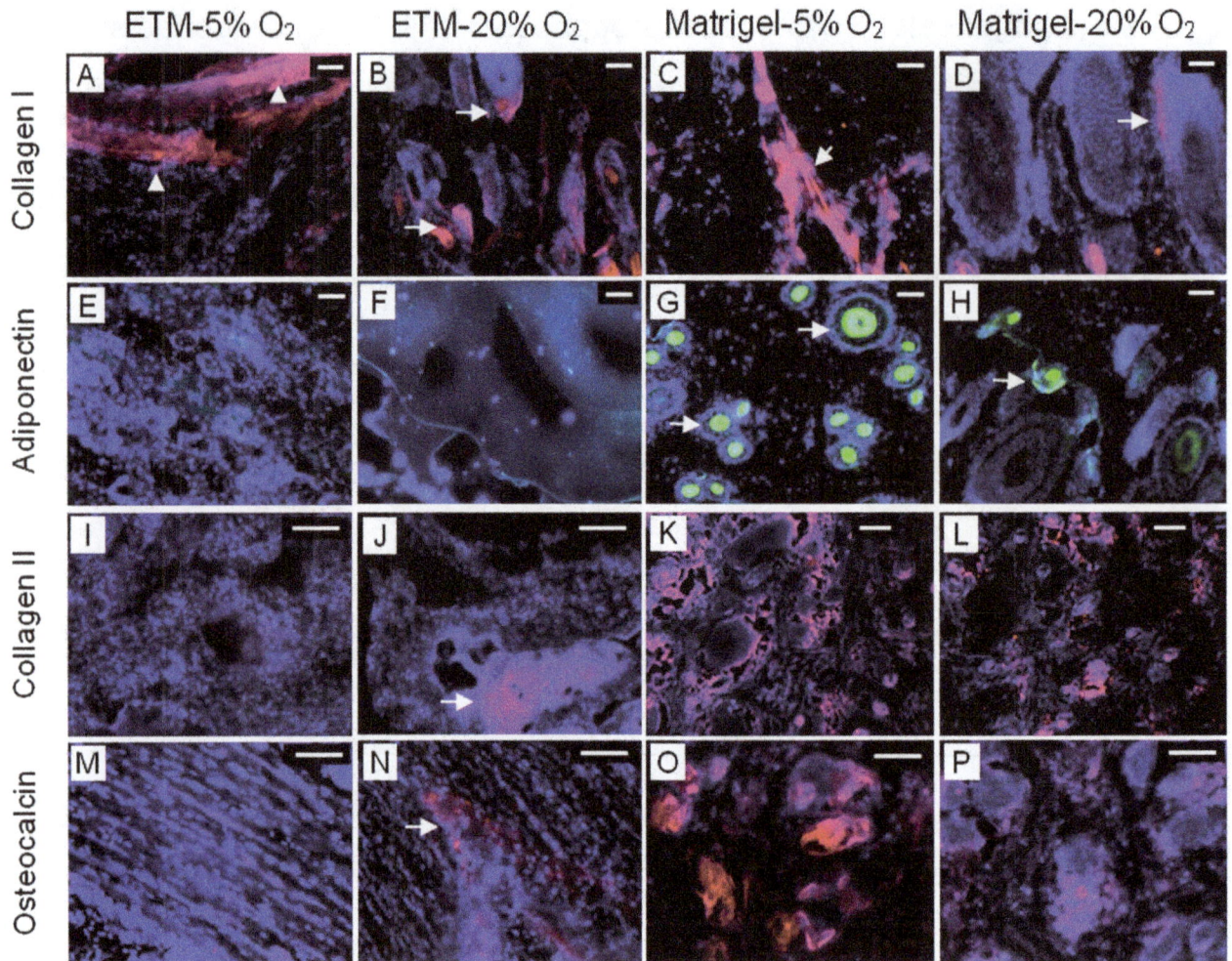

Figure 7. In vivo implantation results of hTSCs after culture in hypoxic and normaxic conditions. hTSCs grown under both hypoxic or normaxic conditions were implanted into nude rats. Implantation of the cells embedded in ETM (ETM-5% O_2) resulted in the formation of tendon-like structures (A, triangles) compared to only spotty areas stained with human collagen type I (B, arrows) when hTSCs treated with 20% O_2 and embedded in the same ETM (ETM-20% O_2) were implanted *in vivo*. In addition, implantation of ETM-5% O_2 hTSCs led to little formation of adiponectin (E), collagen type II (I), and osteocalcin (M); in contrast, implantation of ETM-20% O_2 hTSCs exhibited strong staining for collagen type II (J, arrow) and osteocalcin (N, arrow). When hTSCs were treated with 5% O_2 and embedded in Matrigel, implantation of the cell-Matrigel composites formed more extensive tendinous (C, D, arrows) and non-tendinous tissues (G, green; K, red; O, red) compared to hTSCs treated with 20% O_2 and embedded in Matrigel (H, green; L, red; P, red). Red represents collagen type I (A–D); Green represents adiponectin (E–H); Red represents collagen type II (I–L) and red represents osteocalcin (M–P). In all figures blue represents nuclei stained with Hoechst 33342. Scale bars: 100 μm.G.

possibility is supported by the previous finding that HIF regulates pluripotency and proliferation in hESCs cultured at hypoxic culture conditions [19] by activating the expression of Oct-4, which is known to control the self-renewal and multi-potency of stem cells [30].

In conclusion, using *in vitro* and *in vivo* experimental approaches, we have shown that culture condition using low oxygen level of 5% encourages self-renewal of hTSCs and, as a result, yields more abundant hTSCs than the conventionally used culture condition at 20% oxygen level. Higher number of hTSCs in 5% oxygen conditions will enable the use of these tendon specific stem cells for cell therapy of injured tendons. Future studies are required to investigate the combined effects of low-oxygen culture conditions with other TSC's niche factors, including tendon matrix and mechanical loading, on TSC functions.

Acknowledgments

We thank the Gift of Hope Organ Tissue Donor Network and Drs. Cs-Szabo and Margulis for making human tissues available, and we also extend our appreciation to the tissue donor families who made this study possible. We also thank Dr. Im for her assistance in obtaining human tissues.

Author Contributions

Conceived and designed the experiments: JHW. Performed the experiments: JZ. Contributed reagents/materials/analysis tools: JHW JZ. Wrote the paper: JHW JZ.

References

1. Butler DL, Juncosa-Melvin N, Boivin GP, Galloway MT, Shearn JT, et al. (2008) Functional tissue engineering for tendon repair: A multidisciplinary strategy using mesenchymal stem cells, bioscaffolds, and mechanical stimulation. J Orthop Res 26: 1–9.

2. Awad HA, Boivin GP, Dressler MR, Smith FN, Young RG, et al. (2003) Repair of patellar tendon injuries using a cell-collagen composite. J Orthop Res 21: 420–431.

3. Harris MT, Butler DL, Boivin GP, Florer JB, Schantz EJ, et al. (2004) Mesenchymal stem cells used for rabbit tendon repair can form ectopic bone and express alkaline phosphatase activity in constructs. J Orthop Res 22: 998–1003.

4. Cao Y, Liu Y, Liu W, Shan Q, Buonocore SD, et al. (2002) Bridging tendon defects using autologous tenocyte engineered tendon in a hen model. Plast Reconstr Surg 110: 1280–1289.

5. Bi Y, Ehirchiou D, Kilts TM, Inkson CA, Embree MC, et al. (2007) Identification of tendon stem/progenitor cells and the role of the extracellular matrix in their niche. Nat Med 13: 1219–1227.

6. Zhang J, Wang JH (2010) Characterization of differential properties of rabbit tendon stem cells and tenocytes. BMC Musculoskelet Disord 11: 10.

7. Rui YF, Lui PP, Li G, Fu SC, Lee YW, et al. (2010) Isolation and characterization of multipotent rat tendon-derived stem cells. Tissue Eng Part A 16: 1549–1558.

8. Zhang J, Li B, Wang JH (2011) The role of engineered tendon matrix in the stemness of tendon stem cells in vitro and the promotion of tendon-like tissue formation in vivo. Biomaterials 32: 6972–6981.

9. D'Ippolito G, Diabira S, Howard GA, Roos BA, Schiller PC (2006) Low oxygen tension inhibits osteogenic differentiation and enhances stemness of human MIAMI cells. Bone 39: 513–522.

10. Zhang J, Wang JH (2010) Production of PGE(2) increases in tendons subjected to repetitive mechanical loading and induces differentiation of tendon stem cells into non-tenocytes. J Orthop Res 28: 198–203.

11. Huangfu D, Osafune K, Maehr R, Guo W, Eijkelenboom A, et al. (2008) Induction of pluripotent stem cells from primary human fibroblasts with only Oct4 and Sox2. Nat Biotechnol 26: 1269–1275.

12. Liu H, Fan H, Wang Y, Toh SL, Goh JC (2008) The interaction between a combined knitted silk scaffold and microporous silk sponge with human mesenchymal stem cells for ligament tissue engineering. Biomaterials 29: 662–674.

13. Risbud MV, Guttapalli A, Tsai TT, Lee JY, Danielson KG, et al. (2007) Evidence for skeletal progenitor cells in the degenerate human intervertebral disc. Spine (Phila Pa 1976) 32: 2537–2544.

14. Ivanovic Z (2009) Hypoxia or in situ normoxia: The stem cell paradigm. J Cell Physiol 219: 271–275.

15. Lavrentieva A, Majore I, Kasper C, Hass R (2010) Effects of hypoxic culture conditions on umbilical cord-derived human mesenchymal stem cells. Cell Commun Signal 8: 18.

16. Grayson WL, Zhao F, Izadpanah R, Bunnell B, Ma T (2006) Effects of hypoxia on human mesenchymal stem cell expansion and plasticity in 3D constructs. J Cell Physiol 207: 331–339.

17. Grayson WL, Zhao F, Bunnell B, Ma T (2007) Hypoxia enhances proliferation and tissue formation of human mesenchymal stem cells. Biochem Biophys Res Commun 358: 948–953.

18. Lennon DP, Edmison JM, Caplan AI (2001) Cultivation of rat marrow-derived mesenchymal stem cells in reduced oxygen tension: effects on in vitro and in vivo osteochondrogenesis. J Cell Physiol 187: 345–355.

19. Forristal CE, Wright KL, Hanley NA, Oreffo RO, Houghton FD (2010) Hypoxia inducible factors regulate pluripotency and proliferation in human embryonic stem cells cultured at reduced oxygen tensions. Reproduction 139: 85–97.

20. Lee WY, Lui PP, Rui YF (2012) Hypoxia-mediated efficient expansion of human tendon-derived stem cells in vitro. Tissue Eng Part A 18: 484–498.

21. Tsai RY, McKay RD (2002) A nucleolar mechanism controlling cell proliferation in stem cells and cancer cells. Genes Dev 16: 2991–3003.

22. Pan G, Thomson JA (2007) Nanog and transcriptional networks in embryonic stem cell pluripotency. Cell Res 17: 42–49.

23. Greco SJ, Liu K, Rameshwar P (2007) Functional similarities among genes regulated by OCT4 in human mesenchymal and embryonic stem cells. Stem Cells 25: 3143–3154.

24. Pesce M, Scholer HR (2001) Oct-4: gatekeeper in the beginnings of mammalian development. Stem Cells 19: 271–278.

25. Gang EJ, Bosnakovski D, Figueiredo CA, Visser JW, Perlingeiro RC (2007) SSEA-4 identifies mesenchymal stem cells from bone marrow. Blood 109: 1743–1751.

26. Henderson JK, Draper JS, Baillie HS, Fishel S, Thomson JA, et al. (2002) Preimplantation human embryos and embryonic stem cells show comparable expression of stage-specific embryonic antigens. Stem Cells 20: 329–337.

27. Cui L, Johkura K, Yue F, Ogiwara N, Okouchi Y, et al. (2004) Spatial distribution and initial changes of SSEA-1 and other cell adhesion-related molecules on mouse embryonic stem cells before and during differentiation. J Histochem Cytochem 52: 1447–1457.

28. Grimshaw MJ, Mason RM (2000) Bovine articular chondrocyte function in vitro depends upon oxygen tension. Osteoarthritis Cartilage 8: 386–392.

29. Zhang J, Wang JH (2010) Mechanobiological response of tendon stem cells: implications of tendon homeostasis and pathogenesis of tendinopathy. J Orthop Res 28: 639–643.

30. Keith B, Simon MC (2007) Hypoxia-inducible factors, stem cells, and cancer. Cell 129: 465–472.

11

Both Canonical and Non-Canonical Wnt Signaling Independently Promote Stem Cell Growth in Mammospheres

Alexander M. Many, Anthony M. C. Brown*

Department of Cell & Developmental Biology, Weill Cornell Medical College, New York, New York, United States of America

Abstract

The characterization of mammary stem cells, and signals that regulate their behavior, is of central importance in understanding developmental changes in the mammary gland and possibly for targeting stem-like cells in breast cancer. The canonical Wnt/β-catenin pathway is a signaling mechanism associated with maintenance of self-renewing stem cells in many tissues, including mammary epithelium, and can be oncogenic when deregulated. Wnt1 and Wnt3a are examples of ligands that activate the canonical pathway. Other Wnt ligands, such as Wnt5a, typically signal via non-canonical, β-catenin-independent, pathways that in some cases can antagonize canonical signaling. Since the role of non-canonical Wnt signaling in stem cell regulation is not well characterized, we set out to investigate this using mammosphere formation assays that reflect and quantify stem cell properties. Ex vivo mammosphere cultures were established from both wild-type and *Wnt1* transgenic mice and were analyzed in response to manipulation of both canonical and non-canonical Wnt signaling. An increased level of mammosphere formation was observed in cultures derived from MMTV-*Wnt1* versus wild-type animals, and this was blocked by treatment with Dkk1, a selective inhibitor of canonical Wnt signaling. Consistent with this, we found that a single dose of recombinant Wnt3a was sufficient to increase mammosphere formation in wild-type cultures. Surprisingly, we found that Wnt5a also increased mammosphere formation in these assays. We confirmed that this was not caused by an increase in canonical Wnt/β-catenin signaling but was instead mediated by non-canonical Wnt signals requiring the receptor tyrosine kinase Ror2 and activity of the Jun N-terminal kinase, JNK. We conclude that both canonical and non-canonical Wnt signals have positive effects promoting stem cell activity in mammosphere assays and that they do so via independent signaling mechanisms.

Editor: Yi Li, Baylor College of Medicine, United States of America

Funding: This work was supported by New York Stem Cell Science (http://stemcell.ny.gov/)grant C024286 (AB and AM), National Institutes of Health (http://www.nih.gov/) grant R25 CA105012 (AM), Strang Cancer Prevention Institute, and National Institutes of Health grant R01 CA123238 (AB). The funders had no role in study design, data collection and analysis, decision to publish, or preparation of the manuscript.

Competing Interests: The authors have declared that no competing interests exist.

* Email: amcbrown@med.cornell.edu

Introduction

Stem cells of the adult mammary gland are predicted to have a capacity for self-renewal and to give rise to the two major epithelial cell lineages of mammary ducts: luminal and basal. Substantial progress has been made towards characterizing mouse mammary stem cell populations, both *in vivo* and *in vitro*, but much remains to be determined about the signaling pathways that regulate their behavior. Elucidating the relevant mechanisms is important for understanding normal stem cell and tissue biology, and also because of the potential for developing therapies that can target stem-like cells in cancer.

Evidence that adult mammary tissue contains multipotent self-renewing stem cells was first provided by classical transplantation studies in which a normal epithelial ductal tree, comprising both basal and luminal cell lineages, could be regenerated from small tissue fragments or individual cells [1,2]. Such assays were subsequently used prospectively to identify several combinations of surface markers that enrich for cells with mammary repopu-

lating activity, and indicated that stem cells were distributed within the basal epithelial layer [3,4]. More recently, however, *in vivo* lineage tracing experiments have challenged some of these conclusions [5,6], suggesting that much of the post-natal development of mammary epithelium is dependent on separate luminal and basal progenitors acting in combination with a smaller population of bipotent stem cells [5–7].

Ex vivo assays of mammary epithelial cell sphere formation in suspension culture, mammospheres, offer a complementary approach to stem cell studies that is amenable to signaling pathway analysis. Originally developed for analysis of neuronal precursors, the ability of cells to form spheroids has been used as a stem cell assay for several other tissue types, including prostate and mesenchymal stem cells [8–11]. Mammosphere-forming cell cultures exhibit stem cell properties in their capacity to self-renew and ability to differentiate into committed luminal and basal lineages [12]. In addition, the ability to form mammospheres correlates with the potential to generate epithelial ductal trees in mammary reconstitution assays [13,14]. Thus, mammosphere formation has been used as an indicator of cells with stem cell properties in mouse and human mammary cell lines as well as in primary tissue culture [13–19].

The canonical Wnt/β-catenin signaling pathway is one of the principal signaling mechanisms associated with regulation of stem cell behavior in numerous tissues [20–23]. Canonical Wnt signaling also has well established roles in regulating embryonic development and adult tissue homeostasis, where many of its functions may result from effects on stem or progenitor cells [20–22,24,25]. Similarly, the Wnt/β-catenin pathway is frequently activated in a wide range of human cancers and may regulate neoplasia in part via modulation of cancer cells with stem-like cell properties [22,23,26].

The MMTV-*Wnt1* mouse strain is a well characterized model for the studying the consequences of Wnt signaling in the mammary gland and its effects on stem cells [27–30]. The mouse mammary tumor virus (MMTV) promoter drives expression of the *Wnt1* transgene predominantly in luminal epithelium, and results in activation of canonical Wnt/β-catenin signaling in the basal layer [31–33]. MMTV-*Wnt1* mice display widespread mammary epithelial hyperplasia and are predisposed to carcinomas with nearly 100% penetrance [34]. Notably, the pre-cancerous hyperplastic tissue of MMTV-*Wnt1* was reported to contain larger numbers of $CD24^+ CD29^{HI}$ cells, which are enriched for stem cell activity, in comparison to wild-type glands [3]. In a matrigel-based colony assay, wild-type mouse mammary cells selected for the $CD24^+ CD29^{HI}$ immunophenotype showed increased colony formation in response to purified Wnt3a [35]. Moreover, lineage tracing experiments using Cre-mediated recombination to mark the descendants of Wnt/β-catenin responsive cells expressing Axin2 suggest that such cells contribute to a stem cell population [36]. Collectively these data support a role for Wnt/β-catenin signaling in the growth and/or maintenance of mammary stem cells.

Intracellular signaling elicited by members of the Wnt family of secreted ligands can be broadly classified into two modes: canonical and non-canonical. In the canonical pathway, Wnt ligand binding to receptor complexes containing Frizzled and Lrp5/6 proteins results in stabilization of cytoplasmic β-catenin and transcriptional activation mediated by β-catenin/TCF complexes [22,37]. Wnt1 and Wnt3a are prototypical examples of ligands that consistently activate this pathway [38]. In contrast, non-canonical Wnt signaling is defined as a signaling response to Wnt ligands that is independent of β-catenin stabilization [39,40]. Wnt5a exemplifies a Wnt protein that typically signals in a non-canonical manner [38,41]. Several non-canonical signaling pathways have been proposed and the cognate receptors include Frizzled proteins, and receptor tyrosine kinases such as Ror2, while Lrp5/6 are not required for non-canonical signaling [41–44].

While there is strong support for Wnt/β-catenin signaling promoting mammary stem cell properties, as described above, the roles of non-canonical signaling in the mammary gland are much less clear. Moreover, apparently conflicting data exists concerning functional interactions between canonical and non-canonical Wnt signaling in other experimental systems. Thus Wnt5a has been reported to act either act in opposition to, in concert with, or independently of Wnt/β-catenin signaling [45]. In the mouse mammary gland, Wnt5a overexpression has been shown to inhibit ductal extension during development, and to reduce the growth rate of certain tumors [46,47]. However, due to the multiplicity of non-canonical signaling pathways that Wnt5a may activate depending on the context, it is essential to test its functional consequences empirically in individual assays.

To elucidate the effects of canonical and non-canonical Wnt signaling on stem cell properties of mouse mammary epithelium, here we test the consequence of altered Wnt signaling activity on mammosphere cultures, specifically quantifying the number of secondary mammosphere-forming units (MFUs)[12,13,48]. To include all potential stem cells, including those that may not express the cell surface markers previously used for enrichment, we used unsorted populations of mammary epithelial cells [3,4]. Contrasting the ability of Wnt5a to antagonize canonical Wnt signaling in other systems, we observed that both Wnt3a and Wnt5a promoted mammosphere formation through distinct signaling pathways. Thus both canonical and non-canonical Wnt signaling have independent abilities to promote stem cell capacity.

Methods

Cell culture

Mammospheres culture methods were adapted from Dontu *et al.* [12]. Number 3, 4, 8, and 9 mammary glands were resected from adult mice between 3 and 9 months of age, from a FVB/NJ background. Glands were mechanically minced with a razor blade and digested at 37°C with collagenase (≥250 units/ml) in DMEM/F12 medium for 3 hours with vortexing and pipetting every 30 minutes. The digested tissue was centrifuged at 650×g for 5 minutes. The floating fat layer was removed by aspiration, and the tissue homogenate was digested in 1 mg/ml dispase in DMEM/F12 media with constant pipetting for 3 minutes to generate a single cell suspension. This was washed twice in "mammosphere medium" (DMEM/F12, 20 ng/ml bFGF, 20 ng/ml EGF, 4 µg/ml Heparin, B-27 Supplement, and 1% Penicillin/Streptomycin antibiotic solution) [12] and resuspended in that medium. Remaining clumps of cells were removed by filtration through a 40 µM cell strainer. The cell suspension was enriched for epithelial cells using an Easy-Sep mouse mammary epithelial cell enrichment kit (Stem Cell Technologies) according to manufacturer's instructions. Cells were resuspended in mammosphere medium and plated into 96 well low adherence plates (Corning Costar). Primary mammosphere cultures were fed once by addition of fresh mammosphere medium and the pooled cultures harvested by mild centrifugation and resuspension for assays of secondary sphere formation as described below.

Secondary mammosphere assay

Primary mammosphere cultures were disassociated with trypsin for 30 minutes in the presence of a vital cell labeling dye, (Di-I, or Cell Tracker Red, Invitrogen). Cell suspensions were filtered through a 40 µM cell strainer and then diluted in mammosphere medium with 1% methylcellulose to limit cell aggregation, and 10 nM dexamethasone to maintain transcription from the MMTV promoter. Cells were plated in 96-well low adherence plates using 24 wells per treatment, and were scored for mammosphere formation after one week of growth. In all secondary mammosphere assays, cells were plated at 1000 cells per well except in lentiviral shRNA knockdown experiments, in which they were plated at 2000 cells per well. All exogenous treatments were given as a single dose to the dissociated cells at the time of secondary assay plating. Mammospheres were defined as colonies in methylcellulose suspension culture that contained 10 or more cells visible under phase contrast, of which fewer than 50% still retained the cell tracking dye, consistent with the dye being diluted out upon successive cell divisions.

Immunostaining

Secondary mammosphere cultures were harvested and washed with PBS. Mammospheres were fixed and permeabilized with a 1:1 acetone:methanol fixation solution for 30 minutes at $-20°C$, and PBS was used to rehydrate the mammospheres for 10 minutes at room temperature. Non-specific antibody binding was blocked with a 3% BSA/PBS solution for 1 hour at room temperature. Primary antibodies to cytokeratin 8 (K8; Troma I antibody, developmental studies hybridoma bank, University of Iowa), and cytokeratin 14 (K14; Abcam, catalog #7800), both at a 1:200 dilution in 3% BSA/PBS, were applied overnight at 4°C. Mammospheres were then washed twice in PBS. Cells were incubated in a 1:1000 dilution of Alexafluor-conjugated secondary antibodies (Life technologies) in 3% BSA/PBS for 4 hours. Cells were washed twice with PBS and mounted in Vectashield mounting media (Vector laboratories) including 0.1 µg/ml DAPI.

Mammosphere assay statistical analysis

Numbers of mammospheres reported are the mean of 24 wells per treatment from representative experiments. All experiments were repeated 3 or more times, demonstrating consistent statistical relationship patterns. Comparisons between treatments were made using Student's t test with $p<.05$ required for significance.

Mammosphere samples for qRT-PCR

Mammospheres were prepared as for secondary mammosphere assays, except cells were plated in 6-well low adherence plates. After one week cells were harvested by centrifugation and total RNA was extracted using RNeasy mini kits (Qiagen). cDNA was produced using the iScript cDNA synthesis kit (Bio-Rad). Quantitech primers (Qiagen) were used with SYBR-Green mastermix (Quanta) for quantitative PCR using an MJ Opticon2 system (BioRad) according to manufacturer's instructions.

Viral infection

To infect mammosphere cultures with lentivirus, primary mammospheres were disassociated as for secondary mammosphere assays. Before plating, single cell suspensions in DMEM/F12 media were mixed with lentiviral particles in PBS at a multiplicity of infection of greater than 5:1, plus a 1:200 dilution of Transdux reagent (System Biosciences) for 30 minutes at 37°C. After infection, the cell suspension was plated as for secondary sphere assays. The 7TGC lentiviral reporter vector was obtained from Addgene (Plasmid#24304) [49]. Lentiviral vectors for knockdown of *Ror2* mRNA were obtained from Dr. T. Stappenbeck [50]. For clarity we renamed the knockdown vector shRor2#7 [50] as shRor2 and the control vector SCH002-EGFP as shControl.

Recombinant proteins and small molecule inhibitors

In experiments using inhibitors and recombinant proteins in conjunction with secondary mammosphere assays, agents were added to single cell suspensions before plating. Recombinant Wnt3a (Peprotech), and Recombinant Wnt5a (R&D systems) were used at 200 ng/ml, except when noted. Recombinant Dkk1 (R&D systems) was used at 200 ng/ml. JNK inhibitor SP600125 (Calbiochem, CAS# 129-56-6) was used at 10 µM, and iCRT3 (Calbiochem, pubchem # 126531502) [51] was used at 25 µM.

Ethics Statement

This study was carried out in strict accordance with the recommendations in the Guide for the Care and Use of Laboratory Animals of the National Institutes of Health. The protocol was approved by the Institutional Animal Care and Use Committees of Weill Cornell Medical College (Protocol Number: 0052–11), and the New York Blood Center (Protocol Number: 267).

Results

MMTV-*Wnt1* transgenic mammospheres exhibit similar properties to those from wild-type mice

To investigate the effects of Wnt-induced signaling in mouse mammary stem cells, we employed *ex vivo* mammosphere cultures derived from primary mouse mammary epithelium. Dissociated epithelial cells were obtained from wild-type mice and from MMTV-*Wnt1* transgenic animals. Such mammosphere cultures provide an assay system for stem cell-initiated sphere growth, independent of previously identified stem cell enrichment markers [12,13]. Single cell suspensions being assayed for mammosphere formation were labeled with the lipid-soluble vital dye Di-I in order to track cell division. The majority of cells in all resultant secondary mammospheres exhibited low, or undetectable, fluorescence, the tracking dye having been diluted through multiple cell divisions (Figure 1A–D). In contrast, we observed bright fluorescence in one to two cells per mammosphere, suggesting that individual mammosphere forming cells can divide asymmetrically so as to retain the cell tracking dye in one daughter cell, while the majority of cells within each sphere are derived by serial proliferation. The lineage-specific markers Cytokeratin 8 (K8) and Cytokeratin 14 (K14) were used to identify luminal and basal mammary cells, respectively, in wild-type and MMTV-*Wnt1* derived mammospheres [12,52]. For both genotypes, all mammospheres contained a mixture of cells expressing both K8 and K14, cells expressing K14 alone, and marker-negative cells. Most mammospheres also contained cells that expressed K8 alone (Figure 1E–N). This indicates that MMTV-*Wnt1* and wild-type mammospheres have similar capacity to produce progeny cells expressing differentiation markers during mammosphere growth *in vitro*. In addition, wild-type and MMTV-*Wnt1* mammospheres exhibited similar morphology in both shape and size (Figure 1). To confirm the expected self-renewal capacity of cells with mammosphere forming ability in these cultures, the number of mammosphere-forming units (MFUs) was measured at sequential passages (Figure S1). The continued capacity to form mammospheres was similar to that observed by others using comparable culture systems, indicating that the cultures contained cells with stem cell properties of differentiation and self-renewal [12,53,54]. In subsequent experiments we quantified the number of MFUs in secondary mammosphere assays, reflecting the number of cells with mammary stem cell properties [13,55].

Canonical Wnt Signaling promotes mammosphere formation

In a variety of other stem cell assays systems, Wnt/β-catenin signaling has been associated with stem cell self-renewal or expansion [22,56,57]. Moreover, the lobuloalveolar mammary hyperplasia characteristic of MMTV-*Wnt1* transgenic mice has been reported to contain an increased absolute number of mammary stem cells, and an increased proportion of stem cells as defined by CD24[+] CD29[HI] surface markers [3]. To examine the consequences of canonical Wnt signaling in mammosphere assays, we measured the numbers of secondary MFUs in wild-type and MMTV-*Wnt1* cultures and observed a significantly larger number of MFUs per 1000 cells in MMTV-*Wnt1* cultures compared to wild type (Figure 2A). This indicates a greater

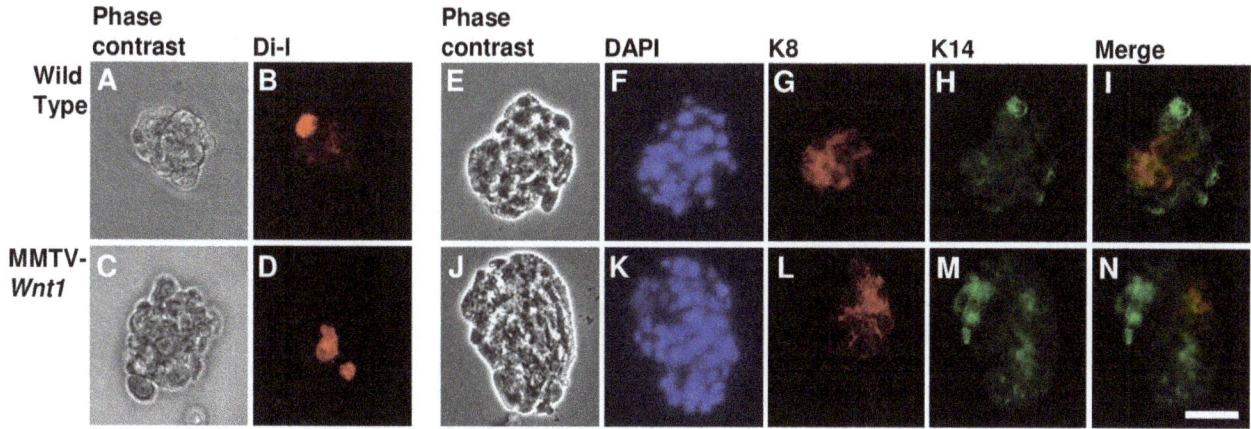

Figure 1. Wild-type and MMTV-*Wnt1* mammospheres contain newly replicated cells and have similar differentiation capacity. Phase contrast (A,C) and Di-I fluorescence images (B,D) of wild-type (A,B) and MMTV-*Wnt1* (C,D) mammospheres. Wild-type (E–I) and MMTV-*Wnt1* (J–N) mammospheres stained with DAPI (F, K), anti-cytokeratin K8 (luminal marker; G, L), anti-cytokeratin K14 (basal marker; H, M), and the merged image of K8 and K14 (I, N). Scale bar =50 μm.

percentage of cells with capacity for stem cell behavior in MMTV-*Wnt1* cultures compared to wild type.

The elevated number of MFUs in MMTV-*Wnt1* ex vivo cultures might depend on a continuous Wnt1 signal, or Wnt1 signaling *in vivo* at an earlier stage might induce a change in cell fate such that mammosphere-forming capacity is permanently altered, irrespective of continued Wnt signaling. To distinguish between these possibilities, we treated dissociated primary mammosphere cells with Dkk1, a specific antagonist of Wnt binding to Lrp5/6, in order to block Wnt1 signaling [58–60]. The addition of Dkk1 to MMTV-*Wnt1* cultures reduced their sphere forming capacity to wild-type levels (Figure 2B). The effect of Wnt1 expression was thus reversible upon blockade of receptors for the canonical Wnt pathway, implying that a continued elevated level of Wnt/β-

catenin signaling is required for the increase in MFU numbers observed in MMTV-*Wnt1* cultures.

We next tested whether acute stimulation of canonical Wnt signaling could substitute for the long term elevated canonical Wnt signaling resulting from the MMTV-*Wnt1* transgene. Disassociated mammospheres were therefore treated with a single dose of recombinant Wnt3a, a Wnt protein that is functionally interchangeable with Wnt1 in canonical signaling assays [35,61]. A single application of Wnt protein was sufficient to induce a two-fold increase in MFUs in wild-type cultures, as assayed by secondary sphere formation (Figure 2C). We also found that addition of Wnt3a to MMTV-*Wnt1* cultures further increased the MFU numbers above the levels observed in MMTV-*Wnt1* cultures or wild-type cultures treated with Wnt3a. As expected, the elevation of MFU numbers induced by Wnt3a was impaired by

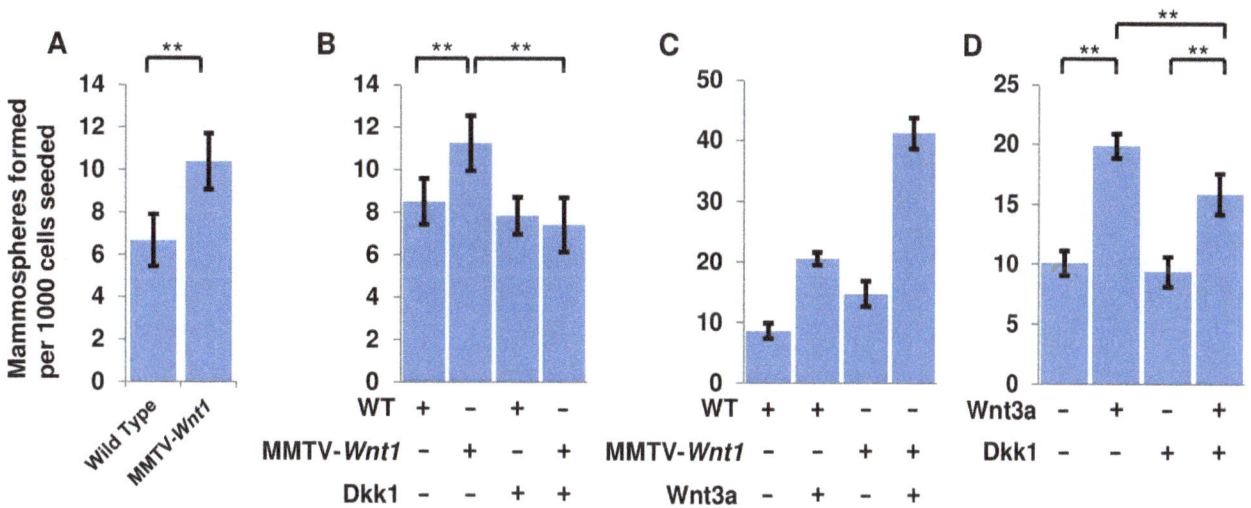

Figure 2. Canonical Wnt signaling increases the number of Mammosphere Forming Units. Secondary mammospheres were counted one week after plating dissociated cells treated as indicated. (A) Increased mammosphere formation in cultures derived from MMTV-*Wnt1* tissue versus wild-type. (B) Mammosphere numbers from wild-type and MMTV-*Wnt1* cultures with or without treatment with 200 ng/ml Dkk1. (C) Mammospheres from wild-type and MMTV-*Wnt1* cultures with or without treatment with 200 ng/ml Wnt3a. (D) Wild-type secondary mammospheres with or without recombinant Wnt3a, Dkk1, or both. **p<.01. All values in panel C were significantly different from one another at p<.01. Error bars show 95% confidence intervals.

pre-treatment of cells with Dkk1 (Figure 2D). Thus, short term stimulation with Wnt3a phenocopies the effects of an MMTV-*Wnt1* transgene in mammosphere cultures.

Wnt5a-mediated signaling promotes mammosphere formation

In several cell systems, non-canonical signaling induced by Wnt5a can result in antagonism of canonical Wnt signaling [62–66]. Moreover, *in vivo* studies of Wnt5a signaling in the mammary gland suggest an antagonistic effect on ductal development [46,47,67,68]. We therefore used recombinant Wnt5a to test the effects of non-canonical Wnt signaling on sphere formation, anticipating a negative effect. Surprisingly, we observed a dramatic increase in MFU number in response to Wnt5a treatment (Figure 3A). Moreover, when added to cultures from MMTV-*Wnt1* mice, recombinant Wnt5a further increased the number of MFUs beyond the elevated level induced by the MMTV-*Wnt1* transgene (Figure 3A). Similarly, when wild-type cultures were treated with recombinant Wnt3a and Wnt5a in combination, we observed an additive elevation of MFU numbers from the two ligands although each was applied at half the concentration used for each Wnt separately (Figure 3B). These results suggest that Wnt3a and Wnt5a independently promote mammosphere formation.

Wnt5a-induced mammosphere formation results from non-canonical Wnt signaling

Signaling induced by Wnt5a typically acts via non-canonical pathways [43,44]. However, under unusual circumstances, such as overexpression of Frizzled4 or 5, a canonical Wnt/β-catenin signal can be induced by Wnt5a [62,69]. To test whether the MFU promoting effect of Wnt5a might be due to such signaling we blocked the canonical-specific Wnt receptors Lrp5 and Lrp6 by

pretreatment of disassociated mammosphere cells with Dkk1 prior to stimulation with Wnt5a. This had no inhibitory effect on the response to Wnt5a, suggesting that Wnt5a does not act via Lrp5/6 in this assay (Figure 3C). To ensure that Wnt5a did not stimulate β-catenin/TCF activity independently of Lrp5/6, we infected secondary mammosphere cultures with a lentivirus (7TGC) that constitutively expresses mCherry, and expresses Green Fluorescent Protein (GFP) only in response to β-catenin-mediated transcriptional activation [49]. These cultures were then challenged with either recombinant Wnt3a or Wnt5a and mCherry-positive mammospheres were examined for GFP expression. While Wnt3a strongly induced the GFP reporter, no such induction was observed in Wnt5a-treated mammospheres or untreated controls (Figure 4). These results demonstrate that, unlike Wnt3a, Wnt5a failed to activate canonical Wnt/β-catenin signaling in mammospheres. We therefore conclude that Wnt5a mediates its effect through a non-canonical signaling pathway.

Wnt5a stimulation of mammospheres requires *Ror2*

One of the Wnt receptors specifically associated with non-canonical signaling is the tyrosine kinase Ror2, which has been shown to bind Wnt5a and to form a ternary complex with Frizzled proteins [42,43,70,71]. To address whether the mammosphere-promoting function of Wnt5a is dependent on *Ror2*, we used a lentiviral *Ror2* shRNA construct (shRor2) to suppress *Ror2* expression, and validated its ability to knock down *Ror2* mRNA in mammospheres by qRT-PCR (Figure 5A) [50]. Secondary mammosphere cultures infected with shRor2 or a non-specific shRNA control vector were then treated with either Wnt5a or Wnt3a. In cultures infected with the control vector, Wnt3a and Wnt5a both increased the number of MFUs as in previous experiments. Infection with shRor2 reduced mammosphere formation compared to shControl. However, while cells infected with shRor2 responded to Wnt3a with increased numbers of

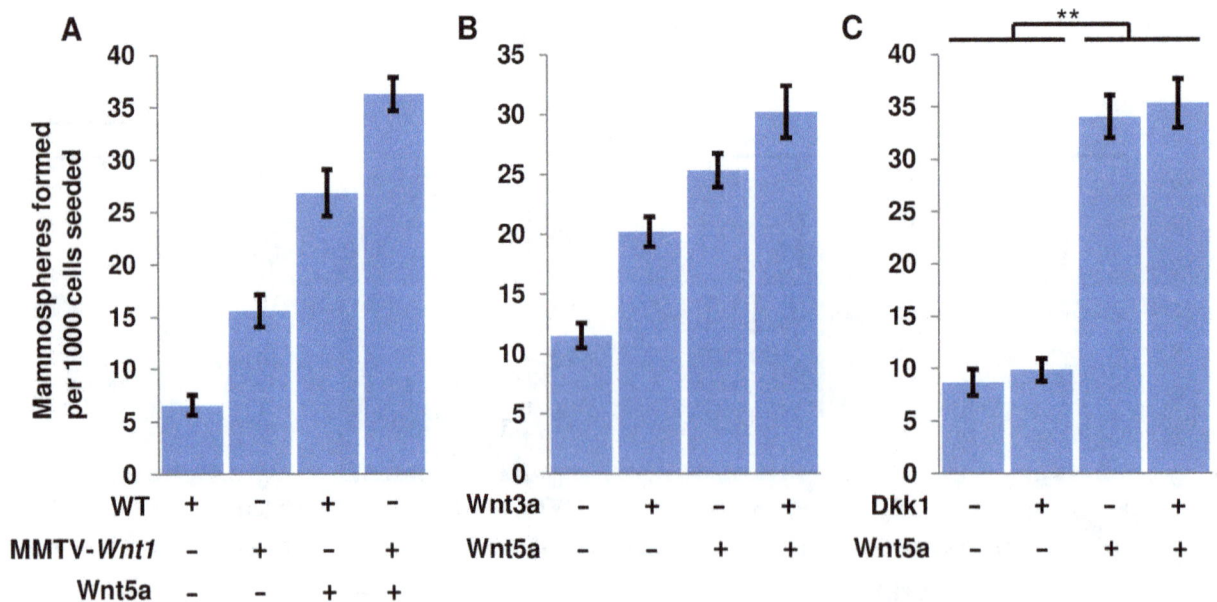

Figure 3. Wnt5a increases MFU number independently of canonical Wnt signaling. Dissociated cells were treated as indicated and secondary mammospheres were counted one week after plating. (A) Numbers of wild-type and MMTV-*Wnt1* secondary mammospheres after treatment with or without 200 ng/ml Wnt5a. (B) Wild-Type mammospheres treated with Wnt3a, Wnt5a or in combination. Wnt3a or Wnt5a were used individually at 400 ng/ml while in combination each was applied at 200 ng/ml. (C) Wild-type secondary mammosphere numbers after treatment with recombinant Wnt5a, Dkk1, or both (200 ng/ml each). In panels A and B all values were significantly different from one another at p<.01; in C, treatments with and without Wnt5a were significantly different at p<.01. Error bars show 95% confidence intervals.

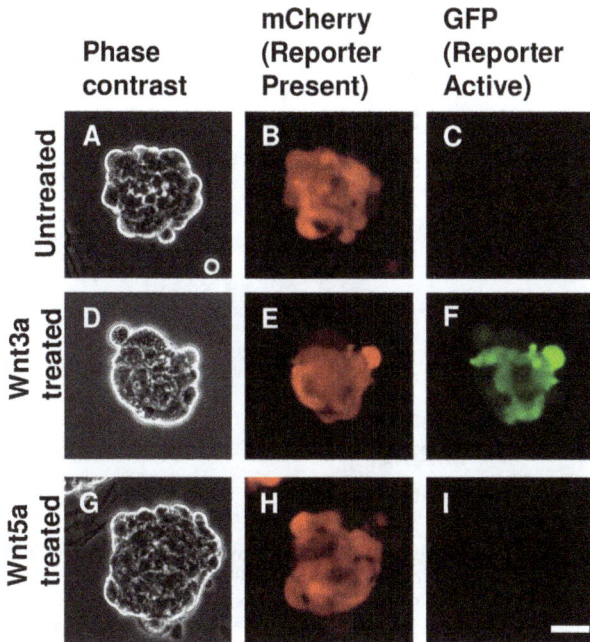

Figure 4. Wnt5a does not induce canonical Wnt signaling in mammospheres. Wild-type mammospheres were infected with the lentiviral reporter 7TGC and untreated (A–C), treated with 200 ng/ml Wnt3a (D–F), or treated with 200 ng/ml Wnt5a (G–I). Representative mammospheres imaged by phase contrast (A, D, and G), mCherry fluorescence (B, E, and H), and GFP fluorescence (C, F, and I). Constitutive mCherry expression indicates presence of the lentiviral reporter, while GFP fluorescence indicates activation of its β-catenin-TCF/LEF responsive promoter. Scale bar = 50 μm. Numbers of spheres imaged: untreated, n = 86; Wnt3a treated, n = 79; Wnt5a treated, n = 103.

MFUs, they failed to respond to Wnt5a in that no significant increase in MFUs was observed (Figure 5B, C). This indicates a requirement for *Ror2* in sphere formation mediated by Wnt5a but not by Wnt3a. It also suggests a basal function for *Ror2* in sphere formation in the absence of exogenous Wnts.

Wnt5a signaling in mammospheres is dependent on JNK but not β-catenin/TCF

Activation of JNK has been reported as an intracellular effector of non-canonical Wnt signaling, and has been specifically implicated downstream of Wnt5a and Ror2 [42,72]. We therefore used a small molecule pan-JNK inhibitor, SP600125, to test whether activation of JNK is required for the effect of Wnt5a in mammosphere cultures [73,74]. Addition of Wnt5a in the presence of JNK inhibitor failed to induce any increase in MFUs, while Wnt3a was still able to induce a two fold increase in MFU numbers in the presence of inhibitor (Figure 6A, B). In a complementary experiment we used iCRT3, a small molecule inhibitor of canonical Wnt signaling which specifically blocks the interaction between β-catenin and TCF/LEF proteins [51]. In secondary mammosphere assays incubated with iCRT3, Wnt3a failed to promote mammosphere formation, while Wnt5a increased the number of MFUs by two fold (Figure 6C, D). Both iCRT3 and JNK inhibitor treatment inhibited mammosphere formation irrespective of exogenous ligands. These results demonstrate a requirement for JNK activity in mediating the effects of Wnt5a, but not Wnt3a, on mammosphere formation. Conversely, they suggest that the increase in MFU numbers

mediated by Wnt3a requires the interaction of β-catenin and TCF/LEF, while Wnt5a acts independently of β-catenin-mediated transcription. Additionally mammosphere formation may be dependent upon a basal level of both canonical and non-canonical Wnt signaling. Together, these results indicate that Wnt5a and Wnt3a promote mammosphere formation through distinct signaling mechanisms.

Discussion

In this study we have examined the effects of Wnt signals on secondary mammosphere formation, an *in vitro* assay that reflects mammary stem cell activity. Using *ex vivo* cultures from both wild-type and MMTV-*Wnt1* transgenic mice, we observed that canonical Wnt/β-catenin signaling stimulates mammosphere formation by primary mouse mammary epithelial cells. This is evident from comparing mammosphere formation by MMTV-*Wnt1* and wild-type *ex vivo* cultures, and from the effect of recombinant Wnt3a on wild-type cells. We also observed that treatment with Wnt5a caused a similar increase in mammosphere formation, although it did not stimulate the Wnt/β-catenin pathway. Instead, the effects of Wnt5a were mediated via a non-canonical Wnt signaling pathway acting via the receptor Ror2 and dependent on activity of the kinase JNK. Our results indicate that both canonical and non-canonical Wnt signals act independently to promote the stem cell properties required for mammosphere formation.

Previous studies of the effects of canonical Wnt signaling on mouse mammary epithelial stem cells, both *in vivo* and *in vitro*, have focused on the sub-population of cells with the immunophenotype $CD24^+ CD29^{hi}$, which are enriched for cells capable of mammary gland repopulation [3,35,75]. Wnt3a promotes the self-renewal of such cells *in vitro* and their abundance *in vivo* is elevated in MMTV-*Wnt1* transgenic mice [3,35]. The present studies constitute a complementary approach in focusing on the mammosphere-forming capacity of mammary stem cells independently of specific cell surface markers. We found that a single dose of Wnt3a protein applied to dissociated cells was sufficient to increase secondary mammosphere formation. Consistent with this, and with the results of Shackleton *et al.* [3], mammary epithelium from MMTV-*Wnt1* mice displayed a greater number of MFUs *in vitro* than equivalent cultures from wild-type mice. This effect was blocked by addition of Dkk1 to the *Wnt1* transgenic cultures. This indicates that elevated canonical Wnt signaling is actively required during *ex vivo* culture in order to produce increased numbers of MFUs, rather than it arising from a permanent change in cell fate mediated by the *Wnt1* transgene during early mammary development. The immediate mammary phenotype of MMTV-*Wnt1* mice is precocious and permanent lobuloalveolar hyperplasia, which imparts significant risk of progression to carcinoma [34]. If the hyperplasia is a consequence of increased numbers of stem-like cells, our data suggest that this would be reversible upon blockade of canonical Wnt signaling. Moreover a continuing effect of canonical Wnt signaling acting on tumor stem cells might account for the suppression of tumor growth in MMTV-*Wnt1* transgenic mice when the Wnt1 signal is antagonized after progression to carcinoma [76–78].

While elevated levels of canonical Wnt/β-catenin signaling are clearly associated with promoting hyperplasia and tumorigenesis in the mammary gland, both in mice and humans [20,23,25], the effects of non-canonical Wnt signaling, e.g. as elicited by Wnt5a, have generally been linked to inhibitory effects. In human breast cancer, for example, loss of Wnt5a expression correlates with poor prognosis, suggesting that the gene acts as a tumor suppressor

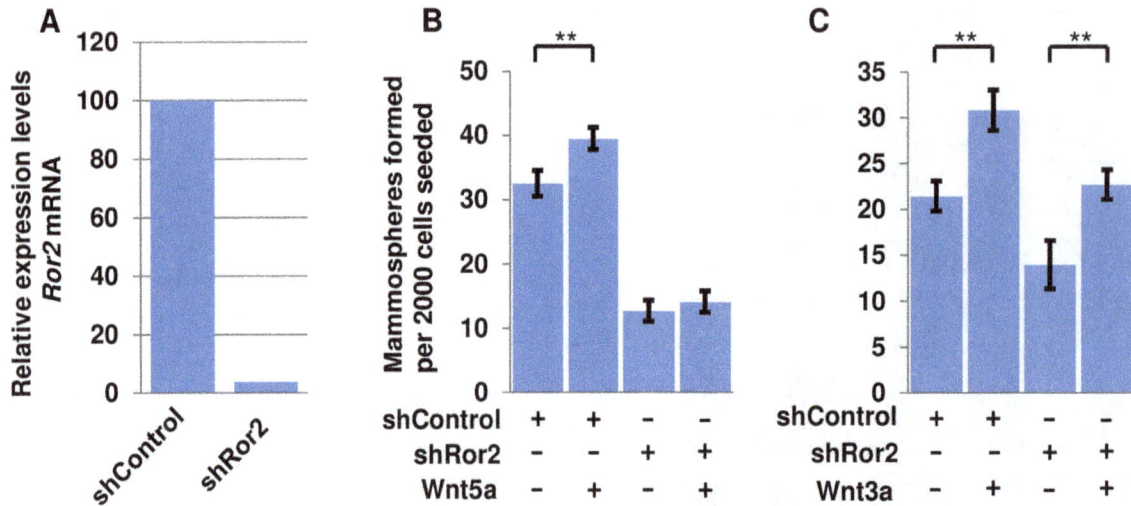

Figure 5. Knockdown of Ror2 by shRNA inhibits the increase in MFU numbers mediated by Wnt5a but not by Wnt3a. (A) Expression of *Ror2* mRNA measured by qPCR in mammospheres infected with control vector shControl or knockdown vector shRor2. (B) Wild-type mammosphere cells were infected with shRor2 or shControl lentiviral vectors and treated with or without 200 ng/ml Wnt5a. Secondary mammospheres were counted one week after plating. (C) Wild-type mammospheres infected with shRor2 or shControl lentiviral vector were treated with and without 200 ng/ml. **p<.01. Error bars show 95% confidence intervals.

[79,80]. Moreover, Wnt5a protein can inhibit ductal proliferation in the mouse mammary gland, loss of *Wnt5a* confers a more aggressive mammary tumor phenotype, and there is evidence that Wnt5a can antagonize the intracellular pathway of canonical Wnt signaling in several settings [25,46,47,62,65]. Given this background, we were surprised to observe a significant increase in MFU numbers upon treatment with Wnt5a in mammosphere assays, an effect comparable to that of Wnt1 or Wnt3a which each stimulate the canonical Wnt pathway [38,62]. Nevertheless our results are consistent with those of Scheel *et al.* (2011), who observed that treatment of human breast epithelial cells with Wnt5a, in conjunction with activation of Wnt/β-catenin and TGFβ pathways, enhanced the efficiency of mammosphere

formation as well inducing the expression of EMT markers [81]. In addition, it has been shown that Wnt5a promotes the self-renewal of spermatogonial stem cells *in vitro*, suggesting that its positive effect on mammospheres is not unique to breast tissues [82]. Given the numerous distinct signaling pathways and receptors through which Wnt5a may act [39,40,83], the positive effects on mammosphere formation observed in our study can perhaps be reconciled with the growth suppressive effects of Wnt5a signaling in mammary tissue [46,47] by invoking distinct signaling responses to Wnt5a in stem cells versus committed progenitors.

Although Wnt5a typically signals via β-catenin-independent mechanisms, there are special circumstances in which it has been

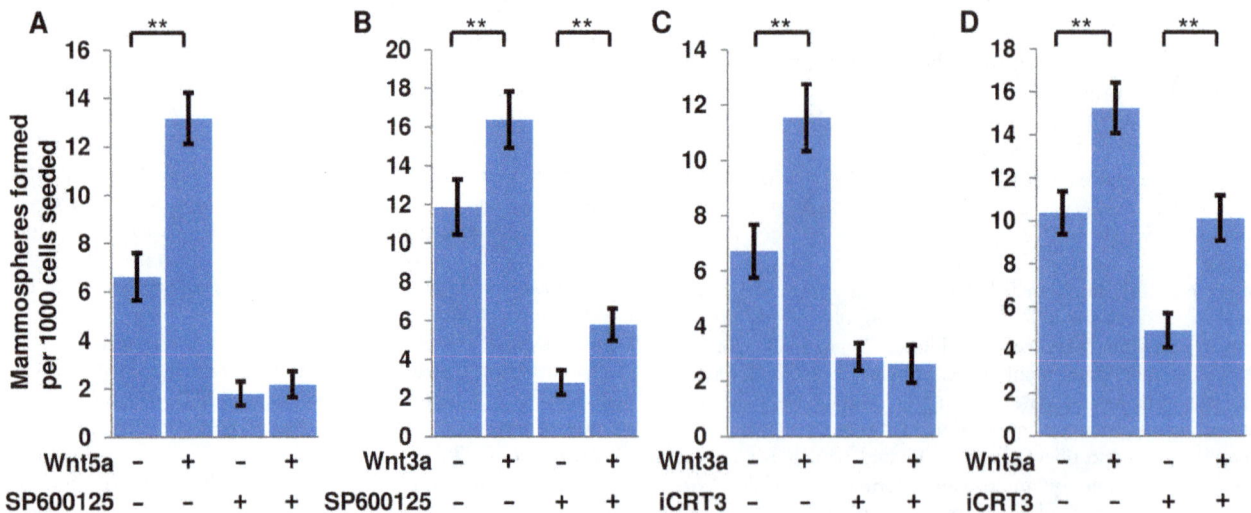

Figure 6. The Wnt5a-induced increase in MFU numbers is mediated via JNK, and not β-catenin/TCF. Secondary mammospheres were counted one week after plating wild-type cells with treatments indicated: (A) Wnt5a at 200 ng/ml, JNK inhibitor SP600125 at 10 μM, or both in combination; (B) Wnt3a at 200 ng/ml, SP600125 at 10 μM, or both in combination; (C) Wnt3a at 200 ng/ml, the β-catenin/TCF antagonist iCRT3 at 25 μM, or both in combination; (D) Wnt5a at 200 ng/ml, iCRT3 at 25 μM, or both in combination. **p<.01. Error bars show 95% confidence intervals.

found to activate the canonical Wnt/β-catenin pathway [62,69,84]. This was not the case in our experiments, however, since Wnt5a treatment of mammospheres failed to activate a transcriptional reporter of Wnt/β-catenin signaling, and the positive effect of Wnt5a was not blocked by either the Lrp5/6 antagonist Dkk1 or the β-Catenin/TCF inhibitor iCRT3. Instead, we provide evidence that stimulation of mammosphere formation by Wnt5a depends on *Ror2*, a receptor tyrosine kinase that binds Wnt5a and transduces a non-canonical Wnt signal that includes activation of JNK [42,62,72,83]. Consistent with *Ror2* involvement, we found that Wnt5a-induced mammosphere formation was abolished by inhibition of JNK. We conclude that while Wnt3a promotes mammosphere formation via canonical Wnt signaling, Wnt5a does so by a non-canonical mechanism. Moreover, our data particularly implicate a *Ror2*-JNK pathway among the numerous other pathways that have been ascribed to non-canonical Wnt signaling [39,83,84].

It remains to be determined whether the independent and additive effects of canonical and non-canonical Wnt signaling on mammosphere formation are caused by both signals operating on the same population of mammosphere-forming cells, or whether they act on distinct target populations. Recent studies aimed at identifying mouse mammary stem cells by lineage tracing *in vivo* have yielded data that may be inconsistent with those from classical mammary reconstitution assays [5–7,85]. Collectively these reports suggest that the capacity to act as multipotent stem cells may reside in several distinct cell types in the developing mammary gland and that they may be activated under different circumstances in response to hormonal signals, pregnancy, tissue damage, or other forms of stress. Thus, the stem cell phenotype may be subject to considerable plasticity in response to extrinsic factors such as Wnts and other putative stem cell niche components [86,87]. Against the growing complexity of mammary stem cell analysis *in vivo*, mammosphere assays provide a promising system for dissecting the responses of individual self-renewing cells to defined factors *in vitro*.

Supporting Information

Figure S1 Wild-type mammospheres can be serially passaged for multiple generations. Wild-type mammosphere cultures were serially passaged weekly. The number of mammospheres formed per 2000 cells plated was assayed at each passage. Passage one represents the number of secondary mammospheres resulting from passage from primary to secondary culture. Error bars show 95% confidence intervals.

Acknowledgments

We thank Dr. T. Stappenbeck (Washington University, St. Louis) for generously sharing *Ror2* knockdown vectors, Dr. R. DasGupta (New York University Medical Center) for reagents, members of the Brown lab for productive discussions, and Dr. Louise Howe for critical reading of the manuscript.

Author Contributions

Conceived and designed the experiments: AM AB. Performed the experiments: AM. Analyzed the data: AM AB. Contributed to the writing of the manuscript: AM AB.

References

1. Daniel CW, Ome KBD, Young JT, Blair PB, Faulkin LJ (1968) The in vivo life span of normal and preneoplastic mouse mammary glands: a serial transplantation study. PNAS 61: 53–60.
2. Kordon EC, Smith GH (1998) An entire functional mammary gland may comprise the progeny from a single cell. Development 125: 1921–1930.
3. Shackleton M, Vaillant F, Simpson KJ, Stingl J, Smyth GK, et al. (2006) Generation of a functional mammary gland from a single stem cell. Nature 439: 84–88.
4. Stingl J, Eirew P, Ricketson I, Shackleton M, Vaillant F, et al. (2006) Purification and unique properties of mammary epithelial stem cells. Nature 439: 993–997.
5. De Visser KE, Ciampricotti M, Michalak E, Wei-Min Tan D, Speksnijder EN, et al. (2012) Developmental stage-specific contribution of LGR5+ cells to basal and luminal epithelial lineages in the postnatal mammary gland. The Journal of Pathology 228: 300–309.
6. Van Keymeulen A, Rocha AS, Ousset M, Beck B, Bouvencourt G, et al. (2011) Distinct stem cells contribute to mammary gland development and maintenance. Nature 479: 189–193.
7. Rios AC, Fu NY, Lindeman GJ, Visvader JE (2014) In situ identification of bipotent stem cells in the mammary gland. Nature 506: 322–327.
8. Méndez-Ferrer S, Michurina TV, Ferraro F, Mazloom AR, Macarthur BD, et al. (2010) Mesenchymal and haematopoietic stem cells form a unique bone marrow niche. Nature 466: 829–834.
9. Shahi P, Seethammagari MR, Valdez JM, Xin L, Spencer DM (2011) Wnt and Notch pathways have interrelated opposing roles on prostate progenitor cell proliferation and differentiation. Stem Cells 29: 678–688.
10. Jensen JB, Parmar M (2006) Strengths and limitations of the neurosphere culture system. Mol Neurobiol 34: 153–161.
11. Ahmed S (2009) The culture of neural stem cells. Journal of Cellular Biochemistry 106: 1–6.
12. Dontu G, Abdallah WM, Foley JM, Jackson KW, Clarke MF, et al. (2003) In vitro propagation and transcriptional profiling of human mammary stem/progenitor cells. Genes Dev 17: 1253–1270.
13. Liao MJ, Zhang CC, Zhou B, Zimonjic DB, Mani SA, et al. (2007) Enrichment of a population of mammary gland cells that form mammospheres and have in vivo repopulating activity. Cancer Res 67: 8131–8138.
14. Liu S, Dontu G, Mantle ID, Patel S, Ahn N, et al. (2006) Hedgehog signaling and Bmi-1 regulate self-renewal of normal and malignant human mammary stem cells. Cancer Res 66: 6063–6071.
15. Zhang M, Behbod F, Atkinson RL, Landis MD, Kittrell F, et al. (2008) Identification of tumor-initiating cells in a p53-null mouse model of breast cancer. Cancer Res 68: 4674–4682.
16. Korkaya H, Paulson A, Iovino F, Wicha MS (2008) HER2 regulates the mammary stem/progenitor cell population driving tumorigenesis and invasion. Oncogene 27: 6120–6130.
17. Dong Q, Wang D, Bandyopadhyay A, Gao H, Gorena KM, et al. (2013) Mammospheres from murine mammary stem cell-enriched basal cells: Clonal characteristics and repopulating potential. Stem Cell Res 10: 396–404.
18. Grimshaw MJ, Cooper L, Papazisis K, Coleman JA, Bohnenkamp HR, et al. (2008) Mammosphere culture of metastatic breast cancer cells enriches for tumorigenic breast cancer cells. Breast Cancer Res 10: R52.
19. Lamb R, Ablett MP, Spence K, Landberg G, Sims AH, et al. (2013) Wnt pathway activity in breast cancer sub-types and stem-like cells. PLoS ONE 8: e67811.
20. Brennan KR, Brown AMC (2004) Wnt proteins in mammary development and cancer. J Mammary Gland Biol Neoplasia 9: 119–131.
21. Many AM, Brown AMC (2010) Mammary stem cells and cancer: roles of Wnt signaling in plain view. Breast Cancer Res 12: 313.
22. Clevers H, Nusse R (2012) Wnt/β-Catenin Signaling and Disease. Cell 149: 1192–1205.
23. Howe LR, Brown AMC (2004) Wnt signaling and breast cancer. Cancer Biol Ther 3: 36–41.
24. Nusse R, Fuerer C, Ching W, Harnish K, Logan C, et al. (2008) Wnt signaling and stem cell control. Cold Spring Harb Symp Quant Biol 73: 59–66.
25. Van Camp JK, Beckers S, Zegers D, Van Hul W (2013) Wnt Signaling and the Control of Human Stem Cell Fate. Stem Cell Rev 10: 207–229.
26. Polakis P (2007) The many ways of Wnt in cancer. Curr Opin Genet Dev 17: 45–51.
27. Li Y, Hively WP, Varmus HE (2000) Use of MMTV-Wnt-1 transgenic mice for studying the genetic basis of breast cancer. Oncogene 19: 1002–1009.
28. Li Y, Welm B, Podsypanina K, Huang S, Chamorro M, et al. (2003) Evidence that transgenes encoding components of the Wnt signaling pathway preferentially induce mammary cancers from progenitor cells. Proc Natl Acad Sci U S A 100: 15853–15858.
29. Liu BY, McDermott SP, Khwaja SS, Alexander CM (2004) The transforming activity of Wnt effectors correlates with their ability to induce the accumulation of mammary progenitor cells. Proc Natl Acad Sci U S A 101: 4158–4163.
30. Incassati A, Chandramouli A, Eelkema R, Cowin P (2010) Key signaling nodes in mammary gland development and cancer: β-catenin. Breast Cancer Res 12: 213.
31. Baker R, Kent CV, Silbermann RA, Hassell JA, Young LJT, et al. (2010) Pea3 transcription factors and wnt1-induced mouse mammary neoplasia. PLoS ONE 5: e8854.

32. Teissedre B, Pinderhughes A, Incassati A, Hatsell SJ, Hiremath M, et al. (2009) MMTV-Wnt1 and ΔN89β-catenin induce canonical signaling in distinct progenitors and differentially activate Hedgehog signaling within mammary tumors. PLoS ONE 4: e4537.

33. Badders NM, Goel S, Clark RJ, Klos KS, Kim S, et al. (2009) The Wnt receptor, Lrp5, is expressed by mouse mammary stem cells and is required to maintain the basal lineage. PLoS ONE 4: e6594.

34. Tsukamoto AS, Grosschedl R, Guzman RC, Parslow T, Varmus HE (1988) Expression of the int-1 gene in transgenic mice is associated with mammary gland hyperplasia and adenocarcinomas in male and female mice. Cell 55: 619–625.

35. Zeng YA, Nusse R (2010) Wnt proteins are self-renewal factors for mammary stem cells and promote their long-term expansion in culture. Cell Stem Cell 6: 568–577.

36. Van Amerongen R, Bowman AN, Nusse R (2012) Developmental stage and time dictate the fate of Wnt/β-catenin-responsive stem cells in the mammary gland. Cell Stem Cell 11: 387–400.

37. Macdonald BT, Semenov MV, He X (2007) SnapShot: Wnt/β-catenin signaling. Cell 131: 1204.

38. Shimizu H, Julius MA, Giarre M, Zheng Z, Brown AMC, et al. (1997) Transformation by Wnt family proteins correlates with regulation of β-catenin. Cell Growth Differ 8: 1349–1358.

39. Semenov MV, Habas R, Macdonald BT, He X (2007) SnapShot: Noncanonical Wnt Signaling Pathways. Cell 131: 1378.

40. Van Amerongen R (2012) Alternative Wnt pathways and receptors. Cold Spring Harb Perspect Biol 4: a007914.

41. Liu G, Bafico A, Aaronson SA (2005) The mechanism of endogenous receptor activation functionally distinguishes prototype canonical and noncanonical Wnts. Mol Cell Biol 25: 3475–3482.

42. Oishi I, Suzuki H, Onishi N, Takada R, Kani S, et al. (2003) The receptor tyrosine kinase Ror2 is involved in non-canonical Wnt5a/JNK signalling pathway. Genes Cells 8: 645–654.

43. Grumolato L, Liu G, Mong P, Mudbhary R, Biswas R, et al. (2010) Canonical and noncanonical Wnts use a common mechanism to activate completely unrelated coreceptors. Genes Dev 24: 2517–2530.

44. Gonzalez-Sancho JM, Brennan KR, Castelo-Soccio LA, Brown AMC (2004) Wnt proteins induce dishevelled phosphorylation via an LRP5/6- independent mechanism, irrespective of their ability to stabilize β-catenin. Mol Cell Biol 24: 4757–4768.

45. McDonald SL, Silver A (2009) The opposing roles of Wnt-5a in cancer. Br J Cancer 101: 209–214.

46. Roarty K, Serra R (2007) Wnt5a is required for proper mammary gland development and TGFβ-mediated inhibition of ductal growth. Development 134: 3929–3939.

47. Roarty K, Baxley SE, Crowley MR, Frost AR, Serra R (2009) Loss of TGF-β or Wnt5a results in an increase in Wnt/β-catenin activity and redirects mammary tumour phenotype. Breast Cancer Res 11: R19.

48. Dontu G, Jackson KW, McNicholas E, Kawamura MJ, Abdallah WM, et al. (2004) Role of Notch signaling in cell-fate determination of human mammary stem/progenitor cells. Breast Cancer Research 6: R605.

49. Fuerer C, Nusse R (2010) Lentiviral Vectors to Probe and Manipulate the Wnt Signaling Pathway. PLoS ONE 5: e9370.

50. Miyoshi H, Ajima R, Luo CT, Yamaguchi TP, Stappenbeck TS (2012) Wnt5a potentiates TGF-β signaling to promote colonic crypt regeneration after tissue injury. Science 338: 108–113.

51. Gonsalves FC, Klein K, Carson BB, Katz S, Ekas LA, et al. (2011) An RNAi-based chemical genetic screen identifies three small-molecule inhibitors of the Wnt/wingless signaling pathway. Proceedings of the National Academy of Sciences 108: 5954–5963.

52. Mani SA, Guo W, Liao MJ, Eaton EN, Ayyanan A, et al. (2008) The epithelial-mesenchymal transition generates cells with properties of stem cells. Cell 133: 704–715.

53. Dey D, Saxena M, Paranjape AN, Krishnan V, Giraddi R, et al. (2009) Phenotypic and functional characterization of human mammary stem/progenitor cells in long term culture. PLoS One 4: e5329.

54. Tao L, Roberts AL, Dunphy KA, Bigelow C, Yan H, et al. (2011) Repression of Mammary Stem/Progenitor Cells by p53 Is Mediated by Notch and Separable from Apoptotic Activity. Stem Cells 29: 119–127.

55. Moraes RC, Zhang X, Harrington N, Fung JY, Wu M-F, et al. (2007) Constitutive activation of smoothened (SMO) in mammary glands of transgenic mice leads to increased proliferation, altered differentiation and ductal dysplasia. Development 134: 1231–1242.

56. Ling L, Nurcombe V, Cool SM (2009) Wnt signaling controls the fate of mesenchymal stem cells. Gene 433: 1–7.

57. Holland JD, Klaus A, Garratt AN, Birchmeier W (2013) Wnt signaling in stem and cancer cells. Current Opinion in Cell Biology 25: 254–264.

58. Semënov MV, Tamai K, Brott BK, Kühl M, Sokol S, et al. (2001) Head inducer Dickkopf-1 is a ligand for Wnt coreceptor LRP6. Curr Biol 11: 951–961.

59. Bafico A, Liu G, Yaniv A, Gazit A, Aaronson SA (2001) Novel mechanism of Wnt signaling inhibition mediated by Dickkopf-1 interaction with LRP6/Arrow. Nat Cell Biol 3: 683–686.

60. Mao B, Wu W, Li Y, Hoppe D, Stannek P, et al. (2001) LDL-receptor-related protein 6 is a receptor for Dickkopf proteins. Nature 411: 321–325.

61. Baljinnyam B, Klauzinska M, Saffo S, Callahan R, Rubin JS (2012) Recombinant R-spondin2 and Wnt3a up- and down-regulate novel target genes in C57MG mouse mammary epithelial cells. PLoS ONE 7: e29455.

62. Mikels AJ, Nusse R (2006) Purified Wnt5a protein activates or inhibits β-catenin-TCF signaling depending on receptor context. PLoS Biol 4: e115.

63. Nemeth MJ, Topol L, Anderson SM, Yang Y, Bodine DM (2007) Wnt5a inhibits canonical Wnt signaling in hematopoietic stem cells and enhances repopulation. Proc Natl Acad Sci USA 104: 15436–15441.

64. Van Amerongen R, Fuerer C, Mizutani M, Nusse R (2012) Wnt5a can both activate and repress Wnt/β-catenin signaling during mouse embryonic development. Dev Biol 369: 101–114.

65. Topol L, Jiang X, Choi H, Garrett-Beal L, Carolan PJ, et al. (2003) Wnt-5a inhibits the canonical Wnt pathway by promoting GSK-3-independent β-catenin degradation. J Cell Biol 162: 899–908.

66. Baksh D, Boland GM, Tuan RS (2007) Cross-talk between Wnt signaling pathways in human mesenchymal stem cells leads to functional antagonism during osteogenic differentiation. J Cell Biochem 101: 1109–1124.

67. Serra R, Easter SL, Jiang W, Baxley SE (2011) Wnt5a as an effector of TGFβ in mammary development and cancer. J Mammary Gland Biol Neoplasia 16: 157–167.

68. Pavlovich AL, Boghaert E, Nelson CM (2011) Mammary branch initiation and extension are inhibited by separate pathways downstream of TGFβ in culture. Exp Cell Res 317: 1872–1884.

69. He X, Saint-Jeannet JP, Wang Y, Nathans J, Dawid I, et al. (1997) A member of the Frizzled protein family mediating axis induction by Wnt-5A. Science 275: 1652–1654.

70. Nishita M, Itsukushima S, Nomachi A, Endo M, Wang Z, et al. (2010) Ror2/Frizzled complex mediates Wnt5a-induced AP-1 activation by regulating Dishevelled polymerization. Mol Cell Biol 30: 3610–3619.

71. O'Connell MP, Fiori JL, Xu M, Carter AD, Frank BP, et al. (2010) The orphan tyrosine kinase receptor, ROR2, mediates Wnt5A signaling in metastatic melanoma. Oncogene 29: 34–44.

72. Nomachi A, Nishita M, Inaba D, Enomoto M, Hamasaki M, et al. (2008) Receptor tyrosine kinase Ror2 mediates Wnt5a-induced polarized cell migration by activating c-Jun N-terminal kinase via actin-binding protein filamin A. J Biol Chem. 283: 27973–27981.

73. Bennett BL, Sasaki DT, Murray BW, O'Leary EC, Sakata ST, et al. (2001) SP600125, an anthrapyrazolone inhibitor of Jun N-terminal kinase. Proc Natl Acad Sci USA 98: 13681–13686.

74. Han Z, Boyle DL, Chang L, Bennett B, Karin M, et al. (2001) c-Jun N-terminal kinase is required for metalloproteinase expression and joint destruction in inflammatory arthritis. J Clin Invest 108: 73–81.

75. Cho RW, Wang X, Diehn M, Shedden K, Chen GY, et al. (2008) Isolation and molecular characterization of cancer stem cells in MMTV-Wnt-1 murine breast tumors. Stem Cells 26: 364–371.

76. Ettenberg SA, Charlat O, Daley MP, Liu S, Vincent KJ, et al. (2010) Inhibition of tumorigenesis driven by different Wnt proteins requires blockade of distinct ligand-binding regions by LRP6 antibodies. Proc Natl Acad Sci USA 107: 15473–15478.

77. DeAlmeida VI, Miao L, Ernst JA, Koeppen H, Polakis P, et al. (2007) The soluble wnt receptor Frizzled8CRD-hFc inhibits the growth of teratocarcinomas in vivo. Cancer Res 67: 5371–5379.

78. Gunther EJ, Moody SE, Belka GK, Hahn KT, Innocent N, et al. (2003) Impact of p53 loss on reversal and recurrence of conditional Wnt-induced tumorigenesis. Genes Dev 17: 488–501.

79. Jonsson M, Dejmek J, Bendahl PO, Andersson T (2002) Loss of Wnt-5a protein is associated with early relapse in invasive ductal breast carcinomas. Cancer Res 62: 409–416.

80. Dejmek J, Dejmek A, Säfholm A, Sjölander A, Andersson T (2005) Wnt-5a protein expression in primary dukes B colon cancers identifies a subgroup of patients with good prognosis. Cancer Res 65: 9142–9146.

81. Scheel C, Eaton EN, Li SH-J, Chaffer CL, Reinhardt F, et al. (2011) Paracrine and autocrine signals induce and maintain mesenchymal and stem cell states in the breast. Cell 145: 926–940.

82. Yeh JR, Zhang X, Nagano MC (2011) Wnt5a is a cell-extrinsic factor that supports self-renewal of mouse spermatogonial stem cells. Journal of Cell Science 124: 2357–2366.

83. Niehrs C (2012) The complex world of WNT receptor signalling. Nat Rev Mol Cell Biol 13: 767–779.

84. Van Amerongen R, Mikels A, Nusse R (2008) Alternative wnt signaling is initiated by distinct receptors. Sci Signal 1: re9.

85. Plaks V, Brenot A, Lawson DA, Linnemann JR, Van Kappel EC, et al. (2013) Lgr5-Expressing Cells Are Sufficient and Necessary for Postnatal Mammary Gland Organogenesis. Cell Rep 3: 70–78.

86. Alexander CM, Goel S, Fakhraldeen SA, Kim S (2012) Wnt Signaling in Mammary Glands: Plastic Cell Fates and Combinatorial Signaling. Cold Spring Harbor perspectives in biology 4: a008037.

87. Marjanovic ND, Weinberg RA, Chaffer CL (2013) Cell Plasticity and Heterogeneity in Cancer. Clinical Chemistry 59: 168–179.

12

Multimodality Molecular Imaging to Monitor Transplanted Stem Cells for the Treatment of Ischemic Heart Disease

Zhijun Pei[1,2], Xiaoli Lan[1]*, Zhen Cheng[3], Chunxia Qin[1], Xiaotian Xia[1], Hui Yuan[1], Zhiling Ding[1], Yongxue Zhang[1]*

1 Department of Nuclear Medicine, Union Hospital, Tongji Medical College of Huazhong University of Science and Technology, Hubei Province Key Laboratory of Molecular Imaging, Wuhan, China, 2 Department of PET Center, Taihe Hospital, Hubei University of Medicine, Shiyan City, Hubei Province, China, 3 Molecular Imaging Program at Stanford and Bio-X Program, Stanford University, Stanford, California, United States of America

Abstract

Purpose: Non-invasive techniques to monitor the survival and migration of transplanted stem cells in real-time is crucial for the success of stem cell therapy. The aim of this study was to explore multimodality molecular imaging to monitor transplanted stem cells with a triple-fused reporter gene [TGF; herpes simplex virus type 1 thymidine kinase (HSV1-tk), enhanced green fluorescence protein (eGFP), and firefly luciferase (FLuc)] in acute myocardial infarction rat models.

Methods: Rat myocardial infarction was established by ligating the left anterior descending coronary artery. A recombinant adenovirus carrying TGF (Ad5-TGF) was constructed. After transfection with Ad5-TGF, 5×10^6 bone marrow mesenchymal stem cells (BMSCs) were transplanted into the anterior wall of the left ventricle (n = 14). Untransfected BMSCs were as controls (n = 8). MicroPET/CT, fluorescence and bioluminescence imaging were performed. Continuous images were obtained at day 2, 3 and 7 after transplantation with all three imaging modalities and additional images were performed with bioluminescence imaging until day 15 after transplantation.

Results: High signals in the heart area were observed using microPET/CT, fluorescence and bioluminescence imaging of infarcted rats injected with Ad5-TGF-transfected BMSCs, whereas no signals were observed in controls. Semi-quantitative analysis showed the gradual decrease of signals in all three imaging modalities with time. Immunohistochemistry assays confirmed the location of the TGF protein expression was the same as the site of stem cell-specific marker expression, suggesting that TGF tracked the stem cells in situ.

Conclusions: We demonstrated that TGF could be used as a reporter gene to monitor stem cells in a myocardial infarction model by multimodality molecular imaging.

Editor: Christoph E. Hagemeyer, Baker IDI Heart and Diabetes Institute, Australia

Funding: This study was supported by the National Natural Science Foundation of China (No. 30830041, 30571816, 30772208, and 30970853) (http://www.nsfc.gov.cn/Portal0/default152.htm), the China Postdoctoral Science Foundation (2005037194) (http://211.166.12.38/V1/Program1/Default.aspx), and Hubei Province Science Fund for Distinguished Young Scholars (2010CDA094) (http://www.hbstd.gov.cn/). The funders had no role in study design, data collection and analysis, decision to publish, or preparation of the manuscript.

Competing Interests: The authors have declared that no competing interests exist.

* E-mail: LXL730724@hotmail.com (XL); zhyx1229@163.com (YXZ)

Introduction

Although significant progress has been made in coronary revascularization and atherosclerosis prevention, cardiovascular diseases are still a major cause of death. Many animal and clinical experiments have demonstrated that treating ischemic heart disease with transplanted bone marrow mesenchymal stem cells (BMSCs) is feasible and promising [1–6]. Although traditional techniques such as in situ hybridization, PCR and immunohistochemistry are widely used to analyze the distribution and migration of transplanted stem cells, they are in vitro or post mortem and obviously not applicable for in vivo studies. Therefore, using non-invasive techniques to monitor the survival

and migration of transplanted stem cells in real-time is crucial for the success of therapy.

In the past decade, techniques to monitor transplanted stem cells have reached a new stage in which the biological progress of transplanted tissues and cells can be monitored in vivo at the molecular level. A variety of cutting edge molecular imaging techniques have been developed [7,8]. Different imaging methods have their advantages and disadvantages. Radionuclide imaging is highly sensitive but suffers low spatial resolution. Magnetic resonance imaging (MRI) shows the highest soft tissue contrast, but has low sensitivity. Bioluminescence and fluorescence imaging have relatively high sensitivity and spatial resolution, but cannot image deep tissues. To address these issues, multimodality molecular imaging has been actively developed in recent years.

Many of these multimodality imaging techniques, such as radionuclide/MRI and optical imaging, are further used for cell trafficking such as stem cell monitoring [9–11]. These studies indicate that multimodality imaging has advantages over single-modality imaging. Furthermore, considering multimodality imaging probes and image fusion techniques have been developed rapidly, multimodality imaging may find many important applications in clinical practice.

The triple fusion gene TGF [herpes simplex virus type 1 thymidine kinase (HSV1-tk), enhanced green fluorescence protein (eGFP) and firefly luciferase (Fluc)] was recently developed and applied to stem cell monitoring [12,13]. We have shown similar results in a previous study [14]. However, TGF has not been used to monitor transplanted stem cells in a myocardial infarction model. Therefore, in our current study, we aimed to determine whether TGF can be expressed in a myocardial infarction model using three imaging methods, and the duration of possible TGF expression. We explored the feasibility of multimodality imaging to monitor the transplanted stem cells in rat models with ischemic heart disease.

Materials and Methods

Ad5-TGF plasmid construction and recombinant adenovirus packaging

The pCDNA3.1 plasmid carrying the TGF fusion gene under the control of the CMV promoter was kindly provided by Dr. Sanjiv Sam Gambhir, Stanford University. The plasmid was repackaged into a recombinant adenovirus (Ad5-TGF) that was amplified and purified. The virus titer of 1.26×10^{10} TU/mL in 7 mL was determined by Vector Gene Technology Co., Ltd (Beijing, China).

Isolation and culture of BMSCs, and Ad5-TGF infection of BMSCs

All studies were approved by the Institutional Animal Care and Use Committee (IACUC) at Tongji Medical College, Huazhong University of Science and Technology (HUST). All the specific animals were maintained in the barrier system room at $22 \pm 2°C$ with an alternating 12 h light/dark cycle, and were given food and water ad libitum throughout the study period. All procedures in the animal experiment were carried out in strict compliance with the Guideline of Laboratory Animals Ethics Committee of Tongji Medical College, HUST. MSCs were isolated from the femur and tibia of a 4-week-old healthy Sprague-Dawley (SD) rat supplied by the Experimental Animal Center of Tongji Medical College, HUST (Wuhan, China). The bone marrow was flushed out with DMEM-F12 (Gibco, Langley, VA) containing 15% fetal bovine serum (FBS) (Gibco, Langley, VA), and then the cells were seeded in six-well plates. Passage 3–5 BMSCs cultured in OPTI-MEM serum-free medium (Gibco, Langley, VA) were infected with Ad5-TGF. The required volume of Ad5-TGF was calculated by the specific multiplicity of infection (MOI) of the BMSC count. Cells were slowly shaken inside a 37°C incubator for 2 h, and they were then transferred into a 37°C/5% CO_2 incubator for 48 h.

Animal model

Thirty specific pathogen-free SD rats (200 ± 25 g) were used to establish the animal model of acute myocardial infarction using previously published methods [15–18]. Rats were anesthetized with Isoflurane (Shanghai Abbott Laboratories, China) and fixed supinely on the operating table. The chest hair was shaved, and the skin was sterilized. A 2 cm incision was made at the left margin of the sternum in the third to fourth intercostal along the direction of the intercostal space. Deep and superficial fasciae were cut. The pectoralis major and serratus anterior muscles were separated to expose the ribs. After the bleeding stopped, the fourth intercostal muscle was separated, and the heart was quickly returned and the left anterior descending artery was ligated with a non-invasive suture. Each layer of the chest wall was sutured after the rat was able to breathe spontaneously.

BMSCs were infected with Ad5-TGF (MOI = 100) for 48 h. Rats (n = 22) that survived for 1 week after surgery were used for experiments. The method described above was used to extrude the heart. Ad5-TGF-infected BMSCs (5×10^6 cells in 100 μl) were slowly injected into the myocardium of the left ventricular anterior wall at the far end of the coronary artery ligation. Rats were transplanted with uninfected BMSCs for the negative control group (n = 8). In the experimental group (n = 14), five rats were performed microPET/CT and bioluminescence imaging continuously, and the other nine rats were used to obtain fluorescence imaging.

All rats for the model preparing were put on heat preservation pads during the surgery and till to fully awake. Surgical wound was prevented to expand during the surgery, which in turn minimized post-operative pain and distress. Penicillin (10^5 U/kg, every 24 hours) was intramuscularly injected for 4–5 days after the surgery in order to prevent infection. All rats were allocated to individual cage post-operation, and bred with nutritious pellets. After surgery, body weight, heart rate, respiration and specific behaviors of rats were recorded every 24 hours for at least one week. Rats were sacrificed after multimodality imaging with overdose anesthetic drug (pentobarbital, 150 mg/kg). All the animal experimental procedures were carried out under the guidance of the veterinaries in Department of Experimental Animals, Tongji Medical College, HUST.

Preparation of ^{18}F-FHBG and microPET/CT imaging

A Siemens Inveon Acquisition Workplace (Siemens Preclinical Solution, Knoxville, TN) was used for microPET/CT imaging. After transplantation, rats were injected with ^{18}F-FHBG (500 ± 51 μCi) via the tail vein at days 2, 3 and 7, and prone position microPET/CT imaging was performed after 1 h for 20 min. The standard ordered-subset expectation maximization method was used for microPET image reconstruction. CT images were used for both attenuation correction of emission data and image fusion. Fusion of microPET/CT images showed horizontal, coronal and sagittal sections. Then, regions of interest (ROIs) were manually drawn. Quantitative analysis of ^{18}F-FHBG uptake in the brain, heart, liver, lung, kidney, stomach, intestine and spleen was performed.

Bioluminescence imaging

A Maestro small animal imaging system (Cambridge Research & Instrumentation) was used for bioluminescence imaging. Rat models received an abdominal injection of 200 μl firefly luciferase substrate (10 mg/ml D-Luciferin; Xenogen, Alameda, CA). Bioluminescence imaging was performed for 5 min after injection. The size of the cooled charge coupled device (CCD) was 4×4 inches, and the operating temperature was –70°C. Bioluminescence imaging was performed at days 2, 3, 5, 7, 10 and 15 after transplantation.

Fluorescence imaging

A fluorescence imaging system for living animals (Roper Scientific, USA) was used. Rats (n = 3) were euthanized at days 2, 3 and 7 after transplantation, respectively. The hair, muscle and ribs were removed to expose the thoracic cavity. The 488-nm

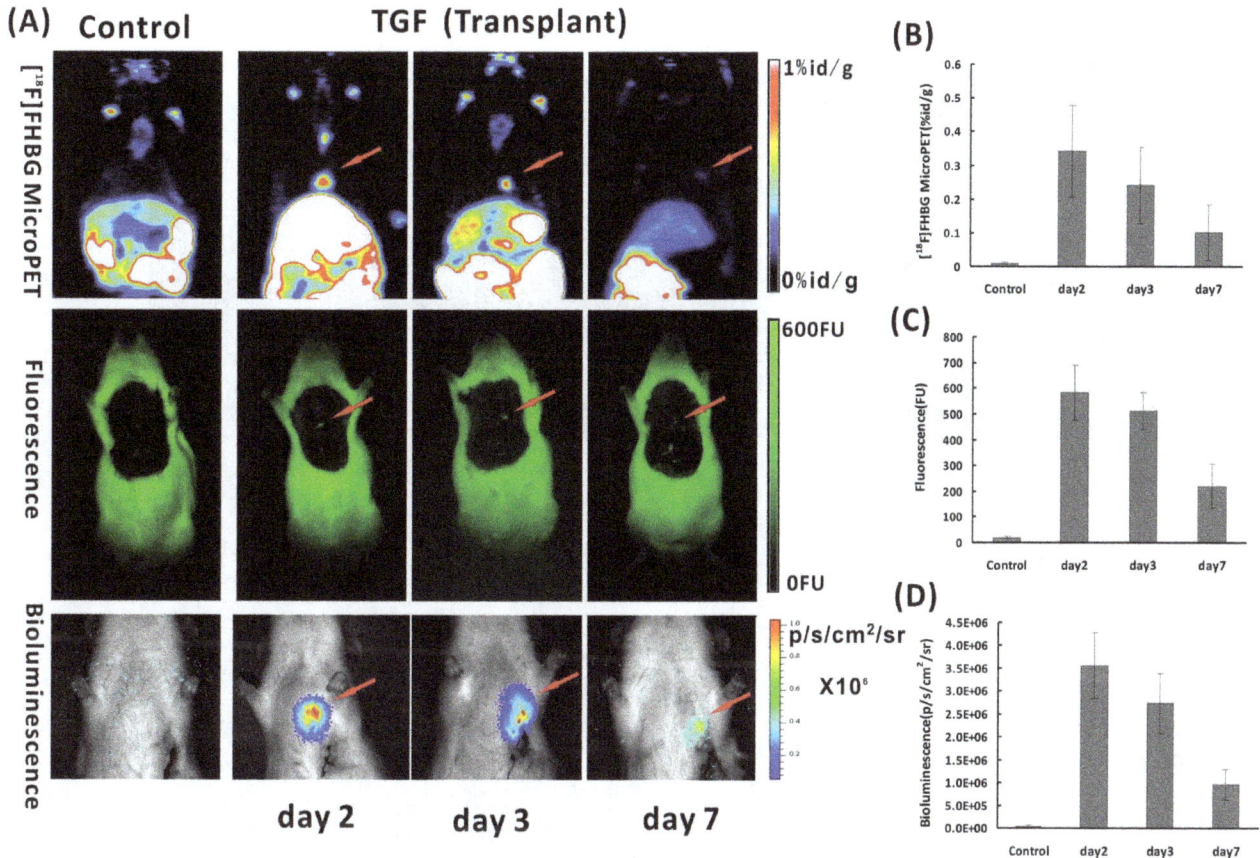

Figure 1. Multimodality molecular imaging in acute myocardial infarction rats after transplanted Ad5-TGF-transfected bone marrow mesenchymal stem cells (BMSCs) into myocardium. A: From images of microPET (upper row), Fluorescence (middle row) and Bioluminescence (lower row), signals in the heart region could be clearly seen in different imaging modalities (indicated by red arrows) at day 2, 3 and 7 after transplantation of Ad5-TGF-transfected BMSCs into myocardium. Semi-Quantitative analysis results obtained by regions of interest (ROIs) analysis of the region of heart from ^{18}F-FHBG microPET (B), Fluorescence (C) and Bioluminescence (D) imaging shows that significant difference could be seen between the experimental group with transplanted Ad5-TGF-BMSCs and the control group with transplanted uninfected BMSCs ($P < 0.05$) in all different imaging modalities.

excitation and 510-nm emission filters were used. The imaging time was 500 ms. Lumazone software was used for imaging and image analysis. Region of interests (ROIs) were manually drawn for quantitative analysis of the heart region. Three control rats were also used in fluorescence imaging.

PCR and immunohistochemistry

Total RNA was extracted from ex vivo myocardial tissue and reverse transcribed into cDNA. PCR primers for TGF were: forward, 5'-ATGCCCACGCTACTGCGG-3'; reverse: 5'-TCAGTTAGCCTCCCCCATCTC-3'. The PCR product size was 837 bp, and PCR conditions were 94°C for 5 min, followed by 35 cycles of 94°C for 30 s, 56°C for 30 s and 72°C for 1 min. The primers were provided by Invitrogen (Shanghai, China).

A portion of the ex vivo myocardial tissue was fixed with 4% formaldehyde, embedded in paraffin, and cut into 3 μm sections for hematoxylin and eosin (H&E) staining and immunohistochemistry analysis. An anti-HSV1-tk antibody (Santa Cruz, CA) was added to sections, followed by incubation at 4°C for 24 h. Then, a secondary antibody was added, followed by DAB staining, hematoxylin counterstaining, dehydration, and mounting. The stained sections were observed under an optical microscope and photographs were obtained.

The remaining ex vivo myocardial tissue was frozen in dry ice and cryosectioned into 15 μm sections. Sections were stained with anti-CD45 and anti-CD90 (eBioscience, San Diego, USA) antibodies at 4°C for 24 h, followed by incubation with a Cy3-labelled secondary antibody in the dark at room temperature for 1 h. Then, sections were counterstained with Hoechst 33258 to visualize cell nuclei, mounted with neutral glycerol, observed under a fluorescence microscope and photographed.

Data analysis

Data were expressed as mean ± standard deviations. SPSS 13.0 software was used to analyze the data. Wilcoxon rank test for two independent samples were used for comparison between two samples. $P < 0.05$ was considered significant.

Results

A serial of images of microPET, Fluorescence and Bioluminescence (Figure 1A) were obtained in acute myocardial infarction rats after transplanted Ad5-TGF-transfected BMSCs into myocardium at day 2, 3 and 7. Signals in the heart region could be clearly seen in different imaging modalities, whereas no signal could be found in the control group which transplanted with uninfected BMSCs (Figure 1A). Semi-Quantitative analysis results

(A)

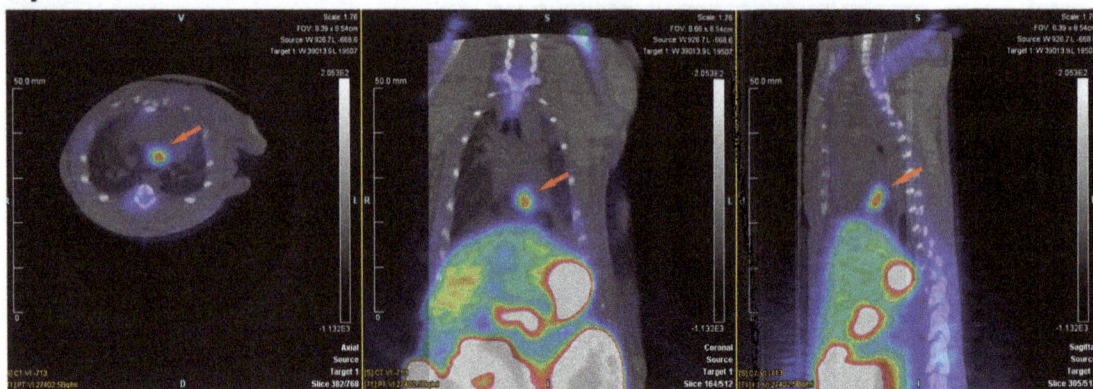

(B)

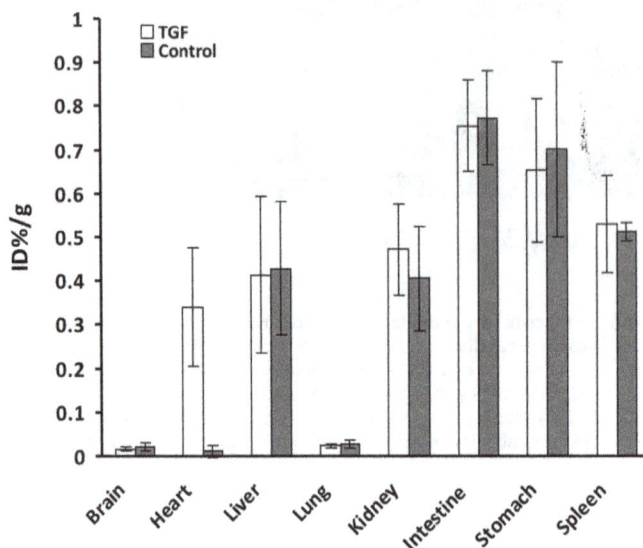

Figure 2. MicroPET/CT fused images in myocardial infarction rats transplanted with Ad5-TGF-transfected BMSCs at 2 days after transplantation (from left to right: horizontal, coronal and sagittal views). High signals (red arrow) could be clearly seen in the region of heart. B: Semi-quantitative analysis of ^{18}F-FHBG uptake in different organs from the images. Significant difference exists in the heart between the experimental rats which were transplanted of Ad5-TGF transfected BMSC into myocardial infarction area and the control rats with non-transfected BMSCs only (n = 5, $P <$ 0.05). However, no significant difference exists in other organs.

(Figure 1B, C and D) obtained by ROIs analysis of the heart region shows significant difference between the experimental group and the control group ($P <$ 0.05) in all different imaging modalities.

MicroPET/CT imaging

MicroPET/CT was performed on modeled rats with transplanted BMSCs at 2 days after transplantation. The fusion image in Figure 2A shows obvious ^{18}F-FHBG uptake in the myocardial infarction region of rats transplanted with Ad5-TGF-infected BMSCs. Next, quantitative analysis was performed. As shown in Figure 2B, ^{18}F-FHBG uptake in the brain, heart, liver, lung, kidney, intestine, stomach and spleen of the modeled group was 0.016±0.005, 0.341±0.136, 0.413±0.179, 0.023±0.004, 0.472±0.104, 0.754±0.104, 0.653±0.164 and 0.531±0.112

percentage of injected dose per gram (%ID/g) (n = 5), respectively, while it showed 0.021±0.008, 0.011±0.014, 0.429±0.151, 0.028±0.009, 0.406±0.119, 0.772±0.107, 0.701±0.201 and 0.512±0.021%ID/g, respectively in the controlled group (n = 5). The heart/lung ratio of ^{18}F-FHBG uptake of the modeled group was 31-fold higher than that of the negative control group ($P =$ 0.043). Next, monitoring was performed for 1 week. As shown in Figure 1A and 1B, ^{18}F-FHBG uptake in the heart region of the modeled group at days 3 and 7 after transplantation was 0.241±0.112 and 0.101±0.082%ID/g, respectively.

Bioluminescence imaging

Continuous monitoring of transfected stem cells was performed for 2 weeks by bioluminescence imaging of the same group of the myocardial infarcted rats that had already been scanned by

A

TGF(Transplant)

day 5   day 10   day 15

B

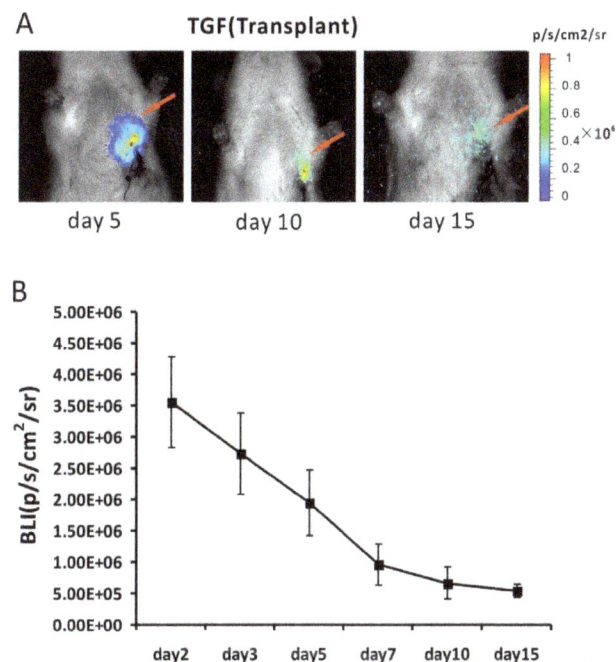

Figure 3. Continuous bioluminescence images after different days of transplanted Ad5-TGF transfected BMSCs into myocardial infarction rats (A). The red arrows indicate the signals in the heart region. B: Semi-quantitative analysis of images shows that the signal decreased with the time after transplantation (n = 5).

microPET/CT (Figure 3A). Quantitative analysis at days 2, 3, 5, 7, 10 and 15 showed that the intensity of the bioluminescence signal in the heart region of rats in the modeled group was $(3.556 \pm 0.725) \times 10^6$, $(2.731 \pm 0.652) \times 10^6$, $(1.946 \pm 0.531) \times 10^6$, $(0.962 \pm 0.326) \times 10^6$, $(0.662 \pm 0.266) \times 10^6$ and $(0.442 \pm 0.126) \times 10^6$ photons/s/cm^2/sr, respectively (Figure 3B) (n = 5). As a comparison, the intensity of the optical signal was only $(0.033 \pm 0.03) \times 10^6$ photons/s/cm^2/sr in the heart region of rats in the negative control group (data not shown).

Fluorescence imaging

Continuous monitoring was also performed for 1 week by fluorescence imaging of transplanted BMSCs in myocardial infarcted rats. Fur, muscle and ribs were removed to expose the thoracic cavity (Figure 1A). Visible green fluorescence was detected in the heart region of rats in the modeled group, whereas no visible fluorescence was detected in the negative control group. Quantitative analysis showed that the fluorescence intensity in the heart region of rats in the modeled group at days 2, 3 and 7 was 582 ± 107, 512 ± 71 and 221 ± 85 FU, respectively (n = 3). However, only 19 ± 5 FU was measured in the heart region of rats in the negative control group (n = 3).

PCR and immunohistochemistry

After PCR and electrophoresis, specific DNA bands were detected between 750 and 1000 bp in DNA obtained from both the rat myocardial tissue and TGF-transfected BMSCs of the modeled group (Figure 4A). The DNA bands of cells were brighter than those of the tissue. No DNA bands were detected in the DNA from myocardial tissue of rats in the negative control group. H&E staining of myocardial tissue showed that Normal myocardial fibers displayed a regular and diffuse distribution (Figure 4B);

however, infarcted myocardial fibers were swollen, disorganized, contained vacuoles and even broken (Figure 4C). Transplanted BMSCs distributed in the myocardial tissue gap were obvious, as shown in Figure 4D and E. eGFP fluorescence from transplanted BMSCs in the infarcted myocardium of rats in the modeled group was observed under a fluorescence microscope (Figure 4F). Positive staining (brown) of transplanted TGF-transfected BMSCs was detected by immunohistochemistry with an anti-HSV1-tk antibody and DAB staining (Figure 4G and H). No obvious staining was detected in normal myocardial cells. Immunofluorescence staining showed that transplanted BMSCs were positive for Cy3-CD90 (red, Figure 4I), but negative for Cy3-CD45 (Fig. 4K). Nuclei were stained and represented in blue (Figure 4J and 4L).

Discussion

This study has shown that microPET/CT, fluorescence and bioluminescence imaging are non-invasive techniques that can be used to repeatedly monitor transplanted stem cells in animal models of myocardial infarction. We performed microPET/CT, fluorescence and bioluminescence imaging on each animal model of myocardial infarction at days 2, 3 and 7 after transplantation. Images of the transplanted region of the heart were even obtained by BLI at 15 days after transplantation. The semi-quantitative analyses of TGF expression obtained by the three imaging techniques were changing at the same trend over time. Finally, we verified the imaging results with the ex vivo assays using PCR and histological identification of the stem cell transplanted heart tissue. This study is the successful application of three different molecular imaging techniques to monitor transplanted stem cells in vivo in a myocardial infarction model.

Because stem cell transplantation is a valid treatment for ischemic heart disease, non-invasive molecular imaging methods have been actively pursued to monitor transplanted stem cells. First, PET reporter gene imaging is one of the most promising non-invasive molecular imaging tools, which is reliable and objective for locating transplanted stem cells in the myocardium of small animals and for quantitative analysis [12,19,20]. Willmann et al applied clinical PET to image large animals such as pigs, in which transplantation of human mesenchymal stem cells into the pig myocardium showed the feasibility of reporter gene imaging [21]. Subsequently, multimodality molecular imaging has been gradually developed and used to monitor transplanted stem cells in the myocardium. Higuchi et al monitored rat cardiac transplantation cell survival and positioning with both PET and MRI [11]. In a study by Wu et al [12], Fluc- and HSV1-sr39tk-transfected embryonic rat H9c2 cardiomyoblasts were transplanted into the myocardium of healthy mice, and in vivo monitoring was performed for 2 weeks using PET and BLI. However, these previous reports all used normal animals and are not an accurate reflection of stem cell survival in a lesioned environment. In this study, the major advantage is the success of continuous multimodality monitoring of stem cells in animal models of myocardial infarction, which is more intuitive and provides a reliable foundation for further applying biological therapy such as stem cells treatment in the future.

Using longitudinal monitoring with the three imaging techniques, we confirmed that BMSCs survived in lesions and did not migrate after transplantation. Based on quantitative analyses, we found that the signals in the heart region decreased as the monitoring time increased using the three imaging techniques. The signal intensity attenuated within 1 week, and by the second week the signal detected by microPET and fluorescence imaging

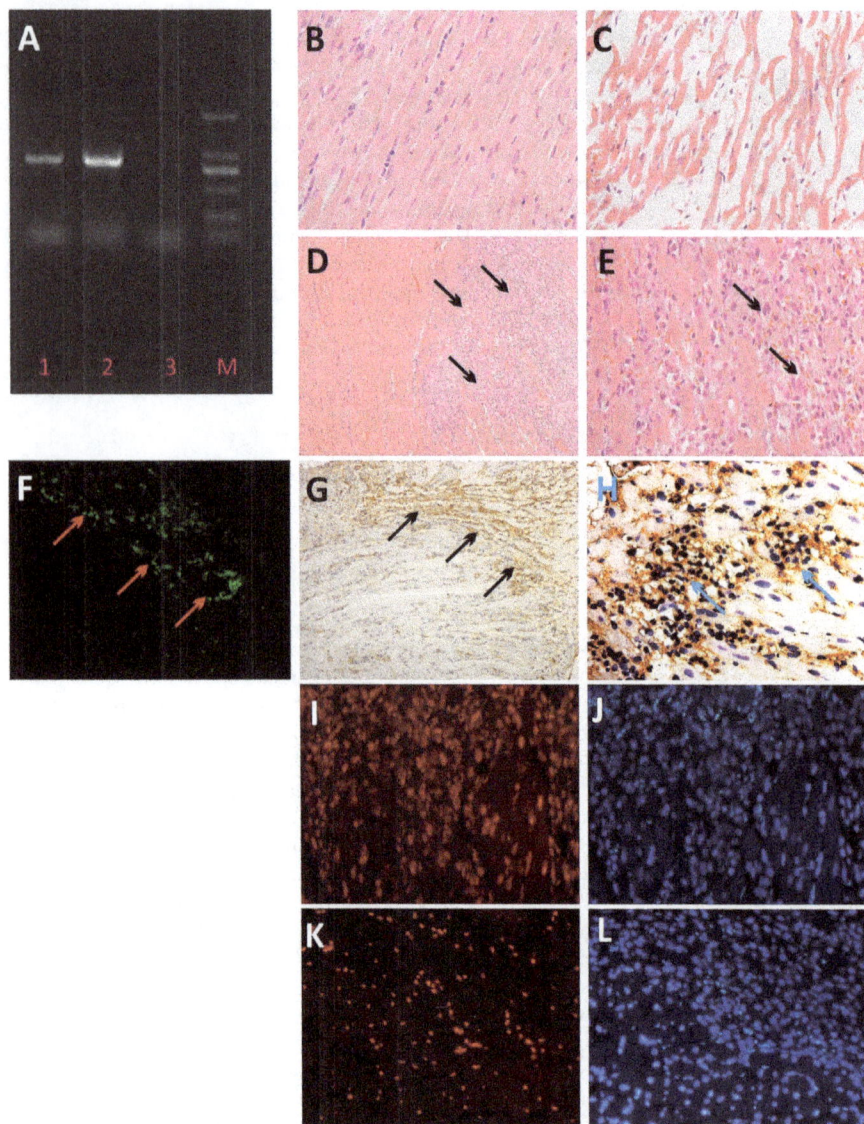

Figure 4. In vitro analysis of the heart tissue after transplanted with Ad5-TGF-transfected BMSCs or non-transfected BMSCs. A: PCR result: 1. rat heart tissue transplanted with Ad5-TGF-transfected BMSCs; 2. Ad5-TGF-transfected BMSCs, 3. the control rat heart tissue with transplanted with non-transfected BMSCs; M, marker (from lower to upper: 100, 250, 500, 750, 1000 and 1500 bp). B: H&E staining of the heart tissue in healthy rats. C: H&E staining of the rat heart tissue with myocardial infarction; D and E: H&E staining of the rat heart tissue with BMSCs transplantation (BMSCs are indicated by black arrows, 100× and 400×, respectively); F: Frozen section of a rat in the modeled group with rBMSC transplantation (eGFP expression is indicated by red arrows); G and H: Immunohistochemistry of HSV1-tk in the heart tissue of rats in the modeled group with BMSC transplantation (brown indicates positive staining, 100× and 400×, respectively). I and K: Immunofluorescence of Cy3-CD90 and Cy3-CD45 in the heart tissue of rats in the modeled group with BMSC transplantation (CD90 is shown as red, CD45 was negative, 400×). J and L: Immunofluorescence of I and K with nuclear counterstaining (blue, 400×).

was significantly reduced and could not be detected. Interestingly, BLI is highly sensitive, and in the second week the signal could only be detected by BLI. In our subsequent analyses, we evaluated the transplanted stem cells in the infarcted myocardium in which the local blood supply mechanism was significantly different from that in the normal myocardium. There may have been insufficient blood supply in the transplantation region, as well as the presence of lesions and inflammation, which could result in the death of some transplanted BMSCs in the infarcted region. The use of this infarction model is also the major difference compared with the normal rat study of Wu et al [12], which indicated that the survival of transplanted stem cells in the infarcted region was affected by

the lesioned environment to a certain extent. One thing to note is that adenovirus was used as the TGF carrier, and it cannot insert the TGF fusion gene into the genome of BMSCs, resulting in the gradual reduction of exogenous proteins as a result of cell metabolism and proliferation.

We used the multi-functional reporter gene TGF for multi-modality molecular imaging to monitor transplanted BMSCs for the treatment of ischemic heart disease. First, we combined microPET and CT technologies in which microPET provided functional imaging and CT provided accurate anatomical localization. As shown in Figure 2B, we precisely located the stem cell transplantation region by coronal, sagittal and cross sections. In the development of molecular imaging, regular conventional imaging has become an inseparable complement. We believe that in the future, PET/CT will be more applicable to clinical development of stem cell tracking techniques in vivo. Second, the sensitivity of BLI reaches a concentration of 10^{-15} mol, which is significantly superior to that of PET [22]. During the 2 weeks of monitoring in our study, PET and fluorescence imaging could only obtain images of the transplanted rats in the first week after cell transplantation, whereas BLI was able to monitor cells for the whole duration. However, the bioluminescence technique is limited in terms of the spatial resolution by the influence of light scattering, and the penetration of the optical signal is only 2 cm [23,24], which is consistent with the images obtained in our study (Figures 1A and 3A). Thus, BLI has limited clinical use, and it is more suitable for small animal studies. Finally, owing to tissue attenuation and refraction, the eGFP of fluorescence imaging is only 2 mm [25,26]. Because of interference by the fur and tissue of rats, thoracotomy is required before fluorescence imaging, as shown in Figure 1A. Fluorescence is an autonomous property of cells, and the generation of fluorescence does not require an exogenous reaction substrate. We directly analyzed the expression of eGFP in tissue under a fluorescence microscope. Moreover, fluorescence imaging is superior to the other two imaging techniques in terms of its use for the in vitro analysis of eGFP.

In our study, although the dynamic observation of survival and migration of stem cells in the myocardium of the infarction model was successful, the duration was relatively short for in vivo monitoring of stem cell proliferation and differentiation as well as evaluation of whether cardiac function improved after stem cell transplantation for treating ischemic heart disease. A recent study has shown that a retrovirus can insert a target gene into the genome of stem cells, which may be advantageous for monitoring stem cell proliferation [27]. Most current studies of in vivo monitoring of transplanted stem cells to treat ischemic heart disease have been focused on cell survival, proliferation and migration [20,28–31]. Further research of stem cell differentiation and evaluation of its treatment efficacy is needed.

BMSCs promote myocardial repair and revascularization, and currently it is one of the promising methods for treating myocardial infarction [32–34]. To improve the repair of infarcted myocardium by transplanted BMSCs, a combination of gene therapy and transplanted BMSCs is used in most cases. For example, after transfection with Bcl-2 [35] or PAI-1 [36], the BMSC survival rate increases. Furthermore, Ang1-tranfected BMSCs provide better remodeling of infarcted myocardium [37]. Integrin-linked kinase promotes the adhesion of BMSCs to the infarcted myocardium [38]. Reporter gene imaging is mature and used for in vivo monitoring regardless of whether a therapeutic gene is expressed or not, the extent of expression and the duration of therapeutic gene expression [39]. In addition, owing to the characteristics the reporter gene technique, namely good specificity and a true reflection of the stem cells, such a technique is relatively mature for in vivo monitoring of stem cell therapy [40]. Therefore, TGF reporter gene imaging is likely to be a comprehensive method not only for tracking stem cells, but also for monitoring the gene expression in combination with gene therapy, which provides a multi-faceted platform for in vivo monitoring of transplanted stem cells for treating ischemic heart diseases.

Conclusion

This is the first application of TGF-transfected BMSC transplantation into the myocardial infarction model. Moreover, it proves that the dynamic situation of BMSCs in vivo can be monitored by microPET/CT, fluorescence and bioluminescence multimodality imaging. This study indicates that TGF can be used for in vivo monitoring of transplanted BMSCs for the treatment of ischemic heart disease as a multimodality reporter gene.

Author Contributions

Conceived and designed the experiments: XL YXZ. Performed the experiments: ZJP XL CXQ XTX HY ZLD. Analyzed the data: ZJP XL. Contributed reagents/materials/analysis tools: ZC ZJP. Wrote the paper: ZJP XL ZC YXZ.

References

1. Clifford DM, Fisher SA, Brunskill SJ, Doree C, Mathur A, et al (2012) Stem cell treatment for acute myocardial infarction. Cochrane Database Syst Rev. 15;2: CD006536.

2. Krause U, Arter C, Seckinger A, Wolf D, Reinhard A, et al (2007) Intravenous delivery of autologous mesenchymal stem cells limits infarct size and improves left ventricular function in the infarcted porcine heart. Stem Cells Dev. 16:31–37.

3. Price MJ, Chou CC, Frantzen M, Miyamoto T, Kar S, et al (2006) Intravenous mesenchymal stem cell therapy early after reperfused acute myocardial infarction improves left ventricular function and alters electrophysiologic properties. Int J Cardiol. 111:231–239.

4. Wolf D, Reinhard A, Krause U, Seckinger A, Katus HA, et al (2007) Stem cell therapy improves myocardial perfusion and cardiac synchronicity: new application for echocardiography. J Am Soc Echocardiogr. 20:512–520.

5. Orlic D, Kajstura J, Chimenti S, Bodine DM, Leri A, et al (2003) Bone marrow cells regenerate infarcted myocardium. Pediatr Transplant. 7:86–88.

6. Plewka M, Krzemińska-Pakuła M, Peruga JZ, Lipiec P, Kurpesa M, et al (2011) The effects of intracoronary delivery of mononuclear bone marrow cells in patients with myocardial infarction: a two year follow-up results. Kardiol Pol. 69:1234–1240.

7. Weissleder R (1999) Molecular imaging: exploring the next frontier. Radiology. 212:609–614.

8. Rodriguez-Porcel M, Wu JC, Gambhir SS (2008-2009) Molecular imaging of stem cells. StemBook [Internet].Cambridge (MA): Harvard Stem Cell Institute.

9. Yaghoubi SS, Creusot RJ, Ray P, Fathman CG, Gambhir SS (2007). Multimodality imaging of T-cell hybridoma trafficking in collagen-induced arthritic mice: image-based estimation of the number of cells accumulating in mouse paws. J Biomed Opt. 12:064025.

10. Gyöngyösi M, Blanco J, Marian T, Trón L, Petneházy O, et al (2008) Serial noninvasive in vivo positron emission tomographic tracking of percutaneously intramyocardially injected autologous porcine mesenchymal stem cells modified for transgene reporter gene expression. Circ Cardiovasc Imaging. 1:94–103.

11. Higuchi T, Anton M, Dumler K, Seidl S, Pelisek J, et al (2009) Combined reporter gene PET and iron oxide MRI for monitoring survival and localization of transplanted cells in the rat heart. J Nucl Med. 50:1088–1094.

12. Wu JC, Chen IY, Sundaresan G, Sundaresan G, Min JJ, et al (2003) Molecular imaging of cardiac cell transplantation in living animals using optical bioluminescence and positronemission tomography. Circulation. 108:1302–1305.

13. Wu JC, Spin JM, Cao F, Lin S, Xie X, et al (2006) Transcriptional profiling of reporter genes used for molecular imaging of embryonic stem cell transplantation. Physiol Genomics. 25:29–38.

14. Pei Z, Lan X, Cheng Z, Qin C, Wang P, et al (2012) A multimodality reporter gene for monitoring transplanted stem cells. Nucl Med Biol.39:813–820.

15. Johns TNP, Olson BJ (1954) Experimental myocardial infarction. A method of coronary occlusion in small animals. Ann Surg. 140:675–682.

16. Fisbein MC, Melecan D, Marko PR (1978) Experimental myocardial infarction in the rat:qualitative and quantitative changes during pathologic evolution. Am J Pathol. 90 :57–70.

17. Pfeffer MA, Pfeffer M, Fisbein MC (1979) Myocardial infarct and ventricular function in rats. Circ Res. 44:503–512.

18. Tarnavski O, McMullen JR, Schinke M, Nie Q, Kong S, et al (2004). Mouse cardiac surgery: comprehensive techniques for the generation of mouse models of human diseases and their application for genomic studies. Physiol Genomics. 16:349–360.

19. Love Z, Wang F, Dennis J, Awadallah A, Salem N, et al (2007). Imaging ofmesenchymal stem cell transplant by bioluminescence and PET. J Nucl Med. 48:2011–2020.

20. Cao F, Lin S, Xie X, Ray P, Patel M, et al (2006) In vivo visualization of embryonic stem cell survival, proliferation,and migration after cardiac delivery. Circulation. 113:1005–1014.

21. Willmann JK, Paulmurugan R, Rodriguez-Porcel M, Stein W, Brinton TJ, et al (2009) Imaging gene expression in human mesenchymal stem cells: from small to large animals. Radiology. 252:117–127.

22. Jang YY, Ye Z, Cheng L (2012) Molecular imaging and stem cell research. Mol Imaging. 11:1–12.

23. Wang HE, Yu HM, Liu RS, Lin M, Gelovani JG, et al (2006) Molecular imaging with ^{123}I-FIAU, ^{18}F-FUDR, ^{18}F-FET, and ^{18}F-FDG for monitoring herpes simplex virus type 1 thymidine kinase and ganciclovir prodrug activation gene therapy of cancer. J Nucl Med. 47:1161–1171.

24. Yaghoubi SS, Couto MA, Chen CC, Polavaram L, Cui G, et al (2006) Preclinical safety evaluation of 18F-FHBG: a PET reporter probe for imaging herpes simplex virus type 1 thymidine kinase (HSV1-tk) or mutant HSV1-sr39tk's expression. J Nucl Med. 47:706–715.

25. Contag CH (2007) In vivo pathology: seeing with molecular specificity and cellular resolution in the living body. Annu Rev Pathol. 2:277–305.

26. Massoud TF, Gambhir SS (2003) Molecular imaging in living subjects: seeing fundamental biological processes in a new light. Genes Dev. 17:545–580.

27. Roelants V, Labar D, Meester C, Havaux X, Tabilio A, et al (2008) Comparison between adenoviral and retroviral vectors for the transduction of the thymidine kinase PET reporter gene in rat mesenchymal stem cells. J Nucl Med. 49:1836–1844.

28. Min JJ, Ahn Y, Moon S, Kim YS, Park JE, et al (2006) In vivo bioluminescence imaging of cord blood derived mesenchymal stem cell transplantation into rat myocardium. Ann Nucl Med. 20:165–170.

29. Inubushi M, Tamaki N (2007) Radionuclide reporter gene imaging for cardiac gene therapy. Eur J Nucl Med Mol Imaging. 34:S27–33.

30. Sheikh AY, Lin SA, Cao F, Cao Y, van der Bogt KE, et al (2007) Molecular imaging of bone marrow mononuclear cell homing and engraftment in ischemic myocardium. Stem Cells. 25: 2677–2684.

31. Hu S, Cao W, Lan X, He Y, Lang J, et al (2011) Comparing study of rNIS and hNIS as reporter genes monitoring rBMSCs transplanted into infarcted myocardium. Mol Imaging. 10:227–237.

32. Picinich SC, Mishra PJ, Mishra PJ, Gold J, Banerjee D (2007) The therapeutic potential of mesenchymal stem cells. Cell- & tissue-based therapy. Expert Opin Biol Ther. 7:965–973.

33. Schafer R, Northoff H (2008) Cardioprotection and cardiac regeneration by mesenchymal stem cells. Panminerva Med. 50:31–39.

34. Copland IB, Lord-Dufour S, Cuerquis J, Coutu DL, Annabi B, et al (2009) Improved autograft survival of mesenchymal stromal cells by plasminogen activator inhibitor 1 inhibition. Stem Cells. 27:467–477.

35. Li W, Ma N, Ong LL, Nesselmann C, Klopsch C, et al (2007) Bcl-2 engineered MSCs inhibited apoptosis and improved heart function. Stem Cells. 25:2118–2127.

36. Deuse T, Peter C, Fedak PW, Doyle T, Reichenspurner H, et al (2009) Hepatocyte growth factor or vascular endothelial growth factor gene transfer maximizes mesenchymal stem cell-based myocardial salvage after acute myocardial infarction. Circulation. 120: S247–254.

37. Sun L, Cui M, Wang Z, Feng X, Mao J, et al (2007) Mesenchymal stem cells modified with angiopoietin-1 improve remodeling in a rat model of acute myocardial infarction. Biochem Biophys Res Commun. 357:779–784.

38. Song SW, Chang W, Song BW, Song H, Lim S, et al (2009) Integrin-linked kinase is required in hypoxic mesenchymal stem cells for strengthening cell adhesion to ischemic myocardium. Stem Cells. 27: 1358–1365.

39. Jaffer FA, Libby P, Weissleder R (2007) Molecular imaging of cardiovascular disease. Circulation. 116:1052–1061.

40. Higuchi T, Anton M, Saraste A, Dumler K, Pelisek J, et al (2009). Reporter gene PET for monitoring survival of transplanted endothelial progenitor cells in the rat heart after pretreatment with VEGF and atorvastatin. J Nucl Med. 50:1881–1886.

13

Influence of Mesenchymal Stem Cell Transplantation on Stereotypic Behavior and Dopamine Levels in Rats with Tourette Syndrome

Xiumei Liu[9], Xueming Wang[9], Lixia Li*, Haiyan Wang, Xiaoling Jiao

Department of Pediatrics, Yuhuangding Hospital of Qingdao University, Yantai, Shandong, China

Abstract

Context: Tourette syndrome (TS) is a heterogeneous neuropsychiatric disorder. Chronic motor and phonic tics are central symptoms in TS patients. For some patients, tics are intractable to any traditional treatment and cause lifelong impairment and life-threatening symptoms. New therapies should be developed to address symptoms and overt manifestations of TS. Transplantation of neurogenic stem cells might be a viable approach in TS treatment.

Objective: We used mesenchymal stem cell (MSC) transplantation to treat TS. We discuss the mechanism of action, as well as the efficiency of this approach, in treating TS.

Settings and Design: An autoimmune TS animal model was adopted in the present study. Forty-eight Wistar rats were randomly allocated to the control group and the 2 experimental groups, namely, TS rats+vehicle and TS rats+MSC. MSCs were co-cultured with 5-bromodeoxyuridine (BrdU) for 24 h for labeling prior to grafting.

Methods: Stereotypic behaviors were recorded at 1, 7, 14, and 28 days after transplantation. Dopamine (DA) content in the striatum of rats in the 3 groups was measured using a high-performance liquid chromatography column equipped with an electrochemical detector (HPLC-ECD) on day 28 after transplantation.

Statistical analysis: Statistical analysis was performed by repeated measurements analysis of variance to evaluate stereotypic behavior counts at different time points.

Results: TS rats exhibited higher stereotypic behavioral counts compared with the control group. One week after transplantation, TS rats with MSC grafts exhibited significantly decreased stereotypic behavior. Rats with MSC grafts also showed reduced levels of DA in the striatum when compared with TS rats, which were exposed only to the vehicle.

Conclusions: Intrastriatal transplantation of MSCs can provide relief from the stereotypic behavior of TS. Our results indicate that this approach may have potential for developing therapies against TS. The mechanism(s) of the observed effect may be related to the suppression of DA system by decreasing the content of DA in TS rats.

Editor: Pranela Rameshwar, University of Medicine and Dentistry of New Jersey, United States of America

Funding: This study was supported by National Natural Science Foundation (81101017), Department of Science and Technology of Shandong Province (BS2010F030), and Yantai Science and Technology Development project (2010148-24). The funders had no role in study design, data collection and analysis, decision to publish, or preparation of the manuscript.

Competing Interests: The authors have declared that no competing interests exist.

* E-mail: dr_llx@sina.com

⑨ These authors contributed equally to this work.

Introduction

Tourette syndrome (TS) is a developmentally regulated neurobehavioral disorder in which chronic motor and phonic tics are central symptoms. The prevalence of TS is estimated to be between 4 and 6 per 1,000 children and adolescents [1]. Epidemiological studies have shown that TS is relatively common and more prevalent in boys. Studies show that approximately 50% of TS patients continue to suffer from tics well into adulthood, and at least a third of patients with TS exhibit tic-related self-injurious behavior [2]. For some individuals, tics can cause lifelong

impairment, and about 5% of TS patients have life-threatening symptoms, which are defined as malignant TS [3].

The pathophysiology and etiology of TS are unclear; however, it is possible that a combination of genetic and environmental factors is involved in it [4]. Structural and functional neuroimaging and neurophysiological and post-mortem studies have shown that the basal ganglia and related cortico-striato-thalamo-cortical circuits, as well as the dopaminergic neuronal system, may be dysfunctional in TS [5,6]. Traditional therapies such as pharmacological treatments, behavioral therapies, and surgical approaches can reduce the frequency and intensity of tics but cannot eliminate them entirely. Therefore, it is recommended that treatments suggested above should be regarded only as symptomatic therapy. Surgical techniques involving deep brain stimulation (DBS) of the thalamus or globus pallidus may also be considered for patients with severe TS. However, side effects of DBS, such as drowsiness, reduced energy, psychosis, and spontaneous tic recurrence, have been reported [7,8]. Furthermore, the use of DBS for the treatment of TS is limited because of post-surgical complications and the necessity of expensive infrastructure to provide this form of therapy [9]. Therefore, new therapeutic options should be explored for TS patients who are significantly impaired by this syndrome.

In recent years, stem cell-based therapy has been perceived as a potential treatment for many neurological disorders. Experiments with animal models suggest that if stem cells are injected into the brain or even the bloodstream, the transplanted cells can survive and migrate to damaged portions of the nervous system, following which they get incorporated into working neural circuits and replace dead neurons. This ability is extremely important for the reconstitutive effects of stem cell treatments. The tested animals display significant improvement in various functions, resulting from stem cell transplants [10]. Stem cell therapy may provide a breakthrough for some of the existing limitations of traditional pharmaceutical approaches, and it is a good choice for the treatment of neural diseases whose exact pathogenesis is unclear [11]. Neural stem cells (NSCs) are considered to be a heterogeneous population of mitotically active, self-renewing, multipotent, and immature progenitor cells [10]. In 2008, we transplanted NSCs into the striatum of TS rats and observed the therapeutic effects of NSCs on the stereotypic behaviors of TS rats. NSCs survived in the brain of TS rats, and a fraction of them differentiated into neurons and gliocytes [12]. Compared with NSCs, mesenchymal stem cells (MSCs) are a better choice for cell transplantation therapy because they are easily accessible, capable of rapid expansion in culture, immunologically inert, and capable of long-term survival and integration with the host tissue. In this study, we propose transplantation of MSCs as a novel therapy for TS and provide insights into the mechanism of action of these cells.

Although the definitive pathophysiological mechanism of TS is not well understood, it is widely believed that abnormalities in the dopaminergic neuronal system play a primary role in the pathophysiology of TS [13]. The efficacy of anti-dopaminergic agents such as haloperidol in treating TS, along with other clinical and basic science findings, have contributed to the concept that abnormal dopamine (DA) signaling and aberrations in basal ganglia processing are important factors contributing to the pathophysiology of TS [14]. In the present study, we transplanted MSCs into the striata of TS rats and subsequently investigated the effect of the transplanted MSCs on stereotypic behaviors and DA levels of TS rats.

Subjects and Methods

Animals

Forty-eight 7–8-week-old Wistar rats (24 of each sex; weight range, 205–220 g) were used for this study. Animals were housed in an environmentally controlled room maintained at a temperature of $21°C \pm 1°C$ with a 12-h light/dark cycle (lights on 0700–1900 hours). The animals had free access to food and water. Experimental procedures were performed in accordance with the NIH Guidelines for the care and use of laboratory animals.

MSC Preparation and Flow Cytometric Analysis

Mononuclear cells were isolated from the long bones of 5 adult Wistar rats. The bones were dissected free, and the proximal and distal ends were removed to reveal the marrow cavity. Cells were cultured in low-glucose Dulbecco's-modified Eagle's medium (Gibco-BRL, Grand Island, NY, USA) supplemented with 15% fetal bovine serum (Gibco-BRL), 100 U/mL penicillin, and 100 µg/mL streptomycin (Gibco-BRL). Cells were incubated in 5% CO_2 at 37°C. Culture medium was replaced every 3–4 days. When the cells reached confluence, adherent cells were harvested with trypsin (Sigma, St. Louis, MO, USA) and subsequently passaged. A small part of the MSC samples from the third to fourth passages was tested utilizing flow cytometric analysis. The remaining MSCs were co-incubated with BrdU for 48 h prior to transplantation. To label MSCs, bromodeoxyuridine (BrdU, 10 µg/mL, Sigma, St. Louis, MO, USA)), a thymidine analog and marker of newly synthesized DNA, was added into the medium for 48 h before transplantation. More than 90% of MSCs selected for immunostaining were immunoreactive for BrdU. Subsequently, cells were harvested and resuspended in PBS at a density of 1×10^5 cells/µL and stored on ice until grafting.

Cells in their third to fourth passage were collected and treated with 0.25% trypsinase. Cells were stained with fluorescein isothiocyanate- or phycoerythrin-conjugated anti-marker monoclonal antibodies in 100 µL PBS for 30 min at 4°C, as suggested by the manufacturer. Antibodies against rat antigens CD29, CD34, CD44, and CD45 were purchased from SeroTec (Raleigh, NC, USA). Appropriate isotype-matched, non-reactive, fluorochrome-conjugated antibodies were used as controls. Cells were analyzed using a flow cytometry system (Cytometer 1.0, CytomicsTM FC500, Beckman Coulter, Los Angeles, CA, USA).

Animal Preparation and in vivo Surgery

A total of 48 Wistar rats were randomly divided into 3 groups: control group, TS+vehicle (PBS) group, and TS+MSC group (n = 16 for each group). Surgery was performed as previously described [15,16]. Briefly, rats were deeply anesthetized with chloral hydrate (400 mg/kg, i.p.) and placed in a stereotaxic apparatus (Stoelting, Wood Dale, Illinois, USA) with the incisor bar set at 3.5 mm below the interaural line. Using aseptic surgical techniques, the skull was exposed, holes were drilled where appropriate, and 28-gauge guide cannulae were implanted into the bilateral striata [15]. Coordinates for cannula placements were 2.0 mm anterior-posterior from the bregma and 4.0 mm medial-lateral and –7.0 mm dorsoventral from the skull. Proper post-surgical care was provided to the animals. The diet of operated rats was supplemented with fresh fruit and egg yolk to maintain body weight.

Rats were allowed to recover for 1 week to re-establish integrity of the blood brain barrier. After the recovery period, osmotic mini-pumps (Alzet, Palo Alto, CA, USA) filled with PBS were connected to each cannula using a polyethylene tube loaded with 50 µL undiluted sera of TS patients using sterile conditions. All sera of TS patients (twelve male subjects and four female subjects, age range = 8–14 years, mean = 11.3) were taken from our serum bank. No subjects were taking psychostimulants at the time that blood was drawn. Collection of sera was performed under a protocol approved by the Institutional Review Board and after consent was obtained. The sera of TS patients bear a high titer of antibasal ganglia antibody, which could induce impairments of the striatum and subsequent stereotypic behavior. The antibasal ganglia antibody ELISA optical density readings of 16 TS subjects selected for microinfusion was 0.952 ± 0.184. Sera were microinfused at a rate of 0.5 µL/h for 72 h. Control surgery rats were microinfused with PBS. After 72 h, animals were again sedated with an intraperitoneal injection of chloral hydrate (100 mg/kg) and placed in the stereotaxic frame. An incision was made along the midline to expose the skull, and the pumps were removed. The MSC suspension (10^5/μL, 5 μL/site) was bilaterally injected into the serum-infusion site. The needle was maintained in place for 5 min before slow removal. The wound was closed with a surgical suture. Each grafted animal received a total of 10^6 MSC. For control grafting, animals underwent the same grafting procedure but received a vehicle infusion of PBS of equal volume. Animals were intramuscularly administered 65,000 units of sodium penicillin and were maintained on a thermal pad until they awoke. The rats were then returned to the home cages.

Assessment of Stereotypic Manifestation of TS

In this study, we established an autoimmune animal model of TS according to protocols established by Hallett and Taylor [15,16]. Stereotypic behaviors in rats were similar to motor and phonic tics in TS patients, and therefore, stereotypy was used as an indicator for successful induction of TS in these animals. Rat movements were video- and audio-taped at the end of the light cycle for 30 min. Stereotypy was recorded, including biting (teeth touching cage or wood chips, vacuous chewing on other objects except the body), taffy pulling (forepaw to the mouth and face), self-gnawing, licking, grooming, head shaking, paw shaking, rearing, and episodic utterances [15,16]. Grooming behavior was recorded by the number of minutes in which grooming occurred. Episodic utterance was defined as repeated medium-pitched sounds of short duration. Stereotypic movements were recorded at 1, 7, 14, and 28 days after transplantation. Animals received a total score equal to the sum of observed movements. A researcher, trained to identify stereotypy of interest and blind to the types of grafts and substance microinfused, quantified the stereotypy by reviewing and listening to the video tapes.

Immunohistochemistry

Twenty eight days after MSC transplantation, rat brains were removed from the cranium and post-fixed in paraformaldehyde prior to sectioning [7]. Rat brain sections were embedded in paraffin and 6-μm thick coronal sections were prepared. All immunostaining processes were performed according to instructions from the HistostainTM-DS kit (Zhongshan, Peking, China). Sections were incubated in 3% H2O2 in methanol for 10 minutes and in 10% goat serum for 10 minutes at room temperature. The sections were incubated in rabbit anti-rat BrdU monoclonal antibody (Zhongshan, Peking, China; 1:250) at 4°C overnight in a humidity chamber and were then treated with biotinylated goat anti-rabbit IgG (Zhongshan; 1:200) for 10 minutes. The remaining primary antibodies, which included mouse anti-rat GFAP monoclonal antibody (Sigma, USA; 1:200, astrocyte marker), mouse anti-rat MAP2 monoclonal antibody (Sigma, USA; 1:200, neuronal marker), and mouse anti-rat Nestin monoclonal antibody (BD Bioscience Pharmingen, Newark, NJ, USA; 1:400, neural precursor cell marker) were added. Subsequently, the sections were treated with biotinylated secondary antibody goat anti-mouse monoclonal IgG (1:100). Finally, horseradish peroxidase-conjugated streptavidin was added and sections were coverslipped in permanent mounting medium.

High-performance Liquid Chromatography with Electrochemical Detection

Twenty-eight days after MSC transplantation, rats in the 3 groups described were killed by decapitation. Both sides of the striatum were isolated and transferred to liquid nitrogen for storage. Samples were homogenized in ice-cold perchloric acid (0.1 mol/L) with 1% ethanol and 0.02% ethylenediamine tetra-acetic acid. DA in the dialysates was measured using a high-performance liquid chromatography column equipped with an electrochemical detector. The HPLC-ECD system comprised a reverse-phase column (MA-5 ODS, 150×4.6-mm ID, Eicom), a model L-6000 pump (Hitachi), and an electrochemical detector (ECD-100, Eiercom). The mobile phase contained 85 mmol/L citric acid–100 mmol/L sodium acetate, 0.2 mmol/L disodium EDTA, 1.2 mmol/L sodium octane sulfonate, and 5% methanol in deionized and distilled water. The flow rate was 1 mL/min, and the pH was adjusted to 3.5. Results were expressed as ng/mg wet weight of brain tissue.

Statistical Analysis

All statistical analyses were carried out using SPSS (version 13.0 for Windows; SPSS Inc, Chicago, IL, USA). Data were reported as mean ± SD. Statistical analysis was performed by repeated measurements analysis of variance to evaluate stereotypy counts at different time point. A P value of <0.05 was considered statistically significant.

Results

Assessment of Stereotypy

Serologic studies of TS have demonstrated that the existence of anti- basal ganglia antibodies induced striatal dysfunction. Stereotypic behaviors were successfully induced in rats by intrastriatal microinfusion of the sera of TS patients. Rats were observed at 1, 7, 14, and 28 days after transplantation. Stereotypic behaviors were recorded and quantified during a 30-min observation period. Using repeated measurements analysis of variance, the overall model exhibited significant group (F = 43.58, P<0.01) and day effects (F = 18.42, P<0.01), as well as (group × day) interactions (F = 14.25, P<0.01), indicating varying degrees of differences among groups and across days. Statistical analysis suggested that TS rats exhibited higher stereotypic behavioral counts compared with the control group. TS+MSC and TS+vehicle injection groups exhibited significant differences between the groups (F = 35.72, P<0.01), as well as a time effect (F = 40.36, P<0.01). TS rats with MSC grafts exhibited significantly decreased stereotypic behaviors 1 week after transplantation. (Figure 1).

Characteristics of MSCs

Bone marrow contains a category of non-hematopoietic multipotent cells that can be cultivated in vitro. They were initially called plastic adherent cells but have recently been renamed as bone marrow stromal cells. In this study, primary MSC cultured as plastic adherent cells were maintained in culture. After 14 days in culture, the attached MSCs developed into an adherent layer with abundant dispersed fibroblast-like cells; each colony was formed by fibroblast-like cells. By day 28, MSCs had proliferated and formed a nearly continuous layer comprising mainly of fibroblast-like cells amongst which a subset of characteristically flattened and spindle-shaped-cells could be recognized. MSCs are mesenchymal elements that provide structural and functional support for hemopoiesis. These cells lack the hematopoietic surface markers but express mesenchymal markers. Our flow cytometry results demonstrate that cultured cells were negative for CD34 and CD45 (hematopoietic lineage markers). These cells showed strong expression of integrin CD29, endothelial progenitor/precursor

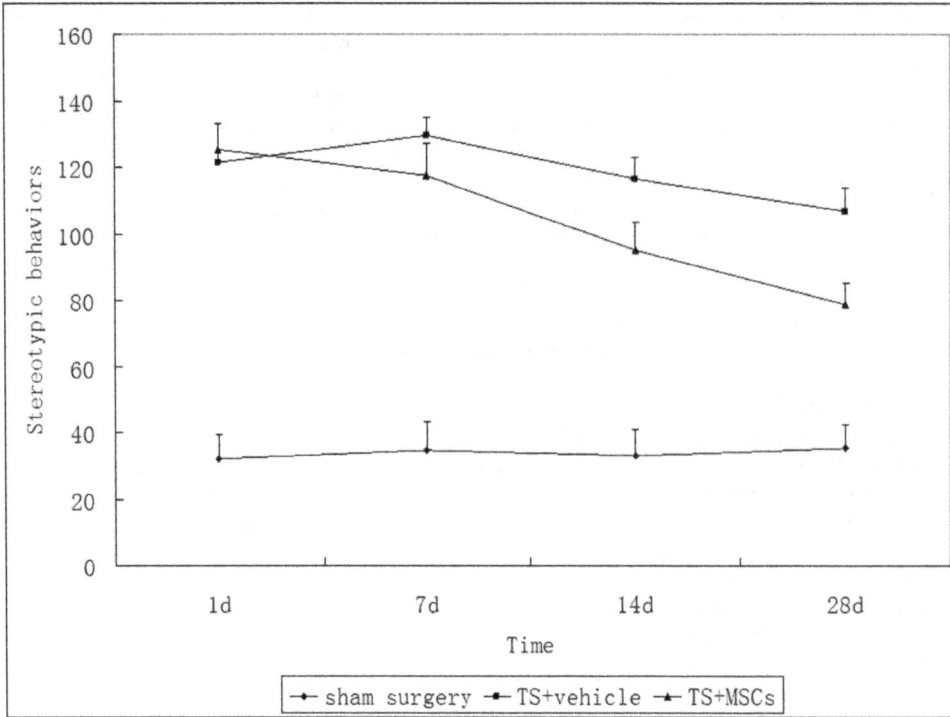

Figure 1. Stereotypy counts were recorded for 30 min at different time points in 3 groups. Scores are higher in the Tourette syndrome (TS) group than those in the control group. TS rats with mesenchymal stem cell (MSC) grafts exhibited decreased stereotypic behaviors compared with TS+vehicle rats. $P<0.05$.

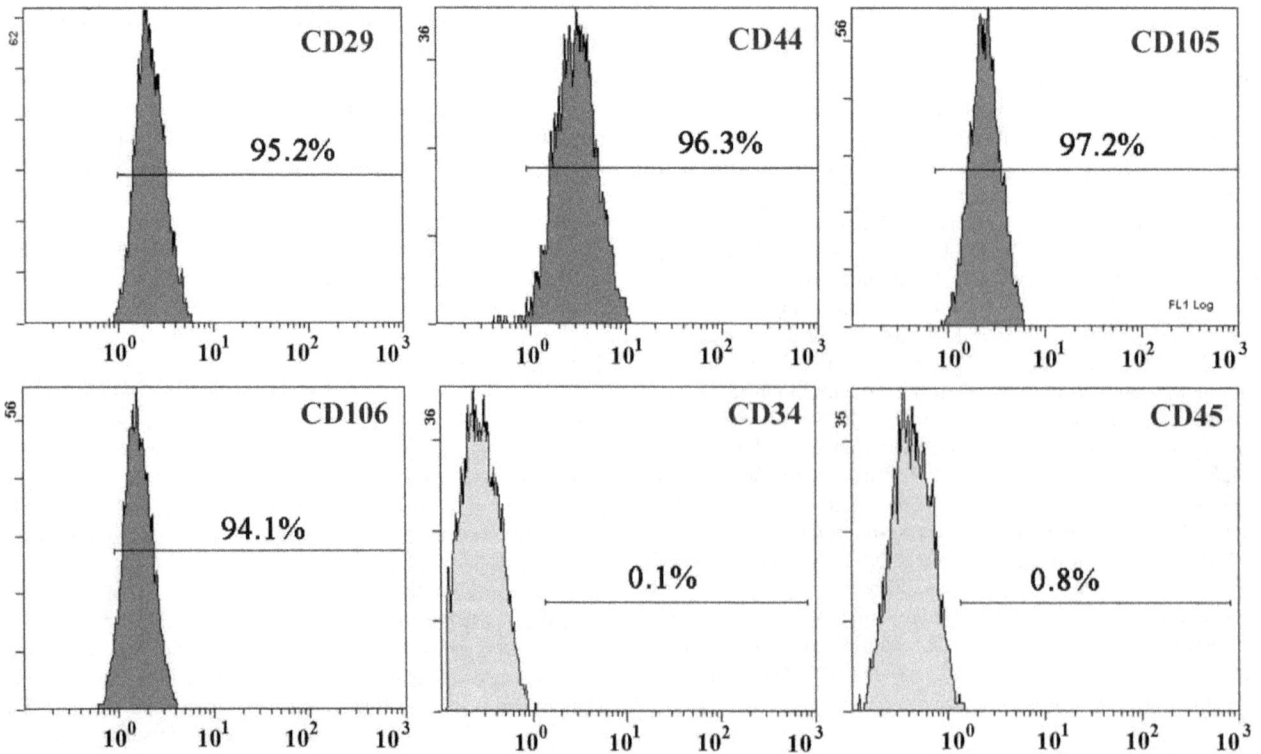

Figure 2. Flow cytometry analysis of MSCs. Cells were cultured for 3–4 passages, harvested, and analyzed by a flow cytometry system. The y-axis represents the number of cells, and the x-axis represents CD29, CD-44, CD105, CD106, CD34, and CD45.

lineage marker CD105, and matrix receptors CD44 and CD106 (Fig. 2).

Immunohistochemistry

In the TS rats with MSC graft, 5-bromodeoxyuridine (BrdU)-positive cells were prevalent in the striatum near the injection site. BrdU and Nestin double-positive cells were also observed, indicating that grafted cells differentiated into neural precursor cells (Figure 3A). Results also showed that a considerable portion of transplanted cells coexpressed BrdU and microtubule-associated protein 2 (MAP-2) (Figure 3B), and several coexpressed BrdU and glial fibrillary acidic protein (GFAP) (Figure 3C).

Levels of DA in the Striatum

Hyperfunction of the DA system plays an important role in the etiopathogenesis of TS. The content and activity of DA in the striatum are closely associated with TS. Some studies have shown that DA levels are increased in TS patients [13]. At the end of our experiments, DA levels in the striatum of TS rats were significantly increased compared with those from the control surgery group (994.8±112.6 ng/g versus 607.3±56.6 ng/g, P<0.01). Reduction in the levels of DA or enhanced degradation of DA is likely to control the stereotypic behavior seen in TS. TS rats that received the MSC grafts had lower DA levels compared with untreated rats in the TS group (757.9±82.4 ng/g versus 994.8±112.6 ng/g, P<0.01). (Figure 4).Our findings demonstrate that transplantation of MSCs can reduce the levels of DA in TS rats and alleviate symptoms of TS.

Discussion

The prevalence of TS is estimated to be approximately between 4 and 6 per 1,000 children and adolescents [17]. In general, tics are self-limited or can be treated by behavioral or pharmacological therapy. It is estimated that approximately 20% of children with TS continue to experience a moderate level of impairment of global functioning by the age of 20 years [18]. Furthermore, severe tics can lead to physical injury and significant functional impairment. Patients with malignant TS have life-threatening and self-injurious symptoms, which are intractable to both conservative treatments and neurosurgical procedures such DBS. In order to find effective new therapies for these unfortunate patients, we have examined the approach of transplanting MSCs into the brain of rats, which serves as an animal model for TS. Our results published earlier show that transplantation of NSCs into

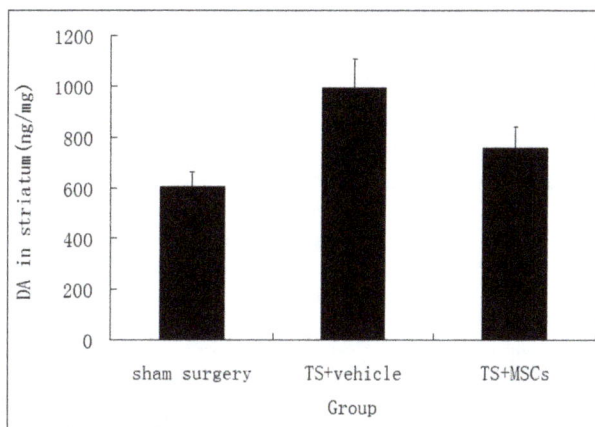

Figure 4. The amounts of dopamine (DA) in striatum homogenate (Figure 4) were analyzed at the end of the test by high-performance liquid chromatography with electrochemical detection in 3 groups (n = 16).

TS rats resulted in improvement of stereotypic behaviors in TS rats [12].

In recent years, MSCs have shown great therapeutic potential in many neurological diseases. Direct or intravascular MSC transplantation has been shown to bring about functional improvements in experimental animal models of central nervous system (CNS) injury and disease [19]. Furthermore, owing to their relatively easy isolation from bone marrow and their extensive capacity for *in vitro* expansion, transplantation of MSCs has been considered as a good choice for cell therapy and tissue re-engineering. In a previous study, we found that TS rats bearing MSC grafts exhibited significantly decreased stereotypic behaviors compared with control animals [20]. Research suggests that under specific experimental conditions, MSC can differentiate into cells with neural phenotypes, which would potentially enable them to replace neural tissue lost after CNS injury. In the present study, BrdU and Nestin double-positive cells were quantified in the rat striatum following MSC transplantation, which demonstrated that grafted MSC differentiated into neural precursor cells. Some of transplanted cells coexpressed BrdU and MAP-2 and several coexpressed BrdU and GFAP, which demonstrated that transplanted MSCs differentiate into neurons and astrocytes. The cell replacement

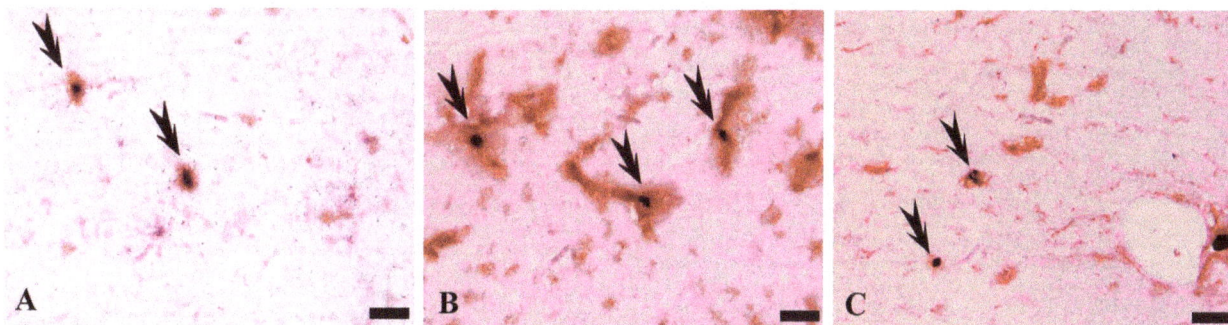

Figure 3. Immunostaining analyses of grafted MSCs in a rat model of Tourette's syndrome at 4 wk after transplantation. Grafted MSCs are identified by BrdU staining. (A) Grafted cells are double-labeled with anti-BrdU (purple) and anti-Nestin (red). (B) Some grafted MSCs differentiated into neurons, as evidenced by double-labeling with BrdU and MAP-2. (C) Grafted cells differentiated into astrocytes, which coexpressed BrdU and GFAP. Scale bar: 20 μm. Arrows represent transplanted MSCs that differentiated into different types of neural cells.

theory was based on replacement of damaged neural cells with alternative functioning cells that induce long-lasting, clinical improvement. It is reasoned that transplanted cells survive, integrate into the endogenous neural network, and lead to functional improvement. These results led us to speculate that replacement of neuronal cells by MSCs contributed to the functional improvement of TS rats. However, the rate of differentiation of MSCs was lower than we expected it to be. This fact makes it difficult to assess evidence for the mechanism of action of MSCs in alleviating symptoms of TS.

The pathophysiology and etiology of TS are not completely understood. Brain imaging, neurophysiology, and post-mortem studies support the involvement of cortical-striatal-thalamo-cortical pathways in the progression of TS. The clinical efficacy of DA antagonists in tic suppression gives rise to speculation that abnormalities of striatal dopaminergic neurotransmission are part of the underlying mechanism that results in TS [21]. Since it is widely believed that abnormalities of DA neuro-transmission play a primary and important role in the pathophysiology of TS, in this study, we examined the levels of DA in TS rats after transplantation of MSC.

DA is a major monoaminergic neurotransmitter released by nerve terminals originating from midbrain neurons. Metabolism of DA in nerve terminals is regulated by multiple processes such as synthesis, storage, reuptake by presynaptic cells, and catabolism. Since these metabolic processes are energy-dependent, DA metabolism in the striatum is disrupted when the striatal region is damaged [22]. It has been demonstrated that excessive release of DA takes place into the extracellular space in brain ischemia animal models [23]. In our study, the autoimmune TS animal model was microinfused with anti-basal ganglia antibody, which induced basal ganglia impairment. Our results show that DA levels were high in the striatum of TS rats. These results concur with those from other studies as well [22,23].

DA modulates the activity of neurons in the striatum by binding to and signaling via DA receptors. There are 2 families

of DA receptors, namely, D1-like receptor (DRD1) and D2-like receptor (DRD2). The content and activity of DA and density and sensitivity of DRD2 in the striatum seemed to be closely associated with TS. DA produced a remarkable effect only after binding to DRD2. Many studies have shown that levels of DA and activity of the DRD2 receptor are increased in TS patients [24]. It is suggested that decreasing the levels of DA or promoting its metabolism and inhibiting the activity of the DRD2 receptor can control the associated stereotypic behavior [25]. In the present study, we demonstrate that following the transplantation of MSC, the levels of DA in TS rats were decreased and the stereotypic behaviors were suspended.

The present study has several limitations. First, this study was an *in vivo* rat study, and caution should be exercised when applying these results to human patients in clinical settings. Second, this research was based on a transient and induced TS model wherein it was difficult to reproduce the full spectrum of pathogenesis of TS as seen in humans. Future studies are needed to examine the effects of MSC transplantation on the dopaminergic nervous system.

Conclusion

In summary, the results of this study provide evidence that transplantation of MSCs effectively inhibits stereotypic TS behaviors. The mechanism of rectification of this pathogenesis is likely to be achieved by downmodulating the activity of dopaminergic neurons.

Acknowledgments

The authors gratefully acknowledge Yuwei Wang, Yu Sun, and Weidong Zhang for their expert technical assistance. We also thank Dr. Xiuli Ju and their laboratory for sharing their protocol for harvesting MSCs.

Author Contributions

Guarantors of integrity of the entire study: XL LL. Made the animal model: XW. Did the statistics: XJ. Analyzed the data: HW.

References

1. Knight T, Steeves T, Day L, Lowerison M, Jette N, et al. (2012) Prevalence of tic disorders: a systematic review and meta-analysis. Pediatr Neurol 47: 77–90.

2. Hassan N, Cavanna AE (2012) The prognosis of Tourette syndrome: implications for clinical practice. Funct Neurol 27: 23–7.

3. Cheung MY, Shahed J, Jankovic J (2007) Malignant Tourette syndrome. Mov Disord 22: 1743–50.

4. Wang Z, Maia TV, Marsh R, Colibazzi T, Gerber A, et al. (2011) The neural circuits that generate tics in Tourette's syndrome. Am J Psychiatry 168: 1326–37.

5. Davila G, Berthier ML, Kulisevsky J, Jurado Chacon S (2010) Suicide and attempted suicide in Tourette's syndrome: a case series with literature review. J Clin Psychiatry 71: 1401–2.

6. Palminteri S, Lebreton M, Worbe Y, Hartmann A, Lehericy S, et al. (2011) Dopamine-dependent reinforcement of motor skill learning: evidence from Gilles de la Tourette syndrome. Brain 134: 2287–301.

7. Neuner I, Halfter S, Wollenweber F, Podoll K, Schneider F (2010) Nucleus accumbens deep brain stimulation did not prevent suicide attempt in tourette syndrome. Biol Psychiatry 68: e 19–20.

8. Albin RL, Mink JW (2006) Recent advances in Tourette syndrome research. Trends Neurosci 29: 175–82.

9. Idris Z, Ghani AR, Mar W, Bhaskar S, Wan Hassan WN, et al. (2010) Intracerebral haematomas after deep brain stimulation surgery in a patient with Tourette syndrome and low factor XIIIA activity. J Clin Neurosci 17: 1343–4.

10. Tang HL, Sun HP, Wu X, Sha HY, Feng XY, et al. (2011) Detection of neural stem cells function in rats with traumatic brain injury by manganese-enhanced magnetic resonance imaging. Chin Med J 124: 1848–53.

11. Notta F, Doulatov S, Laurenti E, Poeppl A, Jurisica I, et al. (2011) Isolation of single human hematopoietic stem cells capable of long-term multilineage engraftment. Science 333: 218–21.

12. Liu X, Wang Y, Li D, Xiuli Ju (2008) Transplantation of rat neural stem cells reduces stereotypic behaviors in rats after intrastriatal microinfusion of Tourette syndrome sera. Behav Brain Res 186: 84–90.

13. Nemoda Z, Szekely A, Sasvari-Szekely M (2011) Psychopathological aspects of dopaminergic gene polymorphisms in adolescence and young adulthood. Neurosci Biobehav Rev 35: 1665–86.

14. Yoon DY, Gause CD, Leckman JF, Singer HS (2007) Frontal dopaminergic abnormality in Tourette syndrome: a postmortem analysis. J Neurol Sci 255: 50–56.

15. Hallett JJ, Harling-Berg CJ, Knopf PM, Stopa EG, Kiessling LS (2000) Antistriatal antibodies in Tourette syndrome cause neuronal dysfunction. J Neuroimmunol 111: 195–202.

16. Taylor JR, Morshed SA, Parveen S, Mercadante MT, Scahill L, et al. (2002) An animal model of Tourette's syndrome. Am J Psychiatry 159: 657–60.

17. Cortese S, Lecendreux M, Bernardina BD, Mouren MC, Sbarbati A, et al. (2008) Attention-deficit/hyperactivity disorder, Tourette's syndrome, and rest-less legs syndrome: the iron hypothesis. Med Hypotheses 70: 1128–32.

18. Rickards H, Cavanna AE, Worrall R (2012) Treatment practices in Tourette syndrome: The European perspective. Eur J Paediatr Neurol 16: 361–4.

19. Li Z, Liu HY, Lei QF, Zhang C, Li SN (2011) Improved motor function in dko mice by intravenous transplantation of bone marrow-derived mesenchymal stromal cells. Cytotherapy 13: 69–77.

20. Liu X, Wang Y, Yi M (2010) Effects of human mesenchymal stem cell transplantation in bilateral corpus striatum in a rat model of Tourette's syndrome. Neural Regeneration Research 5: 1285–90.

21. Jijun L, Zaiwang L, Anyuan L, Shuzhen W, Fanghua Q, et al. (2010) Abnormal expression of dopamine and serotonin transporters associated with the pathophysiologic mechanism of Tourette syndrome. Neurol India 58: 523–9.

22. Palminteri S, Lebreton M, Worbe Y, Hartmann A, Lehéricy S, et al. (2011) Dopamine-dependent reinforcement of motor skill learning: evidence from Gilles de la Tourette syndrome. Brain 134: 2287–301.

23. Arbouw ME, Movig KL, Guchelaar HJ, Neef C, Egberts TC (2012) Dopamine agonists and ischemic complications in Parkinson's disease: a nested case-control study. Eur J Clin Pharmacol 68: 83–8.

24. Herzberg I, Valencia-Duarte AV, Kay VA, White DJ, Müller H, et al. (2010) Association of DRD2 variants and Gilles de la Tourette syndrome in a family-based sample from a South American population isolate. Psychiatr Genet 20: 179–83.

25. Taylor JL, Rajbhandari AK, Berridge KC, Aldridge JW (2010) Dopamine receptor modulation of repetitive grooming actions in the rat: potential relevance for Tourette syndrome. Brain Res 1322: 92–101.

14

Cryopreserved Dental Pulp Tissues of Exfoliated Deciduous Teeth Is a Feasible Stem Cell Resource for Regenerative Medicine

Lan Ma[1,2,◑], Yusuke Makino[1,◑], Haruyoshi Yamaza[2], Kentaro Akiyama[3], Yoshihiro Hoshino[2], Guangtai Song[4], Toshio Kukita[1], Kazuaki Nonaka[2], Songtao Shi[3], Takayoshi Yamaza[1]*

1 Department of Molecular Cell Biology and Oral Anatomy, Graduate School of Dental Science, Kyushu University, Fukuoka, Japan, 2 Department of Pediatric Dentistry, Graduate School of Dental Science, Kyushu University, Fukuoka, Japan, 3 Center for Craniofacial Molecular Biology, Herman Ostrow School of Dentistry of USC, University of Southern California, Los Angeles, California, United States of America, 4 Department of Pedodontics, School of Stomatology, Wuhan University, Wuhan, China

Abstract

Human exfoliated deciduous teeth have been considered to be a promising source for regenerative therapy because they contain unique postnatal stem cells from human exfoliated deciduous teeth (SHED) with self-renewal capacity, multipotency and immunomodulatory function. However preservation technique of deciduous teeth has not been developed. This study aimed to evaluate that cryopreserved dental pulp tissues of human exfoliated deciduous teeth is a retrievable and practical SHED source for cell-based therapy. SHED isolated from the cryopreserved deciduous pulp tissues for over 2 years (25–30 months) (SHED-Cryo) owned similar stem cell properties including clonogenicity, self-renew, stem cell marker expression, multipotency, in vivo tissue regenerative capacity and in vitro immunomodulatory function to SHED isolated from the fresh tissues (SHED-Fresh). To examine the therapeutic efficacy of SHED-Cryo on immune diseases, SHED-Cryo were intravenously transplanted into systemic lupus erythematosus (SLE) model MRL/lpr mice. Systemic SHED-Cryo-transplantation improved SLE-like disorders including short lifespan, elevated autoantibody levels and nephritis-like renal dysfunction. SHED-Cryo amended increased interleukin 17-secreting helper T cells in MRL/lpr mice systemically and locally. SHED-Cryo-transplantation was also able to recover osteoporosis bone reduction in long bones of MRL/lpr mice. Furthermore, SHED-Cryo-mediated tissue engineering induced bone regeneration in critical calvarial bone-defect sites of immunocompromised mice. The therapeutic efficacy of SHED-Cryo transplantation on immune and skeletal disorders was similar to that of SHED-Fresh. These data suggest that cryopreservation of dental pulp tissues of deciduous teeth provide a suitable and desirable approach for stem cell-based immune therapy and tissue engineering in regenerative medicine.

Editor: Niels Olsen Saraiva Câmara, Universidade de Sao Paulo, Brazil

Funding: This work was supported by the grants from Ministry of Education, Culture, Sports, Science and Technology of Japan for Challenging Exploratory Research Project (no. 24659815 to TY and no. 23659967 to KN) and for Young Scientists (B) (no. 20790260 to HY) of Japan Society for Promotion of Science and from National Institute of Dental and Craniofacial Research, National Institutes of Health, Department of Health and Human Services, USA (R01DE17449 and R01DE019156 to SS). The funders had no role in study design, data collection and analysis, decision to publish, or preparation of the manuscript.

Competing Interests: The authors have declared that no competing interests exist.

* E-mail: yamazata@dent.kyushu-u.ac.jp

◑ These authors contributed equally to this work.

Introduction

Mesenchymal stem cells (MSCs) have been isolated from a variety of fetal and adult tissues and considered as an ideal candidate source for cell-based therapy due to their unique properties such as multipotency and immunomodulatory functions [1]. Many researchers have investigated to apply MSCs as progenitors of osteoblasts for bone tissue engineering. Clinical evidences support the efficacy of MSC-based skeletal tissue regeneration [2,3]. On the other hand, MSCs exert striking regulatory effects on immune cells such as T- and B-lymphocytes, dendritic cells and natural killer cells [4,5]. This immunological traits of MSCs lead to take clinical advantages to immune diseases such as acute graft-versus-host-disease (GVHD) [4,6], hematopoietic stem cell (HSC) engraftment [7,8] and systemic lupus erythematosus (SLE) [9].

Recent discovery has evaluated that fresh dental pulp tissues of human exfoliated deciduous teeth preserve MSC population, termed SHED [10]. SHED display typical stem cell properties including clonogenicity, cell proliferation and multipotency to differentiate into odontoblast/osteoblast-, adipocyte-, and neural cell-like cells [10]. SHED also express a unique in vivo tissue regeneration capability of forming dentin/pulp and bone/bone marrow structures when subcutaneously transplanted into immunocompromised mice [10]. SHED implantation govern bone repair in critical-sized bone defects in mouse calvarias [11] and swine mandible [12]. Moreover, systemic SHED-transplantation exhibited effective improvement on SLE-like disorders including hyper-autoantibody levels, renal dysfunction and hyperactivity of interleukin 17 (IL-17)-producing helper T (Th17) cells, in MRL/lpr mice [13]. Therefore SHED are considered to be a feasible and promising cell source for cell-based tissue engineering and immune therapy in regenerative medicine.

Exfoliated deciduous teeth possess advantages of minimal invasiveness and easily accessible tissue source in comparison with other human tissues such as bone marrow and adipose tissue [10]. However, the effective preservation of deciduous teeth has remained a primary concern for clinical applications of SHED. In addition, SHED isolation is impractical immediately after the exfoliation of deciduous teeth because the opportunity of the exfoliation is unpredictable. Recently, cryopreservation of human cells and tissues is amenable to be a reliable and feasible approach for stem cell storage [14]. Herein, we cryopreserved dental pulp tissues of exfoliated deciduous teeth for over 2 years and investigated the effects of the long term cryopreservation on the recovering of SHED properties including *in vitro* and *in vivo* biological and immunological properties. Furthermore, we assessed the therapeutic efficacy of the recovered SHED from the cryopreserved deciduous dental pulp tissues on immune modulation and bone regeneration in SLE model-MRL/*lpr* and bone defect model-immunocompromised mice.

Materials and Methods

Ethics Statement

Procedures using human samples (exfoliated deciduous teeth and peripheral blood) were conducted in accordance with the Declaration of Helsinki and approved by the Kyushu University Institutional Review Board for Human Genome/Gene Research (Protocol Number: 393-01). We obtained written informed consent from all the children's parents on the behalf of the children participants involved in this study. All animal experiments were approved by the Institutional Animal Care and Use Committee of Kyushu University (Protocol Number: A21-044-1) and conformed to all the guidelines outlined in the Guide for the Care and Use of Laboratory Animals by the National Institutes of Health (NIH).

Human Subjects

Human exfoliated deciduous teeth were collected as discarded biological/clinical samples from children (5–7-year-old) at Department of Pediatrics of Kyushu University Hospital. Human peripheral blood mononuclear cells (PBMNCs) were collected from healthy volunteers (28–34 year-old).

Cryopreservation of Deciduous Dental Pulp Tissues of Exfoliated Deciduous Teeth and Isolation and Culture of SHED

A protocol for the cryopreservation of dental pulp tissues of exfoliated deciduous teeth was summarized in **Figure S1**. Dental pulp tissues were separated from exfoliated deciduous teeth. Half of the samples were mixed with a cryopreserved medium at 4°C and kept overnight at −80°C. The cryopreserved medium consisted of 10% dimethyl sulfoxide (DMSO) (Sigma, St Louis, MO) and 90% fetal bovine serum (FBS) (Equitech-Bio, Kerrville, TX). They were transferred into liquid nitrogen and stored for over 2 years (25–30 months). The other half of fresh deciduous dental pulp tissues were treated to isolate SHED (SHED-Fresh). SHED from the cryopreserved deciduous pulp tissues (SHED-Cryo), as well as SHED-Fresh, were isolated by an adherent colony-forming unit fibroblasts (CFU-F) method [10,13,15]. The cryopreserved tissues were quickly thawed at 37°C. Both cryopreserved and fresh dental pulp tissues were digested with 0.3% collagenase type I (Worthington Biochemicals, Lakewood, NJ) and 0.4% dispase II (Sanko Junyaku, Tokyo, Japan) in

phosphate buffered saline (PBS, pH 7.4) for 60 min at 37°C. Single-cell suspensions were obtained through a 70-μm-cell strainer (BD Bioscience, San Jones, CA). The cells (1×10^6) were seeded on T-75 flasks, washed with PBS after 3 hours and cultured at 37°C in 5% CO_2 with a growth medium containing 15% FBS (Equitech-Bio), 100 μM L-ascorbic acid 2-phosphate (WAKO Pure Chemical, Osaka, Japan), 2 mM L-glutamine (Nacalai Tesque, Kyoto, Japan) and 100 U/ml penicillin and 100 μg/ml streptomycin (Nacalai Tesque) in alpha Modification of Eagle's Medium (alphaMEM) (Invitrogen, Carlsbad, CA) and subsequently cultured for 14–16 days until obtaining adherent colonies. The adherent colonies-forming cells were recognized as SHED as described before [10,13]. The colonies-forming cells were passed and sub-cultured in the growth medium. The medium was changed twice a week.

Antibodies

Antibodies used in this study are summarized in the **Table S1**.

Mice

C57BL/6J and Balb/cA-nu/nu mice (female, 6 week-old) were purchased from CLEA Japan. (Tokyo, Japan). C57BL/6J MRL/*lpr* mice (female, 6 week-old) were from Japan SLC. (Shizuoka, Japan). They were housed in temperature- and light-controlled environmental conditions with a 12-hour light and dark cycle, and permitted ad *libitum* consumption of water and standard pellet chow.

Histology of Cryopreserved Dental Pulp Tissues of Exfoliated Deciduous Teeth

Cryopreserved deciduous pulp tissues were fixed with 4% paraformaldehyde (PFA) in PBS and immersed in O.C.T. compound (Sakura Finetek Japan, Tokyo, Japan). The frozen specimens were cut into 6-μm thick sections. Some sections were stained with hematoxylin and eosin (H&E). The others were immunostained with anti-STRO-1 and anti-human CD146 antibodies by using SuperPicture kit (Invitrogen). Subclass-matched antibodies were used for negative controls for immunohistochemistry. The sections were observed under a microscope BIORVO BZ-9000 (Keyence Japan, Osaka, Japan).

CFU-F Assay

Cells (10×10^3) isolated from frozen/fresh deciduous dental pulp tissues were seeded on 100-mm culture dishes and cultured in the growth medium for 16 days. The flasks were treated with 4% PFA and 0.1% toluidine blue in PBS for 18 hours. Colonies containing >50 cells were recognized as single colony clusters under a microscope as previously [10,13]. The numbers of the colonies were counted.

Population Doubling Assay

Cells were seeded on T-75 culture flasks (BD Bioscience). When the cells reached at sub-confluent condition, they were passed. These steps were repeated until the cells lost dividing capability. The population doubling score was calculated at every passage according to the equation: $\log_2$ (number of final harvested cells/number of initial seeded cells) and the total scores were determined by adding up each population doubling score in each sample as described in previous reports [10,13].

Bromodeoxyuridine (BrdU) Incorporation Assay

Passaged 3 (P3) SHED were seeded at 1×10^3 per well on 35-mm dishes and cultured in the growth medium for 2 days. BrdU reagent (Invitrogen) was added at 1:100 in the medium and subsequently cultured additionally for 24 hours. The cells were fixed in 70% ethanol for 15 min, treated with the BrdU staining kit (Invitrogen) and lightly stained with hematoxylin according to the company's instruction. The stained samples were observed under BIORVO microscope (BZ-9000, Keyence Japan, Osaka, Japan). Seven images were randomly selected to calculate BrdU-positive nuclei number in each sample. Cell proliferation capacity was shown as a percentage of BrdU-positive nuclei over total nucleated cells [13].

Telomerase Activity Assay

Telomerase activity was measured by a telomere repeat amplification protocol assay using the quantitative telomerase detection kit (Allied Biotech, Inc., Ijamsville, MD) applied with real-time PCR as referred to our previous reports [13]. As positive control, HEK293T cells were used. Some extracts from each cell were heated at 85°C for 10 min and used as negative control samples. The average starting quantity (SQ) of fluorescence units was used to compare the telomerase activity among the samples.

Flow Cytometric Analysis for SHED

Passaged 3 (P3) SHED were cultured at 50–60% confluent condition. Cells ($100 \times 10^3 / 100$ µl) were stained with primary specific antibodies. Subclass-matched antibodies were used as negative controls. Flow cytometry was analyzed on FACSCalibur flow cytometer (BD Bioscience) [13]. The number (percentage) of positive cells was determined using CellQuest software (BD Bioscience) by comparison with the corresponding control cells stained with the subclass-matched antibody in which a false-positive rate of less than 1% was accepted.

Immunofluorescence for Cultured Cells

The cells were fixed with 4% PFA in PBS and blocked with PBS containing 10% normal serum matched to the secondary antibodies. The samples were incubated with the specific antibodies to cell surface markers or the subclass-matched antibodies overnight at 4°C and treated with CF 633-conjugated-secondary antibodies (Biotium, Hayward, CA). Finally, they were stained with 4', 6-diamidino-2-phenylindole (DAPI) (Dojindo, Kumamoto, Japan) and observed under the BIORVO microscope (Keyence Japan). Numbers of cells positive to the specific antibodies and nuclei stained with DAPI were counted and shown as a percentage of positive cells over total nucleated cells.

Semi-quantitative Reverse Transcription Polymerase Chain Reaction (RT-PCR)

Total RNA was isolated from the cultures using TRIzol (Invitrogen) and digested with DNase I. The cDNA was synthesized from 100 ng of total RNA using Revatra Ace (TOYOBO, Osaka, Japan). The specific primer pairs are listed in **Table S2**. PCR was performed using gene specific primers and RT-PCR Quick Taq HS DyeMix (TOYOBO) at 94°C for 2 min for one cycle and then react for 40 cycles with denature at 94°C for 45 sec, annealing at 56°C for 45 sec, extension at 72°C for 60 sec as one cycles, with a final 10-min extension at 72°C. Finally, 5 µl of each amplified PCR product was analyzed by 2% agarose gel electrophoresis and visualized by ethidium bromide staining. The intensity of bands was measured by using Image-J soft ware and normalized to an internal control gene, *glyceraldehyde 3-phosphate dehydrogenase* (*GAPDH*).

In vitro Multipotent Assay

***In vitro* dentinogenic/osteogenic induction assay.** SHED (P3, 5×10^3) were grown on 60-mm dishes in the growth medium until confluent condition and induced with an dentinogenic/osteogenic medium supplemented with 1.8 mM potassium dihydrogen phosphate (Sigma, St. Louis, MO) and 10 nM dexamethasone (Sigma) in the growth medium. The dentinogenic/osteogenic medium were changed twice a week. One week after the induction, dentinogenic/osteogenic markers were analyzed by semi-quantitative RT-PCR and alkaline phosphatase (ALP) activity test [13]. For mineralized nodule assay, the cultures were stained with 1% Alizarin Red-S (Sigma) at 4 weeks post the osteoinduction. The Alizarin red-positive area was analyzed using NIH image software Image-J and shown as a percentage of Alizarin red-positive area over total area [13].

***In vitro* chondrogenic induction assay.** SHED (P3, 2×10^6) were aggregated in a 15 mL tube and cultured with Dulbecco's modified Eagle's medium (Invitrogen) supplemented with 15% FBS (Equitech-Bio), 2 mM L-glutamine (Nacalai Tesque), 100 µM L-ascorbate-2-phosphate (Wako Pure Chemicals), 2 mM sodium pyruvate (Nacalai Tesque), 1% insulin-transferring-selene mixture (ITS) (BD Biosciences), 100 nM dexamethasone (Sigma), 10 ng/ml transforming growth factor beta$_1$ (TGFbeta$_1$) (PeproTech, Rocky Hill, NJ) and 100 U/ml penicillin and 100 µg/ml streptomycin (Nacalai Tesque). The chondrogenic medium were changed twice a week. After 3-week induction, chondrocyte-specific genes were analyzed by semi-quantitative RT-PCR.

***In vitro* adipogenic induction assay.** Cultured SHED (P3, 5×10^3/dish) were cultured until the confluent condition and induced in an adipogenic medium with the growth medium plus 500 µM isobutyl-methylxanthine (Sigma), 60 µM indomethacin (Sigma), 0.5 µM hydrocortisone (Sigma) and 10 µM insulin (Sigma). After 6-week induction, the cultures were stained with 0.3% Oil red O (Sigma) to detect lipid droplets. The samples stained with Oil red O were treated with isopropanol and the absorbance of the extracts were measured at 520 nm. Adipocyte-specific genes were also analyzed by semi-quantitative RT-PCR.

***In vitro* endothelial cell induction assay.** P3 SHED (5×10^3 cells) were seeded on fibronectin-coated 35-mm dishes (BD Biosciences) and cultured by using a commercial available endothelial cell growth media kit, EGM-2, (Lonza, Basel, Switzerland) with 100 U/ml penicillin and 100 µg/ml streptomycin (Nacalai Tesque) according to the manufacture's instruction. The endothelial growth medium was changed every 2 days. The cultures were fixed 7 days after the induction. Endothelial cell differentiation was determined by immunofluorescence with anti-CD31 and CD34 antibodies.

***In vitro* neuronal cell induction assay.** P3 SHED (5×10^3 cells/well) were plated in laminin-coated 35-mm dishes (BD Biosciences) and cultured in a neurogenic medium containing supplemented with 1xN2 supplement (Invitrogen), 10 ng/ml basic fibroblast growth factor (bFGF) (PeproTech), 10 ng/ml epidermal growth factor (EGF) (PeproTech), 100 U/ml penicillin and 100 µg/ml streptomycin (Nacalai Tesque) in Neurobasal A (Invitrogen) for 21 days according to the recent study [10]. The medium was changed with 50% of fresh medium twice a week. Neural cell differentiation was determined by immunofluorescence with anti-*glial fibrillary acidic protein*, anti-neural filament M and anti-betaIII tubulin antibodies.

***In vitro* hepatic induction assay.** P3 SHED (P3, 5×10^3) were seeded on fibronectin-coated 35-mm dishes (BD Biosciences) and cultured with Iscove's modified Dulbecco's medium (Invitrogen) supplemented with 10 nM dexamethasone (Sigma), 1% ITS

(BD Biosciences), 20 ng/ml EGF (PeproTech), 10 ng/ml bFGF (PeproTech), 20 ng/ml hepatocyte growth factor (PeproTech), 20 ng/ml oncostain M (PeproTech), 100 U/ml penicillin and 100 μg/ml streptomycin (Nacalai Tesque) for 4 weeks. Hepatocyte-specific gene albumin was assayed by semi-quantitative RT-PCR.

In vivo Tissue Regeneration Assay

Xenogenic transplantation was performed as previously [13,16]. Cultured SHED (P3) in the growth medium (4×10^6) were mixed with a carrier, hydroxyapatite tricalcium phosphate (HA/TCP) (40 mg, Zimmer Inc., Warsaw, IN). The mixtures were subcutaneously transplanted into the dorsal surface of 8–10-week-old Balb/c *nude/nude*. The implants were harvested 8 weeks after the surgery, fixed with 4% PFA and decalcified with 10% EDTA. Frozen sections were cut and used for H&E staining or immunofluorescence observed under a BIORVO microscope (BZ-9000, Keyence Japan).

In vivo Self-renewal Assay

In vivo self-renewal assay with a sequential transplantation was referred to a recent study [17]. SHED (P3) (4×10^6) were implanted with HA/TCP carrier (40 mg) (Zimmer) into primary Balb/c *nu/nu* mice. Eight weeks after the transplantation, the primary implants were harvested and treated with 0.4% dispase II (Sanko Junyaku) for 60 min at 37°C to obtain single cells from the implants. The cells were stained with R-PE-conjugated anti-human CD146 and magnetic-beads-conjugated anti-R-PE antibodies (Miltenyi Biotec, Bergisch Gladbach, Germany) and sorted magnetically to obtain human CD146-positive cells. The purity was confirmed by flow cytometry with anti-human CD146 or anti-mouse CD146 antibody. The sorted cells were seeded at low density and cultured to obtain CFU-F. Expanded CFU-F-forming cells (P1) (4×10^6) were subcutaneously transplanted with HA/TCP carriers (40 mg) (Zimmer) into secondary immunocompromised mice for 8 weeks and the secondary implants were assayed by H&E staining and immunofluorescence.

Histological Analysis of Implant Tissues

Implant tissues were fixed with 4% PFA in PBS overnight and decalcified with 5% EDTA solution (pH 7.4). Frozen sections were cut into 8-μm thickness and treated with H&E staining and immunofluorescence with anti-human mitochondria, anti-STRO-1 or anti-human CD146 antibody. The sections were observed under BIORVO microscope (BZ-9000, Keyence Japan). To measure newly-formed area of mineralized tissue, seven fields were randomly selected and the newly-formed area was calculated by NIH image-J software, and shown as a percentage over total tissue area as described before [13,16].

Single Colonies-derived Cell Assay

As referred to a previous study [18], cells isolated from cryopreserved deciduous dental pulp tissues were seeded at 1, 2 or 4 cells per well on 24-well multiplates with the growth medium. The wells containing more than two attached cells were excluded from further analysis. The single cell-attached wells were cultured for 14–16 days and obtained CFU-F-forming cells. The single colony-forming cells were used for population doubling, BrdU incorporation and dentinogenic/osteogenic analyses.

Induction Assay of Th17 cells

Induction of Th17 cells was performed as previously [13]. Human CD4$^+$CD25$^-$ T cells were magnetically sorted from human PBMNCs using CD4$^+$CD25$^+$ regulatory T cell isolation kit (Miltenyi Biotec). They (1×10^6/well) were activated by plate-bounded anti-CD3 (5 μg/ml) and soluble anti-CD28 (1 μg/ml) antibodies for 3 days and loaded on SHED (20×10^3/well) with recombinant human TGFbeta$_1$ (2 μg/ml, PeproTech) and IL-6 (50 μg/ml, PeproTech) for 3.5 days. T cells were stained with anti-PerCP-conjugated CD4 and FITC-conjugated anti-CD8a antibodies, treated with R-PE-conjugated anti-IL-17 and APC-conjugated anti-interferon gamma (IFNgamma) antibodies using Foxp3 staining buffer kit (eBioscience, San Diego, CA) and analyzed on FACSCalibur (BD Biosciences). IL-17 levels in the culture supernatants were analyzed by enzyme linked immunosorbent assay (ELISA) with a commercial available kit (R&D Systems) according to the manufacture's instruction.

Systemic Transplantation of SHED into MRL/lpr Mice

The protocol for systemic transplantation of SHED into MRL/*lpr* mice was referred to our previous reports [9,13]. Briefly, SHED (0.1×10^6/10 g body weight) or PBS were systemically infused into MRL/*lpr* mice via the tail vein at the age of 16 weeks old. Their survival was inspected daily until died. Peripheral blood, urine, kidney, axial lymph nodes and long bones were collected from MRL/*lpr* mice at the age of 20 weeks old according to our previous studies [9,13].

Assay for Autoantibodies, Immunoglobulins, Biomarkers and Cytokines in Peripheral Blood Serum, Urine and Tissue Samples

Biological factors in mouse biological samples were measured by ELISA with commercial available kits (anti-double strand DNA [dsDNA] IgG and IgM, anti-nuclear antibody [ANA], albumin and complement C3: alpha diagnostic [San Antonio, TX]; IL-17 and nuclear factor kappaB ligand [sRANKL] [R&D Systems]; C-terminal telopeptides of type I collagen [CTX]: Nordic Bioscience [Herlev, Denmark]) according to the manufactures' instructions. Creatinine (R&D Systems) and urine protein (Bio Rad, Hercules, CA) were also assayed by colorimetry according to the manufactures' instructions.

Flow Cytometric Analysis for Mouse Peripheral Blood Th17

To measure Th17 cells in mouse peripheral blood, mouse PBMNCs were incubated with PerCP-conjugated anti-CD4, FITC-conjugated anti-CD8a, followed by the treatment with R-PE-conjugated anti-IL-17 and APC-conjugated anti-IFNgamma antibodies using a Foxp3 staining buffer kit (eBioscience). The stained cells were analyzed on FACSCalibur (BD Bioscience).

Histopathological Analysis of Mouse Kidney

Kidney samples were fixed with 4% PFA in PBS overnight. After washed with PBS, some samples were processed for paraffin embedding and cut into 6-μm thick paraffin sections. The others were cut into 8-μm thick cryosections. The paraffin sections were treated with H&E, Gomori trichrome or Periodic Acid Schiff (PAS) staining. The cryosections were stained with anti-mouse complement C3 antibody and treated with CF 633-conjugated-secondary antibodies (Biotium). Finally, they were stained with DAPI (Dojindo) and observed under a BIORVO microscope (BZ-9000, Keyence Japan).

In vivo Tracing Assay of SHED

To trace the trafficking of SHED transfused in MRL/*lpr* mice, SHED were labeled with carboxyfluorescein diacetate succinimidyl ester (CFSE) labeling. Single suspended SHED population (10×10^6/ml) was incubated with CFSE solution (Invitrogen) for 10 min at 37°C. The labeled cells (1×10^6) were intravenously injected into the tail vein of MRL/*lpr* mice. Lymph nodes, kidneys and femurs of the mice were harvested either 24 h or 1 week after the infusion. The tissue samples were fixed with 4% PFA in PBS overnight. Only bone samples were treated with 10% EDTA. All of the samples were cut into 6-μm frozen sections, stained with DAPI (Dojindo) and observed under a microscope.

Bone Phenotype Analysis

The femoral bone samples were fixed with 4% PFA in PBS overnight and analyzed by micro-computed tomography (microCT) with Skyscan 1076 scanner (Skyscan, Kontich, Belgium). Bone mineral density (BMD) and bone structural parameters including bone volume/trabecular volume (BV/TV), trabecular thickness (Tb.Th), trabecular number (Tb.N), trabecular separation (Tb.Sp) and trabecular space (Tb.Spac) were calculated using CT Analyzer software (Skyscan) according to the manufacture's instruction. After microCT analysis, the bone samples were decalcified with 10% EDTA and cut into 6-μm paraffin sections. The paraffin sections were treated with H&E staining. Some sections were stained with a tartrate-resistant acid phosphate (TRAP) staining solution containing 0.01% naphthol AS-MX phosphate (Sigma), 50 mM tartrate (Sigma) and 0.06% fast red violet LB salt (Sigma) in 0.1 M acetate buffer (pH 5.0) for 10 min. TRAP-positive cells were counted and quantified as previously [19].

In vitro Osteoclastic and Osteogenic Assays

Bone marrow were flashed out from the bone cavity of femurs and tibias with heat-inactivated 3% FBS (Equitech-Bio) in PBS. For osteoclastic assay, the bone marrow nucleated cells (BMCs) were seeded at 1×10^6 per well on the 24-well culture plates with 10 ng/ml macrophage colony stimulating factor (M-CSF) (R&D Systems) and 25 ng/ml sRANKL (PeproTech) in alphaMEM for 5–6 days. The medium was changed on Day 3. The cultures were treated with TRAP staining [19] and the TRAP-positive cells with muitinuclei (n>3) were counted [19]. The BMCs (1×10^6) were also seeded on 35-mm culture dishes and cultured in an osteogenic medium containing 10% FBS (Equitech-Bio), 2 mM L-glutamine (Nacalai Tesque), 100 μM L-ascorbic acid 2-phosphate (WAKO pure chemicals), 2 mM beta-glycerophosphate (Sigma), 10 nM dexamethasone (Sigma), 100 U/ml penicillin and 100 μg/ml streptomycin (Nacalai Tesque) in alphaMEM [20]. Four weeks after the induction, the osteogenic cultures were stained with 1% Alizarin Red (Sigma). The mineralized area was quantified by using NIH Image-J [13].

Bone Regeneration in Calvarial Bone Defects

The bone defect models in calvarial bone were generated on immunocompromised mice as described before [11]. Briefly, P3 SHED (4×10^6) were mixed with HA/TCP carriers (40 mg) (Zimmer). Calvarial bones, especially parietal bones, were removed to make a critical defect area. The mixtures of SHED and HA/TCP were placed on the defect area and the calvarial skins were sutured. Twelve weeks after the implantation, the calvariae were fixed with 4% PFA in PBS overnight and imaged by Skyscan 1076 scanner (Skyscan). Then, the samples were decalcified with 10% EDTA and cut into 6-μm thick paraffin sections. The sections were treated with H&E staining and TRAP staining. Some of sections were used for immunofluorescence with anti-human CD146 antibody.

Statistical Analysis

Student's *t*-test was used to analyze significance between 2 groups. A *P* value of less than 0.05 was considered as a significant difference.

Results

SHED-Cryo Possess MSC Properties

SHED have been elucidated to possess MSC characteristics including clonogenicity, self renew, multipotency, *in vivo* regeneration and *in vitro* immunomodulatory functions [10,13]. Here, we demonstrated the effects of cryopreservation of deciduous pulp tissues on these SHED properties. Dental pulp tissues were removed from exfoliated deciduous teeth and frozen in the freezing medium (10% DMSO and 90% FBS) at −80°C overnight followed by the preservation in a liquid nitrogen-filled tank for over 2 years (**Figure S1**). The remaining tooth bodies were returned to the donor children. The long-term cryopreserved tissues were quickly thawed at 37°C before used (**Figure S1**). Histological analysis confirmed that the cryopreserved tissues showed dense connective tissues containing blood vessels and nerve fibers (**Figure 1A**), but not odontoblastic cells. The odontoblast layer may be lost because of mechanical damage or freezing fracture of the pulp samples. Early MSC markers STRO-1 and CD146-positive cells were detected at the perivascular area (**Figure 1B**) as dental pulp stem cells (DPSCs) were localized around the blood vessels in adult dental pulp tissues [21]. These data suggested a possibility of the remaining of MSCs in the cryopreserved deciduous pulp tissues.

Next, we isolated MSCs from the cryopreserved dental pulp tissues of exfoliated deciduous teeth, SHED-Cryo, by CFU-F approach (**Figure S1**), which is a classical and standard method [15]. When seeded at low density, single cells were adhered to the plastic culture dishes and then divided to generate cell clusters (**Figure 1C**). The clonogenic cell clusters exhibited different size and varied density (**Figure 1D**). The colony forming efficiency of SHED-Cryo showed similar to that of SHED-Fresh (**Figure 1E**). BrdU incorporation assay demonstrated that the proliferation capacity of SHED-Cryo maintained at a high level (**Figure 1F**) similar to that of SHED-Fresh (**Figure 1G**). Flow cytometric analysis verified that SHED-Cryo were positive to STRO-1, CD146, CD73 and CD105 but negative to hematopoietic cell markers CD34, CD45 and CD14 (**Figure 1H**). SHED-Cryo were also positive to CD90 (over 95%) and the positive level in SHED-Cryo was similar to that in SHED-Fresh (data not shown). SHED-Cryo shared the immunophenotype with SHED-Fresh (**Figure 1I**). RT-PCR demonstrated that SHED-Cryo expressed genes for embryonic stem cells, *NANOG* and *octamer 4*, and for neural crest cells, *NOTCH1*, *NESTIN* and *low-affinity neural growth factor* (**Figure 1J**), as seen in SHED-Fresh (**Figure 1K**). Expression of a neural crest cell marker, Nestin, was also detected in SHED-Cryo by flow cytometry (**Figure 1L**) and the expression in SHED-Cryo showed a similar level to that in SHED-Fresh (data not shown). Recent studies [22,23] demonstrated the effect of cryopreservation on the expression of NANOG, octamer 4 and Nestin in deciduous teeth stem cells and may support, at least in partially, our cryopreserved effect of deciduous dental pulp tissues on SHED properties. Taken together, these data suggested that SHED-Cryo retained primitive MSC properties likely to SHED-Fresh.

SHED-Cryo Own Multipotency

Four weeks after dentinogenic/osteogenic induction, SHED-Cryo were capable of forming Alizarin Red-positive nodules

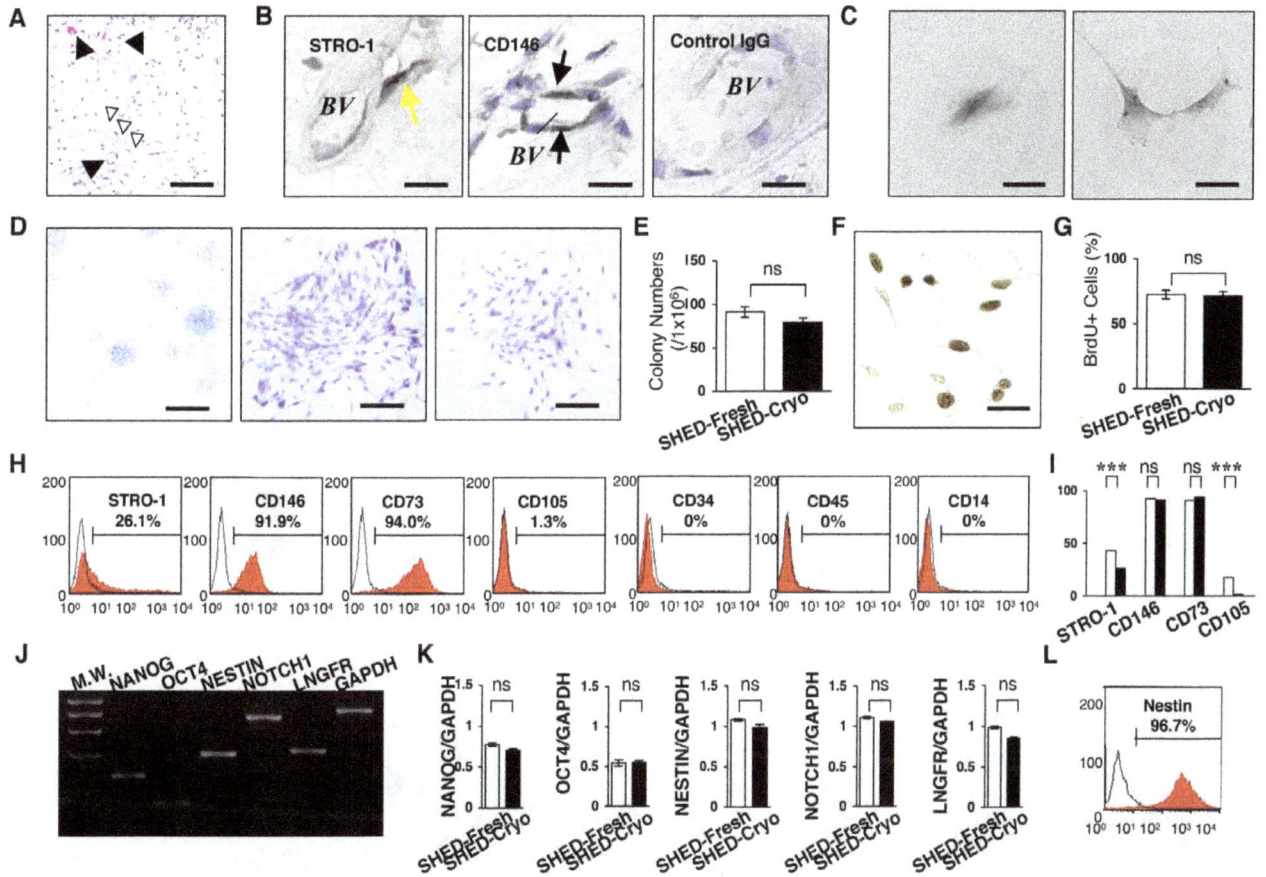

Figure 1. Clonogenicity, cell proliferation capacity and stem cell marker expression of SHED-Cryo. (A) Histology of cryopreserved dental pulp tissue of exfoliated deciduous teeth. Black arrowheads: blood vessel, white arrowheads: nerve fibers. H&E staining. (B) Localization of MSC markers in the cryopreserved deciduous pulp tissues. Yellow arrows: STRO-1-positive cells, black arrows: CD146-positive cells. *BV*: blood vessel, Control: subclass-matched antibody staining. (C–E) CFU-F assay. Formation of a clonogenic cell cluster from a single attached cell (C). Images of attached colonies of SHED-Cryo. Toluidine blue staining (D). Comparison of CFU-F number (E). (F, G) Cell proliferation assay. Immunostaining of BrdU-positive nuclei. (F). Comparison of cell proliferation (G). (H, I) Flow cytometry of MSC markers in SHED-Cryo. Representative histograms (H). Comparison of STRO-1, CD146, CD73 and CD105. Black columns: SHED-Cryo, white columns: SHED-Fresh (I). (J, K) Gene expression of embryonic stem and neural crest cell markers. MW: molecular weight markers (J). Comparative analysis of *NANOG*, *octamer 4 (OCT4)*, *NESTIN*, *NOTCH1* and *low affinity nerve growth factor receptor (LNGFR)* (K). (L) Flow cytometry of Nestin in SHED-Cryo. A, B: n = 3. C–L: n = 5 for all group. A–E: Bar = 30 μm (A), 5 μm (B, C, F), 1 mm (D, left) 25 μm (D, middle and right). G, I, K: ***P<0.005. ns: no significance. The graph bars represent mean±SD.

(**Figure 2A**). SHED-Cryo showed a high ALP activity (**Figure 2B**) and expressed odontoblast/osteoblast-specific genes *runt-related gene 2*, *ALP*, *osteocalcin*, and *dentin sialophosphoprotein* (**Figure 2C**) after the 1-week induction. SHED-Cryo expressed chondrocyte-specific genes for *SOX9*, *aggrecan* and *type X collagen* 3 weeks after chondrogenic culture (**Figure 2D**). SHED-Cryo expressed adipocyte-like phenotypes including accumulation of Oil red-O-positive droplets (**Figure 2E**) and expression of adipocyte-specific genes *lipoprotein lipase* and *peroxisome proliferator activated receptor-gamma2* (**Figure 2F**) 6 weeks after adipogenic induction. SHED-Cryo expressed albumin gene, one of hepatocyte-specific genes, 4 weeks after hepatogenic induction (**Figure 2G**). Immunofluorescence revealed that SHED-Cryo expressed endothelial cell markers CD31 and CD34 1 week after endothelial cell-induction (**Figure 2H**) and neural cell markers neurofilament M and tubulin betaIII 3 weeks after neural cell-induction (**Figure 2I**). SHED-Cryo also exhibited similar capabilities of differentiating into odontoblasts/osteoblasts, adipocytes, chondrocytes, hepatocytes, endothelial cells and neural cells to SHED-Fresh (**Figure 2**).

These data indicated that SHED-Cryo maintained multipotency as MSCs.

SHED-Cryo Showed *in vivo* Mineralized Tissue Regeneration

Eight weeks after subcutaneous transplantation of SHED-Cryo with HA/TCP carriers into immunocompromised mice (**Figure S2**), dentin/pulp-like complex and bone/bone marrow-like structures were formed around the surface of HA/TCP by histological analysis (**Figure 3A**). Immunofluorescence showed that human mitochondria-positive cells were arranged on mineralized matrix (**Figure 3C**). Human specific STRO-1 and CD146 antibodies-positive cells were also detectable on the regenerated mineralized matrix (**Figure 3D**). On the other hand, control transplant-tissues that implanted only HA/TCP carriers without SHED-Cryo did not express any mineralized tissue and human specific antibody-positive cells (**Figure S3**). Therefore these results suggested that SHED-Cryo were responsible cells for mineralized tissue formation in the implant tissues. SHED-Cryo formed similar amount of the regenerated mineralized tissues when compared to

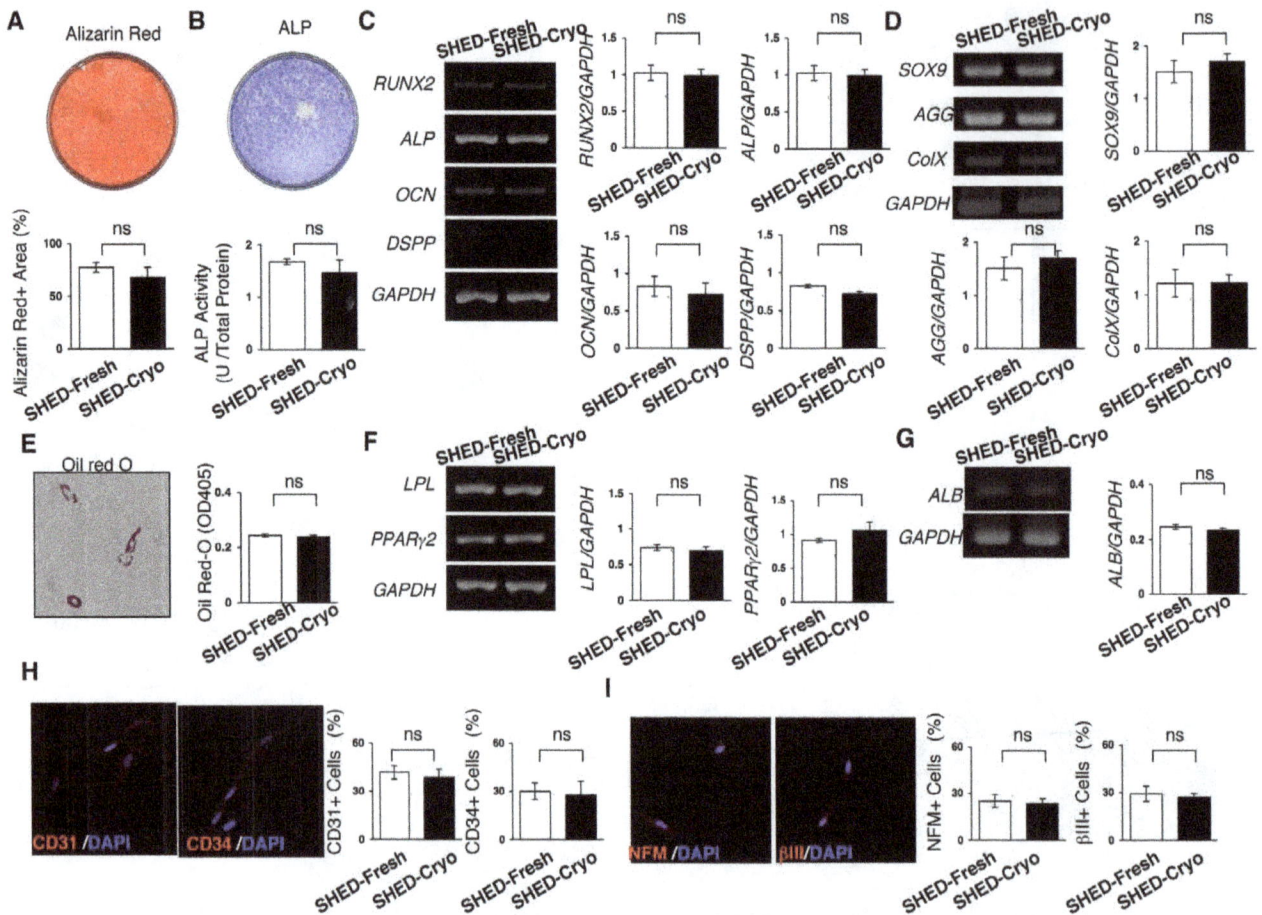

Figure 2. Multipotency of SHED-Cryo. (A–C) Dentinogenic/osteogenic differentiation capacity. Images of Alizarin Red staining (**A**) and alkaline phosphatase (ALP) activity (**B**) of SHED-Cryo. Comparison of Alizarin Red-positive (Alizarin Red+) area (**A**), ALP activity (**B**) and odontoblast/osteoblast-specific genes, *runt-related gene 2 (RUNX2), ALP, osteocalcin (OCN), and dentin sialophosphoprotein (DSPP)* (**C**). (**D**) Chondrogenic differentiation capacity. Comparison of chondrocyte-specific genes, *SOX9, aggrecan (AGG)* and *type X collagen (ColX)*. (**E, F**) Adipogenic differentiation assay. A representative image of Oil Red-O staining and comparison of Oil Red-O accumulation (**E**). Comparison of adipocyte-specific genes *lipoprotein lipase (LPL)* and *peroxisome proliferator activated receptor-gamma2 (PPARgamma2)* (**F**). (**G**) Hepatogenic differentiation capacity. Comparison of hepatocyte-specific gene *albumin (ALB)*. (**H**) Endothelial cell differentiation assay. Comparison of endothelial cell markers CD31 and CD34. (**H**) Neural cell differentiation assay. Comparison of neural cell markers neurofilament M (NFM) and tubulin betaIII (betaIII). **A–I:** n = 5 for all group. ns: no significance. The graph bars represent mean±SD.

SHED-Fresh (**Figure 3B**). These data indicated that SHED-Cryo retained a unique *in vivo* regenerative capacity likely to SHED-Fresh. Although the origin of bone marrow cells are host cells in bone marrow MSC-implants [24,25], the origin of cells in bone marrow and dental pulp in SHED-implants have not been elucidated. Furthermore study will be needed to clarify the origin of the dental pulp cells and bone marrow cells in the SHED-implants in future.

SHED-Cryo Retain Self-renewal Capability

To evaluate self-renewal capability of SHED-Cryo, sequential transplantation, which is one of traditional and gold standard methods to identify the self-renewal capability of stem cells [17], was performed (**Figure S2**). CD146 is considered to be a critical cell surface marker for human MSCs [18]. Cell population was isolated from the primary implants and stained with human CD146 antibody. Then human CD146 antibody-positive cells were purified by a magnet sorting system. Flow cytometric analysis confirmed that the sorted cells were positive to human CD146 (>95%) but negative to mouse CD146 (0%) (**Figure 3E**),

evaluating the high purity of the sorted cells as human stem cells. When the sorted cells were seeded at low density, they were capable of forming CFU-Fs that exhibited positive to human CD146 by immunostaining (**Figure 3F**), indicating that the CD146-positive human cells maintained as intact MSCs in the transplant tissues after the long-term implantation. After the colony-forming cells were secondarily transplanted into immuno-compromised mice for 8 weeks, the secondary implants contained similar dentin/pulp complex-like structures (**Figure 3G**) to the mineralized structural complexes in the primary transplants. Cells positive to anti-human CD146 or anti-human specific mitochondria antibodies were localized on the mineralized matrix (**Figure 3H**). Population doubling and telomerase activity are associated with self-renewal potential of stem cells [26]. Population doubling assay indicated a prolonged and time-limited cell proliferation in SHED-Cryo (**Figure 3I**). Telomerase activity test revealed the lower activity in SHED-Cryo (**Figure 3J**). Collectively, these results verified that SHED-Cryo represented a self-renewal capacity.

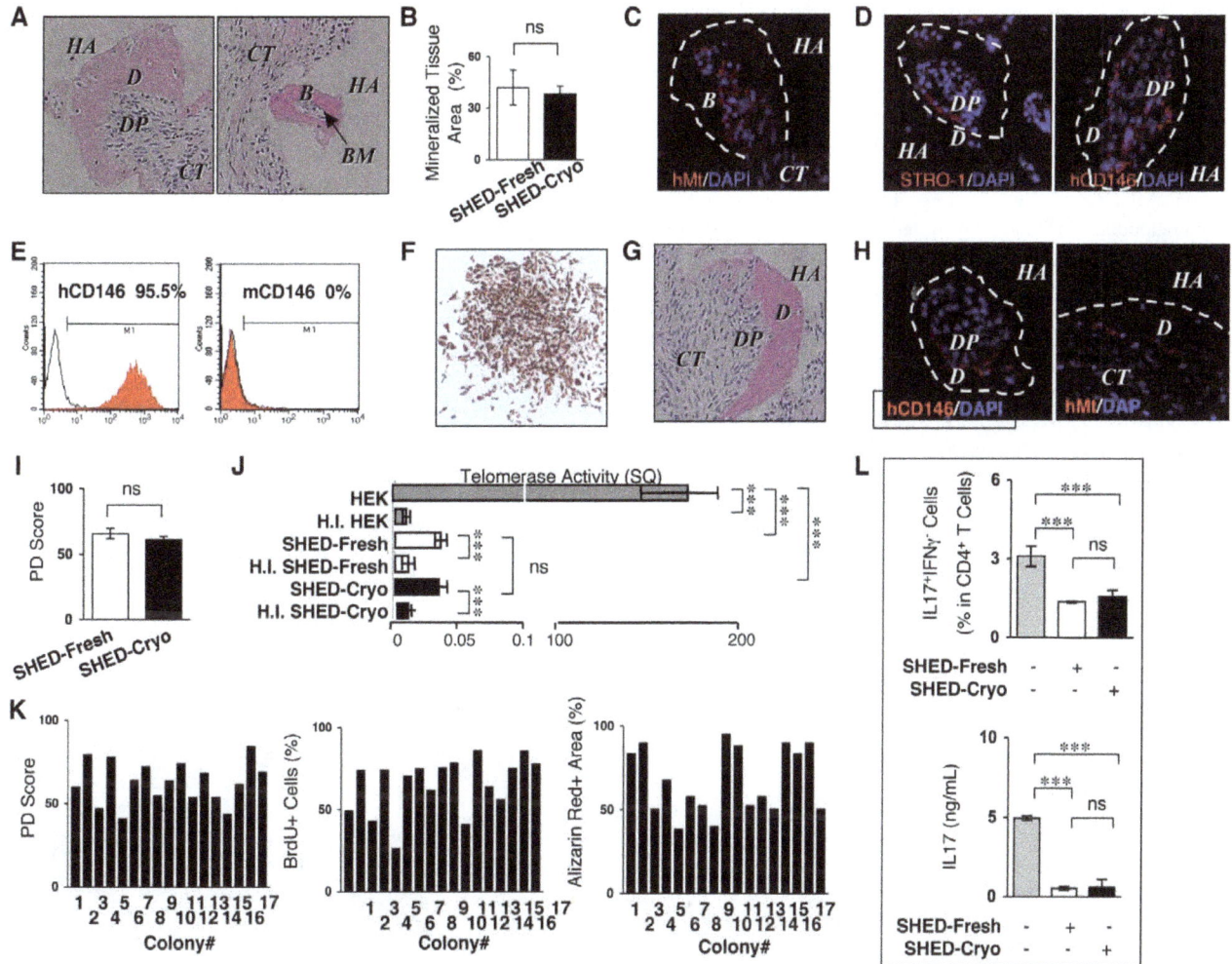

Figure 3. Tissue regeneration capability, self-renewal potency, heterogeneity and *in vitro* immunomodulatory functions of SHED-Cryo. (**A–D**) Images of primary transplant tissues of SHED-Cryo. H&E staining (**A**). Comparison of newly formed-mineralized tissue (B). Immunofluorescence with anti-human specific mitochondria (hMt) (**C**) and anti-STRO-1/human CD146 (hCD146) (**D**) antibodies. (**E, F**) Purity of hCD146 antibody-sorted cells from primary transplants. Flow cytometry with hCD146 and mouse CD146 (mCD146). (**E**). Immunocytochemistry with hCD146 antibody of sorted cell-derived CFU-F (F). (**G, H**) *in vivo* self-renewal assay. Images of secondary transplant tissues. H&E staining (**G**). Immunofluorescence with anti-hCD146/Anti-Mt antibodies. (**H**). (**I**) Comparison of population doubling (PD) scores. (**J**) Comparison of telomerase activity. (**K**) Single-colony-derived cell assay with 17 single cell colonies from a cryopreserved deciduous pulp tissues. (**L**) *In vitro* direct immunosuppressive effects of SHED-Cryo on human Th17 cells. **A–J, L**: n = 5 for all group. **A, C, D, F, H**: *B*: bone, *BM*: bone marrow, *CT*: connective tissue, *D*: dentin, *DP*: dental pulp, *HA*: HA/TCP, **I**: HEK: HEK293 cells, H.I. HEK: heat inactivated HEK, H.I. SHED-Cryo: heat inactivated SHED-Cryo, H.I. SHED-Fresh: heat inactivated SHED-Fresh. **C, D, H**: Dot lined areas: mineralized tissue. Nuclei are counterstained with DAPI. **B, H, I, L**: ***$P < 0.005$, ns: no significance. The graph bars represent mean ± SD.

SHED-Cryo are Heterogeneous Population

To identify the heterogeneity of SHED-Cryo, total 17 clonogenic single-colonies were acquired from a cryopreserved dental pulp tissue of exfoliated deciduous tooth. Each single-colony-derived SHED-Cryo showed various population-doubling score, diverse cell proliferation capacity and varied Alizarin red-positive area (**Figure 3K**). These findings indicated that SHED-Cryo displayed heterogeneous as seen in SHED-Fresh [10].

Immunomodulatory Properties of SHED-Cryo *in vitro*

To explore whether SHED-Cryo display immunomodulatory capacity to Th17 cells as seen in SHED-Fresh (**Figure 3L**) [13], SHED-Cryo were co-cultured with anti-CD3 and anti-CD28 antibodies-activated human CD4⁺CD25⁻ T cells under the stimulation with IL-6 and TGFbeta₁. SHED-Cryo were able to

inhibit both the differentiation of CD4⁺IL17⁺IFNgamma⁻ Th17 cells and the secretion of IL-17 (**Figure 3L**), suggesting that SHED-Cryo maintained *in vitro* inhibitory effect on Th17 cells.

SHED-Cryo Transplantation Prolongs the Life Span and Improves Autoimmune Disorder and Renal Dysfunction in MRL/*lpr* Mice

SLE-like autoimmune disorders appear around age 7–8 weeks in MRL/*lpr* mice. Peripheral levels of autoimmune antibodies increase extremely from about 12 weeks of age in MRL/*lpr* mice. To evaluate the therapeutic potency of SHED-Cryo in severed SLE condition as seen in previous studies [13], SHED-Cryo were intravenously transfused to human SLE model MRL/*lpr* mice at the age of 16 weeks old (**Figure S4**). Kaplan-Meyer curve demonstrated that systemic SHED-Cryo-transplantation significantly prolonged the lifespan of MRL/*lpr* mice (**Figure 4A**). Mantel-Haenszel test evaluated that the median survival time was 155.0, 219.0 and 212.5 days in control, SHED-Fresh-transplanted and SHED-Cryo-transplanted MRL/*lpr* mice, respectively. Elevated serum levels of autoantibodies to ANA and anti-dsDNA IgG and IgM antibodies, which are critical clinical markers in human SLE therapy, were markedly decreased in SHED-Cryo transplanted MRL/*lpr* mice (**Figure 4B**). Increased peripheral immunoglobulins including IgG_1, IgG_{2a}, IgG_{2b} and IgM were also significantly reduced in SHED-Cryo-transplanted mice (**Figure S5**). Histopathological analysis demonstrated that SHED-Cryo-transplantation prevented renal nephritis associated with hypercellularity, mesangial matrix hyperplasia and basal membrane disorder in MRL/*lpr* mice (**Figure 4C**). Immunofluorescence showed that complement C3-deposition in the glomeruli of the kidney in MRL/*lpr* mice was disappeared after SHED-Cryo-transplantation (**Figure 4C**). SHED-Cryo-transplantation elevated serum albumin level and reduced urine protein (**Figure 4D**), meanwhile, decreased serum creatinine in MRL/*lpr* mice (**Figure 4D**). SHED-Cryo-transplantation displayed similar therapeutic effects on the lifespan, autoantibody levels and renal function in MRL/*lpr* mice with SHED-Fresh-trans-

plantation (**Figures 4 and S5**). These findings provided that SHED-Cryo retained therapeutic efficacy on MRL/*lpr* mice.

Transplantation of SHED-Cryo Suppresses Peripheral Th17 Cells in MRL/*lpr* Mice

Th17-cell regulation is a critical therapeutic strategy for SLE treatment [27]. The present study showed that the levels of peripheral Th17 cells and IL-17 were remarkably reduced in SHED-Cryo-received MRL/*lpr* mice at the age of 20 weeks old (**Figures 5A–5C**). This Th17-cell suppressive effect was similar to the effect of SHED-Fresh-transplantation on MRL/*lpr* mice (**Figures 5B and 5C**) [13], suggesting that SHED-Cryo maintained *in vivo* immunomodulatory functions.

Systemically Infused SHED-Cryo Home to Lymph Node and Kidney in MRL/*lpr* Mice and Regulate the Local Immune Microenvironment

To assess the homing ability of SHED-Cryo into sites of injured tissues, CFSE-labeled SHED-Cryo were intravenously injected into MRL/*lpr* mice at the age of 16 weeks old. The high frequency of CFSE-positive SHED-Cryo was observed in the lymph nodes and kidneys, particularly in the glomeruli, on one day after the

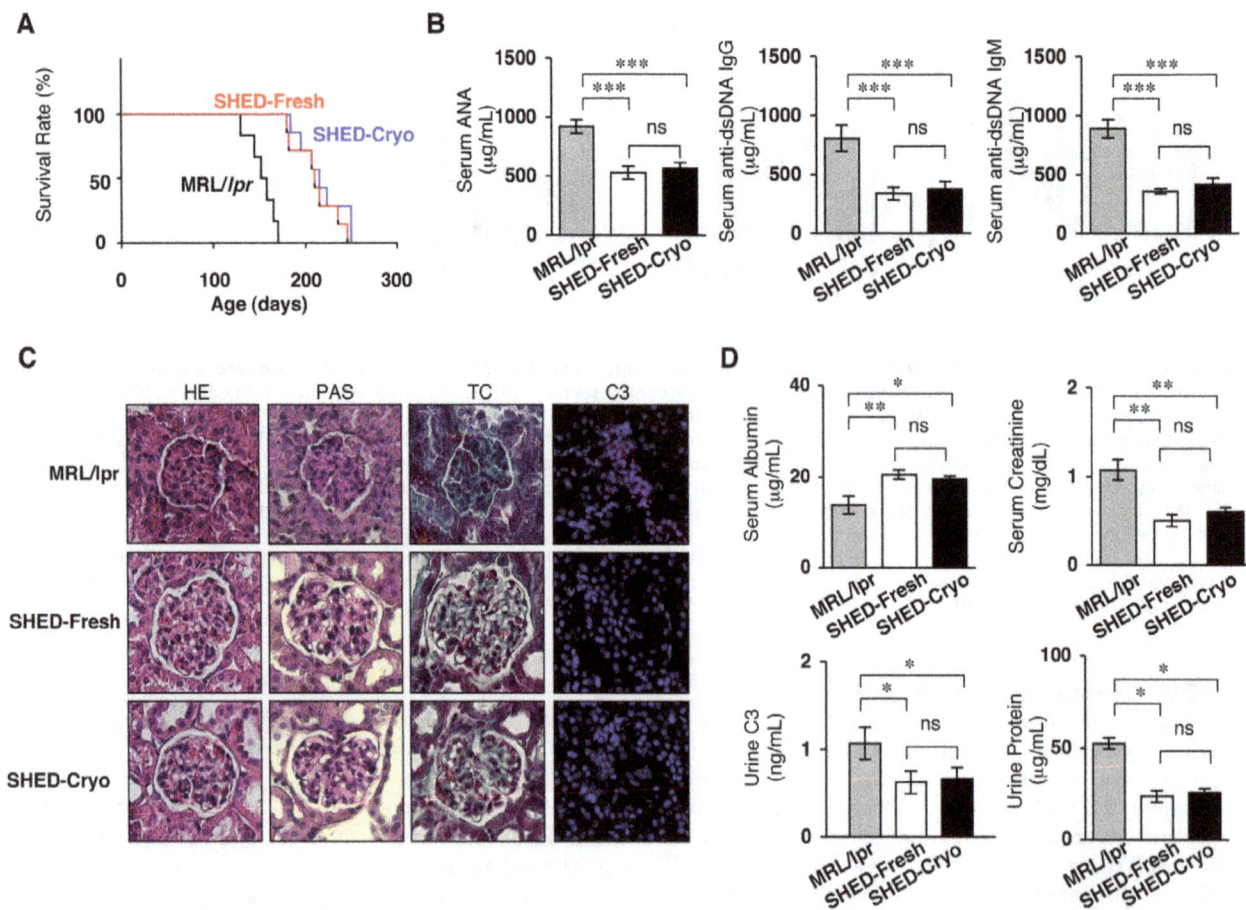

Figure 4. Systemic SHED-Cryo-transplantation improves lifespan and SLE-like disorders in MRL/*lpr* mice. (**A**) Kaplan-Meier survival curve of MRL/*lpr* mice. (**B**) ELISA of serum levels of autoantibodies ANA and anti-dsDNA IgG and IgM antibodies. (**C**) Histopathology of kidneys. *G* and dot-circled area: glomerular. HE: H&E staining, TC: Gomori trichrome staining, PAS: Periodic acid-Schiff staining, C3: Immunofluorescence of Complement C3. DAPI staining. (**D**) Levels of serum albumin and creatinine and urine C3 and protein. **A–D**: MRL/*lpr*: control group, SHED-Cryo: SHED-Cryo-transplant group, SHED-Fresh: SHED-Fresh-transplant group. **A**: n = 7, **B–D**: n = 5 for all group. **B, D**: *$P < 0.05$, **$P < 0.01$, ***$P < 0.005$, ns: no significance. The graph bars represent mean ± SD.

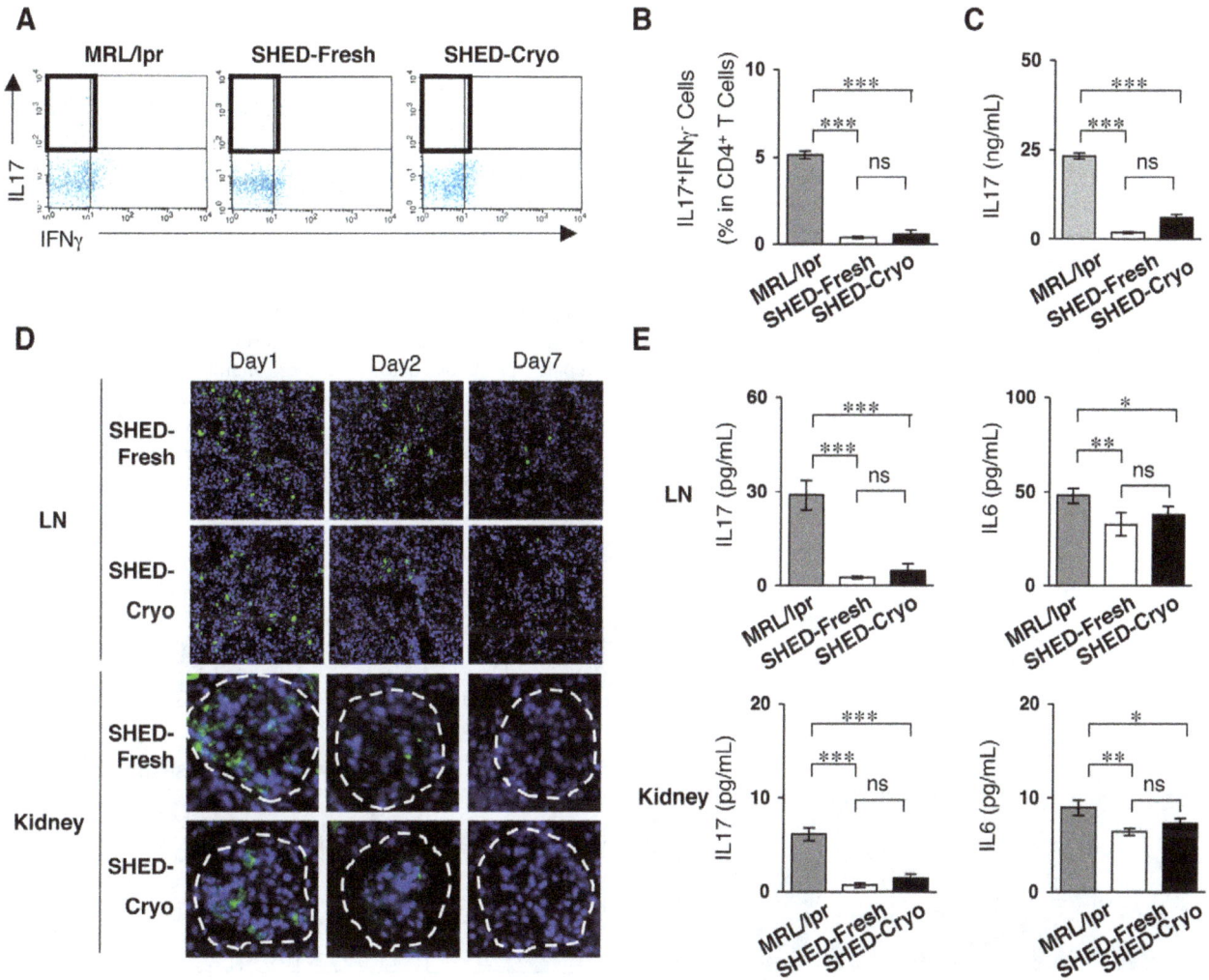

Figure 5. SHED transplantation suppresses circulating and local levels of Th17 cells in MRL/*lpr* mice. (A, B) Flow cytometry of peripheral CD4+IL17+IFNgamma⁻ Th17 cells. **(C)** Serum levels of IL-17. **(D)** Homing of systemically infused CFSE-labeled SHED-Cryo and SHED-Fresh to lymph node (LN) and kidney of MRL/*lpr* Mice after 1- (Day 1), 2- (Day 2) or 7- (Day 7) day transplantation. Dot-circled area: glomerular. **(E)** ELISA of IL-17 an IL-6 levels in lymph node and kidney. **A–E:** n = 5 for all group. MRL/*lpr*: control group, SHED-Cryo: SHED-Cryo-transplant group, SHED-Fresh: SHED-Fresh-transplant group. **B, C, E:** *P<0.05, **P<0.01, ***P<0.005, ns: no significance. The graph bars represent mean±SD.

transplantation (**Figures 5D and S6**). The frequency of CFSE-positive SHED-Cryo decreased gradually in both tissues from the 2nd day to 7th day after the transplantation (**Figures 5D and S6**). The localization of SHED-Cryo were similar to that of SHED-Fresh (**Figures 5D and S6**). The levels of inflammatory cytokines IL-17 and IL-6 in the lymph nodes and kidneys were significantly reduced in SHED-Cryo-transfused MRL/*lpr* mice, as well as SHED-Fresh-transfused mice, when compared to control MRL/*lpr* mice (**Figure 5E**). Taken together, these *in vivo* studies suggested that SHED-Cryo were capable of homing to lesional sites and might improve the pathological environments of damaged tissues.

Systemic SHED-Cryo-transplantation Improves Osteoporotic Skeletal Disorder in MRL/*lpr* Mice

The homing ability of SHED-Cryo into bones was analyzed 1 day after the infusion into MRL/*lpr* mice at the age of 16 weeks old. The CFSE-positive SHED-Cryo was sparsely observed in the bone marrow space of MRL/*lpr* mice 7 days after the infusion (**Figure S7**). The frequency of CFSE-positive SHED-Cryo was

similar to that of SHED-Fresh (**Figure S7**). MRL/*lpr* mice expressed a remarkable osteoporotic bone-loss in their long bones (**Figures 6A–6D**). Systemic SHED-Cryo-transplantation was capable of increasing BMD and recovering trabecular bone structures in the long bones of MRL/*lpr* mice (**Figures 6A and 6B**). TRAP-positive cells were significantly reduced in the long bones of SHED-Cryo-transplanted group compared to the non-transplanted control group (**Figure 6E**). SHED-Cryo-transplantation markedly reduced serum levels of sRANKL and CTX in MRL/*lpr* mice (**Figure 6F**). Flow cytometry revealed that CD4+IL-17+IFNgamma⁻ Th17 cells were significantly reduced in bone marrow cells of SHED-Cryo- and SHED-Fresh-transplanted MRL/*lpr* mice compared to the control MRL/*lpr* mice (**data not shown**), suggesting that immunomodulatory functions of SHED-Cryo and SHED-Fresh may contribute to reduce the bone reduction in MRL/*lpr* mice. To examine *ex vivo* osteoclastogenesis and osteogenesis, BMCs were isolated from control, SHED-Fresh-transplanted and SHED-Cryo-transplanted MRL/*lpr* mice, termed control-, SHED-Fresh- and SHED-Cryo-BMCs, respectively. When BMCs were stimulated with M-CSF and sRANKL, the number of TRAP-positive multinucleated cells was significantly reduced in SHED-Cryo-BMCs than in control-BMCs (**Figure 6G**). Osteogenic analysis showed that Alizarin Red-positive area was shared larger in SHED-Cryo-BMCs than control-BMCs (**Figure 6H**). SHED-Cryo-transplanted MRL/*lpr* mice expressed similar bone regenerative effects to SHED-Fresh-transplanted MRL/*lpr* mice (**Figure 6**). These data indicated that SHED-Cryo-transplantation improved osteoporotic disorder in MRL/*lpr* mice. Furthermore studies will be needed to evaluate the therapeutic mechanism of SHED to osteoporotic disorder in immune diseases likely SLE.

SHED-Cryo-implantation Repairs the Calvarial Bone Defects in Immunocompromised Mice

From the present *in vivo* tissue regeneration capability of SHED-Cryo, we hypothesized that SHED-Cryo could regenerate bone tissues in bone defects, as seen in SHED-Fresh (**Figure 7A–C**) [11]. We generated critical calvarial bone defects on immuno-

compromised mice and implanted SHED-Cryo with HA/TCP carrier onto the defect area (**Figure S8**). SHED-Cryo-implantation was able to regenerate the calvarial defects with a large amount of bone-like structures and bone-marrow-like components compared to implantation with only HA-TCP (**Figure 7A–C**). The amount of regenerated bone and bone marrow at SHED-Cryo-implanted sites was similar to that at SHED-Fresh implanted sites (**Figure 7A–C**). Immunofluorescence with anti-human CD146 antibody revealed that SHED-Cryo were responsible cells for bone regeneration in SHED-Cryo- and SHED-Fresh-implanted group, but not in HA/TCP-implanted group (**Figure 7B**). These findings suggested that SHED-Cryo and SHED-Fresh could be differentiated into bone-forming cells to contribute to repair the bone defect. A large number of TRAP-positive osteoclast-like cells were found in the regenerated bone tissues in SHED-Cryo-implanted sites, as well as in that of SHED-Fresh-implanted sites, but less in HA/TCP-implanted sites (**Figure 7D**), suggesting that the regenerated bone tissues might indicate

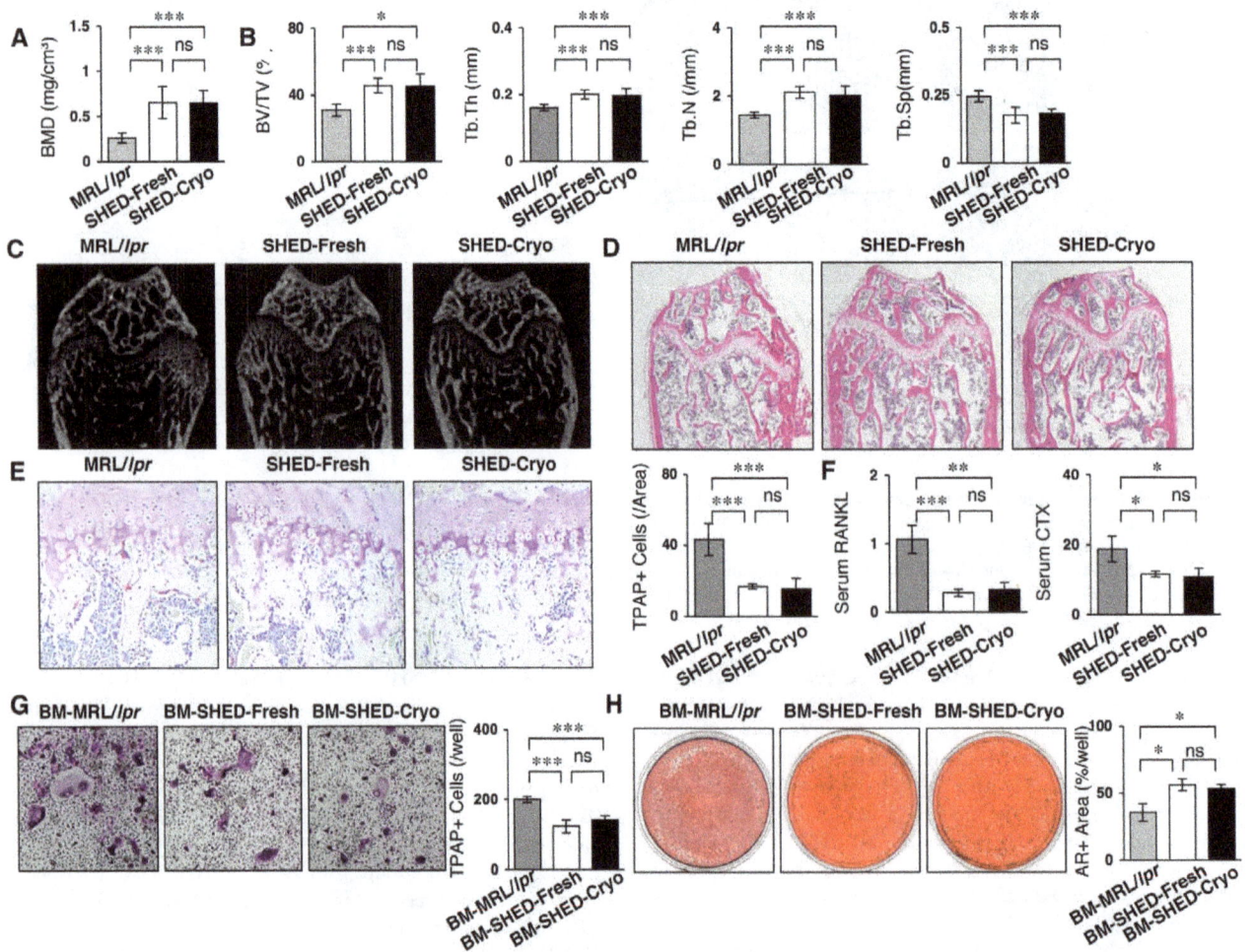

Figure 6. SHED-Cryo transplantation ameliorates osteoporotic bone disorder in MRL/*lpr* mice. (A, B) MicroCT analysis of tibiae. BMD (**A**). Trabecular parameters, bone volume ratio to tissue volume (BV/TV), trabecular thickness (Tb.Th), and trabecular number (Tb.N) along with increased trabecular separation (Tb.Sp) (**B**). (**C, D**) MicroCT (**C**) and histological (**D**) images of trabecular bone structures of tibiae. H&E staining (**D**). (**E, F**) *In vivo* osteoclast activity. TRAP staining (**E**). ELISA of serum sRANKL and C-terminal telopeptides of type I collagen (CTX) (**F**). (**G**) *Ex vivo* sRANKL-induced osteoclastogenesis. TRAP+ cells: TRAP-positive osteoclast-like cells. (**H**) *Ex vivo* osteogenic capacity. Alizarin red-positive (AR+) area after four-week induction. **A–H:** n = 5 for all groups. **A–F:** MRL/*lpr*: control group, SHED-Cryo: SHED-Cryo-transplant group, SHED-Fresh: SHED-Fresh-transplant group, **G, H:** BM-MRL/*lpr*: control MRL/*lpr* mice-derived bone marrow cells, BM-SHED-Cryo: SHED-Cryo-transplanted mice-derived bone marrow cells, BM-SHED-Fresh: SHED-Fresh-transplanted mice-derived bone marrow cells. **A, B, E, F:** *P<0.05, ***P<0.005 (vs. MRL/*lpr*). **G, H:** *P<0.05, **P<0.01, ***P<0.005 (vs. BM-MRL/*lpr*), ns: no significance. **A, B, E–H:** The graph bars represent mean±SD.

a physiological bone-remodeling ability by osteoclasts and osteoblasts. These data implied that SHED-Cryo were a useful cell source for tissue engineered bone regeneration. Implanted MSCs are impaired by host lymphocytes through secreting the pro-inflammatory cytokines IFNgamma and TNFalpha to suppress MSC-mediated bone regeneration [28]. On the other hand, deciduous teeth stem cells are capable of regenerating bone tissues in the calvarial defects of non-immunosuppressed rats [29]. Furthermore study will be necessary to clarify the kinetics of SHED under immunocompetent conditions for future clinical tissue engineering.

Discussion

Since SHED has been identified in dental pulp of deciduous tooth and characterized as MSCs [10], deciduous dental pulp tissues have been considered a promising stem cell source. SHED or deciduous teeth stem cells express multipotency into several lineage cells including dentin/bone-forming cells [10,30], endothelial cells [30], neural cells [10,31] and myocytes [22] *in vitro* and *in vivo*. SHED or deciduous teeth stem cells were also applied for tissue-engineering in large animal models including bone defects, muscular dystrophy and dentin defects [32–34], as well as small animal models including bone defect and spinal cord

injury [11,29,35]. Recent study demonstrates *in vitro* immuno-modulatory functions of SHED and evaluates the immune therapeutic efficacy on SLE-like model mice [13]. Herein, we demonstrated the feasibility of cryopreserved dental pulp tissue of human deciduous teeth in SHED-based bone tissue engineering and immune therapy. Our results indicate that long-term cryopreservation of human dental pulp tissues of deciduous teeth provides a great potential in future translational researches and clinical applications.

Postnatal stem cells have offered great promise to care diverse diseases. Besides HSCs have acquired outstanding and extensive success in a variety of human diseases such as leukemia, aplastic anemia and autoimmune diseases over half century [36,37]. To date, MSCs also admit to regenerative medicine for GVHD [6], skeletal reconstruction [2,3] and SLE [9]. Several challenges have remained to concern quality and safety in respective processes of MSC transplantation such as the cell processing and preserving *ex vivo*. Traditional bone marrow MSCs significantly reduced their frequency and multipotency donor-age-dependently [38–41]. Aspiration of bone marrow might accompany sever invasion to the donors [42]. As these significant disadvantages promote to seek alternative resources with accessible and least invasive approaches, novel MSC populations have been identified from various sources

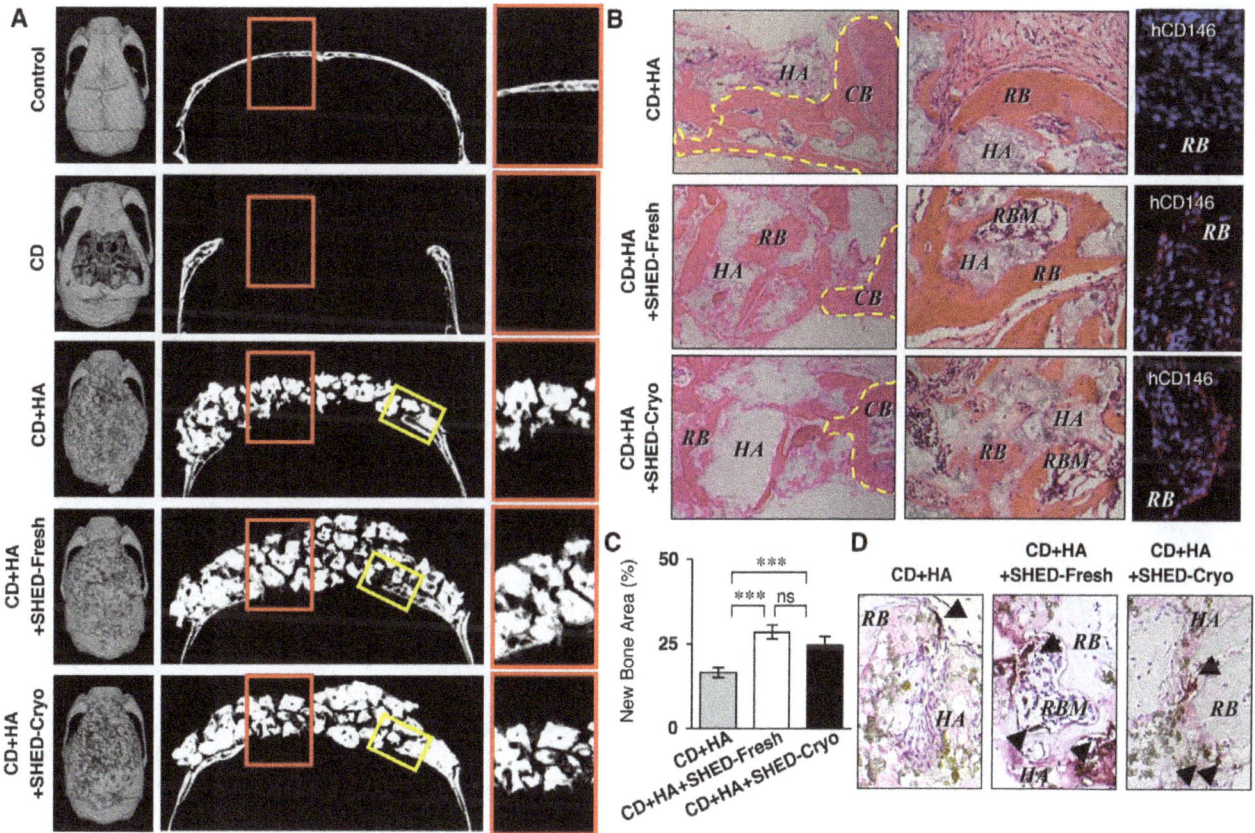

Figure 7. SHED-Cryo are capable of repairing critical calvarial bone defects in immunocompromised mice. (**A**) MicroCT images of mouse calvariae. Left panels: cranial images, middle panels: saggital images, right panels: images of red-bowed area in riddle panels. (**B**) Histology of bone regeneration in mouse calvariae. Left panels: edge parts of the defect area (yellow-boxed area in Figure 7A). H&E staining; middle panels: middle parts of the defect area (red red-bowed area in Figure 7A). H&E staining; right panels: immunofluorescence with anti-human CD146 antibody (hCD146). DAPI staining; right panels: immunofluorescence with anti-human CD146 antibody (hCD146). DAPI staining. CB, yellow dot-circled area: calvarial bone, HA: HA/TCP, RB: regenerated bone, RBM: regenerated bone marrow. (**C**) Regenerated bone area in the defect area. (**D**) Distribution of osteoclasts. Arrowheads: TRAP-positive cells. TRAP staining. **A–D:** n = 5 for all groups. Control: control (non-defect) group, CD: calvarial defect group, CD+HA: HA/TCP-implanted group, CD+HA+SHED-Fresh: SHED-Fresh-implanted group, CD+HA+SHED-Cryo: SHED-Cryo-implanted group. **C:** ***$P<0.005$, ns: no significance. The graph bars represent mean±SD.

such as adipose tissue, cord blood and dental pulp [43–45]. While the isolation process of MSCs is intricate, clinically applicable processing for storing and banking of the resources can offer great advantages for MSC-based therapy as well as the reduction of the number of staffing required for cell processing. Cryopreservation of stem cells has provided several utilities such as long-term storage, adjusting a therapeutic cell dose, reducing contamination for safety and quality in the clinical applications [34,46]. This approach has been used widely and successfully in bone marrow transplantation [47] and HSC transplantation [48]. Therefore, cryopreserved store and banking of MSC resources would be a variable, indispensable and practical approach for stem cell-based therapy.

SHED have been considered to be a primary promising source for regenerative medicine [10]. Exfoliated deciduous teeth represent the most easily, least invasively accessible and feasible resource because of their natural fate (exfoliation) and clinical abolition. SHED offered profound therapeutic efficacy to skeletal defects [11,12] and autoimmune disease [13]. On the other hand, several tasks have remained to be solved in SHED processing and banking. It is generally hard to expect a chance of the exfoliating of deciduous teeth and to maintain the tissue activity of the interests for a while after the harvesting. In addition, isolation process of SHED accompanies with several complicated, arduous and time-consuming steps and attentive operations. Recent discoveries about functional MSCs recovered from cryopreserved intact dental pulp and periodontal ligament (PDL) tissues of adult human third molars suggest that the least minimal processing may be adequate for the banking of samples [49,50]. Recovered MSCs after the cryopreservation can still maintain the immunomodulatory capacity *in vitro* [51,52]. Deciduous teeth stem cells is known to maintain the stem cell property and multipotency after the long-term cryopreservation [22,23]. In the present study, we firstly demonstrate that stem cells are capable of recovering from human dental pulp tissues of exfoliated deciduous teeth after long-term (over 2 years) cryopreservation. The recovered SHED retained superior MSC properties such as self-renew, multipotency, *in vivo* dentin/bone-regeneration and *in vitro* immunomodulatory function. In addition, transplantation of SHED-Cryo showed critical therapeutic efficiency on both immune and skeletal disorders in MRL/*lpr* mice and bone defects in the calvariae of immunocompromised mice. Taken together, these data suggest that long-term cryopreservation of dental pulp tissues of deciduous teeth is an innocuous and designable approach for clinical banking of stem cells and gives great advantages in immune therapy and bone tissue engineering of regenerative medicine. Furthermore, the present pulp tissue banking system could allow children and their parents to return the "baby teeth" bodies as the most precious souvenirs to the children's growing.

Current studies indicate that the biological recovery of the PDL stem cells and adult DPSCs are inferior to the freshly isolated stem cells after the cryopreservation of PDL tissues [50] and adult teeth [49]. Whereas, cryopreserved stem cells from apical papilla (SCAP) show a similar biological and immunomodulatory functions to freshly isolated SCAP [51]. Apical papillae have a responsibility to form and extend tooth roots at the developing stage [53,54], indicating that apical papillae are an active tissue biologically. Cryopreserved deciduous teeth stem cells can also maintain the stem cell property and multipotency [22,23]. The present study demonstrated that cryopreservation of deciduous dental pulp did not affect the biological and immunological properties of SHED. Moreover, SHED show higher cell proliferation capability than adult DPSCs [10], suggesting that deciduous pulp tissues maintain greater biological activity than

adult dental pulp tissues. The discrepancy of the functionally recovery efficiency among deciduous dental pulps, adult PDLs and adult dental pulps after the cryopreservation might depend on the age and/or potential activity of donor samples, supporting the feasibility of the cryopreservation of deciduous dental pulp tissues.

In conclusion, the present cryopreservation of dental pulp tissues of human exfoliated deciduous teeth does not affect on the biological, immunological and therapeutic functions of SHED. Therefore, cryopreserved approach of deciduous dental pulp tissues not only serve as a most clinically desirable banking approach, but also provide sufficient number of SHED for critical therapeutic benefits to stem cell-based immune therapy and tissue engineering in regenerative medicine.

Supporting Information

Figure S1   A scheme of the cryopreservation and isolation of mesenchymal stem cells (MSCs) from dental pulp tissues of exfoliated deciduous teeth. Deciduous dental pulp tissues in the remnant crown (yellow-dot circled region) were removed *en bloc* mechanically, stored in a freezing medium and preserved in a liquid nitrogen tank over 2 years. The frozen tissues were thawed at 37°C and treated with an enzyme solution. SHED from the cryopreserved deciduous dental pulp tissues (SHED-Cryo), as well as SHED from fresh deciduous dental pulp tissues (SHED-Fresh), were obtained by colony forming units fibroblasts (CFU-F) method.

Figure S2   A scheme of *in vivo* tissue regeneration and self-renewal assays of SHED-Cryo. SHED-Cryo were subcutaneously transplanted with HA/TCP carrier into immuno-compromised mice. Eight weeks after the implantation, the primary transplants were harvested. Some transplants were used for histological and immunofluorescent analyses. Cells were isolated from the other primary transplants and stained with human-specific CD146 antibody. Human CD146-positive cells were magnetically sorted. The purity of the cells was confirmed by flow cytometry as described in Materials and Methods. The CD146-positive cells were seeded at low density to obtain CFU-F-forming cells. The colony-forming cells were expanded and transplanted secondarily into immunocompromised mice. The secondary transplants were harvested 8 weeks after the implantation and analyzed morphologically.

Figure S3   Images of primary transplant tissues with HA/TCP alone. (**A**) H&E staining (HE). (**B-D**) Immunofluorescence with anti-human specific mitochondria (hMt) (**B**), anti-STRO-1 (**C**) and human CD146 (hCD146) (**D**) antibodies.

Figure S4   A scheme of the transplantation of SHED-Cryo into MRL/*lpr* mice (*lpr*). SHED-Cryo were infused into MRL/*lpr* mice via the tail vein at the age of 16 weeks. The mice were maintained until died for the survival assay. At 20-week-old, some mice were harvested and biological samples were collected to assess the therapeutic efficacy.

Figure S5   Systemic SHED-Cryo-transplantation improves levels of serum immunoglobulins in MRL/*lpr* mice. n = 5 for all group. *$P<0.05$, **$P<0.01$, ***$P<0.005$, ns: no significance. The graph bars represent mean±SD. MRL/lpr: non-transplanted group, SHED-Fresh: SHED-Fresh-transplanted group, SHED-Cryo: SHED-Cryo-transplanted group.

Figure S6 Homing of systemically infused SHED-Cryo to lymph node and kidney of MRL/*lpr* Mice. Images of CFSE-labeled SHED-Cryo and SHED-Fresh in lymph nodes (LN) and kidneys of MRL/*lpr* mice 1 (Day 1) or 7 (Day 7) days after the transplantation. CFSE: CSFE image, DAPI: DAPI image, SHED-Fresh: SHED-Fresh-infused group, SHED-Cryo: SHED-Cryo-infused group.

Figure S7 Homing of systemically infused SHED-Cryo to bone of MRL/*lpr* Mice. Images of CFSE-labeled cells in bone of MRL/*lpr* mice 7 days after the transplantation. CFSE: CFSE image, DAPI: DAPI image, CFSE/DAPI: Merged image of CFSE and DAPI images, SHED-Fresh: SHED-Fresh-infused group, SHED-Cryo: SHED-Cryo-infused group.

Figure S8 A scheme of the transplantation of SHED-Cryo into calvarial bone defect of immunocompromised mice. SHED-Cryo were expanded and mixed with HA/TCP

carriers. Calvarial bones, especially parietal bone area (P), were removed to generate a bone defect on immunocompromised mice. SHED & HA/TCP mixture were implanted to cover over the defect area. Twelve weeks after the implantation, the samples were harvested and analyzed by microCT and histology.

Table S1 The list of antibodies.

Table S2 The list of primer pairs for RT-PCR.

Author Contributions

Conceived and designed the experiments: TY SS. Performed the experiments: TY LM YM HY KA YH. Analyzed the data: TY LM YM HY KA YH GS KN TK SS. Contributed reagents/materials/analysis tools: TY HY. Wrote the paper: TY.

References

1. Porada CD, Almeida-Porada G (2010) Mesenchymal stem cells as therapeutics and vehicles for gene and drug delivery. Adv Drug Deliv Rev 62: 1156–1166.
2. Kwan MD, Slater BJ, Wan DC, Longaker MT (2008) Cell-based therapies for skeletal regenerative medicine. Hum Mol Genet 17: R93–98.
3. Panetta NJ, Gupta DM, Quarto N, Longaker MT (2009) Mesenchymal cells for skeletal tissue engineering. Panminerva Med 51: 25–41.
4. Aggarwal S, Pittenger MF (2005) Human mesenchymal stem cells modulate allogeneic immune cell responses. Blood 105: 1815–1822.
5. Nauta AJ, Fibbe WE (2007) Immunomodulatory properties of mesenchymal stromal cells. Blood 110: 3499–3506.
6. Le Blanc K, Rasmusson I, Sundberg B, Götherström C, Hassan M, et al. (2004) Treatment of severe acute graft-versus-host disease with third party haploidentical mesenchymal stem cells. Lancet 363: 1439–1441.
7. Koç ON, Gerson SL, Cooper BW, Dyhouse SM, Haynesworth SE, et al. (2000) Rapid hematopoietic recovery after coinfusion of autologous-blood stem cells and culture-expanded marrow mesenchymal stem cells in advanced breast cancer patients receiving high-dose chemotherapy. J Clin Oncol 18: 307–316.
8. Noort WA, Kruisselbrink AB, in't Anker PS, Kruger M, van Bezooijen RL, et al. (2002) Mesenchymal stem cells promote engraftment of human umbilical cord blood-derived CD34 cells in NOD/SCID mice. Exp Hematol 30: 870–878.
9. Sun L, Akiyama K, Zhang H, Yamaza T, Hou Y, et al. (2009) Mesenchymal Stem Cell Transplantation Reverses Multi-Organ Dysfunction in Systemic Lupus Erythematosus Mice and Humans. Stem Cells 27: 1421–1432.
10. Miura M, Gronthos S, Zhao M, Lu B, Fisher LW, et al. (2003) SHED: Stem cells from human exfoliated deciduous teeth. Proc Natl Acad Sci USA 100: 5807–5812.
11. Seo BM, Sonoyama W, Yamaza T, Coppe C, Kikuiri T, et al. (2008) SHED repair critical-size calvarial defects in immunocompromised mice. Oral Dis 14: 428–434.
12. Zheng Y, Liu Y, Zhang CM, Zhang HY, Li WH, et al. (2009) Stem cells from deciduous tooth repair mandibular defect in swine. J Dent Res 88: 249–254.
13. Yamaza T, Akiyama K, Chen C, Liu Y, Shi Y, et al. (2010) Immunomodulatory properties of stem cells from human exfoliated deciduous teeth. Stem Cell Res Ther 1:5.
14. Wood EJ, Benson JD, Agca Y, Crister JK (2004) Fundamental cryobiology of reproductive cells and tissues. Cryobiology 48: 146–156.
15. Friedenstein AJ (1980) Stromal mechanisms of bone marrow: cloning in vitro and retransplantation in vivo. Haematol Blood Transfus 25: 19–29.
16. Shi S, Gronthos S, Chen S, Reddi A, Counter CM, et al. (2002) Bone formation by human postnatal bone marrow stromal stem cells is enhanced by telomerase expression. Nat Biotechnol 20: 587–591.
17. Bi Y, Ehirchiou D, Kilts TM, Inkson CA, Embree MC, et al. (2007) Identification of tendon stem/progenitor cells and the role of the extracellular matrix in their niche. Nat Med. 13: 1219–1227.
18. Gronthos S, Zannettino AC, Hay SJ, Shi S, Graves SE, et al. (2003) Molecular and cellular characterisation of highly purified stromal stem cells derived from human bone marrow. J Cell Sci 116: 1827–1835.
19. Yamaza T, Miura Y, Bi Y, Liu Y, Akiyama K, et al. (2008) Pharmacologic stem cell based intervention as a new approach to osteoporosis treatment in rodents. PLoS ONE 3: e2615.
20. Danjo A, Yamaza T, Kido MA, Shimohira D, Tsukuba T, et al. (2007) Cystatin C stimulated the differentiation of mouse osteoblastic cells and bone formation. Biochem Biophys Res Commun 360: 199–204.
21. Shi S, Gronthos S (2003) Perivascular Niche of Postnatal Mesenchymal Stem Cells Identified in Human Bone Marrow and Dental Pulp. J Bone Miner Res 18: 696–704.
22. Kerkis I, Kerkis A, Dozortsev D, Stukart-Parsons GC, Gomes Massironi SM, et al. (2006) Isolation and characterization of a population of immature dental pulp stem cells expressing OCT-4 and other embryonic stem cell markers. Cells Tissues Organs. 184: 105–116.
23. Lizier NF, Kerkis A, Gomes CM, Hebling J, Oliveira CF, et al. (2012) Scaling-up of dental pulp stem cells isolated from multiple niches. PLoS One 7: e39885.
24. Miura Y, Gao Z, Miura M, Seo BM, Sonoyama W, et al. (2006) Mesenchymal stem cell-organized bone marrow elements: an alternative hematopoietic progenitor resource. Stem Cells 24: 2428–2436.
25. Yamaza T, Miura Y, Akiyama K, Bi Y, Sonoyama W, et al. (2009) Mesenchymal stem cell-mediated ectopic hematopoiesis alleviates aging-related phenotype in immunocompromised mice. Blood 113: 2595–2604.
26. Morrison SJ, Prowse KR, Ho P, Weissman IL (1996) Telomerase activity in hematopoietic cells is associated with self-renewal potential. Immunity 5: 207–216.
27. Yang J, Yang X, Zou H, Chu Y, Li M (2011) Recovery of the immune balance between Th17 and regulatory T cells as a treatment for systemic lupus erythematosus. Rheumatology 50: 1366–1372.
28. Liu Y, Wang L, Kikuiri T, Akiyama K, Chen C, et al. (2011) Mesenchymal stem cell-based tissue regeneration is governed by recipient T lymphocytes via IFN-gamma and TNF-α. Nat Med 17: 1594–1601.
29. de Mendonça Costa A, Bueno DF, Martins MT, Kerkis I, Kerkis A, et al. (2008) Reconstruction of large cranial defects in nonimmunosuppressed experimental design with human dental pulp stem cells. J Craniofac Surg 19: 204–210.
30. Sakai VT, Zhang Z, Dong Z, Neiva KG, Machado MA, et al. (2010) SHED differentiate into functional odontoblasts and endothelium. J Dent Res 89: 791–796.
31. Nourbakhsh N, Soleimani M, Taghipour Z, Karbalaie K, Mousavi SB, et al. (2011) Induced in vitro differentiation of neural-like cells from human exfoliated deciduous teeth-derived stem cells. Int J Dev Biol 55: 189–195.
32. Zheng Y, Liu Y, Zhang CM, Zhang HY, Li WH, et al. (2009) Stem cells from deciduous tooth repair mandibular defect in swine. J Dent Res 88: 249–254.
33. Kerkis I, Ambrosio CE, Kerkis A, Martins DS, Zucconi E, et al. (2008) Early transplantation of human immature dental pulp stem cells from baby teeth to golden retriever muscular dystrophy (GRMD) dogs: Local or systemic? J Transl Med 6: 35.
34. Zheng Y, Wang XY, Wang YM, Liu XY, Zhang CM, et al. (2012) Dentin regeneration using deciduous pulp stem/progenitor cells. J Dent Res 91: 676–682.
35. Sakai K, Yamamoto A, Matsubara K, Nakamura S, Naruse M, et al. (2012) Human dental pulp-derived stem cells promote locomotor recovery after complete transection of the rat spinal cord by multiple neuro-regenerative mechanisms. J Clin Invest 122: 80–90.
36. Korbling M, Estrov Z (2003) Adult stem cells for tissue repair–a new therapeutic concept? N Engl J Med 349: 570–582.
37. Woods EJ, Pollok KE, Byers MA, Perry BC, Purtteman S, et al. (2007) Cord blood stem cell cryopreservation. Transfus Med Hemother 34: 276–285.
38. D'Ippolito G, Schiller PC, Ricordi C, Roos BA, Howard GA (1999) Age-related osteogenic potential of mesenchymal stromal stem cells from human vertebral bone marrow. J Bone Miner Res 14: 1115–1122.
39. Stenderup K, Justesen J, Clausen C, Kassem M (2003) Aging is associated with decreased maximal life span and accelerated senescence of bone marrow stromal cells. Bone 33: 919–926.
40. Stolzing A, Jones E, McGonagle D, Scutt A (2008) Age-related changes in human bone marrow-derived mesenchymal stem cells: consequences for cell therapies. Mech Age Dev 129: 163–173.

41. Garvin G, Connie F, Sharp JG, Berger A (2007) Does the number or quality of pluripotent bone marrow stem cells decrease with age? Clin Orthod Relat Res 465: 202–207.

42. Muschler GF, Boehm C, Easley K (1997) Aspiration to obtain osteoblast progenitor cells from human bone marrow: The influence of aspiration volume. J Bone Joint Surg Am 79: 1699–1709.

43. Zuk PA, Zhu M, Mizuno H, Huang J, Futrell JW, et al. (2001) Multilineage cells from human adipose tissue: implications for cell-based therapies. Tissue Eng 7: 211–228.

44. Erices A, Conget P, Minguell JJ (2000) Mesenchymal progenitor cells in human umbilical cord blood. Br J Haematol 109: 235–242.

45. Gronthos S, Mankani M, Brahim J, Robey PG, Shi S (2000) Postnatal human dental pulp stem cells (DPSCs) in vitro and in vivo. Proc Natl Acad Sci USA 97: 13625–13630.

46. Hubel A (1997) Parameters of cell freezing: implications for the cryopreservation of stem cells. Transfus Med Rev 11: 224–233.

47. Areman E, Sacher R, Deeg H (1990) Processing and storage of human bone marrow: A survey of current practices in North America. Bone Marrow Transplant 6: 203–209.

48. Watt SM, Austin E, Armitage S (2007) Cryopreservation of hematopoietic stem/progenitor cells for therapeutic use. Methods Mol Biol 368: 237–259.

49. Perry BC, Zhou D, Wu X, Yang FC, Byers MA, et al. (2008) Collection, cryopreservation, and characterization of human dental pulp-derived mesenchymal stem cells for banking and clinical use. Tissue Eng Part C. 14: 149–156.

50. Seo BM, Miura M, Sonoyama W, Coppe C, Stanyon R, et al. (2005) Recovery of stem cells from cryopreserved periodontal ligament. J Dent Res 84: 907–912.

51. Ding G, Wang W, Liu Y, An Y, Zhang C, et al. (2010) Effect of cryopreservation on biological and immunological properties of stem cells from apical papilla. J Cell Physiol 223: 415–422.

52. Zhao ZG, Li WM, Chen ZC, You Y, Zou P (2008) Hematopoiesis capacity, immunomodulatory effect and ex vivo expansion potential of mesenchymal stem cells are not impaired by cryopreservation. Cancer Invest 26: 391–400.

53. Sonoyama W, Liu Y, Fang D, Yamaza T, Seo BM, et al. (2006) Mesenchymal stem cell-mediated functional tooth regeneration in swine. PLoS One 1: e79.

54. Sonoyama W, Liu Y, Yamaza T, Tuan RS, Wang S, et al. (2008) Characterization of the Apical Papilla and Its Residing Stem Cells from Human Immature Permanent Teeth: A Pilot Study. J Endod 34: 166–171.

15

Cobalt and Nickel Stabilize Stem Cell Transcription Factor OCT4 through Modulating Its Sumoylation and Ubiquitination

Yixin Yao[1], Yinghua Lu[1], Wen-chi Chen[1], Yongping Jiang[2], Tao Cheng[3], Yupo Ma[4], Lou Lu[5], Wei Dai[1,6]*

1 Department of Environmental Medicine, New York University Langone Medical Center, Tuxedo, New York, United States of America, **2** Biopharmaceutical Research Center, Chinese Academy of Medical Sciences & Peking Union Medical College, Suzhou, China, **3** Institute of Hematology & Blood Disease Hospital, Chinese Academy of Medical Sciences & Peking Union Medical College, Tianjin, China, **4** Yupo Ma, Department of Pathology, The State University of New York at Stony Brook, Stony Brook, New York, United States of America, **5** Department of Medicine, David Geffen School of Medicine, University of California Los Angeles, Torrance, California, United States of America, **6** Department of Biochemistry and Molecular Pharmacology, New York University Langone Medical Center, Tuxedo, New York, United States of America

Abstract

Stem cell research can lead to the development of treatments for a wide range of ailments including diabetes, heart disease, aging, neurodegenerative diseases, spinal cord injury, and cancer. OCT4 is a master regulator of self-renewal of undifferentiated embryonic stem cells. OCT4 also plays a crucial role in reprogramming of somatic cells into induced pluripotent stem (iPS) cells. Given known *vivo* reproductive toxicity of cobalt and nickel metals, we examined the effect of these metals on expression of several stem cell factors in embryonic Tera-1 cells, as well as stem cells. Cobalt and nickel induced a concentration-dependent increase of OCT4 and HIF-1α, but not NANOG or KLF4. OCT4 induced by cobalt and nickel was due primarily to protein stabilization because MG132 stabilized OCT4 in cells treated with either metals and because neither nickel nor cobalt significantly modulated its steady-state mRNA level. OCT4 stabilization by cobalt and nickel was mediated largely through reactive oxygen species (ROS) as co-treatment with ascorbic acid abolished OCT4 increase. Moreover, nickel and cobalt treatment increased sumoylation and mono-ubiquitination of OCT4 and K123 was crucial for mediating these modifications. Combined, our observations suggest that nickel and cobalt may exert their reproductive toxicity through perturbing OCT4 activity in the stem cell compartment.

Editor: Xianglin Shi, University of Kentucky, United States of America

Funding: This study was supported in part by US Public Service Awards to WD (CA090658 and ES019929), an NIEHS Center grant Nickel and cobalt stabilize (ES000260) and a key State project of China on iPS and stem cell research (2011ZX09102-010-04). The funders had no role in study design, data collection and analysis, and decision to publish, or preparation of the manuscript.

Competing Interests: The authors have declared that no competing interests exist.

* E-mail: wei.dai@nyumc.org

Introduction

Cobalt [Co(II)] and Nickel [Ni(II)] are capable of crossing the placenta barrier and exerting their toxicity on the animal reproduction system, thus affecting embryonic development [1,2]. Exposure of Ni(II) and Co(II) at a high concentration (100 µM) significantly reduced proliferation of inner cell mass and trophoblast cells [3]. The reduced proliferative ability of trophoblast cells compromises invasiveness of the embryo [3]. Intriguingly, exposure of Co(II) at a low concentration (1 µM) induces a highly organized inner cell mass with an abnormally large size [2]. Human exposure to cobalt and nickel occur environmentally and occupationally. It has been reported that there is a correlation between occupational exposure to nickel (refinery female workers) and delivery of newborns small-for-gestational-age [4]. Both soluble and insoluble nickel can potentially pose threat to human health. It has been reported that potential intracellular concentrations of nickel ion can reach the molar range after cell phagocytizes a crystalline NiS particle [5].

Octamer binding protein 4 (OCT4), SOX2, Krüppel-like factor 4 (KLF4), and MYC are important transcription factors that are capable of reprogramming somatic cells into pluripotent stem cells [6–8]. Induced pluripotent stem (iPS) cells possess the capacity of developing into an entire organism [9]. Hypoxia improves the rate of reprogramming differentiated cells into iPS cells [10–14]. Consistent with these findings, bovine blastocysts produced under a reduced oxygen tension exhibit significantly more inner cell mass (consisting of embryonic stem cells) than those maintained at a normal oxygen tension [15].

OCT4 is a stem cell transcription factor that activates or represses target gene expression depending on cellular context [16–18]. OCT4 and other stem cell factors including NANOG and SALL4 form a transcriptional network that controls pluripotency in ES cells [19]. *OCT4* mRNA and its protein are present in unfertilized oocytes; OCT4 protein is localized to pronuclei following fertilization [20]. *OCT4* mRNA levels drop dramatically after fertilization albeit OCT4 protein remains detectable in the nuclei of 2-cell embryos [20]. Zygotic *OCT4* expression is activated prior to the 8- cell stage, leading to the increase of both mRNA and protein [20].

OCT4 is subject to post translational modifications including phosphorylation [21–23], poly-ubiquitination [24,25] and sumoylation [26–28]. For example, AKT1 phosphorylates OCT4 at threonine 235 (T235) in embryonic carcinoma cells [22]. The

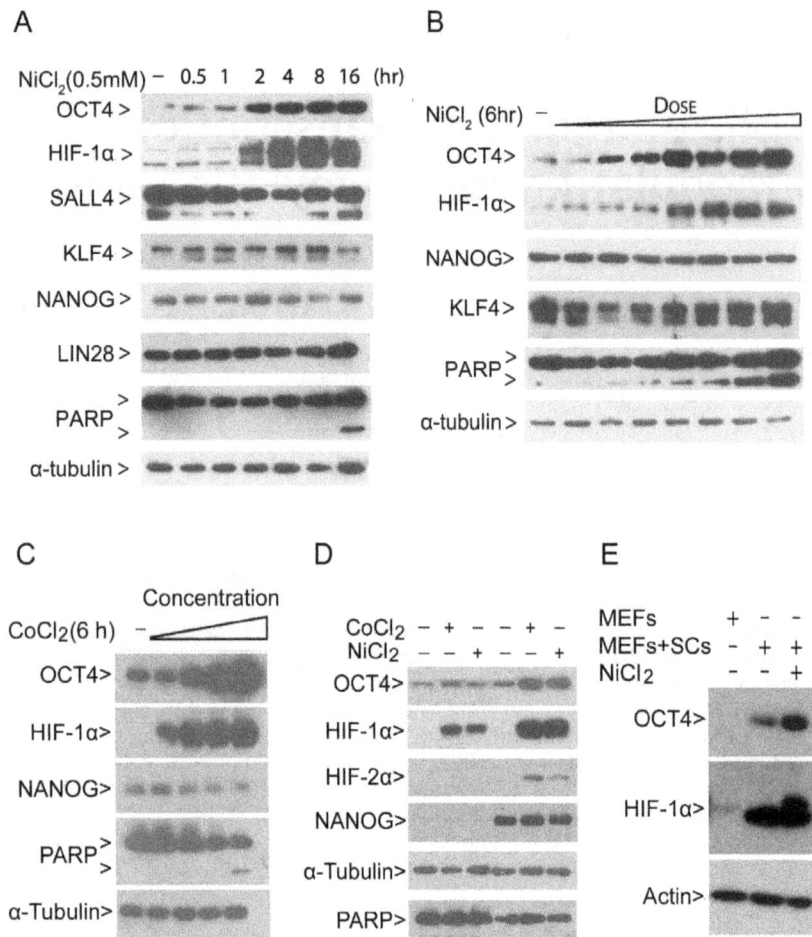

Figure 1. Induction of OCT4 expression by nickel and cobalt. A, Tera-1 cells treated with or without NiCl$_2$ (0.5 mM) for various times were lysed and equal amounts of cell lysates were blotted with indicated antibodies. B, Tera-1 cells treated with various concentrations of nickel (e.g., 15.6, 31.3, 62.5, 125, 250, 500, and 1000 µM) were lysed and equal amounts of cell lysates were blotted with antibodies as indicated. C, Tera-1 cells were treated with CoCl$_2$ at various concentrations (e.g., 125, 250, 500, and 1000 µM) and equal amounts of cell lysates were blotted with antibodies as indicated. D, Tera-1 and NT2/D1 cells were treated with NiCl$_2$ or CoCl$_2$ (0.5 mM) for 6 h and equal amounts of cell lysates were blotted with antibodies as indicated. E, Human embryonic stem cells (H1) grown on MEFs were treated with or without nickel (0.25 mM) for 24 h, after which cells were collected and lysed. MEFs cells grown separately without stem cells were also collected as control. Equal amounts of cell lysates were then blotted for OCT4.

phosphorylation increases the stability of OCT4 and facilitates its nuclear localization and interaction with SOX2. OCT4 is also modified by sumoylation, which positively regulates its stability, chromatin binding, and transcriptional activity [26].

To understand whether toxicity of nickel and cobalt on embryonic development is partly mediated by their effect on stem cell transcription factors, we studied OCT4 expression in both primary stem cells and stem cell-derived cell lines treated with nickel or cobalt ions. We observed that Ni(II) and Co(II) significantly increased expression of OCT4 in a time- and concentration-dependent manner. Ni(II)- or Co(II)-induced OCT4 expression is primarily due to protein stabilization. Our further studies reveal that ROS produced as the result of Ni(II) and Co(II) exposure is responsible for OCT4 stabilization partly via modulating post-translational modifications.

Results

Ni(II) and Co(II) Induce OCT4

To determine if expression of key stem cell transcription factors was affected by metal-induced stresses, Tera-1 cells (embryonic carcinoma origin) were treated with nickel chloride (NiCl$_2$) for various times. Equal amounts of cell lysates were blotted with antibodies to a panel of transcription factors including OCT4, NANOG, KLF4, SALL4, and HIF-1α. As expected, HIF-1α levels were stabilized by Ni(II) (Fig. 1A and 1B) as the metal is known to be a hypoxic mimetic [29]. Interestingly, OCT4 protein levels, but not other key stem cell factors including SALL4, NANOG, and KLF4, also exhibited a time- and concentration-dependent increase (Fig. 1A and 1B). Cobalt, a metal with many overlapping properties with nickel, also induced the increase of OCT4, but not NANOG, in Tera-1 cells in a concentration-dependent manner (Fig. 1C). As expected, it induced HIF-1α as well given its known property as a hypoxic mimetic [29]. Ni(II) and Co(II) also induced OCT4 in NT2 cells (embryonic origin) although the magnitude of induction was not as great as seen in Tera-1 cells (Fig. 1D),

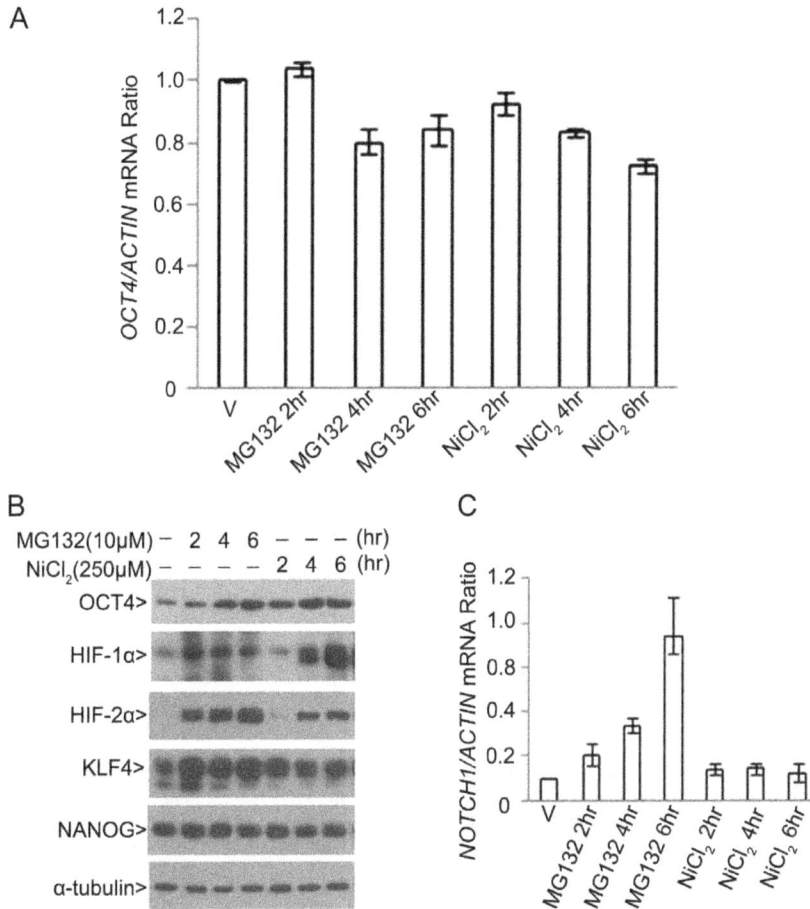

Figure 2. OCT4 induction by nickel or cobalt is not due to increased transcription. A, Tera-1 cells treated with MG132 (10 μM) or NiCl₂ (0.25 mM) for various times were collected and total RNAs were extracted. OCT4 specific mRNA was measured using quantitative PCR and specific signals were normalized by β-actin mRNA levels. B, Tera-1 cells treated with MG132 or NiCl₂ (0.25 mM) for various times were lysed and equal amounts of lysates were blotted with indicated antibodies. C, Quantification of expression of *NOTCH-1* mRNA in cells treated with MG132 or NiCl₂ for various times.

suggesting that cell lines with different genetic backgrounds may respond to the metal stress differently. Supporting this, HIF-2α was not inducible in NT2 cells by either Ni(II) or Co(II) whereas it was induced in Tera-1 cells (Fig. 1D).

To further confirm that induction of OCT4 occurs in primary stem cells, we treated feeder-dependent human embryonic stem cells (H1, WiCell) with NiCl₂. We observed that there is a basal level of OCT4 expression in H1 stem cells but not in feeder cells (murine embryonic fibroblasts). Nickel treatment significantly elevated the level of OCT4 (Fig. 1E). As expected, nickel induced expression of HIF-1α as well. In addition, we observed that nickel (or cobalt) treatment of human iPS cells could induce expression of OCT4 (data not shown). Moreover, chromium, another environmental metal toxicant, did not induce expression of OCT4 (data not shown). Combined, our observations are consistent with the notion that the steady-state level of OCT4 can be perturbed by exposure to nickel or cobalt ions.

OCT4 Induction by Ni(II) or Co(II) was Not Due to Transcriptional Activation

To determine whether increased expression of OCT4 by Ni(II) or Co(II) was due to transcriptional activation, RNA samples extracted from Tera-1 cells treated with Ni(II) or MG132 were

analyzed by quantitative polymerase chain reaction (qPCR). There was no increase in *OCT4* mRNA in cells treated with Ni(II) and/or MG132 whereas Ni(II) or MG132 greatly stimulated the accumulation of OCT4 and HIF-1α protein levels (Fig. 2A and 2B). As control, we analyzed *NOTCH1* mRNA levels via qPCR as it was under control of SALL4, also a stem cell transcription factor [30]. We observed that *NOTCH1* mRNA was significantly increased in cells treated with MG132 but not with Ni(II) (Fig. 2C).

Cobalt and Nickel Prolong the Half-life of OCT4 in Tera-1 Cells

To confirm that Ni(II) or Co(II) affects OCT4 protein stability, Tera-1 cells treated with cycloheximide (CHX), a chemical that blocks new protein synthesis, in the presence or the absence of Ni(II). At various times of treatment, cells were collected and equal amounts of cell lysates were blotted for OCT4, as well as other transcription factors. Ni(II) significantly stabilized the level of OCT4, but not NANOG and KLF4, in cells treated with CHX and prolonged its half-life (Fig. 3A and 3B). As expected, Ni(II) treatment also greatly stabilized HIF-1α. In addition, Co(II) significantly prolonged the half-life of both OCT4 and HIF-1α in cells treated with CHX (Fig. 3C and 3D). Combined, these studies

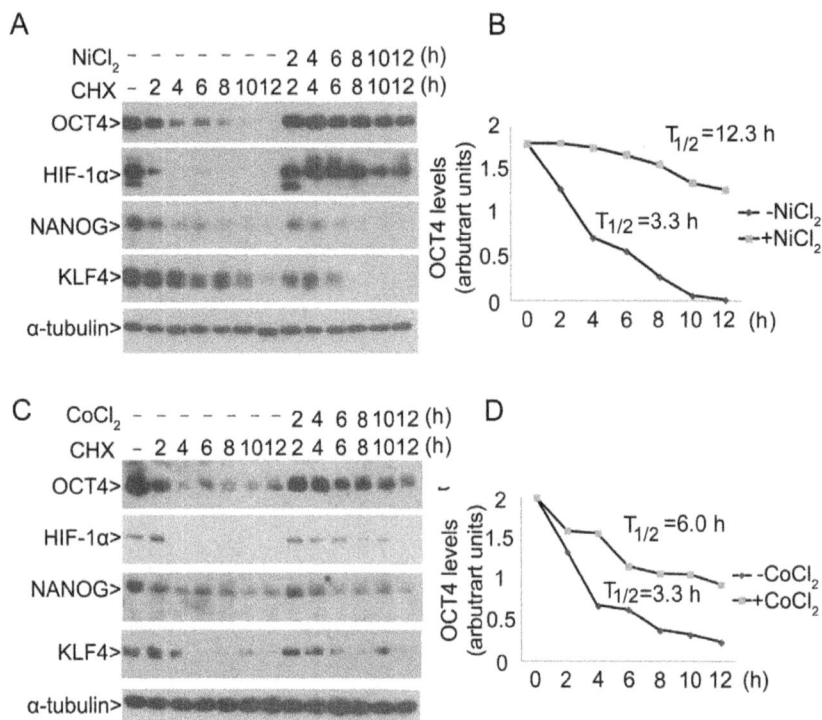

Figure 3. Cobalt and nickel prolong the half-life of OCT4. A. Tera-1 cells were treated with cycloheximide (CHX) in the presence of absence of NiCl₂ for various times. Equal amounts of cell lysates were blotted with indicated antibodies. B. OCT4 signals shown in *A* were quantified and plotted. $T_{1/2}$ denotes an estimated half-life. C. Tera-1 cells were treated with CHX in the presence of absence of CoCl₂ for various times. Equal amounts of cell lysates were blotted with indicated antibodies. D. OCT4 signals shown in *C* were quantified and plotted. $T_{1/2}$ denotes an estimated half-life.

indicate that OCT4 increase after Ni(II) or Co(II) treatment is primarily due to an increased protein stability.

Post-translational Modifications of OCT4 are Enhanced by Co(II)

OCT4 protein stability is modulated by ubiquitination and sumoylation [26–28]. To test whether Co(II) or Ni(II) stabilizes OCT4 through affecting post translational modifications including ubiquitination and/or sumoylation, His₆-OCT4 ectopically expressed in HEK293T cells was pulled down by Ni-NTA resin. We used ectopic expression system in HEK293 cells partly because endogenous OCT4 in Tera-1 migrated at or near 55 kDa position, which interfered with various biochemical studies (e.g., co-immunoprecipitation). Western blotting analysis showed that many slow mobility bands of OCT4 were detected in pull-down samples that these bands were induced/enhanced after treatment with Co(II) or MG132 (Fig. 4A). Moreover, major bands that were modified by SUMO-1 (~75 kDa) and ubiquitin (55 kDa) co-migrated with slow mobility bands of OCT4 (marked by *), indicating that these bands are OCT4-specific. Although pull-down samples were also positive for SUMO-2 modification its level appeared to be much lower than that of SUMO-1 modification (Fig. 4A, SUMO-1 and SUMO-2 blots). Enhanced modifications of OCT4 were also demonstrated with cells treated with Ni(II) (data not shown).

K123 is Important for Mono-sumoylation and Mono-ubiquitination of OCT4

To identify potential lysine residues that were modified by ubiquitination, we analyzed OCT4 amino acid sequences for optimal ubiquitination sites using the criteria available (www.ubpred.org). Four lysines sites (K123, K126, K128, and K140) with the highest scores along with a low score lysine site K222 were subjected to mutagenic analysis. The relative position of these sites to OCT4 domains is shown in Fig. 4B. K123 appeared to be critical for mediating mono-ubiquitination of OCT4 as its mutation into R largely abolished 55 kDa band (Fig. 4C, Upper panel). Neither Co(II) nor MG132 induced ubiquitin-modified OCT4 in K123 mutant. Blotting with antibody to Flag (part of ectopic ubiquitin) confirmed the importance of K123 in mediating ubiquitination of OCT4 although other mutants appeared to have a negative effect on OCT4 sumoylation (Fig. 4C, Lower panel).

K123 but not K222 of OCT4 is modified by sumoylation [26,27]. To identify potential site(s) whose SUMO-modification can be affected by Ni(II) or Co(II) treatment, we co-transfected HEK293 cells with Flag-tagged *SUMO-1* (or *SUMO-2*) and OCT4 (or its mutant) expression constructs. Pull-down analysis coupled with immunoblotting confirmed that K123 was indeed a site that could be modified by SUMO-1 and SUMO-2 (Fig. 5A and 5B). SUMO modification was greatly enhanced/induced after Co(II) and MG132 treatment. Blotting with the antibody against the FLAG tag confirmed that modification by SUMO-1 was much more pronounced than that by SUMO-2 (Fig. 5A and 5B), which is consistent with the early observation (Fig. 4A).

K123 is Important for OCT4 to Bind to Chromatin after Co(II) Exposure

OCT4 functions are primarily mediated through binding to promoters of target genes, thereby regulating their expression [31]. To determine whether OCT4 modifications on K123 were important for its induction by Co(II), HEK293 cells were

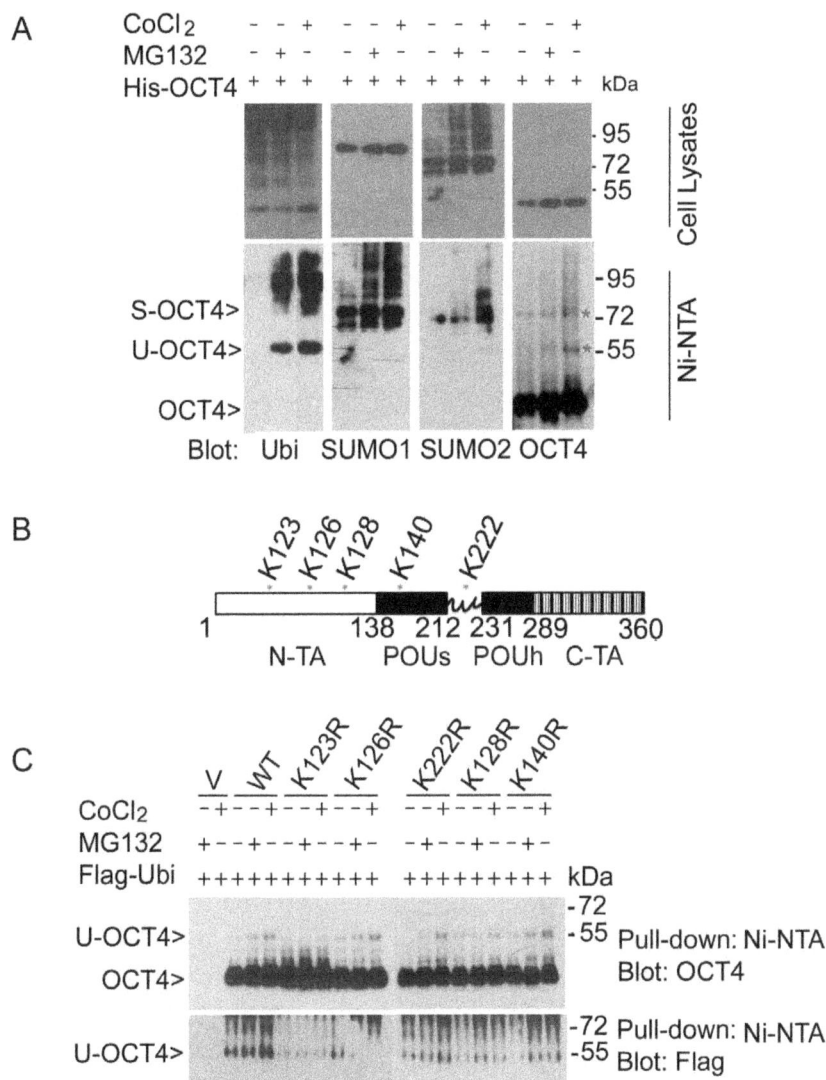

Figure 4. OCT4 is post-translationally modified. A. HEK293T cells transfected with a His$_6$-OCT4 expression plasmid for 48 h were treated with vehicle, MG132 or CoCl$_2$ for 3 h. Equal amounts of cell lysates were incubated with Ni-NTA resin. Proteins specifically bound to the resin, along with the lysate inputs, were blotted with antibodies to OCT4, SUMO-1, SUMO-2, and ubiquitin. S-OCT4 and U-OCT4 denote sumoylated and ubiquitinated OCT4, respectively. B. Schematic presentation of OCT4 domain, as well as lysine residues potential for ubiquitination. C. HEK293T cells were co-transfected with various OCT4 expression plasmids and a Flag-tagged ubiquitin expression plasmid for 48 h. After transfection, cell lysates of various treatments were incubated with Ni-NTA resin. Proteins specifically bound to the resin, along with total cell lysates, were blotted with antibodies to Flag and OCT4.

transfected with a wild-type (WT) construct of OCT4 or its mutant OCT4^{K123R} and treated with Co(II) for various times, after which cell lysates were blotted for OCT4. In contrast to WT OCT4, OCT4^{K123R} expression was not induced by Co(II) (Fig. 6A). We then asked whether K123 mutation affected its subcellular localization. Immunoblot analysis of fractionated cell lysates revealed that both WT OCT4 and OCT4^{K123R} were associated with chromatin in untreated cells; however, Co(II) exposure stimulated the increase of WT OCT4 but not OCT4^{K123R} on chromatin (Fig. 6B). Intriguingly, OCT4^{K222R} remained elevated after Co(II) treatment, which behaved in a manner similar to that of wild-type OCT4.

As both Co(II) and Ni(II) are capable of generating reactive oxygen species (ROS) [29], we asked whether induction of OCT4 by these metal was partly mediated through ROS. Tera-1 cells treated with Co(II) were supplemented with various concentrations

of ascorbic acid (AA) that inhibits ROS production. Co(II) was used as it is a stronger ROS inducer than Ni(II) [29]. Whereas AA alone did not significantly modulate OCT4 levels it repressed Co(II)-induced OCT4 (Fig. 6C). Significantly, in combination with Co(II), AA destabilized the steady-state level of OCT4 in a concentration-dependent manner. Our further analysis revealed that AA treatment also reduced OCT4 modifications (sumoylation and monoubiquitination) mediated by K123 (Fig. 6D).

Co(II) Increases OCT4 Activity by Modulating SUMO-1-modification on K123

To determine whether K123 was important for transcriptional functions of OCT4, we co-transfected HEK293T cells with an OCT4 [or mutant (OCT4^{K123R}) expression construct and a luciferase reproter construct driven by a thymidine kinase

A

B

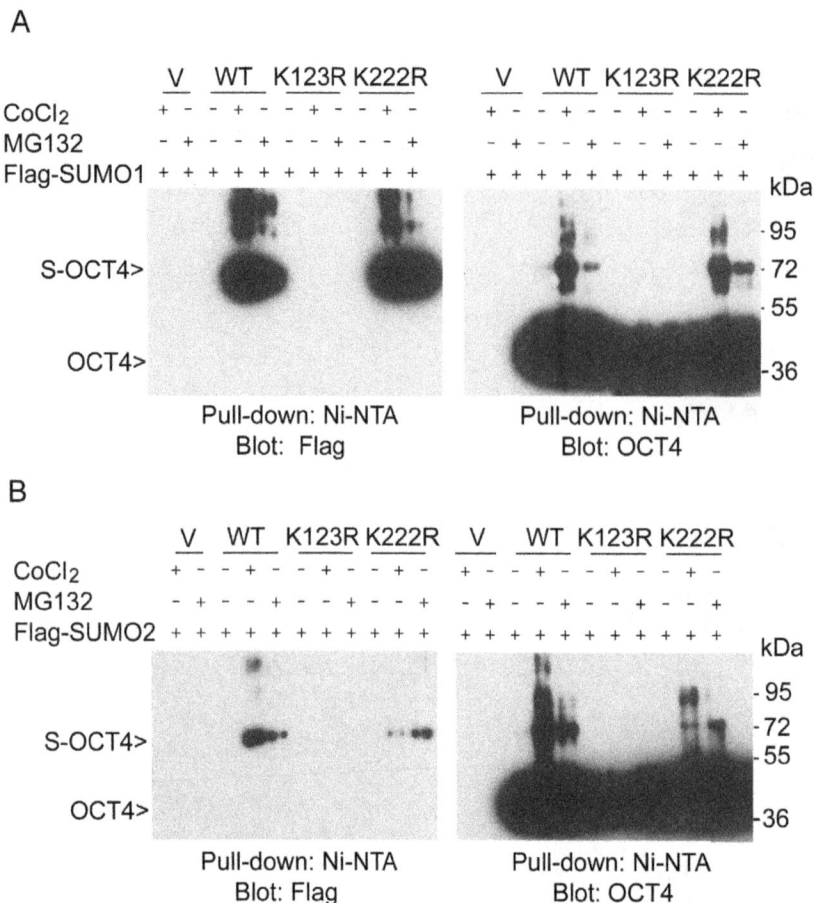

Figure 5. Cobalt enhances sumoylation of OCT4 at K123. A. HEK293T cells were co-transfected with various OCT4 expression plasmids and a FLAG-tagged SUMO-1 expression plasmid for 48 h followed by treatment with CoCl₂ or MG132 for 3 h. Cell lysates of various treatments were incubated with Ni-NTA resin. Proteins specifically bound to the resin, along with total cell lysates, were blotted with antibodies to Flag and OCT4. B. HEK293T cells were co-transfected with various OCT4 expression plasmids and a FLAG-tagged SUMO-2 expression plasmid for 48 h followed by treatment with CoCl₂ or MG132 for 3 h. Cell lysates of various treatments were incubated with Ni-NTA resin. Proteins specifically bound to the resin, along with total cell lysates, were blotted with antibodies to Flag and OCT4.

promoter fused to six copies of octamer (OCT4 monomer binding sequence]. Flag-tagged-SUMO-1, SUMO-2, or ubiquitin expression construct was also used for co-transfection. After transfection for 48 h, equal amounts of cell lysates were analyzed for luciferase activities. Ectopical expression of SUMO-1 greatly enhanced reporter gene activities in cells expressing WT OCT4 (Fig. 7A), which was further boosted by Co(II) treatment. However, the reporter gene activities were not significantly increased when cells were co-transfected with constructs expressing SUMO-1 and mutant OCT4 in the presence or absence of Co(II). Moreover, co-transfection with plasmid constructs expressing SUMO-2 and OCT4 (WT or mutant) did not significantly modulate the reporter gene activities (Fig. 7B), suggesting that SUMO-2 may not be used as a major modification *in vivo* and/or that SUMO-2 modification does not significantly affect OCT4 activity. Interestingly, co-expression of ubiquitin and WT-OCT4, but not OCT4[K123R], significantly boosted the reporter gene activities although Co(II) treatment did not significantly increase the activity (Fig. 7C). Expression of various OCT4 constructs was monitored by immunoblotting (Fig. 7, right panels).

Discussion

OCT4 is a master regulator of proliferation and self-renewal of embryonic stem cells [6,32]. *OCT4* mRNA and protein are present in unfertilized oocytes, acting as an important maternal factor to regulate embryonic development [20]. The inner cell mass and trophoblast layer regulated by OCT4 are crucial because both contribute to the normal development of healthy embryos [33]. Given its importance, OCT4 expression is tightly controlled and any perturbations of its expression are expected to have an adverse effect on cell proliferation and differentiation [32].

Nickel and cobalt are both belong to group VII in the periodic chart, thus having similar chemical properties. Cobalt also shares similar features with nickel on iron regulation [34]. An earlier *in vivo* study showed Ni(II) reduced mouse embryo implantation frequency significantly when it was injected to mice during the pre-implantation stage [35]. The size and weight of mouse litters were reduced in the treated groups as compared with that of control group. In a separate study, it has been shown that Ni(II) treated mice exhibit a high rate of embryo resorption, abnormal fetuses, and stillborn [36]. Nickel exposure also causes a significant reduction in the trophoblast area and inner cell mass [2]. Reduced proliferative ability of trophoblast cells appears to be associated

A

WT   K123R
CoCl₂ 0 2 4 6 8   0 2 4 6 8 (h)

OCT4>

Tubulin>

B

V   WT   V   K123R   V   K222R
Cy N Ch Cy N Ch Cy N Ch Cy N Ch Cy N Ch Cy N Ch
CoCl₂ - - - + + + - - - + + +   - - - + + +

OCT4>

PARP>

Tubulin>

C

CoCl₂ - - - - -   + + + + + (50 µM)
AA

OCT4>

Tubulin>

D

V   WT K123R K222R
CoCl₂ - + + + + + + + +
AA   - - + - + - + - +

OCT4>   Short
                        Exposure
                        kDa
S-OCT4>   -72
U-OCT4>   -55   Long
                              Exposure
OCT4>   -36

GFP>

Tubulin>

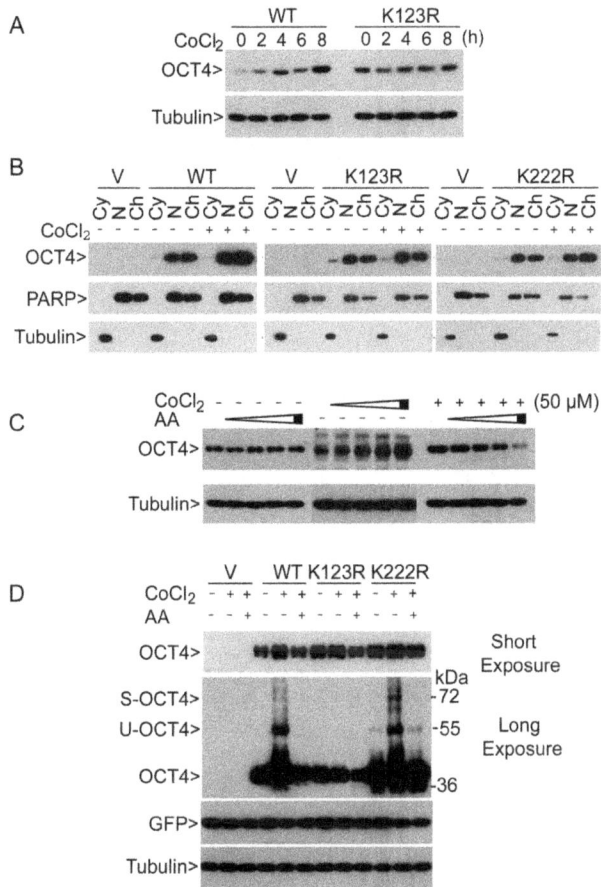

Figure 6. K132 is important for OCT4 chromatin association after CoCl₂ treatment. A. HEK293 cells were transfected with WT OCT4 or OCT4^{K123R} for 48 h and then treated with CoCl₂ for various times. Equal amounts of cell lysates were then blotted for OCT4 and α-tubulin. B. HEK293 cells were transfected with WT OCT4 or various mutant constructs for 48 h followed by treatment with CoCl₂ for 3 h. Cells with various treatments were lysed and separated into cytoplasmic (Cy), soluble nuclear (N) and chromatin (Ch) fractions. Equal amounts of proteins from each fraction were blotted with antibodies to OCT4 and α-tubulin. C. HEK293 cells transfected with WT OCT4 for 48 h were treated with CoCl₂ (0, 50, 100, 150, and 200 µM) and/or ascorbic acid (AA) (0, 50, 100, 150, and 200 µM) for 3 h. At the end of treatments, cells were lysed and equal amounts of proteins were blotted for OCT4 and α-tubulin. D. HEK293 cells were transfected with WT OCT4 or various mutant constructs for 48 h followed by treatment with CoCl₂ or ascorbic acid (AA) for 3 h. GFP expression plasmid was used for co-transfection to monitor transfection efficiency. Cell lysates of various treatments were blotted with antibodies to OCT4, GFP and α-tubulin. Both short and long exposures of OCT4 blots were shown. S-OCT4 and U-OCT4 denote sumoylated and ubiquitinated OCT4, respectively.

with compromised invasiveness of the embryo [3]. Our current studies strongly suggest that embryonic toxicity caused by nickel or cobalt exposure is likely due, at least impart, to altered expression and activity of OCT4.

It has been shown that nickel and cobalt toxicity and carcinogenicity are mediated through ROS production [29]. Using the electron paramagnetic resonance spin trapping approach, Hanna et al. have shown that various Co(II) complexes generate ROS from the reaction of hydrogen peroxide under physiological conditions [37]. Moreover, it has been suggested that depletion of glutathione may be a possible mechanism of oxidative

stress induced by nickel [38,39]. Many stem cell transcription factors function as onco-proteins, thus promoting cell proliferation and facilitating malignant transformation when their expression and activities are deregulated [40–43]. Given that OCT4 controls expression of many transcription factors including NANOG, SALL4, Myc and SOX2 [19,44], it is tempting to speculate that Co(II) or Ni(II) carcinogenesis in the stem cell compartment may be partly due to an enhanced activities of OCT4 and its downstream targets.

OCT4 has two distinct DNA binding domains, POU domain (a.a.138–212 in human) and homeobox (a.a. 231-189) which independently bind half-sites of the canonical octamer motif [45]. This flexibility allows OCT4 to form heterodimers with other transcription factors, as well as to form homodimers [46]. Post-translational modifications are known to impact on protein conformation. In fact, it has been shown that OCT4 protein stability and transcriptional activities are subjected to the regulation by post-translational modifications including phosphoylation [21,22], sumoylation [26–28] and poly-ubiquitination [24,25]. Here we showed that OCT4 exhibits multiple modifications including ubiquitination and sumoylation, levels of which appear to correlated with OCT4 stability. Moreover, modifications of OCT4 can be induced by exposure to Co(II) or Ni(II). We have observed that OCT4 can be modified by SUMO-1 and SUMO-2. We have also demonstrated that Ni(II) and Co(II) enhance SUMO-modification of OCT4, leading to its stabilization. These observations are consistent with early reports that SUMO-1-modification of OCT4 affects its stability, as well as its transcriptional activity [26,27]. In this study, we also showed that OCT4 can be modified by SUMO-2 albeit its level appears to be lower than that of SUMO-1. Our luciferase assays suggest that SUMO-2 modification does not seem to be important for OCT4 transcriptional activities.

Experimental Procedures

Cell Lines and Antibodies

HEK293T, TERA-1 and NT2/D1 cell lines were obtained from the American Type Culture Collection (ATCC, Manassas, VA). Anti-HIF-1α antibody was purchased from Bethyl Laboratories (Montgomery, TX). Antibodies to α-tubilin, PARP, and HIF-2α were purchased from Cell Signaling Technology (Danvers, MA). Antibodies against GFP, NANOG, and OCT4 were purchased from Santa Cruz Biotechnology (Santa Cruz, CA). SALL4 antibody was purchased from Abcam (Cambridge, MA). Human embryonic stem cells (H1, WiCell, WI) were cultured using a feeder-dependent culture condition. These cells were maintained in DMEM-F12 (Invitrogen, USA) medium which was supplemented with 20% KSR, 10 ng/mL bFGF, 2mM Gluta-MAX™-I, 0.1 mM MEM Non-Essential Amino Acids Solution, 1×β-mercaptoethanol. Cells were passed every other day after trypsinization. Mitomycin C treated murine embryonic fibroblasts (MEFs) were prepared as feeder cells.

Site-directed Mutagenesis

Mutant OCT4 with lysine 123 (K123), K126, K128, K140, and/or K222 replaced with arginine (R) were carried out using the QuikChange Lightning Site-directed Mutagenesis kit from Stratagene (Santa Clara, CA) according to the instruction provided by the supplier. Individual mutations were confirmed by DNA sequencing service from Seqwright (Houston, TX). Plasmid transfection was carried out using Lipofectamine reagents from Life Technology according to manufacturer's instruction.

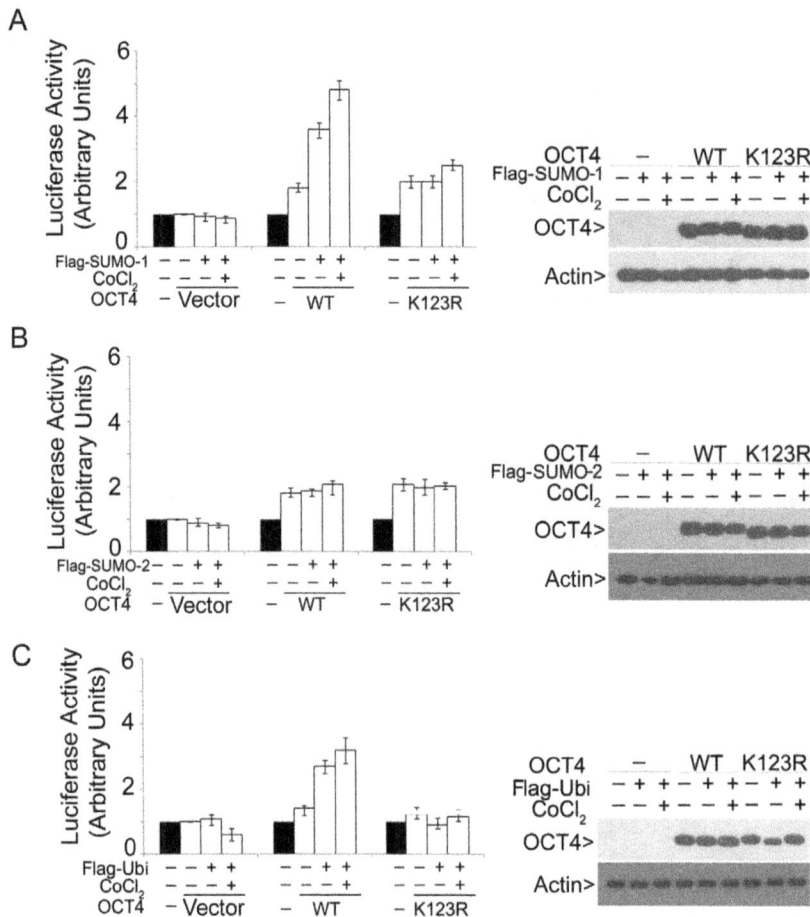

Figure 7. Cobalt stimulates SUMO-1-mediated transcriptional activity of OCT4. A. HEK293T cells were seeded in triplicate and co-transfected for 42 h with 6W-37tk-luc reporter, Flag-SUMO-1 (or Flag-SUMO-2 or Flag-ubiquitin), and WT OCT4 (or OCT4^{K123R}) expression plasmids. Transfected cells were then treated with CoCl$_2$ (50 μM) for 6 h, after which cells were lysed. Equal amounts of cell lysates were assayed for firefly luciferase activities as described in Materials and Methods. Data are expressed as fold-changes after normalization by the renilla luciferase activity. Each experiment was repeated for at least three times. Samples from each transfection were also blotted with antibodies to OCT4 and α-tubulin.

RNA Isolation and qPCR

Total RNA was isolated from cells with various treatments using TriZol reagent (Life Technology) and converted into cDNA using SuperScript III First-Strand Synthesis Supermix for qRT-PCR (Life technology). Briefly, cells were immediately lysed in the TRIZOL reagent. RNA (1 μg) was reverse transcribed using oligo dT by reverse transcriptase. The synthesized cDNA was then used for quantitative real-time PCR (qPCR), which was carried out using ABI 7300 Real-Time PCR System (Life Technology). Eexpression levels of various genes were normalized to the levels of *ACT-B* mRNA, and expressed as fold induction relative to the untreated control.

Immunoprecipitation and Pulldown Assays

For OCT4 pulldown assay, HEK293T cells overexpressed with various OCT4 constructs were lysed in 8 M urea buffer (100 mM NaH$_2$PO$_4$; 10 mM Tris-HCl ,pH 8.0; 8 M urea). After extensive washing, proteins on the resin were eluted in the SDS sample buffer and subjected to analysis by SDS-PAGE followed by Western blotting with appropriate antibodies.

Cell Fractionation

Cytoplasmic, nuclear and chromatin fractions were obtained using a modification of the procedure of Jin and Felsenfeld [47]. Cells were washed 3 times with PBS, suspended in the hypotonic buffer (10 mM Tris-HCl, pH 7.4; 10 mM KCl; 1.5 mM MgCl$_2$; and 1 mM Dithiothreitol) supplemented with inhibitor cocktails (10 mM Na-butyrate, 0.5 μg/mL aprotinin, 0.5 μg/mL leupeptin, and 1 μg/mL aprotinin). Cells were disrupted using a 25 gauge needle. Nuclei were pelleted and resuspended in a low salt buffer (20 mM Tris-HCl ,pH 7.4; 20 mM KCl, 1.5 mM MgCl$_2$; 25% glycerol and 1 mM DTT). Nuclei were homogenized with a 25 gauge needle followed by the addition of an equal volume of a high salt buffer (20 mM Tris-HCl, pH 7.4; 1.2 M KCl; 1.5 mM MgCl$_2$; 25% glycerol 0.2 mM EDTA and 1 mM DTT). Soluble nuclear fraction and insoluble materials were seperated by centrifugation (14000 g×15 min) at 4°C. Pellets were resuspended in Tris saline magnesium buffer (20 mM Tris-HCl, ,pH, 7.4; 150 mM NaCl; 2 mM CaCl$_2$; 2 mM MgCl$_2$). The resuspended nuclei were digested with 120 U/μL micrococcal nuclease (Fisher) for 12 min at 37°C. The reaction was stopped by adding EDTA (pH 8.0) to a final concentration of 10 mM. After centrifugation (2500 rpm×5 min), the supernatant S1 was collected. After passing four times through a 20-gauge needle followed by four passes through a 25-gauge needle, the pellets were resuspended in the lysis buffer plus with 0.25 mM EDTA and incubated on ice for 15 min followed by centrifugation (10,000 rpm×10 min). The supernatant S2 was then collected and combined with S1 as the chromatin binding fraction.

Half-life Study

After transfection of HEK293T cells with either wild-type or various mutant His_6-OCT4 expression plasmid constructs for 24 h, cycloheximide (CHX) was added at a final concentration of 50 μg/ml to block new protein synthesis. Cells were harvested at various times post CHX treatment (2, 4, 6, 8, 10, and 12 h). Equal amounts of cell lysates were blotted for OCT4.

Luciferase Reporter Gene Assays

Plasmid construct expressing firefly (Photinus pyralis) luciferase gene driven by the OCT4 promoter was kindly provided by Dr. Yupo Ma (SUNY Stony Brook). Additional plasmid constructs (6XW, PORE, and MORE reporters) were gifts from Dr. Michael Atchison (University of Pennsylvania). HEK293T cells seeded in 12-well plate for 16 h. The total amount of DNA per well was equalized to 1.6 μg with carrier plasmid. Cells were co-transfected with the firefly reporter plasmid (0.3 μg/well), Flag tagged ubiquitin (or SUMO-1 and SUMO-2) plasmid (0.6 μg/well) RL

Renilla luciferase reporter plasmid (Promega, Madison, WI; for monitoring transfection efficiency), and an OCT4 expression plasmid (0.6 μg). Cells were then lysed and luciferase activities were measured using the Dual-Luciferase Reporter Assay System (Promega). Cell lysates were also blotted with antibodies to OCT4.

Statistical Analysis

Data were represented as the mean ± SD. Differences between mean values of various samples were compared by Statistical Package for the Social Sciences (SPSS) software by two tailed Student t test. The differences were considered significant at P value ≤ 0.05.

Acknowledgments

We thank coworkers in the laboratory for valuable discussions and suggestions. We also thank Dr. Michael Atchison for 6XW, PORE, MORE reporter constructs.

Author Contributions

Conceived and designed the experiments: YM WD. Performed the experiments: YY WC YL. Analyzed the data: YY WD YL. Contributed reagents/materials/analysis tools: YJ TC YM LL. Wrote the paper: YY WD.

References

1. Clarson LH, Roberts VH, Hamark B, Elliott AC, Powell T (2003) Store-operated Ca2+ entry in first trimester and term human placenta. J Physiol 550: 515–528.
2. Paksy K, Forgacs Z, Gati I (1999) In vitro comparative effect of Cd2+, Ni2+, and Co2+ on mouse postblastocyst development. Environ Res 80: 340–347.
3. Forgacs Z, Massanyi P, Lukac N, Somosy Z (2012) Reproductive toxicology of nickel - review. J Environ Sci Health A Tox Hazard Subst Environ Eng 47: 1249–1260.
4. Vaktskjold A, Talykova LV, Chashchin VP, Odland JO, Nieboer E (2007) Small-for-gestational-age newborns of female refinery workers exposed to nickel. International Journal Of Occupational Medicine And Environmental Health 20: 327–338.
5. Cangul H, Broday L, Salnikow K, Sutherland J, Peng W, et al. (2002) Molecular mechanisms of nickel carcinogenesis. Toxicol Lett 127: 69–75.
6. Pantazis P, Bollenbach T (2012) Transcription factor kinetics and the emerging asymmetry in the early mammalian embryo. Cell Cycle 11: 2055–2058.
7. Takahashi K, Yamanaka S (2006) Induction of pluripotent stem cells from mouse embryonic and adult fibroblast cultures by defined factors. Cell 126: 663–676.
8. Zhao H, Sun N, Young SR, Nolley R, Santos J, et al. (2013) Induced Pluripotency of Human Prostatic Epithelial Cells. PLoS One 8: e64503.
9. Boland MJ, Hazen JL, Nazor KL, Rodriguez AR, Gifford W, et al. (2009) Adult mice generated from induced pluripotent stem cells. Nature 461: 91–94.
10. Yoshida Y, Takahashi K, Okita K, Ichisaka T, Yamanaka S (2009) Hypoxia enhances the generation of induced pluripotent stem cells. Cell Stem Cell 5: 237–241.
11. Bae D, Mondragon-Teran P, Hernandez D, Ruban L, Mason C, et al. (2012) Hypoxia enhances the generation of retinal progenitor cells from human induced pluripotent and embryonic stem cells. Stem Cells Dev 21: 1344–1355.
12. Mathieu J, Zhang Z, Nelson A, Lamba DA, Reh TA, et al. (2013) Hypoxia Induces Re-Entry of Committed Cells into Pluripotency. Stem Cells.
13. Heddleston JM, Li Z, McLendon RE, Hjelmeland AB, Rich JN (2009) The hypoxic microenvironment maintains glioblastoma stem cells and promotes reprogramming towards a cancer stem cell phenotype. Cell Cycle 8: 3274–3284.
14. Forristal CE, Christensen DR, Chinnery FE, Petruzzelli R, Parry KL, et al. (2013) Environmental Oxygen Tension Regulates the Energy Metabolism and Self-Renewal of Human Embryonic Stem Cells. PLoS One 8: e62507.
15. Harvey AJ, Kind KL, Pantaleon M, Armstrong DT, Thompson JG (2004) Oxygen-regulated gene expression in bovine blastocysts. Biol Reprod 71: 1108–1119.
16. Liu L, Roberts RM (1996) Silencing of the gene for the beta subunit of human chorionic gonadotropin by the embryonic transcription factor Oct-3/4. J Biol Chem 271: 16683–16689.
17. Botquin V, Hess H, Fuhrmann G, Anastassiadis C, Gross MK, et al. (1998) New POU dimer configuration mediates antagonistic control of an osteopontin preimplantation enhancer by Oct-4 and Sox-2. Genes Dev 12: 2073–2090.
18. Zhang ZN, Chung SK, Xu Z, Xu Y (2013) Oct4 maintains the pluripotency of human embryonic stem cells by inactivating p53 through Sirt1-mediated deacetylation. Stem Cells.

19. Tan MH, Au KF, Leong DE, Foygel K, Wong WH, et al. (2013) An Oct4-Sall4-Nanog network controls developmental progression in the pre-implantation mouse embryo. Mol Syst Biol 9: 632.
20. Ovitt CE, Scholer HR (1998) The molecular biology of Oct-4 in the early mouse embryo. Mol Hum Reprod 4: 1021–1031.
21. Saxe JP, Tomilin A, Scholer HR, Plath K, Huang J (2009) Post-translational regulation of Oct4 transcriptional activity. PLoS One 4: e4467.
22. Lin Y, Yang Y, Li W, Chen Q, Li J, et al. (2012) Reciprocal regulation of Akt and Oct4 promotes the self-renewal and survival of embryonal carcinoma cells. Mol Cell 48: 627–640.
23. Campbell PA, Rudnicki MA (2013) Oct4 interaction with Hmgb2 regulates Akt signaling and pluripotency. Stem Cells 31: 1107–1120.
24. Liao B, Jin Y (2010) Wwp2 mediates Oct4 ubiquitination and its own auto-ubiquitination in a dosage-dependent manner. Cell Res 20: 332–344.
25. Xu H, Wang W, Li C, Yu H, Yang A, et al. (2009) WWP2 promotes degradation of transcription factor OCT4 in human embryonic stem cells. Cell Res 19: 561–573.
26. Wei F, Scholer HR, Atchison ML (2007) Sumoylation of Oct4 enhances its stability, DNA binding, and transactivation. J Biol Chem 282: 21551–21560.
27. Zhang Z, Liao B, Xu M, Jin Y (2007) Post-translational modification of POU domain transcription factor Oct-4 by SUMO-1. FASEB J 21: 3042–3051.
28. Wu YC, Ling TY, Lu SH, Kuo HC, Ho HN, et al. (2012) Chemotherapeutic sensitivity of testicular germ cell tumors under hypoxic conditions is negatively regulated by SENP1-controlled sumoylation of OCT4. Cancer Res 72: 4963–4973.
29. Salnikow K, Su W, Blagosklonny MV, Costa M (2000) Carcinogenic metals induce hypoxia-inducible factor-stimulated transcription by reactive oxygen species-independent mechanism. Cancer Res 60: 3375–3378.
30. Yang J, Aguila JR, Alipio Z, Lai R, Fink LM, et al. (2011) Enhanced self-renewal of hematopoietic stem/progenitor cells mediated by the stem cell gene Sall4. J Hematol Oncol 4: 38.
31. Zhao XM, Du WH, Hao HS, Wang D, Qin T, et al. (2012) Effect of vitrification on promoter methylation and the expression of pluripotency and differentiation genes in mouse blastocysts. Mol Reprod Dev 79: 445–450.
32. Niwa H, Miyazaki J, Smith AG (2000) Quantitative expression of Oct-3/4 defines differentiation, dedifferentiation or self-renewal of ES cells. Nat Genet 24: 372–376.
33. Oda M, Shiota K, Tanaka S (2006) Trophoblast stem cells. Methods Enzymol 419: 387–400.
34. Jungwirth U, Kowol CR, Keppler BK, Hartinger CG, Berger W, et al. (2011) Anticancer activity of metal complexes: involvement of redox processes. Antioxid Redox Signal 15: 1085–1127.
35. Storeng R, Jonsen J (1981) Nickel toxicity in early embryogenesis in mice. Toxicology 20: 45–51.
36. Storeng R, Jonsen J (1980) Effect of nickel chloride and cadmium acetate on the development of preimplantation mouse embryos in vitro. Toxicology 17: 183–187.

37. Hanna PM, Kadiiska MB, Mason RP (1992) Oxygen-derived free radical and active oxygen complex formation from cobalt(II) chelates in vitro. Chem Res Toxicol 5: 109–115.

38. Liu CM, Zheng GH, Ming QL, Chao C, Sun JM (2013) Sesamin Protects Mouse Liver against Nickel-Induced Oxidative DNA Damage and Apoptosis by the PI3K-Akt Pathway. J Agric Food Chem 61: 1146–1154.

39. Rodriguez RE, Misra M, North SL, Kasprzak KS (1991) Nickel-induced lipid peroxidation in the liver of different strains of mice and its relation to nickel effects on antioxidant systems. Toxicol Lett 57: 269–281.

40. Qinyu L, Long C, Zhen-dong D, Min-min S, Wei-ze W, et al. (2013) FOXO6 promotes gastric cancer cell tumorigenicity via upregulation of C-myc. FEBS Lett 587: 2105–2111.

41. Dhodapkar KM, Gettinger SN, Das R, Zebroski H, Dhodapkar MV (2013) SOX2-specific adaptive immunity and response to immunotherapy in non-small cell lung cancer. Oncoimmunology 2: e25205.

42. Ma W, Ma J, Xu J, Qiao C, Branscum A, et al. (2013) Lin28 regulates BMP4 and functions with Oct4 to affect ovarian tumor microenvironment. Cell Cycle 12: 88–97.

43. Li W, Reeb AN, Sewell WA, Elhomsy G, Lin R-Y (2013) Phenotypic Characterization of Metastatic Anaplastic Thyroid Cancer Stem Cells. PLoS One 8: e65095.

44. Loh YH, Wu Q, Chew JL, Vega VB, Zhang W, et al. (2006) The Oct4 and Nanog transcription network regulates pluripotency in mouse embryonic stem cells. Nat Genet 38: 431–440.

45. Rosner MH, Vigano MA, Ozato K, Timmons PM, Poirier F, et al. (1990) A POU-domain transcription factor in early stem cells and germ cells of the mammalian embryo. Nature 345: 686–692.

46. Remenyi A, Lins K, Nissen LJ, Reinbold R, Scholer HR, et al. (2003) Crystal structure of a POU/HMG/DNA ternary complex suggests differential assembly of Oct4 and Sox2 on two enhancers. Genes Dev 17: 2048–2059.

47. Jin C, Felsenfeld G (2007) Nucleosome stability mediated by histone variants H3.3 and H2A.Z. Genes Dev 21: 1519–1529.

16

Embryonic Carcinoma Cells Show Specific Dielectric Resistance Profiles during Induced Differentiation

Simin Öz[1], Christian Maercker[2,3]*, Achim Breiling[1]*

1 Division of Epigenetics, DKFZ-ZMBH Alliance, German Cancer Research Center, Heidelberg, Germany, 2 Mannheim University of Applied Sciences, Mannheim, Germany, 3 Genomics and Proteomics Core Facilities, German Cancer Research Center, Heidelberg, Germany

Abstract

Induction of differentiation in cancer stem cells by drug treatment represents an important approach for cancer therapy. The understanding of the mechanisms that regulate such a forced exit from malignant pluripotency is fundamental to enhance our knowledge of tumour stability. Certain nucleoside analogues, such as 2'-deoxy-5-azacytidine and 1β-arabinofuranosylcytosine, can induce the differentiation of the embryonic cancer stem cell line NTERA 2 D1 (NT2). Such induced differentiation is associated with drug-dependent DNA-damage, cellular stress and the proteolytic depletion of stem cell factors. In order to further elucidate the mode of action of these nucleoside drugs, we monitored differentiation-specific changes of the dielectric properties of growing NT2 cultures using electric cell-substrate impedance sensing (ECIS). We measured resistance values of untreated and retinoic acid treated NT2 cells in real-time and compared their impedance profiles to those of cell populations triggered to differentiate with several established substances, including nucleoside drugs. Here we show that treatment with retinoic acid and differentiation-inducing drugs can trigger specific, concentration-dependent changes in dielectric resistance of NT2 cultures, which can be observed as early as 24 hours after treatment. Further, low concentrations of nucleoside drugs induce differentiation-dependent impedance values comparable to those obtained after retinoic acid treatment, whereas higher concentrations induce proliferation defects. Finally, we show that impedance profiles of substance-induced NT2 cells and those triggered to differentiate by depletion of the stem cell factor OCT4 are very similar, suggesting that reduction of OCT4 levels has a dominant function for differentiation induced by nucleoside drugs and retinoic acid. The data presented show that NT2 cells have specific dielectric properties, which allow the early identification of differentiating cultures and real-time label-free monitoring of differentiation processes. This work might provide a basis for further analyses of drug candidates for differentiation therapy of cancers.

Editor: Austin John Cooney, Baylor College of Medicine, United States of America

Funding: This work was supported by the Ministerium für Wissenschaft, Forschung und Kunst, Baden Württemberg (Kooperatives Promotionskolleg Krankheitsmodelle und Wirkstoffe) (http://mwk.baden-wuerttemberg.de/). The funders had no role in study design, data collection and analysis, decision to publish, or preparation of the manuscript.

Competing Interests: The authors have declared that no competing interests exist.

* E-mail: a.breiling@dkfz.de (AB); c.maercker@hs-mannheim.de (CM)

Introduction

The induction of differentiation by treatment with natural ligands and synthetic drugs represents an important approach for cancer therapy [1,2]. Tumours are thought to originate from cells with stem cell characteristics that have acquired aberrant gene expression patterns, mostly due to genetic and/or epigenetic mutations, which destabilise the homeostasis of cellular proliferation and differentiation [1,3]. Cancer is thus characterised by a block in differentiation and by the induction of uncontrolled proliferation [3]. The identification and characterisation of substances that induce differentiation in human cancer cells therefore represents an important aspect in the development of novel cancer therapies.

A prominent example for a differentiation inducing drug is 2'-deoxy-5-azacytidine (decitabine, DAC), that has been suggested to induce differentiation by DNA demethylation [4]. A compound closely related to decitabine, 1β-arabinofuranosylcytosine (cytarabine, araC), induces differentiation without inhibiting DNA methylation [5]. DAC, araC and the structurally related drug 5-

azacytidine (AZA), are used for the treatment of myeloid leukaemias, a group of diseases that is characterised by a differentiation block of precursor cells [6,7]. While the precise molecular modes of action of these drugs are still not well understood, nucleoside analogues can be incorporated into DNA and thereby trigger DNA damage or other stress response pathways [8]. Indeed, we have recently shown that both DAC and araC induce neuronal differentiation in the embryonal carcinoma (EC) cell line NTERA2 D1 (NT2) by triggering degradation of OCT4 and other stem cell proteins via DNA damage pathways [9].

NT2 EC cells express high levels of stem cell specific transcription factors (especially OCT4 and NANOG), Polycomb Group (PcG) proteins and DNA methyltransferases. The cells also show significant levels of non-CpG methylation, a DNA mark restricted to pluripotent cells that is strongly reduced upon differentiation induction with all-trans-retinoic acid (RA), a conserved intercellular signaling molecule found in most vertebrates [10]. NT2 cells have not only been shown to differentiate along the neuronal lineage, but also show mesodermal and ectodermal lineage potential and thus represent a valuable human cancer stem

A

B

C

D

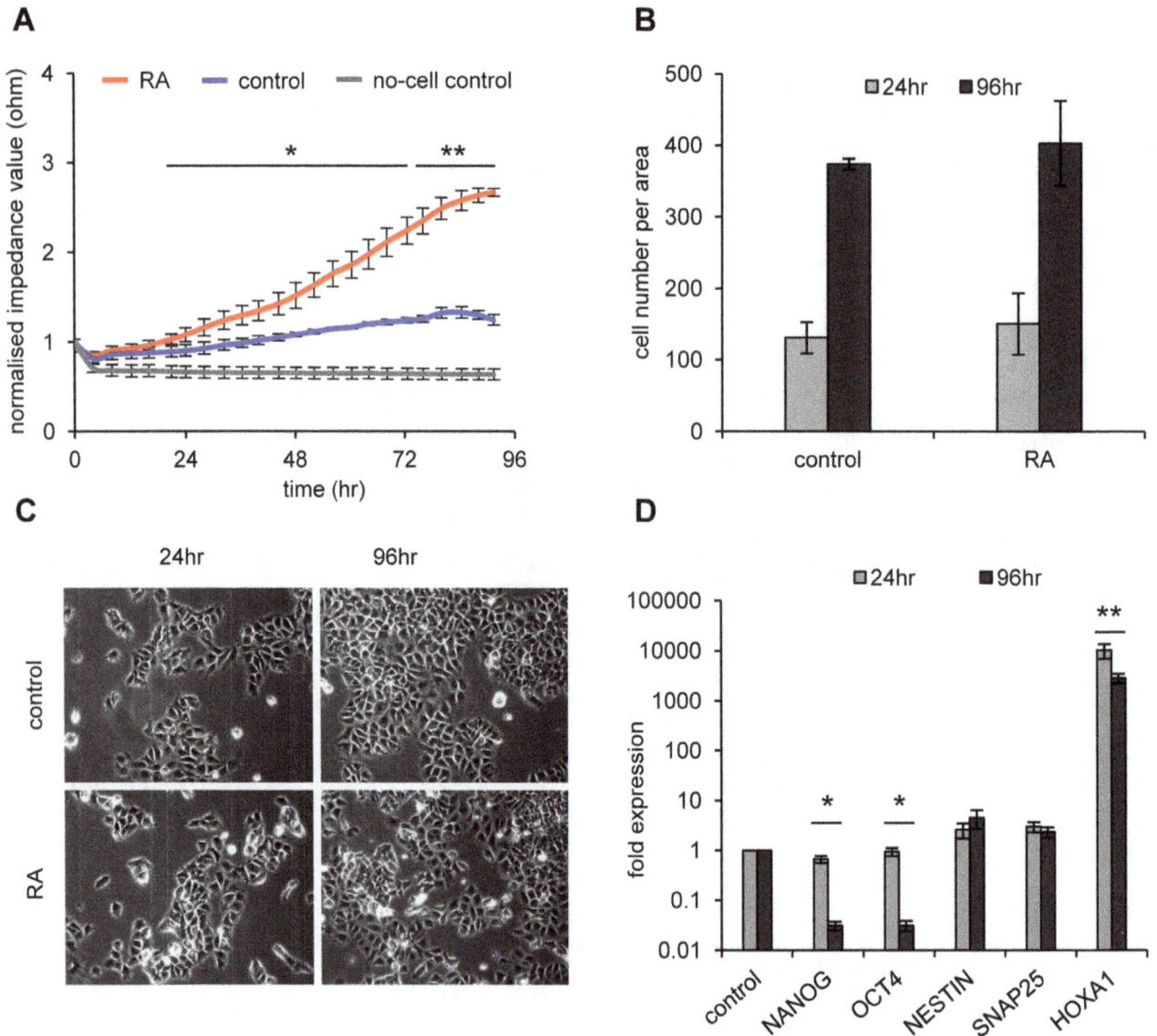

Figure 1. Retinoic acid induced neuronal differentiation of NT2 EC cells. (**A**) Impedance profiles comparing RA-induced (10 µM - red) and untreated NT2 cells (blue) during a 4 day period. The mean of three independent experiments is shown. Standard deviations are indicated by error bars every four hours. Measurements were executed at 45 kHz in 5-minute intervals for 96 hours. Normalised resistance values were compared by two-tailed Student's t-test. After 20 hours of RA treatment differences in impedance values start to become statistically significant (*p<0.05, **p<0.005). Black lines show regions with significant differences in respect to the untreated cell control. (**B**) Average cell numbers of three replicates of untreated and RA-treated NT2 cells after 24 and 96 hours do not differ significantly. Standard deviations are indicated by error bars. (**C**) Microscopic images (10× magnification) of NT2 control cells and NT2 cells treated with RA for 24 and 96 hours. No clear differentiation phenotype becomes apparent for the RA treatment. (**D**) qRT-PCR expression analysis of stem cell factors *NANOG*, *OCT4* and the differentiation markers *NESTIN*, *SNAP25* and *HOXA1* in RA- treated and control cells after 24 and 96 hours of treatment. Data is shown in logarithmic scale. Only *HOXA1* is prominently induced by retinoic acid at both time points. The stemness genes are only found reduced after 96 hours of RA treatment. All qRT-PCR measurements were repeated at least three times and internally normalised to the corresponding *β-actin* values. Standard deviations are indicated by error bars. Two-tailed student's t-test showed significant differences when comparing expression levels of *OCT4*, *NANOG* and *HOXA1* at 24 hours with the expression levels at 96 hours. (*p<0.05, **p<0.005).

cell model system [11,12]. Cultures exposed to differentiation-inducing substances are usually rather heterogeneous and show a mixture of neuronal, ectodermal and mesodermal features [11–14]. Induction of differentiation with the natural ligand retinoic acid results in visible morphological changes only after prolonged treatment of at least three days [9,15]. Changes in marker gene expression are even more delayed. Efficient reduction of stem cell factors or induced expression of neuronal markers becomes apparent only after several days of RA treatment [9,13,15]. In order to screen drug libraries for differentiation-inducing substances a fast method for early-identification of cellular differentiation is thus desirable.

Electrical cell-substrate impedance sensing (ECIS) is a label-free, non-invasive monitoring technique to study the formation of cell-matrix as well as cell-cell contacts during cell proliferation, cell migration, metastasis, wound healing, cellular differentiation and cancer development [16–18]. The method is based on the phenomenon that living cells behave as dielectric particles and thus alter the electrode impedance after attachment to a microelectrode surface. Impedance measurements at the electrode-cell interface are influenced by increasing cell number, increased adhesion, morphological changes and cell spreading [19]. We have previously used this non-invasive assay to measure impedance profiles of differentiating mesenchymal stem cells [20]. Mesenchymal stem cells (MSCs) induced for adipogenesis or osteogenesis *in vitro*, showed characteristic changes in dielectric properties, that were already visible within 24 hours. ECIS is thus a reliable tool for real-time monitoring of stem cell differentiation [20,21].

To study immediate effects on impedance values, we analysed the onset of drug-induced differentiation in NT2 cells by ECIS. Already after 20 hours of retinoic acid induction we found a significant increase of impedance values. The slope/time ratios of the dielectric resistance profiles positively correlated with the employed concentration of RA. Further experiments determined the concentrations of nucleoside drugs that induced impedance changes with slope/time ratios comparable to those obtained with retinoic acid. These differentiation-specific effects could be separated from cytotoxicity. Finally, we show that differentiation induction by nucleoside drugs and retinoic acid is mainly caused by the reduction of the levels of stemness factors, in particular OCT4. Taken together, our work provides a basis for further real-time studies in living cells evaluating drug candidates as differentiation inducing agents for cancer therapy.

Materials and Methods

Cell Culture and Drug Treatment

The human cell line NT2 D1 [22,23] was a kind gift from Peter W. Andrews (University of Sheffield). Cell line authentication was provided by LGC Standards (Teddington, report tracking no. 71008933). Cells were maintained in Dulbecco's Modified Eagle Medium (DMEM) supplemented with 10% FCS (Invitrogen), 200 U/ml penicillin (Gibco) and 200 µg/ml streptomycin (Gibco). NT2 cells were induced to differentiate (if not mentioned otherwise) with 10 µM all-trans retinoic acid (Sigma), 1 µM deoxycytidine (Sigma), 1 µM azacytidine (Sigma), 1 µM decitabine (Sigma), 1 µM cytarabine (Sigma), 5 mM hexamethylene bisacetamide (Sigma) and 50 µM fibroblast growth factor 2 (Novitec) in 5% CO_2 at 37°C.

RNA Isolation and qRT PCR

Total RNA was isolated from NT2 D1 cells using the Trizol reagent (Invitrogen) or the RNeasy kit (Qiagen), following the manufacturer's recommendations. Total RNA (500 ng) was reverse transcribed using Superscript III (Invitrogen). Quantitative RT-PCR was performed utilising the LightCycler 480 System (Roche). 1 µl of cDNA was used for 10 µl PCR reaction using Absolute QPCR SYBR Green Mix (Thermo Scientific) under following conditions: 1 cycle at 95°C for 15 min followed by 50 cycles at 95°C for 15 s, at 60°C for 40 s. All samples were measured in triplicates. Cycle threshold numbers for each amplification were measured with the LightCycler 480 software, and relative expression values were calculated and normalised using β-actin as an internal standard. For RT-primer sequences see Table S5.

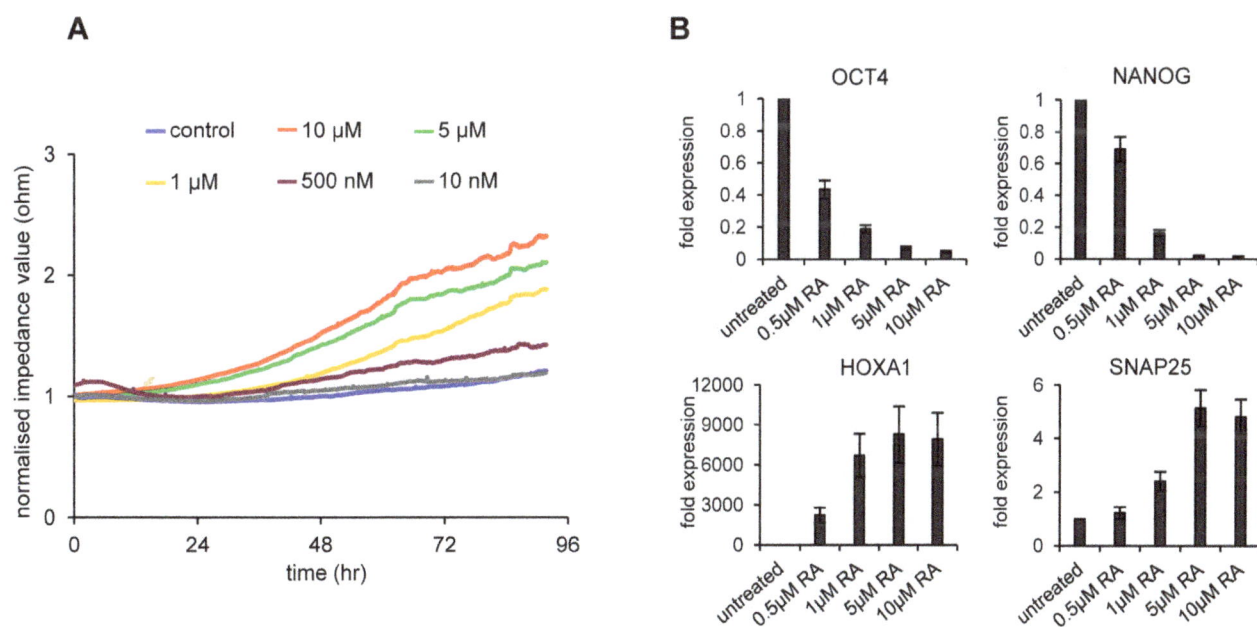

Figure 2. Induced concentration-dependent differentiation of NT2 cells by RA. (A) Impedance profiles comparing induction profiles of different RA concentrations during a 4 day period. Measurements were executed at 45 kHz in 5-minute intervals for 96 hours. The mean of three independent experiments is shown. Standard deviations are not shown to avoid crowding of the diagram. For single diagrams including standard deviations and statistical tests for these data sets see Fig. S1. (B) qRT-PCR expression analysis of stem cell factors *NANOG*, *OCT4* and the differentiation markers *HOXA1* and *SNAP25* and in RA- treated and control NT2 cells after 96 hours of treatment. The concentration of RA employed correlates negatively with the expression of stem cell factors, but positively with the expression of differentiation markers. All qRT-PCR measurements were repeated at least three times and internally normalised to the corresponding *β-actin* values. Standard deviations are indicated by error bars.

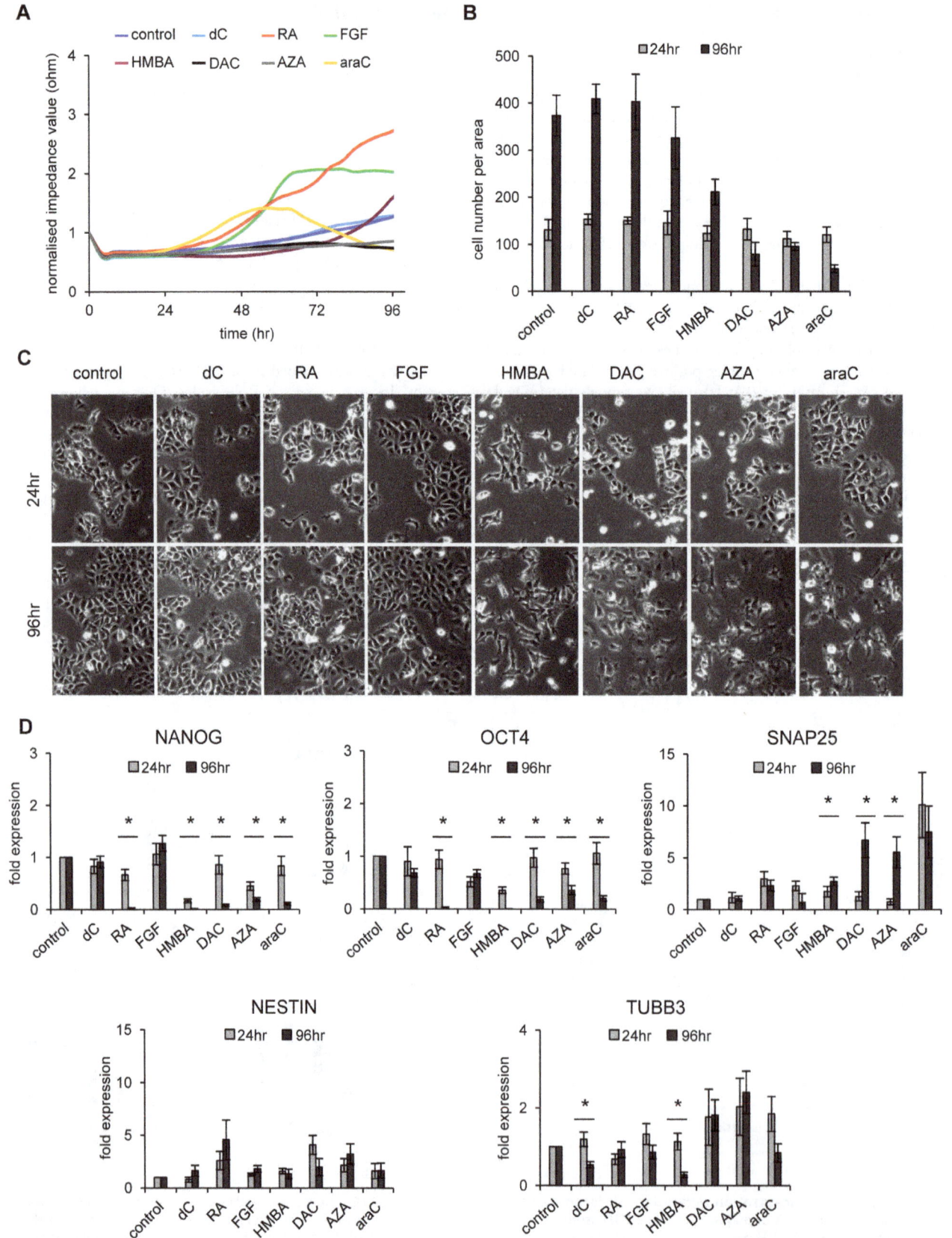

Figure 3. Induced differentiation by a defined panel of drugs. (**A**) Impedance profiles comparing NT2 cells treated with retinoic acid (RA, 10 μM), hexamethylene bisacetamide (HMBA, 5 mM), 5-azacytidine (AZA, 1 μM), deoxycytidine (dC, 1 μM), fibroblast growth factor 2 (FGF, 50 μM), 2'-deoxy-5-azacytidine (DAC, 1 μM) and 1β-arabinofuranosylcytosine (araC, 1 μM) during a 4 day period. Measurements were executed at 45 kHz in

5-minute intervals for 96 hours. The mean of three independent experiments is shown. Standard deviations are not shown to avoid crowding of the diagram. For single diagrams including standard deviations and statistical tests for these data sets see Fig. S2. (**B**) Average cell numbers of three replicates of untreated and treated NT2 cells after 24 and 96 hours. Nucleoside drugs are cytotoxic at the concentrations used and show significant growth inhibition. Standard deviations are indicated by error bars. (**C**) Microscopic images (10× magnification) of NT2 control cells and NT2 cells treated with the various substances mentioned in (A) after 24 and 96 hours of treatment. (**D**) qRT-PCR expression analysis of stem cell factors NANOG, OCT4 and the differentiation markers NESTIN, SNAP25 and TUBB3 in treated and NT2 control cells after 24 and 96 hours of treatment. All qRT-PCR measurements were repeated at least three times and internally normalised to the corresponding β-actin values. Standard deviations are indicated by error bars. Treatments showing significant differences comparing expression levels at 24 hours with those at 96 hours are marked with an asterisk (two-tailed student's t-test; $p < 0.05$).

Impedance Measurements

NT2 D1 cells (2×10^4 per well) were seeded in triplicates or quadruplicates and grown in 400 μl DMEM supplemented with 10% FCS (Invitrogen), 200 U/ml penicillin (Gibco) and 200 μg/ml streptomycin (Gibco) on 8W10E+ ECIS Cultureware arrays (Applied Biophysics) that contain 40 250-μm gold electrodes per well. Cells were then treated as indicated. Measurements were carried out in cell culture medium (without medium changes) and the arrays were kept in the incubator with 5% CO_2 at 37°C. The arrays were measured on an ECIS™ Model 1600 (Applied Biophysics) at 45 kHz in 5-minute intervals for 96 hours. Data were normalised to their starting values. Analysis was done on ECIS software based on the model developed by Giaever and Keese [19].

Staining and Microscopy

At ECIS endpoints, cells attached to the arrays were washed with 1X Phosphate Buffered Saline (PBS) and fixed with 4% paraformaldehyde. Cells were permeabilised with 0.1% TritonX-100, stained with Phalloidin Tritc (Sigma) and DAPI (Invitrogen) in PBS for 1 hour and washed. After staining, 8-well chamber tops were removed from the base slides and mounting media and cover slips were added. Fluorescence images were taken on Zeiss Axioskop 2 Plus. Phase contrast images of living cells were taken on a Leica DM-IRBE inverse microscope.

Protein Depletion via siRNAs

For the depletion of OCT4, specific ON-TARGETplus SMARTpool siRNAs (Dharmacon) were used as described [9]. In brief, NT2 cells were seeded in quadruplicate into 8W10E+ ECIS Cultureware arrays (Applied Biophysics) at a density of 2×10^4 per well, with 400 μl of medium (see above). Cells were transfected with siRNAs using the DharmaFECT1 transfection reagent (Dharmacon) according to the manufacturer's instructions. The final siRNA concentration was 50 nM. Scrambled siRNAs for negative control experiments were also obtained from Dharmacon. Impedance measurements were started immediately after transfection, as described above and carried out during a 6 day period. Medium was changed once after 3 days. Impedance peaks caused by the medium change were equalised during data analysis with the ECIS software.

Statistical Analysis and Bioinformatics

Two-tailed student's t-test was used for statistical analysis of ECIS and qRT-PCR data. To determine the slope maxima of the differentiation-induced impedance data, we applied a cubic smoothing spline with 10 degrees of freedom to the resistance values and performed a generalised cross validation to the data [24]. The steepest rise in impedance was determined as the time point with the maximum slope.

Results

Monitoring RA-induced Differentiation using Electric Impedance Sensing

RA-induced neuronal differentiation of NT2 cells [15,25] is a comparably slow process that usually requires several weeks of treatment before morphological changes become visible, even if gene expression patterns change more rapidly [15]. In order to use a non-invasive method and to directly monitor differentiation induction in vitro, we seeded NT2 cells (2×10^4 per well) into eight-well ECIS-arrays (8WE10+) and measured resistance changes at 45 kHz every 5-minutes in the absence or presence of RA over a four day period (96 hours). As shown in Figure 1A, untreated NT2 cells showed only a weak increase of frequency dependent impedance values over four days, which was most likely mainly caused by the increasing cell number. In striking contrast, treatment with 10 μM RA led to a significant increase of impedance values starting with 20 hours of treatment (Fig. 1A). Total cell numbers did not differ significantly between untreated and RA-treated cells at 24 or 96 hours of growth (Fig. 1B), indicating that impedance differences were not due to differing growth rates. Also, overall the morphology of both cell populations was very similar (Fig. 1C). Early onset of differentiation is usually monitored by marker gene expression using quantitative reverse transcription PCR (qRT-PCR) on total RNA isolated from growing cells. As shown in Fig. 1D, transcription of the stem cell factors NANOG and OCT4 was significantly down-regulated in RA-treated cells, but only after 96 hours of RA treatment, whereas specific differentiation markers NESTIN, SNAP25 and HOXA1 were induced. As expected, HOXA1, a very early and prominent marker of differentiation, showed a strong increase of expression within 24 hours (see also ref. 9). However, expression differences of the other genes investigated were barely visible by qRT-PCR within the first day after start of treatment (Fig. 1D, light grey bars). Thus, impedance measurement is a highly sensitive and robust method to follow the early onset of RA-induced differentiation.

In order to monitor the effect of RA concentration on differentiation induction, we treated NT2 cells with different concentrations of retinoic acid and registered the dielectric resistance profiles (Fig. 2A, Fig. S1). Increasing RA concentrations lead to increased resistance (as indicated by the time point of first statistical significant difference in impedance values of control experiments versus RA treatment, see Fig. S1) and also a steeper slope of the impedance profiles. This correlated with the state of differentiation, as confirmed by the measurement of marker gene expression after 96 hours of treatment (Fig. 2B). We then chose two parameters that characterise RA-induced resistance changes: the slope of the curve obtained when joining the single points of resistance measurements and the time point when the maximum slope is reached. The higher the slope and the earlier the slope maximum, the faster and stronger are the resistance changes that reflect ongoing differentiation. We therefore analysed the dataset shown in Fig. 2A by applying a cubic smoothing spline and

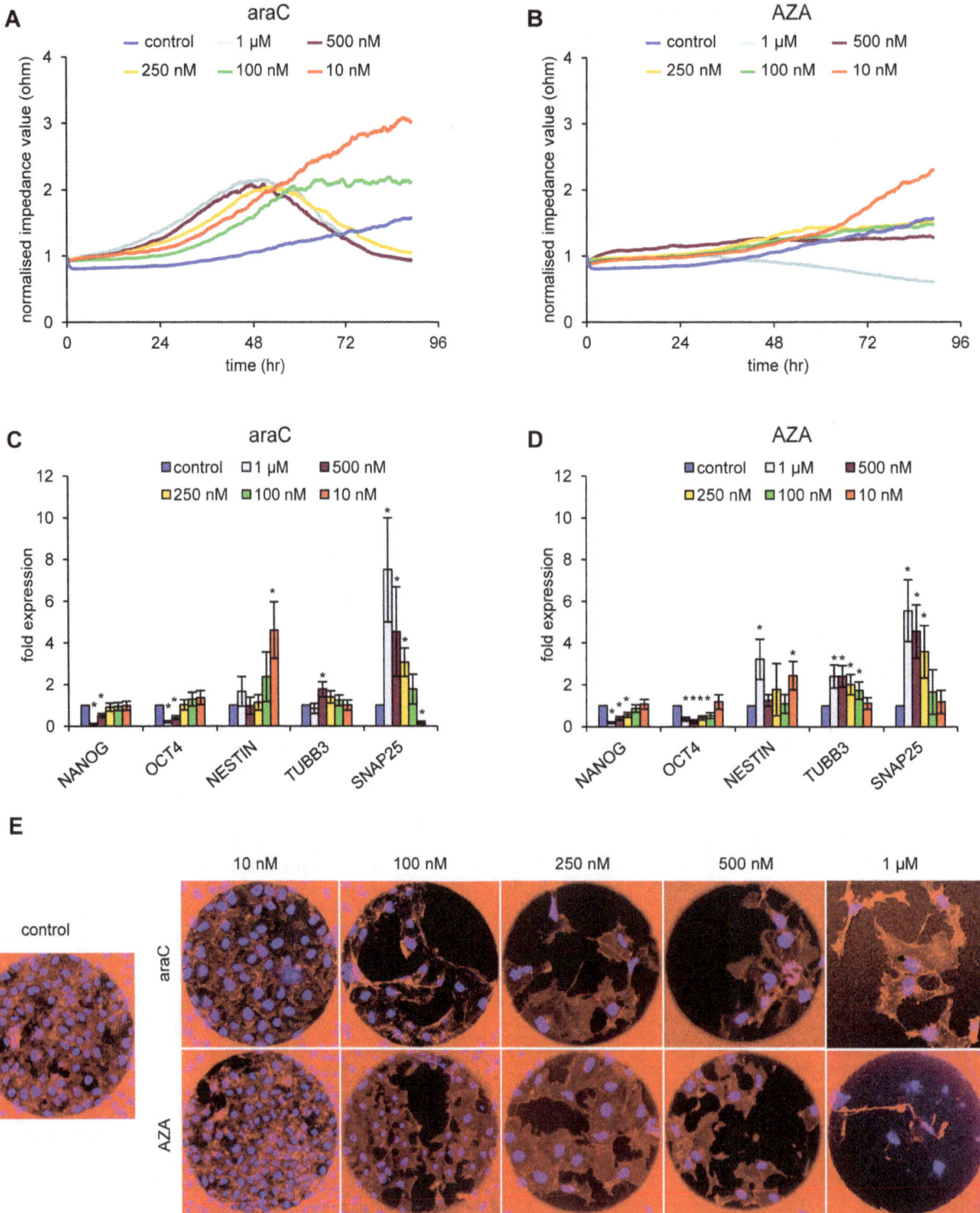

Figure 4. Induced concentration-dependent differentiation by araC and AZA. (**A**) Impedance profiles of NT2 cells treated with different concentrations of 1β-arabinofuranosylcytosine (araC) during a 4 day period. Concentrations above 100 nM are severely cytotoxic, which leads to a drastic drop in impedance values after 48 hours. Measurements were executed at 45 kHz in 5-minute intervals for 96 hours. One representative experiment is shown. For single diagrams showing the mean of at least three experiments including standard deviations and statistical tests see Fig. S3A. (**B**) Impedance profiles of NT2 cells treated with different concentrations of 5-azacytidine (AZA) during a 4 day period. Concentrations above 100 nM strongly induce proliferative defects, which prevents the increase of impedance values. Measurements were executed at 45 kHz in 5-minute intervals for 96 hours. One representative experiment is shown. For single diagrams showing the mean of at least three experiments including standard deviations and statistical tests see Fig. S3B. (**C**) qRT-PCR expression analysis of stem cell factors *NANO*G, *OCT4* and the neuronal differentiation markers *NESTIN*, *SNAP25* and *TUBB3* in NT2 cells treated with different concentrations of araC and control cells after 96 hours of treatment. All qRT-PCR measurements were repeated at least three times and internally normalised to the corresponding *β-actin* values. Standard

deviations are indicated by error bars. Expression levels of the respective genes showing significant differences compared with the untreated control are marked an asterisk (two-tailed student's t-test; p<0.05). (**D**) qRT-PCR expression analysis of stem cell factors *NANOG*, *OCT4* and the neuronal differentiation markers *NESTIN*, *SNAP25* and *TUBB3* in NT2 cells treated with different concentrations of AZA and control cells after 96 hours of treatment. All qRT-PCR measurements were repeated at least three times and internally normalised to the corresponding *β-actin* values. Standard deviations are indicated by error bars. Expression levels of the respective genes showing significant differences compared with the untreated control are marked an asterisk (two-tailed student's t-test; p<0.05). (**E**) Phalloidin staining of growing cultures. Flourescence images (10× magnification) of NT2 control cells and NT2 cells treated with the indicated concentrations of AZA and araC. The circular dark region is the electrode measuring area covered by the cells. Cells were stained with Phalloidin TRITC (red) and DAPI (blue).

determined for each treatment the maximum slope and the time point the maximum was reached. Then, by calculating the slope/time ratio (in order to compensate for low, but early slope maxima), a clear positive correlation between RA concentration, marker gene expression and maximum slope and time was found (Table S1). Thus the slope/time ratio can be used as an early marker for differentiation.

Electric Impedance Sensing of NT2 Cells Treated with a Panel of Differentiation Inducing Drugs

In order to expand our analyses to other differentiation-inducing substances and to monitor NT2 cells induced to differentiate into other lineages, we treated the cells with the nucleoside analogues araC, DAC and AZA and measured the impedance profiles (Fig. 3A, Fig. S2). In parallel experiments, differentiation was induced by Fibroblast Growth Factor 2 (FGF2; bFGF) and hexamethylene bisacetamide (HMBA). Expression of bFGF increases during retinoic acid induced differentiation of NT2 cells and bFGF treatment of floating spheres of NT2 cells has been shown to trigger terminal differentiation into neurons [14,25–27]. In addition, mesodermal features in aggregated NT2 cells after prolonged treatment with bFGF have also been reported [12]. HMBA treatment of NT2 cells has been shown to result in the expression of marker genes usually associated with epithelial

structures, suggesting that HMBA triggers differentiation into epidermal ectoderm [11,13].

As shown in Figure 3A (see also Fig. S2), araC treatment led to an early increase of resistance, indicating onset of differentiation. This effect was followed by a significant drop below the values of the deoxycytidine (dC) control (Fig. 3A, Fig. S2F). All three nucleoside drugs induced strong proliferation defects after 24 hours of growth (Fig. 3B), most likely by triggering DNA-damage dependent apoptotic pathways [9], which led to low cell numbers and reduced impedance values.

bFGF induced an increase in resistance after 48 hours of treatment (Fig. 3A, Fig. S2B). Cells treated with HMBA, however, showed values in the range of the untreated control. HMBA treatment also reduced the cell number in the culture, but in contrast to araC, an increase in resistance was observed after 72 hours (Fig. 3A, Fig. S2C).

The analysis of the slope/time ratios reveals similar values for RA and HMBA, reflecting a similar potential to induce differentiation (Table S2). Nevertheless, after HMBA treatment the maximum slope was reached much later, indicating a distinct differentiation pathway. bFGF and araC show higher slope/time ratios, due to an earlier onset of increasing resistance for araC and a steeper slope for bFGF, again reflecting different modes of differentiation induction. Taken together, RA, HMBA, bFGF and araC showed significant induction potentials, resulting in specific

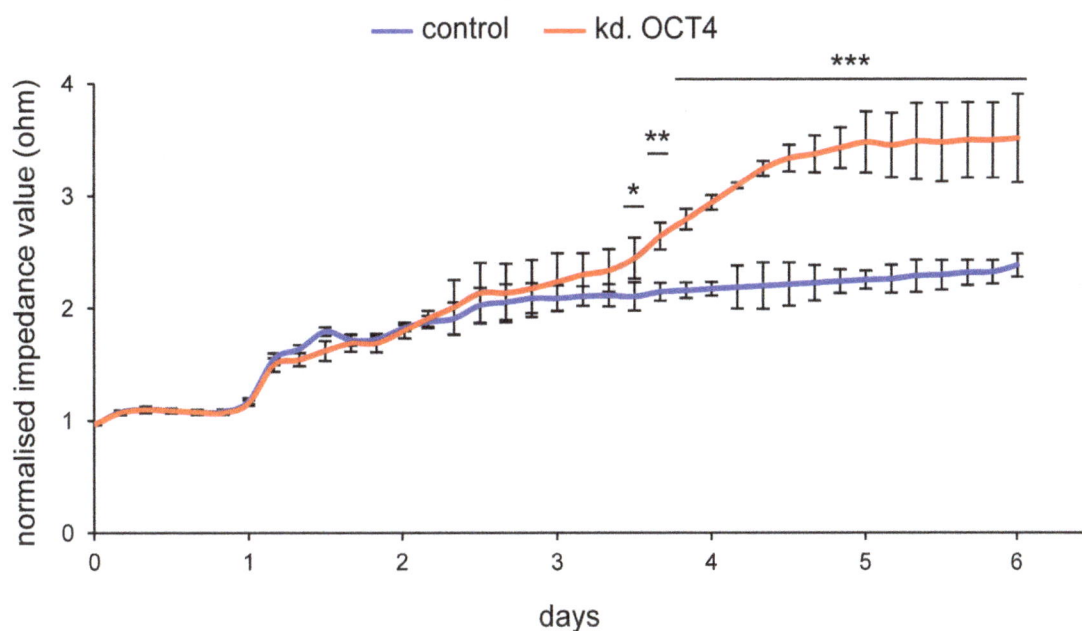

Figure 5. Induced differentiation by RNAi-mediated depletion of OCT4. Impedance profiles of control NT2 cells (blue - scrambled knock down) and NT2 cells depleted for OCT4 (red) during a 6 day period. Measurements were executed at 45 kHz in 5-minute intervals for 6 days. The mean of three independent experiments is shown. Standard deviations are indicated by error bars every four hours. Student's t-test was used for statistical analysis. Differences between control and knock down experiments in the indicated regions have been found to be statistically significant (*p<0.05, **p<0.005, ***p<0.001).

dielectric resistance profiles and slope/time ratios, whereas AZA and DAC treated cells had a similar profile as the controls after 24 hours of treatment. Nevertheless, for araC, DAC and AZA the strong reduction of surviving cells lead to declining impedance values at later time points (Fig. 3A, Fig. 3B, Fig. S2D, S2E and S2F).

As already observed for retinoic acid (Fig. 1C), the majority of differentiation-inducing factors did not induce any significant morphological differences after 24 hours of treatment (Fig. 3C). However, HMBA treated cells, which had shown a delayed increase in dielectric resistance, appeared morphologically different already after 24 hours of treatment (Fig. 3C). After 96 hours of incubation, onset of differentiation was visible for all differentiation-inducing factors (Fig. 3C). In addition, araC, DAC and AZA treated cells showed clear reduction of cell numbers due to the cytotoxicity of these compounds.

As shown by qRT-PCR in Figure 3D, stem cell factors (*OCT4* and *NANOG*) were less expressed after 96 hours of treatment with all substances (except bFGF), indicating ongoing differentiation and loss of pluripotency. Consistent with the morphological alterations, expression of both genes was only weakly reduced after 24 hours. Furthermore, differentiation markers (*SNAP25, NESTIN, TUBB3*) were only moderately increased after 96 hours of treatment (Fig. 3D). At 24 hours, PCR-based expression analysis of differentiation marker genes as well as phase contrast microscopy failed to clearly indicate onset of differentiation (Fig. 3C, 3D, light grey bars). However, the drug-specific impedance values suggest that differentiation already starts within the first day of treatment, especially with RA, bFGF and araC (Fig. 3A, Fig. S2). Thus treatment-induced early differentiation steps obviously trigger changes in cell-extracellular matrix contacts, leading to increased resistance (Fig. 1A, Fig. 2A, Fig. 3A, Table S2). This finding underscores the value of ECIS analysis, especially to analyse early differentiation states, and is also in accordance with recent *in vitro* differentiation data of MSCs, which revealed changes in impedance profiles already within the first hours of adipogenic or osteogenic differentiation [20,21]. Our data further suggest that lineage specific morphological changes influence impedance values in different ways, leading to characteristic resistance profiles and slope/time ratios.

Concentration Dependence of Drug Induced Impedance Curves

As shown in Figures 3A and 3B, the nucleoside drugs were cytotoxic at the concentrations used initially (1 μM), leading to reduced cell numbers after prolonged treatment, therefore impeding the monitoring of differentiation-dependent impedance changes. We addressed the possibility to separate their influence on differentiation induction from cytotoxic side effects by lowering the concentration of these compounds. As shown in Figs. 4 and S3, 10 nM araC induced impedance values that were similar to RA-treated cells, without triggering cell death, with a slope/time ratio comparable to the one obtained with RA at 10 μM (Table S2 and Table S3). Up to 48 hours, also araC concentrations higher than 10 nM induced increased dielectric resistance, leading to positively correlated slope maxima (Table S3). However, at later time points cytotoxicity lead to significantly reduced growth and a drop in impedance (Fig. 4A, Fig. S3). Treatment with AZA at 10 nM lead to a similar, but retarded increase in resistance, again comparable to RA at 10 μM, which is also reflected by the slope/time ratio (Fig. 4B; Fig. S3, Table S3). Thus, as with araC, 10 nM AZA triggered differentiation of NT2 cells without inducing proliferation defects. Higher concentrations of AZA were mostly toxic for the cells.

End point qPCR of stem cell and differentiation markers showed only moderate changes for low concentrations (100 nM and 10 nM) of araC and AZA (Fig. 4C and 4D). Only *NESTIN* expression was significantly increased in araC-treated cells. It should also be noted that the moderate differentiation phenotypes obtained with 10 nM araC or AZA were morphologically similar to those of RA-induced cultures (see Fig. 1C). Stronger differentiation phenotypes and gene expression changes were only obtained with higher, cytotoxic drug concentrations (Fig. 4C–E). These results further underscore the sensitivity of the ECIS method in detecting early onset of differentiation.

Electric Impedance Sensing of OCT4-depleted NT2 Cells

We have recently shown that siRNA-mediated depletion of the stem cell specific protein OCT4 induces neuronal differentiation in NT2 cells [9]. In order to analyse if reduction of OCT4 levels alone will lead to an increase of impedance levels in a similar way as retinoic acid or drug treatment, we seeded NT2 cells into ECIS-arrays and depleted OCT4 by siRNA transfection, using conditions that cause more than 90% reduction of *OCT4* mRNA levels [9]. As shown in Figure 5 increased resistance became apparent for the OCT4 depleted population after 2–3 days, reaching levels comparable to RA treatment around day 4. The delay of 2–3 days is caused by the knock down procedure, as efficient turnover of OCT4 protein is only achieved after 3 days [9], which also explains the observed increase in impedance over this period, as the cells continued to grow (Fig. 5). After substantial depletion of OCT4 was achieved at day 3, the cells started to differentiate, leading to a subsequent increase in resistance values (Fig. 5). Interestingly, when calculating the slope/time ratio (Table S4), we observed very similar values to the ones found in RA treated cells (compare Tables S1, S2 and S4). These findings show that differentiation induction of NT2 cells is triggered by the reduction of OCT4 and provide important confirmation for the argument that the observed changes in resistance were indeed caused by the onset of cellular differentiation.

Discussion

In this study we demonstrate that treatment with well characterised differentiation triggering substances induces distinct dielectric changes in differentiating NT2 cell populations. We were able to generate impedance profiles during the first days of differentiation that allow to monitor the onset of differentiation very early (after 20 hours), when other phenotypic changes or differentiation specific marker gene expression patterns are not yet apparent. Further, by calculating slope/time ratios of each data set, we obtained a measure for the degree of induced differentiation. Impedance analysis also seems to allow the correlation of lineage choices with specific resistance profiles that could enable the prediction of differentiation pathways induced by specific drugs (RA, bFGF - early max. slope) from epidermal differentiation (HMBA - late max. slope).

MSCs show specific impedance profiles during adipogenic or osteogenic differentiation, caused by the modulation of cellular contacts with the gold electrodes or the extracellular matrix [20,21]. As shown in this work, RA-induced neuronal differentiation of NT2 cells is also accompanied by specific interactions with the extracellular matrix (ECM) that change during later stages. Untreated NT2 cells express *alpha5beta1 integrin*, which specifically interacts with fibronectin, whereas NT2 neuron-like cells express *alpha3beta1 integrin* with LAMIN-5 as a ligand in the extracellular matrix [28]. During epithelial transition induced by HMBA, NT2 cells also show specific morphological characteristics. The nuclear to cytoplasm ratio is increased and the actin cytoskeleton is reorganised [29]. Impedance sensing thus seems to discriminate between these different modes of interaction with the extracellular matrix, which are specific for each differentiation pathway. Extracellular matrix molecules can also directly influence the efficiency of differentiation *in vitro* [20]. Besides cell-ECM interactions, also cell-cell contacts are of outmost importance during differentiation. Changes in cell membrane capacitance also can be described by an ECIS scan with different frequencies [21]. For example, connexins, expressed during neuronal differentiation [28] or desmosomes, expressed during epithelial differentiation [29] might come into focus in this respect.

Further we show that the previously described drug-induced differentiation using nucleoside analogues [9] induces similar impedance profiles and slope/time ratios as the natural ligand retinoic acid. The activation of differentiation and the maintenance of differentiation-specific gene expression patterns require substantial epigenetic modulation, especially changes in PcG presence, histone modification patterns and also DNA methylation [30–33]. We have previously used the DNMT inhibitor 2′-deoxy-5-azacytidine to induce hypomethylation and differentiation in NT2 cells [9]. Nevertheless, we observed even stronger differentiation induction by cytarabine, a drug that has no epigenetic modulatory potential and does not trigger any changes in DNA methylation [9]. This suggested a mechanism of drug-dependent differentiation that does not interfere directly with the epigenetic maintenance system.

RA- and nucleoside-drug induced NT2 cells showed very similar early impedance profiles. At higher concentrations of the drugs, cytotoxicity became predominant, resulting in reduced impedance. The ECIS assay allowed to determine the concentration thresholds of the tested drugs that were sufficient for the induction of differentiation without triggering cytotoxic side effects. At concentrations as low as 10 nM, araC and AZA induced differentiation-specific impedance profiles very similar to RA treatment that were stably increasing over more than three days. At concentrations above 100 nM, cytotoxicity became prominent, although surviving cells showed strong differentiation phenotypes. For both DAC and AZA, dose dependent dual mechanisms have been described, with cytotoxic and anti-proliferative effects at high doses and DNA hypomethylation at low doses [34]. Our impedance data further support this concept and may indicate that low doses of these drugs, including araC, can specifically induce differentiation in cancer stem cell populations.

Depletion of OCT4 by RNA interference induced similar resistance profiles and slope/time ratios as RA and low concentrated nucleoside drugs. This confirms the hypothesis that induced differentiation is caused predominantly by the reduction of OCT4 levels, either by proteolytic degradation (as for the nucleoside drugs) or by the transcriptional down-regulation of the gene (as for the natural ligand retinoic acid) [9]. Since cell membrane capacitance obviously is an early marker of differentiation, lower (1 kHz to 8 kHz) and higher frequencies (62.5–64 kHz) to measure the multifrequency complex impedance (Z*) might improve the characterisation of the drug induced cell inherent dielectric properties [21]. We so far only monitored impedance at 45 kHz. Measurements at other frequencies possibly will help to describe cell-cell vs. cell-matrix interactions in more detail. Nevertheless, our work shows that impedance sensing is a robust and sensitive method to describe the effects of differentiation inducing drugs by measuring the dielectric properties of cells in real-time. Using this method, very early differentiation processes could be followed within the first day after drug treatment in non-invasive conditions. Therefore, our work can serve as a basis for a more detailed analysis of molecular effects of signalling pathways involved in cellular differentiation and for the screening for drugs that modulate cellular phenotypes.

Supporting Information

Figure S1 Induced concentration-dependent differentiation by RA. Impedance profiles comparing untreated NT2 with cells treated with 10 nM RA (**A**), 500 nM RA (**B**), 1 μM RA (**C**), 5 μM RA (**D**) and 10 μM RA (**E**) are shown. Measurements were executed at 45 kHz in 5-minute intervals for 96 hours. Each experiment was repeated at least three times. Standard deviations are indicated by error bars every four hours. Student's t-test was used for statistical analysis (*p<0.05. **p<0.005). Black lines show regions with significant differences in respect to the untreated control.

Figure S2 Induced differentiation by a panel of drugs. Impedance profiles comparing untreated NT2 with cells treated with 1 μM dC (**A**), 50 μM bFGF (**B**), 5 mM HMBA (**C**), 1 μM DAC (**D**), 1 μM AZA (**E**) and 1 μM araC (**F**) are shown. Measurements were executed at 45 kHz in 5-minute intervals for 96 hours. Each experiment was repeated at least three times. Standard deviations are indicated by error bars every four hours. Student's t-test was used for statistical analysis (*p<0.05. **p<0.005). Black lines show regions with significant differences in respect to the dC control.

Figure S3 Induced concentration-dependent differentiation by araC and AZA. (**A**) Impedance profiles comparing untreated NT2 cells (dark blue) and cells treated with 1 μM (light blue), 500 nM (purple), 250 nM (yellow), 100 nM (green) and 10 nM (red) araC. (**B**) Impedance profiles comparing untreated NT2 cells (dark blue) and cells treated with 1 μM (light blue), 500 nM (purple), 250 nM (yellow), 100 nM (green) and 10 nM (red) AZA. Measurements were executed at 45 kHz in 5-minute intervals for 96 hours. Each experiment was repeated at least three times. Standard deviations are indicated by error bars every four hours. Student's t-test was used for statistical analysis (*p<0.05. **p<0.005). Black lines show regions with significant differences in respect to the control.

Table S1 Slope maxima of RA-treated NT2 cells.

Table S2 Slope maxima of drug-treated NT2 cells.

Table S3 Slope maxima of araC- and AZA-treated NT2 cells.

Table S4 Slope maxima of OCT4-depleted NT2 cells.

Table S5 RT-Primer pairs used in this study.

Acknowledgments

We thank Francesca Tuorto for help with microscopy, Fabian Graf for help with impedance measurements and cell staining and Sebastian Bender for bioinformatic assistance. S.Ö. is PhD student of the HBGIS graduate school, Heidelberg, and member of the joint PhD program "disease models and drugs" between Heidelberg University and Mannheim University of Applied Sciences.

Author Contributions

Conceived and designed the experiments: CM AB SÖ. Performed the experiments: AB SÖ. Analyzed the data: CM AB SÖ. Contributed reagents/materials/analysis tools: CM AB. Wrote the paper: CM AB SÖ.

References

1. Sell S (2004) Stem cell origin of cancer and differentiation therapy. Crit Rev Oncol Hematol 51: 1–28.
2. Degos L (1999) Differentiating agents in the treatment of leukemia. Leuk Res 14: 717–719.
3. von Wangenheim KH, Peterson HP (2008) The role of cell differentiation in controlling cell multiplication and cancer. J Cancer Res Clin Oncol 134: 725–741.
4. Jones PA, Taylor SM (1980) Cellular differentiation, cytidine analogs and DNA methylation. Cell 20: 85–93.
5. Hatse S, De Clercq E, Balzarini J (1999) Role of antimetabolites of purine and pyrimidine nucleotide metabolism in tumor cell differentiation. Biochem Pharmacol 58: 539–555.
6. Burnett A, Wetzler M, Löwenberg B (2011) Therapeutic Advances in Acute Myeloid Leukemia. J Clin Oncol 29: 487–494.
7. Robak T, Wierzbowska A (2009) Current and emerging therapies for acute myeloid leukemia. Clin Ther 31: 2349–2370.
8. Ewald B, Sampath D, Plunkett W (2008) Nucleoside analogs: molecular mechanisms signalling cell death. Oncogene 27: 6522–6537.
9. Musch T, Oz Y, Lyko F, Breiling A (2010) Nucleoside drugs induce cellular differentiation by caspase-dependent degradation of stem cell factors. PLoS One 5: e10726.
10. Bocker MT, Tuorto F, Raddatz G, Musch T, Yang FC, et al. (2012) Hydroxylation of 5-methylcytosine by TET2 maintains the active state of the mammalian HOXA cluster. Nat Commun 3: 818.
11. Andrews PW (2002) From teratocarcinomas to embryonic stem cells. Philos Trans R Soc Lond B Biol Sci 357: 405–417.
12. Pal R, Ravindran G (2006) Assessment of pluripotency and multilineage differentiation potential of NTERA-2 cells as a model for studying human embryonic stem cells. Cell Prolif 39: 585–598.
13. Przyborski SA, Christie VB, Hayman MW, Stewart R, Horrocks GM (2004) Human embryonal carcinoma stem cells: models of embryonic development in humans. Stem Cells Dev 13: 400–408.
14. Marchal-Victorion S, Deleyrolle L, De Weille J, Saunier M, Dromard C, et al. (2003) The human NTERA2 neural cell line generates neurons on growth under neural stem cell conditions and exhibits characteristics of radial glial cells. Mol Cell Neurosci 24: 198–213.
15. Pleasure SJ, Page C, Lee VM (1992) Pure, postmitotic, polarized human neurons derived from NTera 2 cells provide a system for expressing exogenous proteins in terminally differentiated neurons. J Neurosci 12: 1802–1815.
16. Wegener J, Keese CR, Giaever I (2000) Electric cell-substrate impedance sensing (ECIS) as a noninvasive means to monitor the kinetics of cell spreading to artificial surfaces. Exp Cell Res 259: 158–166.
17. Keese CR, Bhawe K, Wegener J, Giaever I (2002) Real-time impedance assay to follow the invasive activities of metastatic cells in culture. Biotechniques 33: 842–850.
18. Hong J, Kandasamy K, Marimuthu M, Choi CS, Kim S (2011) Electrical cell-substrate impedance sensing as a non-invasive tool for cancer cell study. Analyst 136: 237–245.
19. Giaever I, Keese CR (1991) A morphological biosensor for mammalian cells. Nature 366: 591–592.
20. Angstmann M, Brinkmann I, Bieback K, Breitkreutz D, Maercker C (2011) Monitoring human mesenchymal stromal cell differentiation by electrochemical impedance sensing. Cytotherapy 13: 1074–1089.
21. Bagnaninchi PO, Drummond N (2011) Real-time label-free monitoring of adipose-derived stem cell differentiation with electric cell-substrate impedance sensing. Proc Natl Acad Sci USA 108: 6462–6467.
22. Andrews PW, Damjanov I, Simon D, Banting GS, Carlin C, et al. (1984) Pluripotent embryonal carcinoma clones derived from the human teratocarcinoma cell line Tera-2. Differentiation in vivo and in vitro. Lab Invest 50: 147–162.
23. Andrews PW (1984) Retinoic acid induces neuronal differentiation of a cloned human embryonal carcinoma cell line in vitro. Dev Biol 103: 285–293.
24. Green PJ, Silverman BW (1994) Nonparametric Regression and Generalized Linear Models: A Roughness Penalty Approach. London: Chapman & Hall. 182 p.
25. Schuldiner M, Yanuka O, Itskovitz-Eldor J, Melton DA, Benvenisty N (2000) Effects of eight growth factors on the differentiation of cells derived from human embryonic stem cells. Proc Natl Acad Sci USA 97: 11307–11312.
26. Zhang RL, Zhang L, Zhang ZG, Morris D, Jiang Q, et al. (2003) Migration and differentiation of adult rat subventricular zone progenitor cells transplanted into the adult rat striatum. Neuroscience 116: 373–382.
27. Coyle DE, Li J, Baccei M (2011) Regional differentiation of retinoic acid-induced human pluripotent embryonic carcinoma stem cell neurons. PLoS One 6: e16174.
28. Meland MN, Herndon ME, Stipp CS (2010) Expression of alpha5 integrin rescues fibronectin responsiveness in NT2N CNS neuronal cells. J Neurosci Res 88: 222–232.
29. Simoes PD, Ramos T (2007) Human pluripotent embryonal carcinoma NTERA2 cl.D1 cells maintain their typical morphology in an angiomyogenic medium. J Negat Results Biomed 6: 5.
30. Fouse SD, Shen Y, Pellegrini M, Cole S, Meissner A, et al. (2008) Promoter CpG methylation contributes to ES cell gene regulation in parallel with oct4/Nanog, PcG complex, and histone H3 K4/K27 trimethylation. Cell Stem Cell 2: 160–169.
31. Mohn F, Weber M, Rebhan M, Roloff TC, Richter J, et al. (2008) Lineage-specific polycomb targets and de novo DNA methylation define restriction and potential of neuronal progenitors. Mol Cell 30: 755–766.
32. Bracken AP, Dietrich N, Pasini D, Hansen KH, Helin K (2006) Genome-wide mapping of Polycomb target genes unravels their roles in cell fate transitions. Genes Dev 20: 1123–1136.
33. Sessa L, Breiling A, Lavorgna G, Silvestri L, Casari G, et al. (2007) Noncoding RNA synthesis and loss of Polycomb group repression accompanies the colinear activation of the human HOXA cluster. RNA 13: 223–239.
34. Stresemann C, Lyko F (2008) Modes of action of the DNA methyltransferase inhibitors azacytidine and decitabine. Int J Cancer 123: 8–13.

17

Therapeutic Interaction of Systemically-Administered Mesenchymal Stem Cells with Peri-Implant Mucosa

Ryosuke Kondo[1], Ikiru Atsuta[1]*, Yasunori Ayukawa[1], Takayoshi Yamaza[2], Yuri Matsuura[1], Akihiro Furuhashi[1], Yoshihiro Tsukiyama[1], Kiyoshi Koyano[1]

1 Section of Implant and Rehabilitative Dentistry, Division of Oral Rehabilitation, Faculty of Dental Science Kyushu University, Fukuoka, Japan, 2 Department of Molecular Cell Biology and Oral Anatomy, Graduate School of Dental Science Kyushu University, Fukuoka, Japan

Abstract

Objectives: The objective of this study was to investigate the effect of systemically transplanted mesenchymal stem cells (MSCs) on the peri-implant epithelial sealing around dental implants.

Materials and Methods: MSCs were isolated from bone marrow of donor rats and expanded in culture. After recipient rats received experimental titanium dental implants in the bone sockets after extraction of maxillary right first molars, donor rat MSCs were intravenously transplanted into the recipient rats.

Results: The injected MSCs were found in the oral mucosa surrounding the dental implants at 24 hours post-transplantation. MSC transplantation accelerated the formation of the peri-implant epithelium (PIE)-mediated mucosa sealing around the implants at an early stage after implantation. Subsequently, enhanced deposition of laminin-332 was found along the PIE-implant interface at 4 weeks after the replacement. We also observed enhanced attachment and proliferation of oral mucous epithelial cells.

Conclusion: Systemically transplanted MSCs might play a critical role in reinforcing the epithelial sealing around dental implants.

Editor: Zoran Ivanovic, French Blood Institute, France

Funding: This work was supported by a grant-in-aid for Scientific Research (C) No. 23592888 (to I. Atsuta) from the Ministry of Education, Culture, Sports, Science and Technology of Japan. The funders had no role in study design, data collection and analysis, decision to publish, or preparation of the manuscript.

Competing Interests: The authors have declared that no competing interests exist.

* E-mail: atyuta@dent.kyushu-u.ac.jp

Introduction

Dental implant therapy is one of the most important and effective prosthodontic therapy options for partially and completely edentulous patients. Dental implants based on the concept of "osseointegration", a term explaining the fixation of a titanium implant in the bone [1], have resulted in dramatic therapeutic success and clinical improvement. However, the peri-implant tissue is always exposed to the possibility of inflammation because the titanium body penetrates the surrounding oral mucosa. Although the mucosal structure around the dental implant shows similarities to normal/healthy gingiva with innate defense mechanisms [2–4], many researchers have described the biological weakness of the peri-implant epithelium (PIE)-mediated sealing against the oral environment [5,6]. Therefore, improvement of the tight PIE-mediated sealing around dental implants is strongly desired to enable clinical success and improve outcomes for oral implant therapy.

Mesenchymal stem cells (MSCs) were first identified as postnatal stem cells in bone marrow by Friedenstein and colleagues [7], and were subsequently found in several human tissues, including adipose tissue, the umbilical cord, and dental pulp [8–10]. Recently, the minimum criteria to define MSCs was proposed by the Mesenchymal and Tissue Stem Cell Committee of the International Society for Cellular Therapy [11] as follows: (1) a capacity for adherence; (2) typical immunophenotypes including positivity for CD105, CD73, and CD90, and negativity for CD45, CD34, CD14, and CD11b; (3) multipotency including cell types of at least three lineages, such as osteoblasts, chondroblasts, and adipocytes. Furthermore, MSCs exhibit anti-inflammatory functions toward diverse immune cell types including lymphocytes, macrophages, and natural killer cells [12]. Therefore, many researchers have a great deal of interests in the therapeutic potential of human MSCs to treat a variety of human diseases [13,14].

In this study, it was investigated that the MSCs potential was applied for implant treatment with some troubles, delayed healing and mucosa inflammation based on the low sealing around implant. A few studies have reported that epithelial healing after implant placement is very similar to mucosa wound healing [15]. Wound healing progresses through a genetically programmed repair process that involves inflammation, cell proliferation, re-epithelialization, formation of granulation tissue, angiogenesis, interactions between various cells, and matrix and tissue remodeling [16]. Additionally, bacteria can accumulate around the implant circumference and induce inflammatory destruction more easily than around the natural tooth [17]. Under such abnormal situations, the PIE structure is formed along the implant surface. In all situations, the aim of treatment is to provide soft tissue regeneration to restore the structure, function, and physiology of the damaged tissues. Thus, it is critical to stabilize the epithelial soft tissue seal by promotion of epithelial cell adherence [18].

The relationship between MSC-based therapy and PIE-implant interface sealing is not well understood. The hypothesis of the present study was that systemic MSCs accumulate around the implant in the early stage and promote PIE formation and soft tissue attachment to the implant surface.

Materials and Methods

1. Animals

Male Wistar rats (4- and 6-weeks-old) and GFP-transgenic SD-Tg (CAG-EGFP) rats were purchased from Kyudo Lab (Tosu, Japan) and Japan SLC (Shizuoka, Japan), respectively. These animal experiments were performed under an institutionally approved guideline for animal care established by Kyushu University (approval number: A24-237-0).

2. Isolation and culture of MSCs

MSCs were isolated from the bone marrow of Wistar or GFP-transgenic rats based on a colony forming unit-fibroblast (CFU-F) assay [19]. Briefly, bone marrow cells were flushed out of the bone cavities of rat femurs and tibias, and then treated with a 0.85% NH_4Cl solution for 10 minutes to lyse red blood cells. The cells were passed through a 70-μm cell strainer to obtain a single cell suspension. The single cells were seeded at 1×10^6 cells/dish in 100-mm culture dishes. At 1 day after seeding, the cells were washed with phosphate buffered saline (PBS) and cultured in growth medium consisting of alpha minimum essential medium (Invitrogen, Grand Island, NY) containing 20% fetal bovine serum (Equitech-Bio, Kerrville, TX), 2 mM L-glutamine (Invitrogen,), 55 μM 2-mercaptoethanol (Invitrogen), 100 U/ml penicillin, and 100 μg/ml streptomycin (Invitrogen). After 1 week of culture, the CFU-Fs had formed colonies. The adherent mesenchymal cells in these colonies (referred to as "MCs" hereafter) were detached by trypsin, reseeded as new cultures, and expanded for further studies.

3. CFU-F assay

The CFU-F assay was performed as described previously [20]. Passage one MSCs were seeded at appropriate cell numbers in 100-mm dishes (Nalge Nunc, Rochester, NY). After 16 days, the cells were stained with a mixture of 0.1% toluidine blue (Merck, Darmstadt, Germany) and 2% paraformaldehyde (PFA; Merck) solution. Clusters containing >50 cells were considered as colonies. Total colony numbers were counted per dish. The CFU-F assay was repeated in independent experiments.

4. Immunofluorescence

Passage two MSCs (2×10^4 cells/dish) were seeded on 35-mm dishes and incubated for 12 hours at 37°C under 5% CO_2. Then, the slides were fixed in 4% PFA for 5 minutes and blocked with normal serum matched to the secondary antibodies for 1 hour followed by incubation with the mouse anti-rat CD44, CD90, and CD105 antibodies (1:100, Sigma–Aldrich) overnight at 4°C. Then, the slides were treated with FITC-conjugated secondary antibodies (1:200, Jackson Immuno Research, West Grove, PA) for 1 hour at room temperature (RT) and mounted with VECTASHIELD Mounting Medium containing 4'6-diamidino-2-phenylindole (DAPI) (Vector Laboratories, Burlingame, CA).

5. Flow cytometric analysis of cell surface markers

Passage two MSCs were collected and incubated with mouse anti-rat CD44, CD90, and CD105 antibodies (2 μg/ml, Chemicon International, Temecula, CA) for 60 minutes at 4°C and then an allophycocyanin-labeled secondary antibody (2 μg/ml, Vender, City, State/Country) for 30 minutes at 4°C. The analysis was then carried out using a FACS Calibur system (BD Biosciences) [21].

6. Osteogenic differentiation assay

Passage two MSCs (5×10^5 cells/dish) were grown on 35-mm dishes to confluency in growth medium and then cultured in osteogenic culture medium [growth medium containing 1.8 mM KH_2PO_4 (Sigma-Aldrich, St. Louis, MO) and 10 nM dexamethasone (Sigma-Aldrich)]. After 28 days of osteogenic induction, the cultures were stained with a 1% Alizarin Red S solution (Sigma-Aldrich). The expression of osteogenic markers including alkaline phosphatase (ALP), Runx2 and osteocalcin (OCN) was determined by western blot analysis.

7. Adipogenic differentiation assay

Passage two MSCs (5×10^5 cells/dish) were grown on 35-mm dishes to confluency in growth medium and then cultured in adipogenic culture medium [growth medium containing 0.5 mM isobutylmethylxanthine (Sigma–Aldrich), 60 μM indomethacin (Sigma–Aldrich), 0.5 μM hydrocortisone (Sigma–Aldrich) and 10 μg/ml insulin (Sigma–Aldrich)]. After 14 days of adipogenic induction, the cultures were stained with Oil Red O. The expression of adipogenic markers including lipoprotein lipase (LPL) and peroxisome proliferator-activated receptor γ (PPARγ) was determined by western blot analysis.

8. Chondrogenic differentiation assay with the pellet culture technique

Passage two MSCs (5×10^5 cells/dish) were collected in 5-ml conical polypropylene tubes and then pelleted for 6 minutes at 1,600 rpm. The cell pellet was incubated in chondrogenic culture medium [growth medium containing 50 μg/ml ascorbic phosphate (Sigma-Aldrich), 2 mM pyruvate (Sigma-Aldrich), 10 mg/ml transforming growth factor-β1 (Sigma-Aldrich), and 10 μg/ml insulin (Sigma-Aldrich)]. After 21 days, the pellets were washed with PBS and then fixed with 4% PFA for 4 hours.

9. Western blot analysis

Proteins were separated by sodium dodecyl sulfate polyacrylamide gel electrophoresis on 7.5% gels and transferred to polyvinylidene fluoride membranes (Bio-Rad Laboratories, Hercules, CA). The blots were probed for 24 hours at 4°C with primary polyclonal antibodies against rat osteogenic markers (ALP, Runx2, and OCN) and adipogenic markers (PPARγ and LPL) (all diluted at 1:100; Cell Signaling, Beverly, MA). The membranes were then incubated with secondary antibody for 60 minutes at RT, and visualized using an ECL Western Blotting Analysis System (GE Healthcare, Little Chalfont, UK).

10. Experimental dental implants

Similar to previously described designs [15,22], one-piece, screw-type pure titanium implants (Japan Industrial Standards Class 1 equivalent to ASTM grade 1) (Sky blue, Fukuoka, Japan) with machine-polished surfaces were used in this study (Fig. 1A). The roughness of the implant surface was measured with a laser scanning microscope (VK-9710, Keyence, Osaka, Japan) and the arithmetic mean roughness (Ra) was 0.16 μm. Before use, the

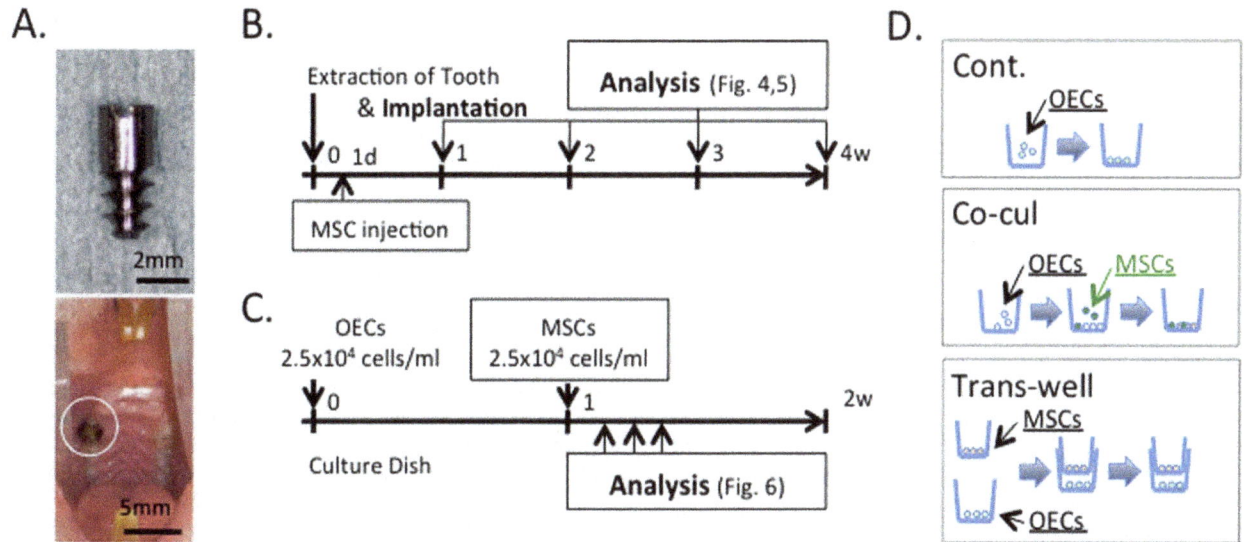

Figure 1. Design of the *in vivo* and *in vitro* experiments. (A) Photograph of the experimental implants (upper panel). Photograph of the implant in the rat oral cavity (lower panel). There is no apparent inflammation in the oral mucosa around the implant. (B) Experimental protocol for the *in vivo* study. Implantation was immediately performed after tooth extraction. Then, 24 hours after implantation, mesenchymal stem cells (MSCs) were injected via the tail vein. The structure of the epithelial tissue around the tooth or implant was observed after 1, 2, 3 and 4 weeks. (C) Experimental protocol for the *in vitro* study. Rat oral epithelial cells (OECs) were analyzed for changes in cell morphology 7 days after seeding OECs with MSCs under various culture conditions. (D) Experimental methods used for the *in vitro* study.

implants were treated with 100% acetone and distilled water and subsequently sterilized by autoclaving.

11. Oral implantation

The oral implantation procedure was completed according to our previous immediate-implantation study [23] (Fig. 1B). Briefly, maxillary right first molar was extracted from 27 male Wistar rats (6-weeks-old, 150–180 g), and the experimental implant was screwed into the cavity under systemic chloral hydrate and local lidocaine hydrochloride (Abbott Laboratory, North Chicago, IL) anesthesia. Following surgery, the rats were administered buprenorphine (0.05 mg/kg, i.m.) to relieve pain.

12. MSC transplantation

Passage three MSCs (1×10^6 cells) were injected into the rats with or without experimental dental implants via the tail vein (n = 5 in each group) at 24 hours post-implantation under anesthesia. As controls, rats received PBS (n = 5) or rat skin-derived fibroblasts (n = 5) in place of MSCs. All rats were sacrificed at 1, 2, 3 and 4 weeks post-transplantation.

13. Immunohistochemistry

According to our previous report [22], at the end of each experimental period, rats were deeply anesthetized and perfused intracardially with heparinized phosphate buffered saline, followed by 4% PFA (pH 7.4). The maxillae were dissected and demineralized in 5% tetrasodium ethylenediaminetetraacetate for 4 days at 4°C. The oral mucosa surrounding the implant and tooth site was carefully removed from the bone, implant or tooth, and then immersed in a 20% sucrose solution. The samples were embedded into O.C.T compound (Sakura, Tokyo, Japan) and cut into 10 μm-thick bucco-palatal sections with a cryostat at −20°C. These sections were then immunostained using an avidin-biotin complex (ABC) procedure (Vectastain ABC, Vector Laboratories, Burlingame, CA), as described previously [24,25]. Briefly, after treatment with 10% normal goat serum for 30 minutes at RT,

samples were incubated overnight at 4°C with a rabbit polyclonal anti-rat laminin-332 antibody (Santa Cruz Biotechnology, Santa Cruz, CA), then treated with biotinylated goat anti-rabbit IgG (1:200) for 45 minutes at RT, and finally reacted with the ABC reagent (1:100) for 60 min at RT. Immuno-positive reactions were visualized by treatment for 5 minutes in 0.02% 3,3′-diaminobenzidine tetrahydrochloride (Dojin Laboratories, Kumamoto, Japan) and 0.006% H_2O_2, at RT and the sections were counterstained lightly with hematoxylin.

14. Isolation and culture of oral mucous epithelial cells (OECs)

OECs were cultured according to a previous report [26] (Fig. 1C). Briefly, oral mucosa derived from 4-day-old Wistar rats was incubated with dispase (1×10^3 IU/ml) in Mg^{2+}- and Ca^{2+}-free PBS for 12 hours at 4°C. The oral epithelium (OE) was then peeled from the connective tissue using two pairs of tweezers. The epithelium was dispersed by pipetting 10 times and seeded onto dishes. OECs were cultured in defined keratinocyte serum free medium (Invitrogen) containing gentamicin (50 μg/ml) on plastic in a humidified atmosphere of 95% air and 5% CO_2 at 37°C.

Additionally, to determine the *in vitro* cellular effects of MSCs on OECs, the epithelial cells were co-cultured with MSCs directly or indirectly (Fig. 1D). For direct co-culture, OECs (2.5×10^4/ml: show the number per well) were first plated followed by MSCs (2.5×10^4/mL: show the number per well). For indirect co-culture using Transwell plates (BD Biosciences), OECs (2.5×10^4 per well) were plated in the lower chambers and MSCs (2.5×10^4 per well) were seeded in the upper chambers.

15. Cell adhesion assay

OEC adhesion assays were conducted according to previously published methods [27]. Twenty-four hours after co-culture with OECs and MSCs, non-adherent or weakly attached cells were removed by shaking (3×5 min at 75 rpm) on a rotary shaker (NX-20, Nissin, Tokyo, Japan). Adherent cells were then counted and calculated as a percentage of the initial count, which was used to define the adhesive strength of the cells.

16. Cell proliferation assay

Cell proliferation was assayed using cell proliferation kits (Molecular Probes Inc., Eugene, OR). Cultured cells were exposed to 5-bromo-2'-deoxyuridine (BrdU) in culture medium for 1 hour and then fixed in 70% methanol for 30 minutes. Fixed cells were incubated with mouse anti-BrdU monoclonal antibody (1:100 dilution) for 1 hour and then with FITC-conjugated anti-mouse IgG (Chemicon International; 1:100 dilution) for 30 minutes.

17. Statistical analysis

Data are expressed as the mean ± SD. One-way analysis of variance with Fisher's least significant difference tests was performed. P-values of less than 0.05 were considered significant. Experiments were performed using triplicate samples and were repeated three or more times to verify their reproducibility.

Results

1. Isolation and characterization of MSCs from rat bone marrow

The colony formation rate was 4–6% per bone marrow-derived cell. Immunofluorescence and flow cytometric analyses showed that passage two MCs were positive for CD44, CD90, and CD105, and negative for CD45 and CD11b (Fig. 2A).

Next, when the MSCs were cultured under an osteogenic condition for 4 weeks, the cultures showed accumulation of calcium by Alizarin Red staining. Furthermore, western blotting revealed expression of osteoblast-specific molecules including ALP, Runx2, and OCN (Fig. 2B). Under adipogenic induction conditions for 2 weeks, the MCs were capable of storing intracellular lipid droplets as shown by Oil Red O staining. Additionally, western blotting confirmed expression of adipocyte-specific markers including LPL and PPARγ (Fig. 2B). Finally, the MCs were able to differentiate into chondrocytes using the pellet culture technique (Fig. 2B). These findings indicated that our isolated cell population contained MSCs according to the criteria by the Mesenchymal and Tissue Stem Cell Committee of the International Society for Cellular Therapy [11]. Figure 2C shows that single colony-derived rat stem cells represented a putative MSC population with clonogenic renewal properties.

2. Homing of transplanted GFP-labeled MSCs to the surrounding tissues of dental implants (Fig. 3)

To determine whether transplanted MSCs are able to home to the oral mucosa around dental implants, we intravenously infused bone marrow MSCs derived form GFP-transgenic rats and analyzed their localization in the surrounding mucosa of dental implants at 3 days after infusion. GFP-positive cells were observed in the connective tissues of the peri-implant mucosa beneath the dental implants (Fig. 3A, S1). Interestingly, the localization of GFP-positive MSCs was limited to the connective tissue around the apical portion of the PIE. GFP-positive MSCs were not found in the naïve gingivae around natural teeth. However, GFP-positive MSCs were observed at the wound site after tooth extraction (Fig. 3A). Additionally, these double-positive cells remained around the extraction and implantation site for about 1 to 2 weeks (Fig. 3B). Transplanted fibroblasts isolated from the back skin of GFP-transgenic rats were not found at any site. Almost all injected cells were detected in the lung or peripheral blood (Fig. S2).

3. Healing of the oral mucosa around dental implants (Fig. 4, 5)

First, to evaluate the effect of MSC transplantation on mucosal repair, we observed the healing process of the oral mucosa at 1, 2, 3, and 4 weeks after tooth extraction (Fig. 4). Interestingly, in the MSC-transplanted group, the healing epithelium bulged vertically and completely covered the extraction cavity at 1 week after the extraction. The regenerated oral epithelium (OE) became more mature until week 4. In contrast, in the non-transplanted group, a thin epithelial layer extended horizontally from the wounded edges of the oral sulcular epithelium (OSE) to the scar at 1 week after the extraction. The regenerating epithelium finally covered the extraction wound. These histological findings indicate that MSC transplantation accelerates the repair process of the oral mucosa after tooth extraction, and that MSC transplantation is capable of inducing the formation of peri-implant mucosa.

Next, we examined the effect of MSC transplantation on formation of the peri-implant mucosa. Rat MSCs were intravenously injected into rats that received dental implants immediately after tooth extraction. Development of the peri-implant mucosa was then analyzed by histochemical and immunohistochemical analyses at 1, 2, 3, and 4 weeks post-implantation (Figs. 4 and 5). In the non-transplanted group, hypertrophic OE/OSE with a corneous layer began to extend along the surface of the dental implant and present at the upper-middle portion of the implant until week 2. Moreover, a thin non-keratinized epithelium had extended from the keratinized epithelium and finally formed non-keratinized PIE and keratinized peri-implant sulcular epithelium (PISE) at week 4 as reported previously [15,22]. However, the MSC-transplanted group exhibited more accelerated formation of the PIE around the dental implant compared with that in the non-transplanted group (Fig. 5). At week 2, a thin epithelium without keratinization extending from the OSE had spread further along the implant. The non-keratinized epithelium, PIE, and PISE covered the surface of the dental implants at week 3.

Laminin-332 is a component of the basement membrane (BM), which is considered to be involved in migration and adhesion of PIE cells, indicating that laminin-332 plays an important role in PIE formation [19,22,24]. To confirm the efficacy of MSC transplantation to promote formation of the PIE around dental implants, we examined the distribution of laminin-332 during the PIE development process (Fig. 5). In the non-transplanted group, laminin-332 was expressed initially along the BM under the OSE and OE in both groups and sparsely in the connective tissue until week 2. At week 3, laminin-332 was strongly expressed in the connective tissues around the apical portion of the immature PIE, but not in the inner interface between the PIE and dental implant. Laminin-332 was also expressed weakly at the BM under the PISE. At week 4, laminin332 was distributed as a band along the implant-PIE interface and PIE-connective tissue interface. However, the MSC-transplanted group showed earlier deposition of lamin-332 around the PIE. Additionally, at week 4, the laminin-332-positive structure at the PIE-implant interface extended to a more upper portion compared with that in the non-transplanted group. Furthermore, the laminin-332-positive structure at the PIE-implant interface in the MSC-transplanted group showed stronger laminin-332 expression than that in the non-transplanted group.

4. Direct and indirect interaction between co-cultured MSCs and OECs (Fig. 6)

OECs were detected and quantified by adhesion and proliferation assays after 24 hours in co-culture with or without MSCs in co-culture and transwell groups (Fig. 6A). Many OECs adhered

A. MSC marker

Immunofluorescence

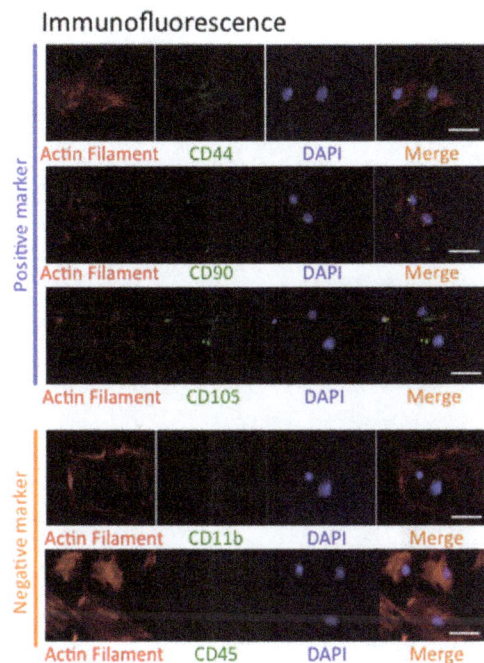

B. Differentiation

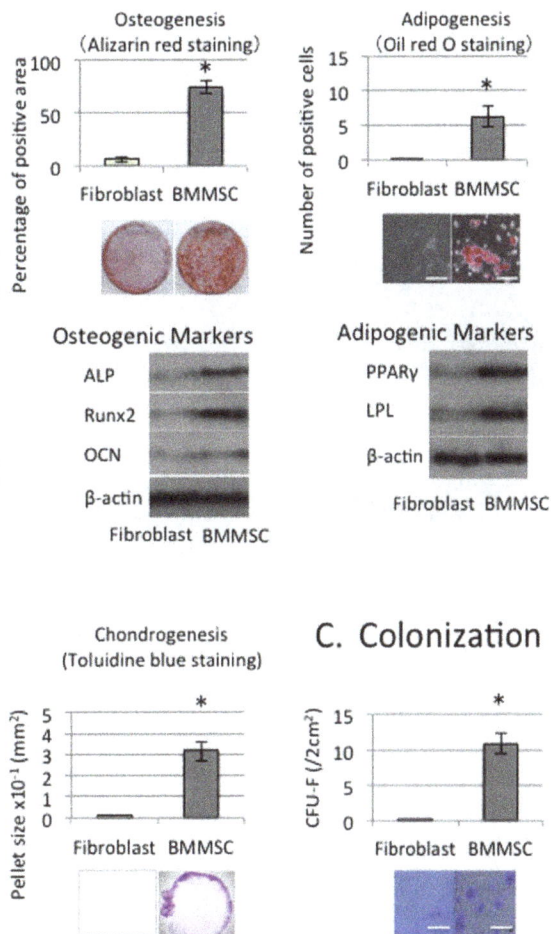

C. Colonization

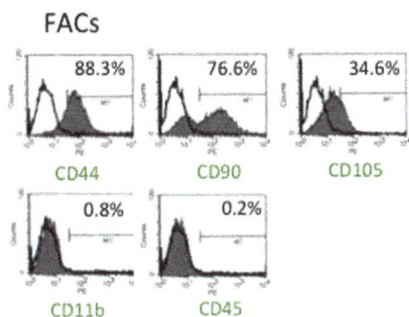

Figure 2. Multipotential differentiation of rat MSCs. (A) Expression of stem cell markers in rat BMMSCs. Cells cultured in 100 mm culture dishes were fixed and immunostained with specific Abs for rat CD-44, CD-90, CD-105 or CD-11b. Cells were incubated with rhodamine- or FITC-conjugated secondary Abs. (A1) Under a fluorescence microscope, positive signals were quantified in five random fields and expressed as the percentage of total DAPI-positive cells (bar = 100 μm) (mean ± SD). (A2) Expression of cell surface markers on MSCs as determined by flow cytometry. (B) (B1) Osteogenic differentiation of MSCs. After culture under osteogenic differentiation conditions for 4 weeks, osteogenic differentiation was determined by Alizarin Red S staining and western blot analysis of specific proteins (ALP, Ranx-2, OCN). The graph shows the quantification of the Alizarin Red S dye content in differentiated osteocytes from independent experiments (mean ± SD). Scale bar, 50 μm. (B2) Adipogenic differentiation of MSCs. After culture under adipogenic differentiation conditions for 2 weeks, adipocyte differentiation was determined by Oil Red O staining and western blot analysis of specific proteins (PPAR-γ, LPL). The graph shows the quantification of the Oil Red O dye content in differentiated adipocytes from independent experiments (mean ± SD). *$P<0.05$ (C) Single colony-derived rat stem cells represented a putative MSC population with clonogenic renewal properties.

when cultured with MSCs indirectly in transwells, but fewer cells adhered when co-cultured directly with MSCs (Fig. 6B). OEC proliferation was significantly higher in the indirect co-culture in transwells than that in direct co-culture (Fig. 6C).

Discussion

1. Isolation and characterization of MSCs from bone marrow (BMMSCs)

A variety of adult stem cells and/or precursor cells have been reported in several complex tissues or organs, including the dental tissues; however, few studies have confirmed which population of precursor cells exists in rat MSCs. In this study, MSCs from rat gingival tissue displayed CD44, CD90 and CD105, but were

negative for CD11b and CD45, as reported previously [28]. To validate our method of rat BMMSC isolation, Figure 2 shows that the cells isolated from the bone marrow of healthy donor rats had CD44 (88.3%), CD90 (76.6%), and CD105 (34.6%) positive signals by immunofluorescence and FACS (Fig. 2A), thus validating the MSC isolation method.

In this study, we used colony formation assays to show the presence of cells exhibiting the functional capacities of MSCs as described elsewhere [20,29]. Phenotypically defined rat MSCs adhered to culture dishes and 4–6% of these cells gave rise to a single colony (CFU-F) (Fig. 2C).

We next examined the multi-differentiation potential of phenotypically defined MSCs. Under adipogenic and osteogenic induction conditions, single colony-derived MSCs could differen-

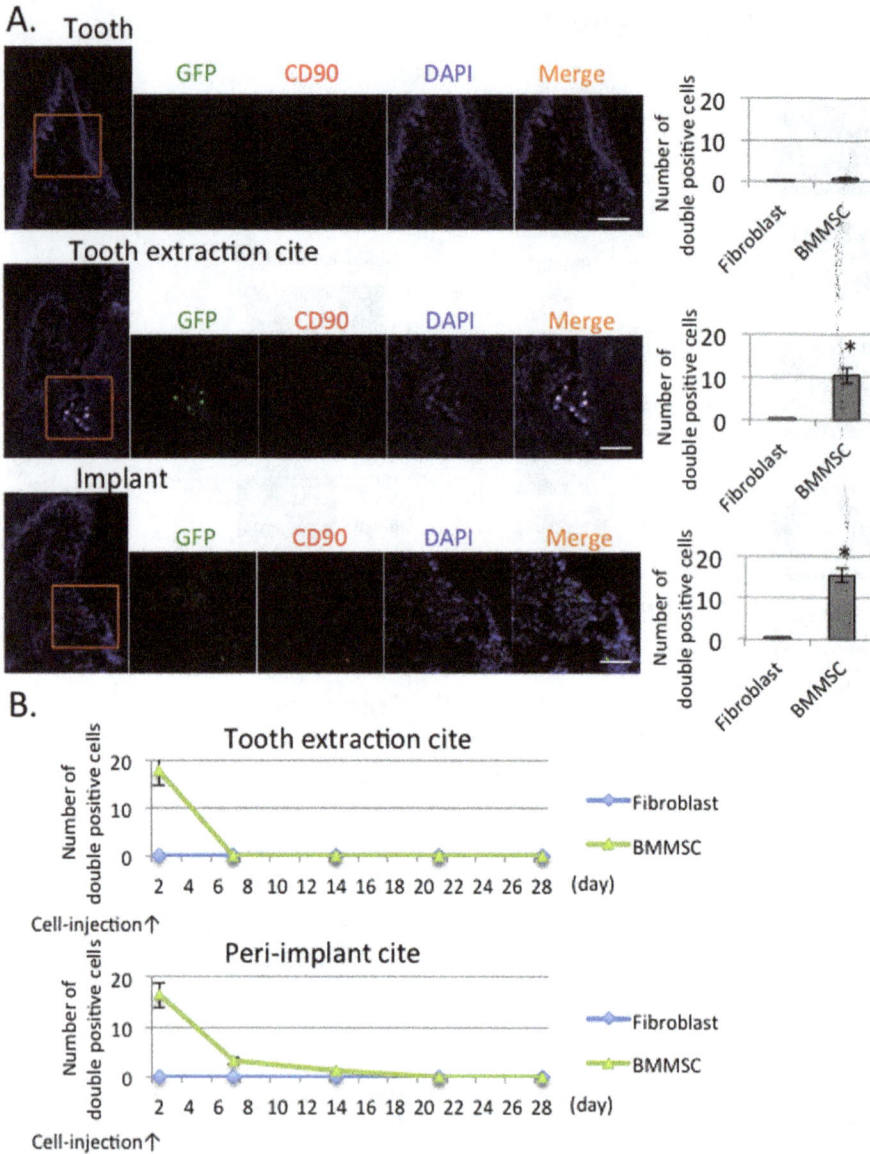

Figure 3. Accumulation of GFP-transgenic injected MSCs after tooth extraction or implantation. (A) Around the experimental implants, injected MSCs selectively accumulated at the extraction site or peri-implant tissue. However, no double-positive MSCs were observed in gingival mucosa. Fibroblasts isolated from back skin as negative controls did not accumulate at any sites. Bar = 100 μm. (B) The accumulated MSCs remained around the extraction and implantation sites for approximately 1 to 2 weeks.

tiate into adipocytes and osteoblasts as determined by Oil Red O and Alizarin Red S staining, respectively (Fig. 2B). Additionally, adipogenic (PPARγ and LPL) and osteogenic (ALP, Runx2 and OCN) marker expression was demonstrated by western blotting (Fig. 2B). These results were consistent with mesenchymal stem cell properties reported in other tissues [29]. We therefore isolated a putative population of stem cells with the methods employed in the present study, and demonstrated that the stem cells represented a putative MSC population with clonogenic renewal and multipotent differentiation capacities.

2. Accumulation of MSCs at the peri-implant tissue

MSCs injected into the rat tail vein were traced by the observation of samples harvested from all rats at various time points. In the present study, the injected MSCs were obtained

from GFP animals [30] Thus, the transplanted MSCs could be distinguished from recipient cells. The injected MSCs were observed to accumulate at peri-implant mucosa, while no cells accumulated at gingival mucosa around natural teeth 3 days after MSC injection. Previous studies showed that MSCs specifically accumulated in injured sites with inflammation [29]. Furthermore, it was reported that soft tissue around a titanium dental implant exhibited chronic inflammation [31]. It is well known in rodents that a majority of injected MSCs were trapped in the lungs [32] and very few re-circulate. In particular, cultured MSCs attach easily to any tissue including blood vessels and lungs. Indeed, because our data and those of many previous studies show local effects of MSCs via systemic injection, this method is considered to be effective [12,33,34]. In this study, our data suggested that peri-

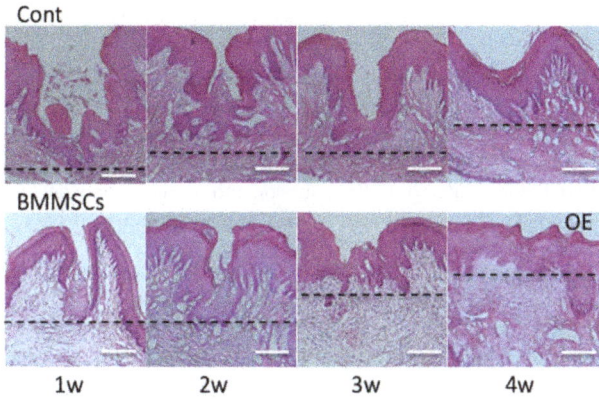

Figure 4. Formation of oral mucosae by injected rat MSCs after tooth extraction. (1w) One week after extraction, a thin epithelial layer extended horizontally on the wound site in the MSC group (B), but not in the control group (A). (2, 3w) After 2 and 3 weeks, in both groups, the OE became mature in structure. In the MSC group, the wound site exhibited the same horizontal height as normal OE. (4w) After 4 weeks, both groups exhibited consolidated OE. Hematoxylin and eosin staining. Bar = 100 μm.

implant soft-tissue inflammation strongly induced the accumulation of MSCs at an early stage.

Moreover, the location of accumulated MSCs was limited to around the apical portion of the PIE, close to connective tissue, in the early stages of wound healing. The localized MSCs were also observed in the connective tissue after tooth extraction (Fig. 3A). Because connective tissue and alveolar bone have more blood vessels than epithelial tissue [35], the inflammatory cells and the transplanted MSCs may easily accumulate in these regions.

Additionally, we determined how long MSCs remained at the inflammation site. The results showed that the isolated MSCs remained for 1 week in the tooth extraction model or 2 weeks in the implantation model. However, MSCs were reported to be present at such sites in the host mouse body for 3 days [36]. This discrepancy may be due to differences in the MSC isolation methods: Wang et al. counted the number of cells in blood, while we analyzed frozen tissue sections directly [36]. Cells accumulating at the injured site may continue to function over a period of time. Figure 3 shows that the transplanted MSCs around the implant remained at the local site twice as long as those around the tooth extraction socket. As shown in our previous study [15], because the recovery of soft tissue around the implant takes much longer than around the tooth extraction socket, the MSCs were maintained for longer in the implantation group compared with the extraction group (Fig. 3B). Therefore, the disappearance of inflammation may release the accumulated MSCs from the titanium surface or the tooth extraction site.

3. Distribution of laminin-332 in the peri-implant oral mucosa

Laminin-332, which mediates the adhesion of basal cells via integrin α6β4, is expressed at the interface between junctional epithelium and natural tooth [22,37] and is predicted to be critical for the attachment of the gingival epithelial cells to substrates [38]. In our previous study, laminin-332 was implicated in the adhesion of the PIE to the dental implant [22]. Therefore, we observed the distribution of laminin-332 during PIE formation around the implant to eliminate the influence of transplanted MSCs on OE.

As previously reported, laminin-332-positive staining was apparent as a band along the implant-PIE interface in most areas except the upper portion [18]. However, the injection of MSCs extended the positive band into this upper portion. Systemic MSC application induced laminin-332 expression by the epithelial cells

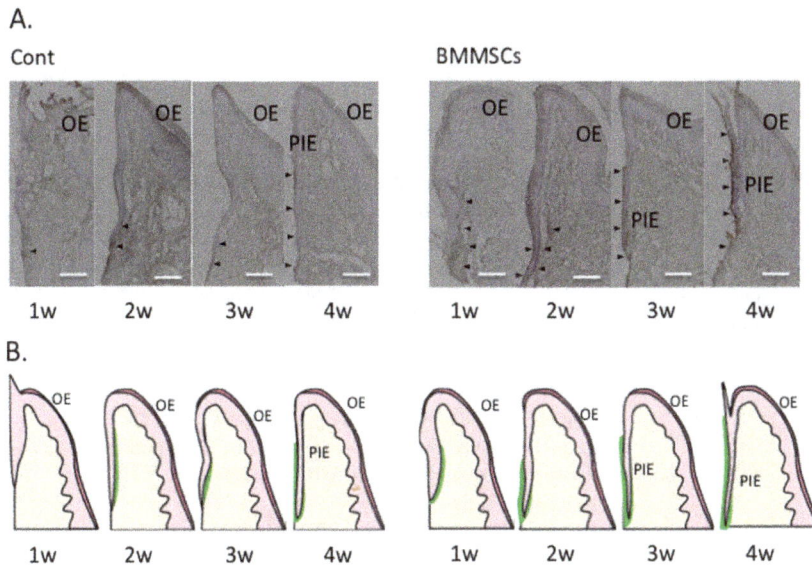

Figure 5. Laminin-332 distribution during the formation of peri-implant epithelium (PIE) following MSC injection. (A) (1w) After 1 week, laminin-332 was expressed in the BM of the new epithelium and in the FC, but not in the epithelial layer facing the implant. After 2 weeks, laminin-332 was intensely expressed in the CT, and the innermost cells of the PIE facing the implant were positive for laminin-332 in the MSC group. (3w) After 3 weeks, a weak positive reaction for laminin-332 was observed in the PIE as a thin band along the implant-PIE interface in the MSC group only. (4w) After 4 weeks, the PIE was completely formed in both groups. Laminin-332 was scarcely expressed along the upper portion of the implant-PIE interface. Laminin-332 was weakly expressed along the BM. Hematoxylin staining. Bar = 100 μm. (B) Lower panels show schematics of these tissue arrangements in the gingiva around the implant. Green lines show laminin-332-positive areas.

A.

Cont   Co-cul   Trans-well

B. Adhesion

C. Proliferation

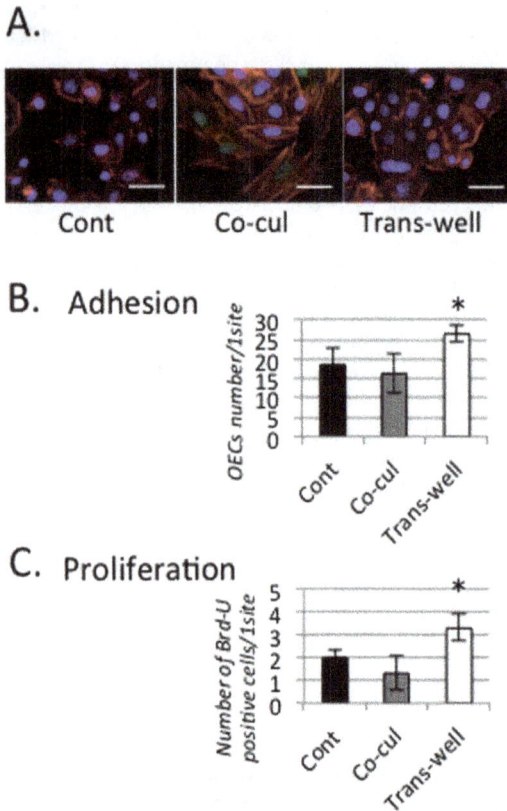

Figure 6. Relationship between MSCs and OECs in co-culture.
(A) Rate of proliferation as determined by Brd-U assay. The rate of OEC proliferation in the trans-well condition was significantly lower than in cells grown with C-medium, but the proliferation rate was markedly increased by indirect culture with MSCs. (bar = 100 μm) (B) Apoptosis analysis by FACS. Apoptosis of OECs was detected and quantified by FACS after annexin V and 7AAD staining. The lower table was arranged from the upper data. The rate of apoptosis in OECs in the Co-cul condition was much higher than in the trans-well condition. Each data point represents the mean ± SD of two parallel experiments. #; $P<0.05$ versus trans-well.

on the dental implant (Fig. 5). MSCs promote tissue regeneration, as indicated in Figure 4, not only by multi differentiation and proliferation, but also by cytokine expression (IGF-1, FGF, PDGF, etc.) to activate cells involved in wound healing [39]. The effect of MSCs on the epithelium after implantation or extraction was long lasting, as shown in Figure 2, although there were a small number of exotic MSCs around the wounded site in the early stages. In contrast, our previous report showed that the expression of laminin-332 on the titanium implant surface was accelerated by the specific growth factor, IGF-1, which promoted PIE formation and improved epithelial sealing around the dental implant [25]. Therefore, we suggest that the MSCs were induced to stimulate tissue regeneration by growth factors in the MSC-injected group at 4 weeks (Fig. 5).

4. Relationship between MSCs and OECs in co-culture conditions

Although it has been shown that MSCs clearly promoted epithelium wound healing and epithelial attachment to the implant surface (Figs. 4 and 5), the mechanism by which the transplanted MSCs activated the epithelial cells is not clear. In Figure 6, the changes in OEC adhesion and proliferation when co-cultured directly or indirectly with MSCs were shown to have a strong relationship with MSCs. As a result, in assays to evaluate the strength of OEC adherence, the adhesion of OECs was accelerated under the indirect co-culture condition (Fig. 6A). Similarly, proliferation of OECs was markedly increased only by indirect co-culture of MSCs and OECs using trans-wells (Fig. 6B). These results appeared to be inconsistent with *in vivo* data showing that injected MSCs accumulated around the implant and promoted epithelial cell attachment to the implant surface. Considering their indirect culture in trans-wells only, MSCs promoted the adhesion and proliferation of OECs on the titanium surface.

Conclusion

In this study, systemic MSC application accelerated OE healing and PIE formation after tooth extraction and implantation, respectively. Additionally, laminin-332 expression at the adhesion structures along the implant-PIE interface was improved by MSC injection. Therefore, systemically applied MSCs may significantly improve the protection of the PIE from peri-implant inflammation.

Supporting Information

Figure S1 Accumulation of GFP-transgenic injected MSCs after implantation. Around the experimental implants, there were many CD-90/CD-44 or CD-90/CD-105 double-positive cells in the mucosa. The location of accumulated MSCs was limited to around the apical portion of the PIE-like epithelial structure. However, fibroblasts isolated from GFP-transgenic rat back skin, which were injected via the tail vein similar to MSCs, did not accumulate at any site. Bar = 100 μm.

Figure S2 Accumulation of GFP-transgenic injected MSCs at various organs. The location of accumulated MSCs, CD-90/GFP double-positive cells, was limited to around the experimental implants. However, almost all injected cells were detected in lung and a few cells were in heart, liver. Bar = 20 μm.

Author Contributions

Conceived and designed the experiments: IA YA TY YT KK. Performed the experiments: IA YA. Analyzed the data: IA RK. Contributed reagents/materials/analysis tools: IA YM. Wrote the paper: IA AF.

References

1. Brånemark PI, Adell R, Albrektsson T, Lekholm U, Lundkvist S, et al. (1983) Osseointegrated titanium fixtures in the treatment of edentulousness. Biomaterials 4: 25–28.
2. Buser D, Weber HP, Donath K, Fiorellini JP, Paquette DW, et al. (1992) Soft tissue reactions to non-submerged unloaded titanium implants in beagle dogs. J Periodontol 63: 225–235.
3. Gould TR, Westbury L, Brunette DM (1984) Ultrastructural study of the attachment of human gingiva to titanium in vivo. J Prosthet Dent 52: 418–420.
4. Schroeder A, van der Zypen E, Stich H, Sutter F. (1981) The reactions of bone, connective tissue, and epithelium to endosteal implants with titanium-sprayed surfaces. J Maxillofac Surg 9: 15–25.

5. Furuhashi A, Ayukawa Y, Atsuta I, Okawachi H, Koyano K (2012) The difference of fibroblast behavior on titanium substrata with different surface characteristics. Odontology 100: 199–205.

6. Ikeda H, Yamaza T, Yoshinari M, Ohsaki Y, Ayukawa Y, et al. (2000) Ultrastructural and immunoelectron microscopic studies of the peri-implant epithelium-implant (Ti-6Al-4V) interface of rat maxilla. J Periodontol 71: 961–973.

7. Friedenstein AJ, Deriglasova UF, Kulagina NN, Panasuk AF, Rudakowa SF, et al. (1974) Precursors for fibroblasts in different populations of hematopoietic cells as detected by the in vitro colony assay method. Exp Hematol 2: 83–92.

8. Gronthos S, Mankani M, Brahim J, Robey PG, Shi S. (2000) Postnatal human dental pulp stem cells (DPSCs) in vitro and in vivo. Proc Natl Acad Sci U S A. 97: 13625–13630.

9. Lee OK, Kuo TK, Chen WM, Lee KD, Hsieh SL, et al. (2004) Isolation of multipotent mesenchymal stem cells from umbilical cord blood. Blood 103: 1669–1675.

10. Zuk PA, Zhu M, Mizuno H, Huang J, Futrell JW, et al. (2001) Multilineage cells from human adipose tissue: implications for cell-based therapies. Tissue Eng 7: 211–228.

11. Dominici M, Le Blanc K, Mueller I, Slaper-Cortenbach I, Marini F, et al. (2006) Minimal criteria for defining multipotent mesenchymal stromal cells. The International Society for Cellular Therapy position statement. Cytotherapy 8: 315–317.

12. Zhang R, Liu Y, Yan K, Chen L, Chen XR, et al. (2013) Anti-inflammatory and immunomodulatory mechanisms of mesenchymal stem cell transplantation in experimental traumatic brain injury. J Neuroinflammation 10: 106.

13. Le Blanc K, Frassoni F, Ball L, Locatelli F, Roelofs H, et al. (2008) Mesenchymal stem cells for treatment of steroid-resistant, severe, acute graft-versus-host disease: a phase II study. Lancet 371: 1579–1586.

14. García-Olmo D, García-Arranz M, Herreros D, Pascual I, Peiro C, et al. (2004) A phase I clinical trial of the treatment of Crohn's fistula by adipose mesenchymal stem cell transplantation. Dis Colon Rectum. 48: 1416–1423.

15. Atsuta I, Yamaza T, Yoshinari M, Mino S, Goto T, et al. (2005) Changes in the distribution of laminin-5 during peri-implant epithelium formation after immediate titanium implantation in rats. Biomaterials 26: 1751–1760.

16. Tarnawski AS (2005) Cellular and molecular mechanisms of gastrointestinal ulcer healing. Dig Dis Sci 50 Suppl 1 S24–33.

17. Lindhe J, Berglundh T (1998) The interface between the mucosa and the implant. Periodontology 2000 17: 47–54.

18. Atsuta I, Ayukawa Y, Ogino Y, Moriyama Y, Jinno Y, et al. (2012) Evaluations of epithelial sealing and peri-implant epithelial down-growth around "step-type" implants. Clin Implant Dent Relat Res 223: 459–466.

19. Takano T, Li YJ, Kukita A, Yamaza T, Ayukawa Y, et al. (2014) Mesenchymal stem cells markedly suppress inflammatory bone destruction in rats with adjuvant-induced arthritis. Lab Invest in press

20. Yamaza T, Miura Y, Bi Y, Liu Y, Akiyama K, et al. (2008) Pharmacologic stem cell based intervention as a new approach to osteoporosis treatment in rodents. PloS one 3: e2615.

21. Métrailler-Ruchonnet I, Pagano A, Carnesecchi S, Ody C, Donati Y, et al. (2007) Bcl-2 protects against hyperoxia-induced apoptosis through inhibition of the mitochondria-dependent pathway. Free Radic Biol Med 42: 1062–1074.

22. Atsuta I, Yamaza T, Yoshinari M, Goto T, Kido MA, et al. (2005) Ultrastructural localization of laminin-5 (gamma2 chain) in the rat peri-implant oral mucosa around a titanium-dental implant by immuno-electron microscopy. Biomaterials 26(32):6280–6287.

23. Ikeda H, Shiraiwa M, Yamaza T, Yoshinari M, Kido MA, et al. (2002) Difference in penetration of horseradish peroxidase tracer as a foreign substance into the peri-implant or junctional epithelium of rat gingivae. Clin Oral Implants Res 13: 243–251.

24. Atsuta I, Ayukawa Y, Furuhashi A, Ogino Y, Moriyama Y, et al. (2013) In Vivo and In Vitro Studies of Epithelial Cell Behavior around Titanium Implants with Machined and Rough Surfaces. Clin Oral Implants Res.

25. Atsuta I, Ayukawa Y, Furuhashi A, Yamaza T, Tsukiyama Y, et al. (2013) Promotive effect of insulin-like growth factor-1 for epithelial sealing to titanium implants. J Biomed Mater Res A. In press

26. Shiraiwa M, Goto T, Yoshinari M, Koyano K, Tanaka T (2002) A study of the initial attachment and subsequent behavior of rat oral epithelial cells cultured on titanium. J Periodontol 73: 852–860.

27. Okawachi H, Ayukawa Y, Atsuta I, Furuhashi A, Sakaguchi M, et al. (2012) Effect of titanium surface calcium and magnesium on adhesive activity of epithelial-like cells and fibroblasts. Biointerphases 7: 27.

28. Egusa H, Sonoyama W, Nishimura M, Atsuta I, Akiyama K (2012) Stem cells in dentistry–part I: stem cell sources. J Prosthodont Res 56: 151–165.

29. Sonoyama W, Liu Y, Yamaza T, Tuan RS, Wang S, et al. (2008) Characterization of the apical papilla and its residing stem cells from human immature permanent teeth: a pilot study. J Endod 34: 166–171.

30. Belema-Bedada F, Uchida S, Martire A, Kostin S, Braun T (2008) Efficient homing of multipotent adult mesenchymal stem cells depends on FROUNT-mediated clustering of CCR2. Cell Stem Cell. 2: 566–575.

31. Esposito M, Thomsen P, Molne J, Gretzer C, Ericson LE, et al. (1997) Immunohistochemistry of soft tissues surrounding late failures of Branemark implants. Clin Oral Implants Res 8: 352–366.

32. Lee RH, Pulin AA, Seo MJ, Kota DJ, Ylostalo J, et al. (2009) Intravenous hMSCs improve myocardial infarction in mice because cells embolized in lung are activated to secrete the anti-inflammatory protein TSG-6. Cell Stem Cell 5: 54–63.

33. Akiyama K, Chen C, Gronthos S, Shi S (2012) Lineage differentiation of mesenchymal stem cells from dental pulp, apical papilla, and periodontal ligament. Methods Mol Biol 887: 111–121.

34. Atsuta I, Liu S, Miura Y, Akiyama K, Chen C, et al. (2013) Mesenchymal stem cells inhibit multiple myeloma cells via the Fas/Fas ligand pathway. Stem Cell Res Ther 4: 111.

35. Schroeder HE (1986) Healing and regeneration following periodontal treatment. Dtsch Zahnarztl Z 41: 536–538.

36. Wang L, Zhao Y, Liu Y, Akiyama K, Chen C, et al. (2013) IFN-gamma and TNF-alpha Synergistically Induce Mesenchymal Stem Cell Impairment and Tumorigenesis via NFkappaB Signaling. Stem cells in press

37. Shimono M, Ishikawa T, Enokiya Y, Muramatsu T, Matsuzaka K, et al. (2003). Biological characteristics of the junctional epithelium. J Electron Microsc (Tokyo) 52: 627–639.

38. Tamura RN, Oda D, Quaranta V, Plopper G, Lambert R, et al. (1997) Coating of titanium alloy with soluble laminin-5 promotes cell attachment and hemidesmosome assembly in gingival epithelial cells: potential application to dental implants. J Periodontal Res 32: 287–294.

39. Chen WL, Chang HW, Hu FR (2008) In vivo confocal microscopic evaluation of corneal wound healing after epi-LASIK. Invest Ophthalmol Vis Sci 4: 2416–2423.

18

Hematopoietic Stem Cells in Neonates: Any Differences between Very Preterm and Term Neonates?

Lukas Wisgrill[1], Simone Schüller[1], Markus Bammer[1], Angelika Berger[1], Arnold Pollak[1], Teja Falk Radke[2], Gesine Kögler[2], Andreas Spittler[3], Hanns Helmer[4], Peter Husslein[4], Ludwig Gortner[5]*

1 Dept. of Pediatrics and Adolescent Medicine, Division of Neonatology, Paediatric Intensive Care & Neuropaediatrics, Medical University of Vienna, Vienna, Austria, 2 Institute for Transplantation Diagnostics and Cell Therapeutics, Heinrich Heine University Medical Center, Duesseldorf, Germany, 3 Department of Surgery, Research Labs & Core Facility Flow Cytometry, Medical University of Vienna, Vienna, Austria, 4 Dept. of Obstetrics and Gynecology, Medical University of Vienna, Vienna, Austria, 5 Dept. of Pediatrics and Neonatology, Saarland University, Homburg, Saar, Germany

Abstract

Background: In the last decades, human full-term cord blood was extensively investigated as a potential source of hematopoietic stem and progenitor cells (HSPCs). Despite the growing interest of regenerative therapies in preterm neonates, only little is known about the biological function of HSPCs from early preterm neonates under different perinatal conditions. Therefore, we investigated the concentration, the clonogenic capacity and the influence of obstetric/perinatal complications and maternal history on HSPC subsets in preterm and term cord blood.

Methods: CD34+ HSPC subsets in UCB of 30 preterm and 30 term infants were evaluated by flow cytometry. Clonogenic assays suitable for detection of the proliferative potential of HSPCs were conducted. Furthermore, we analyzed the clonogenic potential of isolated HSPCs according to the stem cell marker CD133 and aldehyde dehydrogenase (ALDH) activity.

Results: Preterm cord blood contained a significantly higher concentration of circulating CD34+ HSPCs, especially primitive progenitors, than term cord blood. The clonogenic capacity of HSPCs was enhanced in preterm cord blood. Using univariate analysis, the number and clonogenic potential of circulating UCB HSPCs was influenced by gestational age, birth weight and maternal age. Multivariate analysis showed that main factors that significantly influenced the HSPC count were maternal age, gestational age and white blood cell count. Further, only gestational age significantly influenced the clonogenic potential of UCB HSPCs. Finally, isolated CD34+/CD133+, CD34+/CD133– and ALDH^{high} HSPC obtained from preterm cord blood showed a significantly higher clonogenic potential compared to term cord blood.

Conclusion: We demonstrate that preterm cord blood exhibits a higher HSPC concentration and increased clonogenic capacity compared to term neonates. These data may imply an emerging use of HSPCs in autologous stem cell therapy in preterm neonates.

Editor: Jörn-Hendrik Weitkamp, Vanderbilt University, United States of America

Funding: This study was funded by the University Saarland (HOMFOR2011). The funders had no role in study design, data collection and analysis, decision to publish, or preparation of the manuscript. No additional external funding received for this study.

Competing Interests: The authors have declared that no competing interests exist.

* Email: Ludwig.Gortner@uniklinikum-saarland.de

Introduction

Umbilical cord blood (UCB) is a rich source of hematopoietic stem and progenitor cells (HSPCs). During the past decades, human cord blood of term neonates has been established as a potential source for HSPC transplantation [1]. Therefore, many studies focus on hematopoietic and immunological features of HSPCs in term newborns.

In normal human development, fetuses gradually mature to adapt to extrauterine conditions. Premature birth is often associated with obstetric and perinatal complications, interrupting the physiologic development process and resulting in organ damage or dysfunction. The role of HSPCs in physiologic and pathophysiologic human development still remains uncertain. Hematopoietic stem and progenitor cells, as assessed by the expression of CD34, are capable of differentiating into non-hematopoietic cells such as microglia [2], hepatocytes [3], and type II alveolar pneumocytes [4]. These findings may indicate a supporting role of HSPCs in the intrauterine development.

The utilization of term UCB has widely become an easily available and acceptable alternative for stem cell transplantation of hematological and non-hematological disorders. Preterm birth is a major determinant of neonatal mortality and morbidity and is associated with severe complications including bronchopulmonary dysplasia (BPD), white matter injury and intracranial hemorrhage [5]. In the last decades, the potential of non-oncologic stem cell and mononuclear cell therapies have been investigated for the regeneration of impaired organ development and tissue regeneration [6–10]. In clinical settings, the infusion of autologous UCB in infants with neurologic disorders seems feasible and safe [11,12].

Double-blind randomized studies are needed to evaluate the therapeutic benefit of autologous UCB transfusion in neonates.

Despite the growing interest of regenerative medicine in preterm neonates [13], less is known about the biological properties of HSPCs obtained from preterm cord blood (PCB). Thus, we aimed to investigate the number and clonogenic capacity of circulating CD34+ HSPCs subsets in PCB and term cord blood (TCB) and the influence of obstetric and maternal history on these subsets. Further, we determine the clonogenic capacity of isolated HSPC subsets of PCB and TCB.

Materials and Methods

Study population

Sixty newborns were enrolled in the study between February and August 2013. Very preterm infants (n = 30; 24–32 weeks of gestational age (GA)) were compared with term newborns (n = 30; 38–42 weeks of GA). The study was approved by the ethics committee of the Medical University of Vienna and written informed consent was obtained from the parents before birth. Heparinized whole blood obtained from umbilical cord was collected immediately after cesarean section and processed within six hours after collection. Maternal history as well perinatal and neonatal variables were documented until discharge from hospital.

Definition of clinical parameters

Prolonged rupture of membranes (PROM) was defined as rupture of membranes ≥18 h prior to delivery. Chorioamnionitis was defined as described previously [14], based on the presence of maternal fever ≥38°C with two or more of the following criteria: maternal leucocytosis (>15 000 cells/mm^3), maternal (>100 beats/min) or fetal (>160 beats/min) tachycardia, uterine tenderness and/or foul odor of the amniotic fluid. Early labour was defined as onset of labour before caesarean section. Small-for-gestational-age (SGA) was defined as birth weight below the 10th percentile using Fenton growth charts [15]. Antenatal corticosteroid administration was defined as any intramuscular administration of betamethasone ≥24 h prior to delivery. Tocolysis was defined by the administration of the oxytocin antagonist Atosiban or the β_2 adrenergic agonist Hexoprenalin within 12 hours of delivery. Gestational diabetes mellitus (GDM) was diagnosed by blood sugar levels of >126 mg/dl after fasting, or a pathological oral glucose tolerance test. Preeclampsia was defined by high maternal blood pressure (>140 mmHg systolic or >90 mmHg diastolic) and proteinuria (>300 mg/24 h).

Enumeration of umbilical cord HSPCs

The enumeration of HSPC was conducted in whole UCB using BD Trucount tubes (BD Biosciences, San Jose, CA, USA), according to the manufacturer's instruction. 100 µl whole UCB were added by reverse pipetting to BD Trucount tubes and stained with anti- CD45− FITC, anti- CD34− PE, anti- CD38− PE-Cy7, anti- CD10− PE- CF594, 7-AAD (all BD Biosciences) and anti- CD133− APC (Miltenyi Biotechnology, Bergisch, Gladbach, Germany). After immunofluorescence staining, erythrocytes were lysed with ammonium chloride lysis solution and analyzed within one hour by flow cytometry (LSRII, BD Biosciences) and analysed using FlowJo software (TreeStar Inc., Ashland, OR, USA). For enumeration of HSPC in UCB, CD34+ cells were gated according to the modified ISHAGE criteria (CD45^{dim}/7-AAD-/CD34+ cells). CD10 was added to exclude B- lymphoid progenitors. The number of cells/µl was calculated as [(# gated cells/# acquired beads) * (# of beads per test/test volume).

Progenitor cell isolation according to common stem cell markers or aldehyde dehydrogenase activity

UCB mononuclear cells (MNCs) were isolated from fresh heparinized cord blood by Ficoll gradient centrifugation (Ficoll-Paque PLUS; Amersham, GE Healthcare Life Sciences, Little Chalfont, Buckinghamshire, UK) and remaining erythrocytes were lysed with ammonium chloride lysis solution (BD Biosciences). 1×10^6 MNCs were subsequently stained with anti- CD45− FITC, anti- CD34− PE, 7-AAD (BD Biosciences) and anti- CD133− APC (Miltenyi Biotec) for isolation of CD34+/CD133+ and CD34+/CD133− HSC subpopulations.

Another MNC aliquot was assayed for aldehyde dehydrogenase (ALDH) activity using Aldefluor reagent (StemCell Technologies, Marseille, France) according to the manufacturer's instructions. Isolated cells were resuspended in aldefluor assay buffer and an appropriate amount of aldefluor substrate was added to 1×10^6 MNCs. Cells were incubated for 30 minutes at 37°C in a waterbath for conversion of substrate to a fluorescent product. An aliquot of Aldefluor- stained cells were immediately treated with diethylaminobenzaldehyde (DEAB), a specific ALDH inhibitor, to serve as negative control. For evaluating clonogenic capacity, SSC^{low}/ALDH^{high} and SSC^{low}/ALDH^{low} cells were selected by FACS. For all samples, 400 cells were directly sorted into 100 µl IMDM using a MoFlo Astrios flow cytometer cell sorter (Beckman Coulter, Brea, CA, USA). For phenotypic analysis of SSC^{low}/ALDH^{high} cells, aldefluor- substrate labeled MNCs were costained with anti- CD34 PE and anti- CD133 APC (Miltenyi Biotechnology) and subsequently analyzed on a LSR II (BD Biosciences).

Clonogenic progenitor assay

Purified HSC population were resuspended into 900 µl semisolid methylcellulose medium (Methocult H4334, StemCell Technologies) and plated in triplicates on 24- well plates with a seeding density of 100 HSPC/well. For whole blood progenitor assays, 100 µl UCB was lysed with ammonium chloride lysis solution, centrifuged and resuspended in IMDM. An appropriate amount of lysed whole blood was added to 900 µl methylcellulose medium and plated in triplicates reaching a seeding density of 100 HSPC/well. Cells were incubated in humidified atmosphere (37°C/5% CO$_2$) for 14 days and colonies were evaluated for number and morphology by light microscopy.

Statistical analysis

Statistical analysis was performed with IBM SPSS Statistics 21. Kolmogorov- Smirnov test was performed to prove normal distribution. Student's T- Test was used to analyze normal distributed data. In case of non- normal distribution, data were further analyzed using Mann- Whitney- U test. Bonferroni correction for multiple testing was used. Differences of quantitative variables were compared by Mann- Whitney U test and qualitative variables with Fisher's exact test. Correlation between variables was determined using Spearman's rank correlation coefficient. Multiple linear backward regression was performed to evaluate the effects of obstetric, perinatal and neonatal factors on the UCB HSC count. A p<0.05 was considered as statistically significant.

Results

Clinical characteristics of preterm and term newborns

Obstetric history as well perinatal and neonatal clinical variables of preterm and term newborns are summarized in Table 1. Mean difference in GA between preterm and term neonates was 9.53 weeks. Preterm neonates had a significantly lower mean birth

Table 1. Obstretic, perinatal and neonatal clinical parameters.

Parameter	Preterm newborns (n = 30)	Term newborns (n = 30)	p-value
Maternal age	32.73±7.09 (32)	34.03±6.48 (33.5)	0.403
No. of Gravidity	2.20±2.79 (1)	2.17±0.98 (2)	0.189
No. of Parity	1.57±0.81 (1)	1.8±0.84 (2)	0.619
Maternal diabetes	3 (10%)	5 (16.7%)	0.706
Smoking	1 (3.3%)	4 (13.3%)	0.353
Preeclampsia	4 (13.3%)	0 (0%)	0.112
Chorioamnionitis	9 (30%)	0 (0%)	0.002
Gestational age (wks)	29.63±2.70 (30.14)	39.16±1.17 (38.92)	<0.0001
Birth weight (g)	1354.77±553,80 (1318)	3392.77±489.06 (3360)	<0.0001
SGA	5 (16.7%)	0 (0%)	0.052
Male Gender	16 (53.3%)	19 (63.3%)	0.601
5- min APGAR	8.47±0.93 (8)	9.87±0.43 (10)	<0.0001
10- min APGAR	8.87±0.90 (9)	9.97±0.18 (10)	<0.0001
Antenatal corticosteroids	27 (90%)	0 (0%)	<0.0001
Labour before cesarean section	13 (43.3%)	2 (6.7%)	0.002
Tocolysis	13 (43.3%)	0 (0%)	<0.0001
PROM	14 (46.7%)	0 (0%)	<0.0001

Data are shown as mean ± SD (median) or n (%), SGA: small-for-gestational-age; PROM: prolonged rupture of membranes.

weight (1354.77±553.80 g vs. 3392.77±489.06 g) including 10 infants of extremely low birth weight (ELBW) (755.80±163.61 g), 7 of very low birth weight (VLBW) (1224.14±113.45 g), and 13 of low birth weight (LBW) (1885.85±316.56 g) compared to term newborns (3392.77±489.06 g). Early labour, chorioamnionitis, tocolysis, antenatal steroids, PROM and APGAR scores at 5 and 10 min were statistically significantly different between both groups, whereas maternal age, number of pregnancies, GDM, preeclampsia, SGA and smoking were not statistically different. Both groups were matched by gender. The most abundant obstetric complication in the preterm group was PROM (46.7%) followed by chorioamnionitis (16.7%) and preeclampsia (13.3%). One preterm neonate was diagnosed with meconium peritonitis.

Preterm newborns display a higher number of circulating cord blood CD34+ HSPC than term newborns

The number of WBC/ml in UCB was significantly higher in TCB than in PCB. However, the concentration of circulating CD34+ HSPCs in UCB was significantly higher in PCB than in TCB (Figure 1A). The proportion of CD34+/CD133– HSPCs was significantly higher in PCB than in TCB, whereas the proportion of CD34+/CD133+ HSPCs was higher in TCB. Furthermore, PCB showed a higher percentage of CD34+/CD38– HSPCs compared to TCB (Table 2). The percentage of CD34+/CD38−/CD133+ HSPCs was not significantly higher in PCB than TCB. CD34+ HSPC count was not statistically different between ELBW, VLBW and LBW infants, but was higher in ELBW infants compared to term newborns (Figure 1B). In the univariate analysis maternal age, gestational age and birth weight significantly correlated with the number of CD34+ HSPCs in both groups. Maternal age positively correlated with circulating CD34+ HSPCs in PCB (Fig. 1C), whereas significant inverse correlations were found in TCB (Fig. 1D). Furthermore, GA (Fig. 1E) and birth weight (Fig. 1F) negatively correlated with CD34+ HSPC in UCB. The percentage of CD34+/CD133+ cells in CB positively correlated with GA and birth weight of infants. Multiple linear

Table 2. HSPC subsets and WBC count of preterm and term cord blood.

	PCB (n = 30)	TCB (n = 30)	p-value
Number of CD45+ WBC ($\times10^6$/ml)	6.5±3.1	9.4±2.8	<0.0001
Number of CD34+ ($\times10^4$/ml)	4.3±3.1	2.3±1.1	<0.01
Percent of CD34+/CD38+/CD133+ (%)	62.18±10.21	73.14±5.68	<0.0001
Percent of CD34+/CD38+/CD133– (%)	37.82±10.21	26.86±5.68	<0.0001
Percent of CD34+/CD38– (%)	17.08±6.52	12.78±6.18	0.008
Percent of CD34+/CD38−/CD133+ (%)	80.32±12.10	70.38±22.76	0.051
Percent of CD34+/CD38−/CD133– (%)	19.68±12.10	29.62±22.76	0.051

PCB: preterm cord blood; TCB: term cord blood; WBC: white blood cells; HSPC: hematopoietic stem and progenitor cells.

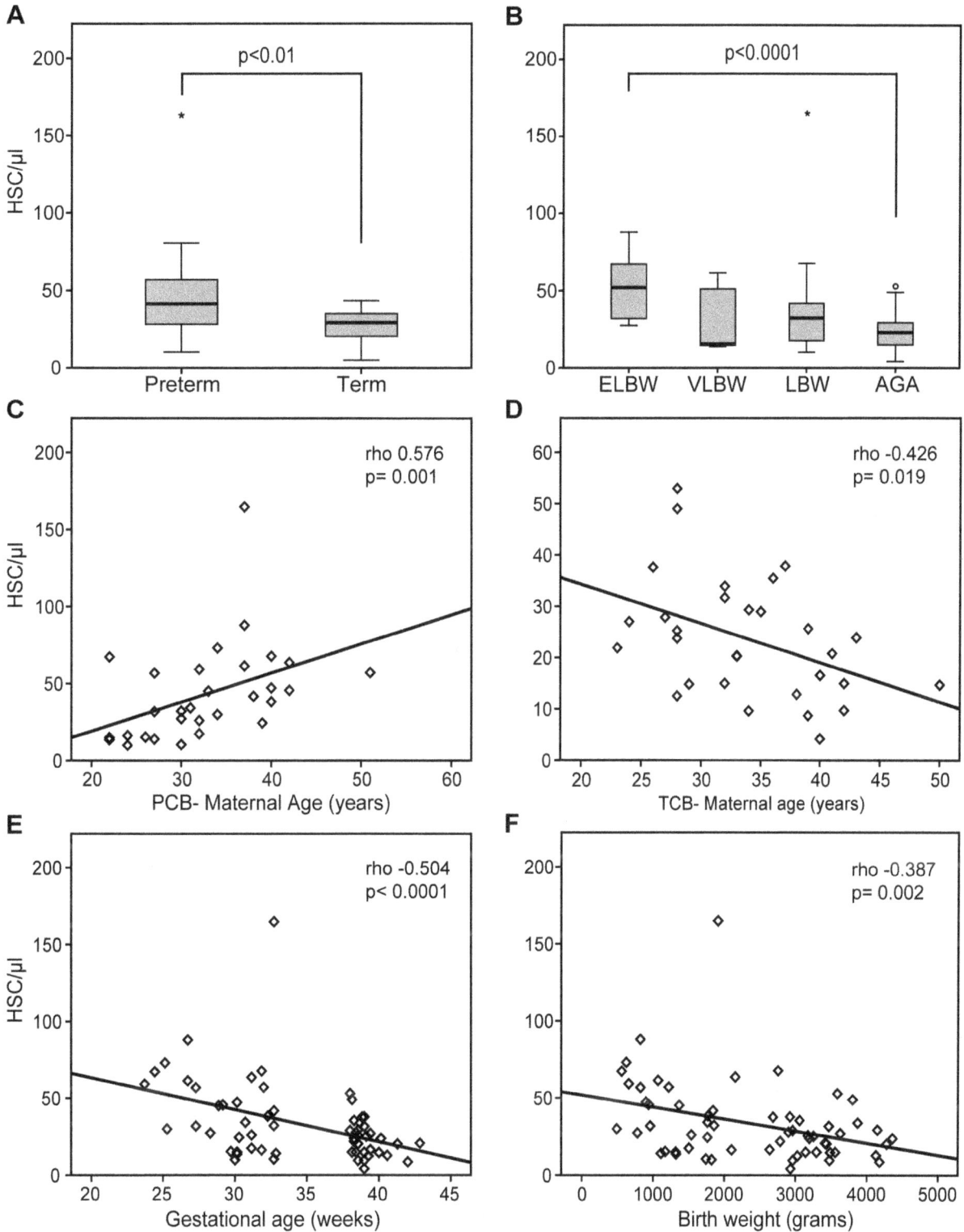

Figure 1. HSPC count and correlation with clinical parameters. Number of circulating HSPCs in umbilical cord blood of preterm and term neonates (A) and in ELBW, VLBW, LBW and appropriate for gestational age (AGA) infants (B). Correlation between number of HSPCs and maternal age of PCB (C) and TCB (D), gestational age (E) and birth weight (F).

Table 3. Clonogenic capacity of HSPCs of preterm and term cord blood.

		PCB	TCB	p-value
LWBA	No. of colonies	96.13±10.81	59.80±11.49	<0.0001
	No. of BFU- E	67.27±10.05	38.33±11.97	<0.0001
	No. of CFU- GM	28.87±12.11	21.47±8.21	0.086
CD34+/CD133+	No. of colonies	101.33±3.51	51.67±16.50	0.007
	No. of BFU- E	58.00±7.54	31.67±7.64	0.013
	No. of CFU- GM	43.34±6.43	20.00±8.89	0.021
CD34+/CD133–	No. of colonies	57.00±6.00	21.33±4.51	0.001
	No. of BFU- E	44.67±4.16	14.33±2.08	<0.0001
	No. of CFU- GM	12.33±4.16	7.00±2.65	0.134
ALDH^{high}	No. of colonies	85.34±5.51	53.34±7.09	0.004
	No. of BFU- E	61.00±5.00	35.00±5.57	0.004
	No. of CFU- GM	24.34±5.03	18.34±1.53	0.119

Clonogenic potential of HSCs in the lysed whole blood assay (LWBA, n = 15) and sorted HSC subpopulations (n = 3).

backward regression showed that main factors that significantly influence the HSPC count were maternal age ($p<0.01$), gestational age ($p<0.001$) and white blood cell count ($p<0.0001$).

Higher in- Vitro clonogenic capacity of PCB derived HSPCs compared to TCB

In our lysed whole blood clonogenic progenitor assay (LWBA), the clonogenic capacity of CD34+ HSPCs was higher in PCB than TCB (Fig. 2A). The number of erythrocyte burst- forming units (BFU- E) was significantly higher in PCB than in TCB. However, there was no statistically significant difference in the number of granulocyte-macrophage colony-forming units (CFU-GM) of PCB and TCB (Fig. 2B) (Table 3). In the univariate analysis gestational age, birth weight and maternal age correlated with the clonogenic capacity of UCB HSPCs. GA (Fig. 2C) and birth weight (Fig. 2D) negatively correlated with the total number of colonies and BFU-E. Furthermore, maternal age positively correlated with clonogenic capacity in PCB (Fig. 2E), whereas significant inverse correlations were found in TCB (Fig. 2F). Interestingly, the number of BFU-E in PCB (rho 0.755; p = 0.001) also positively correlated with the maternal age. Using multiple linear backward regressions, only gestational age ($p<0.012$) significantly influenced the clonogenic capacity of UCB HSPCs.

Isolated CD34+/CD133+, CD34+/CD133– and ALDH^{high} HSPCs of PCB exhibit higher clonogenic capacity than in TCB

To evaluate the higher clonogenic capacity of PCB, we isolated HSPCs according to common stem cell markers (CD34+/CD133+, CD34+/CD133–) and ALDH activity (Fig. 3). Isolated CD34+/CD133+ HSPCs of PCB displayed a significant higher clonogenic capacity compared to TCB. The percentage of CFU- GM and BFU-E did not differ significantly between PCBs and TCBs. Furthermore, CD34+/CD133– HSPCs exhibited a higher clonogenic capacity when isolated from PCB as opposed to TCB. Interestingly, CD34+/CD133– HSPCs had a higher potential to differentiate into BFU- E than CFU- GM in both groups. UCB HSPCs sorted according to their ALDH activity displayed a similar clonogenic capacity and BFU-E/CFU- GM differentiation pattern as the lysed whole blood assay. The clonogenic capacity of ALDH^{high} HSPCs was higher in PCB than TCB. ALDH^{high} cells of PCB and TCB showed a similar

proportion of CD133+ HSPCs. ALDH^{low} cells, serving as negative control, showed hardly any clonogenic potential in both groups. Although the number of BFU- E was significantly higher in PCB than in TCB, there was no significant difference in the number of CFU- GM (Table 3).

Discussion

Many other studies examined the impact of obstetric, perinatal and neonatal factors in term newborns, showing an influence of gestational age, birth weight, maternal age, small for-gestational-age, preeclampsia, delivery mode and fetal distress [16–23]. However, those studies investigated interfering factors on term cord blood HSPC count. In our study, we found differences in CD34+ HSPC concentrations and in vitro clonogenic capacity in preterm compared with term cord blood cells and, using multiple linear backward regression, three predictive factors influencing the count and in vitro clonogenic capacity: (1) gestational age, (2) white blood cell count and (3) maternal age.

The gestational age of neonates is a consistent predictor for both HSPC concentration and clonogenic capacity. Gasparoni et al. [18] reported a gestational age-dependent decrease of concentration and clonogenic potential of CD34+ HSPCs, which is in line with our results. The HSPC concentration in preterm cord blood samples might be influenced by individual occurring fetal stress during premature birth. Acute fetal stress seems to induce the release of cytokines resulting in higher HSPC counts [22,23]. An approach to explain the higher clonogenic potential of HSPCs from PCB might be the fluctuating cytokine and chemokine levels in neonates that reflect the transfer of hematopoiesis from liver to bone marrow, which may vary in certain preterm newborns [17].

Cord blood from term newborns exhibit a higher WBC count than those from preterm newborns [24]. Although TCB possesses a higher WBC count, we measured a fewer HSPC concentration compared to preterm cord blood, which is in contrast to other studies [21,25]. This might be explained by different used enumeration methods. In our study, we used a lyse-no-wash whole blood assay without prior cell separation or enrichment step minimizing cell loss.

To date, the influence of maternal age on cord blood HSPCs is disputable. As described by others [16,24,26], maternal age had no impact on the HSPC concentration. In contrast, other groups

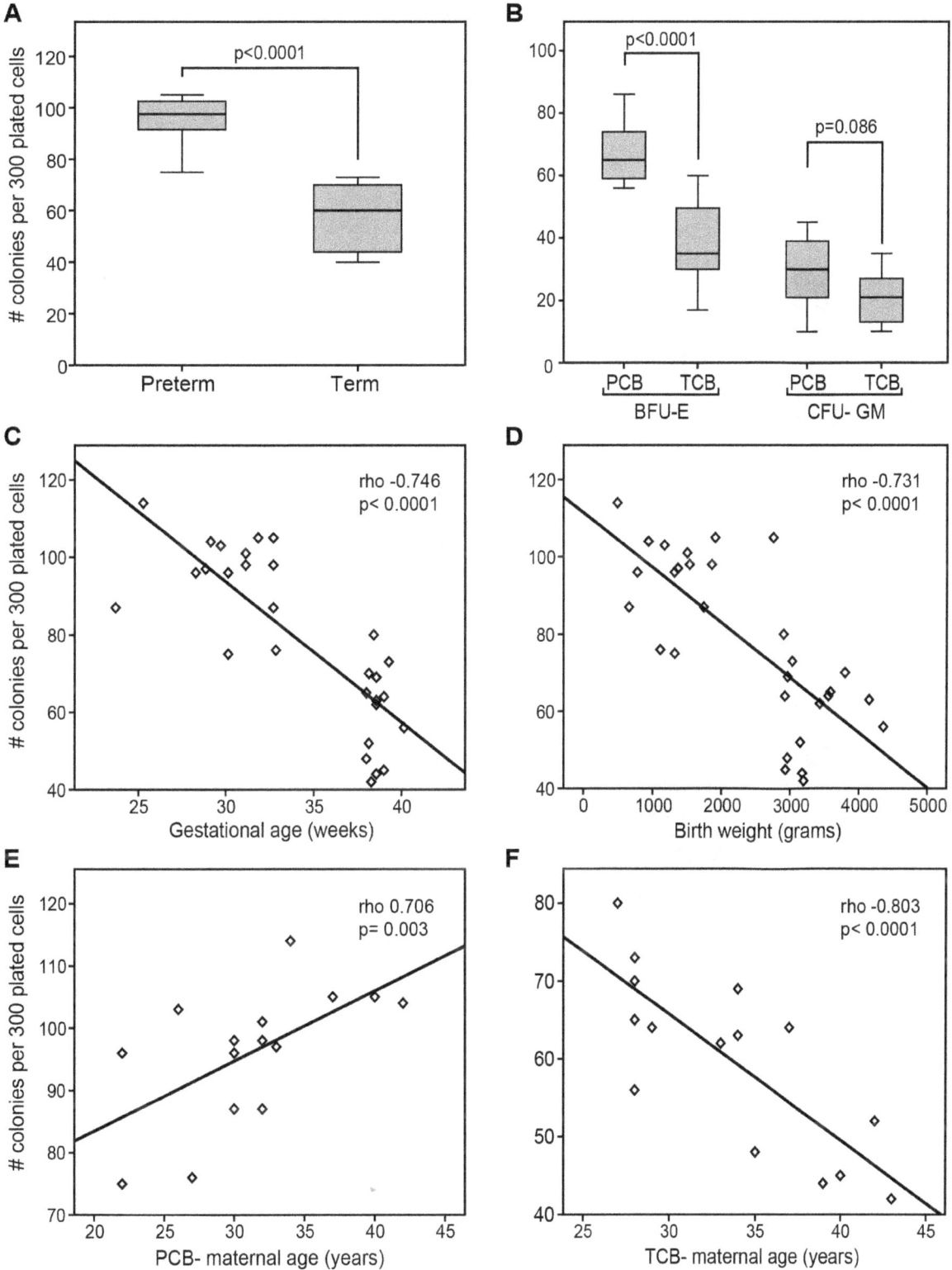

Figure 2. Clonogenic capacity and correlation with clinical parameter. Total clonogenic capacity of PCB and TCB (A) and potential to differentiate into CFU- GM and BFU-E (B). Correlation between total number of colonies and gestational age (C), birth weight (D), maternal age of PCB (E) and TCB (F). Data are presented as number of colonies per 300 plated HSPCs.

reported an influence of maternal age on HSPC concentration, which is in line with our results [19,27]. In addition, the univariate analysis shows a maternal age-dependent influence on the

clonogenic capacity. The etiology for this finding is unclear. Speculative hypotheses include alternating fetal hormone levels during pregnancy [28]. Interestingly, we found no clinical

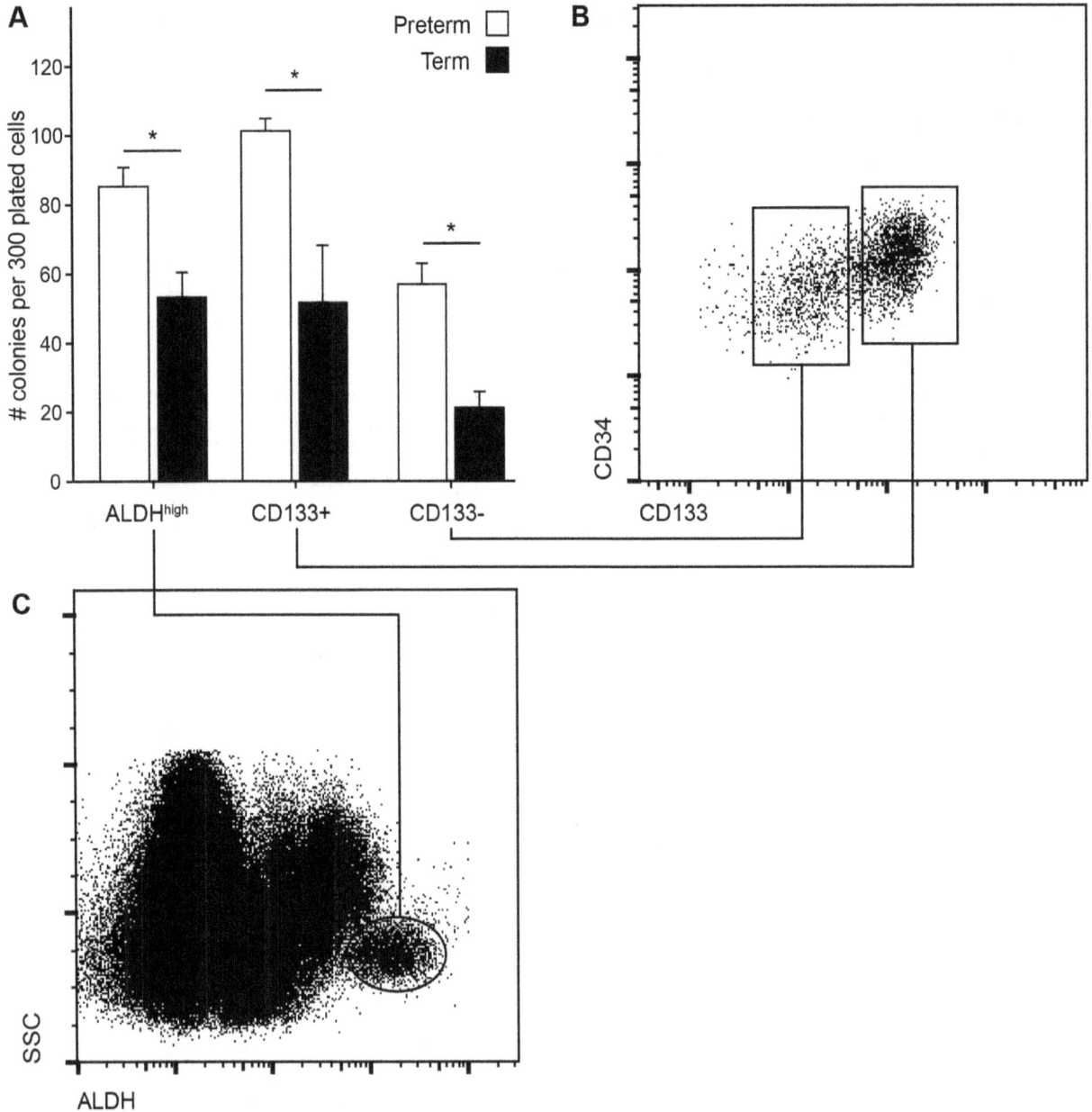

Figure 3. Clonogenic capacity of isolated HSPC subsets. Isolated HSPCs from preterm neonates displayed higher clonogenic capacity compared to term neonates (A). UCB MNCs were either stained to stem cell markers CD34 and CD133 (B) or ALDH activity (C). Data are presented as number of colonies per 300 plated HSPCs.

correlation of premature birth associated morbidities on HSPC count and clonogenic capacity. In our study, preeclampsia and small-for-gestational-age did not affect the cord blood HSPC concentration as described by other groups [20,29] and may be due to our small sample size. Although tocolysis with Atosiban had no effect on the HSPC population, the influence of magnesium sulfate on CD34+ cells, which is widely used as tocolytic drug in the U.S., is still unknown.

In 1997, Yin et al. [30] identified the novel stem cell marker CD133 restricted to a subset of CD34+ HSPC with long-term repopulating ability [31]. The indicated role of HSPC subsets in tissue repair and the sufficient isolation from UCB may suggest a therapeutic capacity of different HSPC subsets in regenerative medicine. Taguchi et al. [32] showed that the systemic adminis-

tration of CD34+ HSPCs promote the neovascularization and enhance the neurogenesis in a mouse stroke model. Especially, CD133+ HSPCs exhibit a high potential to regenerate ischemic tissue by promoting local angiogenesis [33,34]. They are considered to possess both hematopoietic and endothelial lineage differentiation potential [35]. The potential use of CD133+ HSPCs in regenerative stem cell therapy is under investigation [36,37]. Although CD34+/CD133+ HSPCs from preterm cord blood showed higher clonogenic potential, the regenerative potential needs to be evaluated. The capability of CD133– HSPCs, especially from preterm cord blood, for regenerative purposes is still unknown. In both groups, CD34+/CD133– HSPCs had a lower clonogenic capacity, which is in consent with other studies [30,38].

Isolating UCB HSPCs according to high ALDH activity has been an efficient and reproducible method without excessive cell manipulation. As reported by Hess et al., ALDH^{high} HSPCs from TCB demonstrated reliable NOD/SCID repopulating function [39]. Furthermore, UCB ALDH^{high} cells are considered to be proangiogenic progenitors that integrate into ischemic tissue and promote vascular regeneration [40]. Isolated HSPCs according to their enhanced ALDH activity in PCB demonstrated a HSPC phenotype and higher in-vitro clonogenic potential than in TCB. Our data indicate that HSPCs can be effectively isolated from PCB using ALDH activity.

Umbilical cord blood is a robust source of different stem and progenitor cells with lower risk of severe acute graft versus host disease when used for allogeneic transplant [41]. Despite the higher CD34+ HSPC counts in PCB, some UCB samples of preterm neonates displayed a lower HSPC concentration compared to term neonates. Another critical point is the lower obtainable blood volume from preterm cord blood. The mean volume of the collected PCB samples (n = 6; mean GA 29±1.5 weeks; range 27+1–31+1 weeks) was 15.5±7.8 ml (range 8.5 ml to 30 ml). Consequently, several techniques have been established for the expansion of UCB-derived HSPCs [42,43]. Expansion with cytokine mixtures in-vitro showed a 1000- fold expansion of HSPCs of PCB [42]. To our knowledge, no study evaluated the regenerative potential in vivo after PCB-derived HSPC expansion.

In conclusion, we found that cord blood of preterm infants contains a variety of HSPC subsets and they possess higher in-vitro clonogenic potential when compared to HSPCs isolated from cord blood of term infants. In our study, obstetric and perinatal complications did not influence the number and clonogenic capacity of HSPC subset in PCB and TCB. It seems feasible to isolate a sufficient cell count and fully functional HSPCs from PCB. Further analyses need to be done, especially the ex vivo expansion of PCB derived HSPCs and their in vivo regenerative potential. However, our findings suggest that PCB might be a potential source of HSPCs for stem cell therapy in preterm neonates.

Acknowledgments

The authors would like to thank Günther Hofbauer for technical assistants and Annica Bee for statistical analysis.

Author Contributions

Conceived and designed the experiments: LW TFR GK LG. Performed the experiments: LW AS. Analyzed the data: LW AS TFR GK LG. Contributed reagents/materials/analysis tools: LG AB AP AS HH PH. Wrote the paper: LW AB AP LG. Collection of blood samples: SS MB HH PH. Obtained parental consent: SS MB. Evaluation of clinical data: LW SS MB.

References

1. Gluckman E (2000) Current status of umbilical cord blood hematopoietic stem cell transplantation. Exp Hematol 28: 1197–1205.
2. Eglitis MA, Mezey E (1997) Hematopoietic cells differentiate into both microglia and macroglia in the brains of adult mice. Proc Natl Acad Sci U S A 94: 4080–4085.
3. Ishikawa F, Drake CJ, Yang S, Fleming P, Minamiguchi H, et al. (2003) Transplanted human cord blood cells give rise to hepatocytes in engrafted mice. Ann N Y Acad Sci 996: 174–185.
4. Krause DS, Theise ND, Collector MI, Henegariu O, Hwang S, et al. (2001) Multi-organ, multi-lineage engraftment by a single bone marrow-derived stem cell. Cell 105: 369–377.
5. Horbar JD, Carpenter JH, Badger GJ, Kenny MJ, Soll RF, et al. (2012) Mortality and neonatal morbidity among infants 501 to 1500 grams from 2000 to 2009. Pediatrics 129: 1019–1026.
6. Kogler G, Critser P, Trapp T, Yoder M (2009) Future of cord blood for non-oncology uses. Bone Marrow Transplant 44: 683–697.
7. Donega V, van Velthoven CT, Nijboer CH, van Bel F, Kas MJ, et al. (2013) Intranasal mesenchymal stem cell treatment for neonatal brain damage: long-term cognitive and sensorimotor improvement. PLoS One 8: e51253.
8. Zhang X, Wang H, Shi Y, Peng W, Zhang S, et al. (2012) Role of bone marrow-derived mesenchymal stem cells in the prevention of hyperoxia-induced lung injury in newborn mice. Cell Biol Int 36: 589–594.
9. Chang YS, Oh W, Choi SJ, Sung DK, Kim SY, et al. (2009) Human umbilical cord blood-derived mesenchymal stem cells attenuate hyperoxia-induced lung injury in neonatal rats. Cell Transplant 18: 869–886.
10. Dalous J, Pansiot J, Pham H, Chatel P, Nadaradja C, et al. (2013) Use of human umbilical cord blood mononuclear cells to prevent perinatal brain injury: a preclinical study. Stem Cells Dev 22: 169–179.
11. Sun J, Allison J, McLaughlin C, Sledge L, Waters-Pick B, et al. (2010) Differences in quality between privately and publicly banked umbilical cord blood units: a pilot study of autologous cord blood infusion in children with acquired neurologic disorders. Transfusion 50: 1980–1987.
12. Cotten CM, Murtha AP, Goldberg RN, Grotegut CA, Smith PB, et al. (2014) Feasibility of autologous cord blood cells for infants with hypoxic-ischemic encephalopathy. J Pediatr 164: 973–979 e971.
13. Gortner L, Felderhoff-Muser U, Monz D, Bieback K, Kluter H, et al. (2012) Regenerative therapies in neonatology: clinical perspectives. Klin Padiatr 224: 233–240.
14. Polin RA (2012) Management of neonates with suspected or proven early-onset bacterial sepsis. Pediatrics 129: 1006–1015.
15. Fenton TR, Kim JH (2013) A systematic review and meta-analysis to revise the Fenton growth chart for preterm infants. BMC Pediatr 13: 59.
16. Cervera A, Lillo R, Garcia-Sanchez F, Madero L, Madero R, et al. (2006) Flow cytometric assessment of hematopoietic cell subsets in cryopreserved preterm and term cord blood, influence of obstetrical parameters, and availability for transplantation. Am J Hematol 81: 397–410.
17. Clapp DW, Baley JE, Gerson SL (1989) Gestational age-dependent changes in circulating hematopoietic stem cells in newborn infants. J Lab Clin Med 113: 422–427.
18. Gasparoni A, Ciardelli L, Avanzini MA, Bonfichi M, di Mario M, et al. (2000) Immunophenotypic changes of fetal cord blood hematopoietic progenitor cells during gestation. Pediatr Res 47: 825–829.
19. Mohyeddin Bonab MA, Alimoghaddam KA, Goliaei ZA, Ghavamzadeh AR (2004) Which factors can affect cord blood variables? Transfusion 44: 690–693.
20. Wahid FS, Nasaruddin MZ, Idris MR, Tusimin M, Tumian NR, et al. (2012) Effects of preeclampsia on the yield of hematopoietic stem cells obtained from umbilical cord blood at delivery. J Obstet Gynaecol Res 38: 490–497.
21. Nakajima M, Ueda T, Migita M, Oue Y, Shima Y, et al. (2009) Hematopoietic capacity of preterm cord blood hematopoietic stem/progenitor cells. Biochem Biophys Res Commun 389: 290–294.
22. Lim FT, Scherjon SA, van Beckhoven JM, Brand A, Kanhai HH, et al. (2000) Association of stress during delivery with increased numbers of nucleated cells and hematopoietic progenitor cells in umbilical cord blood. Am J Obstet Gynecol 183: 1144–1152.
23. Manegold G, Meyer-Monard S, Tichelli A, Pauli D, Holzgreve W, et al. (2008) Cesarean section due to fetal distress increases the number of stem cells in umbilical cord blood. Transfusion 48: 871–876.
24. Ballen KK, Wilson M, Wuu J, Ceredona AM, Hsieh C, et al. (2001) Bigger is better: maternal and neonatal predictors of hematopoietic potential of umbilical cord blood units. Bone Marrow Transplant 27: 7–14.
25. Hassanein SM, Amer HA, Shehab AA, Hellal MM (2011) Umbilical cord blood CD45(+)CD34(+) cells coexpression in preterm and full-term neonates: a pilot study. J Matern Fetal Neonatal Med 24: 229–233.
26. Omori A, Takahashi K, Hazawa M, Misaki N, Ohba H, et al. (2008) Maternal and neonatal factors associated with the high yield of mononuclear low-density/CD34+ cells from placental/umbilical cord blood. Tohoku J Exp Med 215: 23–32.
27. McGuckin CP, Basford C, Hanger K, Habibollah S, Forraz N (2007) Cord blood revelations: the importance of being a first born girl, big, on time and to a young mother! Early Hum Dev 83: 733–741.
28. Baik I, Devito WJ, Ballen K, Becker PS, Okulicz W, et al. (2005) Association of fetal hormone levels with stem cell potential: evidence for early life roots of human cancer. Cancer Res 65: 358–363.
29. Surbek DV, Danzer E, Steinmann C, Tichelli A, Wodnar-Filipowicz A, et al. (2001) Effect of preeclampsia on umbilical cord blood hematopoietic progenitor-stem cells. Am J Obstet Gynecol 185: 725–729.
30. Yin AH, Miraglia S, Zanjani ED, Almeida-Porada G, Ogawa M, et al. (1997) AC133, a novel marker for human hematopoietic stem and progenitor cells. Blood 90: 5002–5012.
31. Handgretinger R, Kuci S (2013) CD133- Positive Hematopoietic Stem Cells: From Biology to Medicine. Adv Exp Med Biol 777: 99–111.
32. Taguchi A, Soma T, Tanaka H, Kanda T, Nishimura H, et al. (2004) Administration of CD34+ cells after stroke enhances neurogenesis via angiogenesis in a mouse model. J Clin Invest 114: 330–338.

33. Salven P, Mustjoki S, Alitalo R, Alitalo K, Rafii S (2003) VEGFR-3 and CD133 identify a population of CD34+ lymphatic/vascular endothelial precursor cells. Blood 101: 168–172.

34. Peichev M, Naiyer AJ, Pereira D, Zhu Z, Lane WJ, et al. (2000) Expression of VEGFR-2 and AC133 by circulating human CD34(+) cells identifies a population of functional endothelial precursors. Blood 95: 952–958.

35. Bailey AS, Jiang S, Afentoulis M, Baumann CI, Schroeder DA, et al. (2004) Transplanted adult hematopoietic stems cells differentiate into functional endothelial cells. Blood 103: 13–19.

36. Elkhafif N, El Baz H, Hammam O, Hassan S, Salah F, et al. (2011) CD133(+) human umbilical cord blood stem cells enhance angiogenesis in experimental chronic hepatic fibrosis. APMIS 119: 66–75.

37. Franceschini V, Bettini S, Pifferi S, Rosellini A, Menini A, et al. (2009) Human cord blood CD133+ stem cells transplanted to nod-scid mice provide conditions for regeneration of olfactory neuroepithelium after permanent damage induced by dichlobenil. Stem Cells 27: 825–835.

38. de Wynter EA, Buck D, Hart C, Heywood R, Coutinho LH, et al. (1998) CD34+AC133+ cells isolated from cord blood are highly enriched in long-term culture-initiating cells, NOD/SCID-repopulating cells and dendritic cell progenitors. Stem Cells 16: 387–396.

39. Hess DA, Meyerrose TE, Wirthlin L, Craft TP, Herrbrich PE, et al. (2004) Functional characterization of highly purified human hematopoietic repopulating cells isolated according to aldehyde dehydrogenase activity. Blood 104: 1648–1655.

40. Putman DM, Liu KY, Broughton HC, Bell GI, Hess DA (2012) Umbilical cord blood-derived aldehyde dehydrogenase-expressing progenitor cells promote recovery from acute ischemic injury. Stem Cells 30: 2248–2260.

41. Grewal SS, Barker JN, Davies SM, Wagner JE (2003) Unrelated donor hematopoietic cell transplantation: marrow or umbilical cord blood? Blood 101: 4233–4244.

42. Wyrsch A, dalle Carbonare V, Jansen W, Chklovskaia E, Nissen C, et al. (1999) Umbilical cord blood from preterm human fetuses is rich in committed and primitive hematopoietic progenitors with high proliferative and self-renewal capacity. Exp Hematol 27: 1338–1345.

43. Kogler G, Radke TF, Lefort A, Sensken S, Fischer J, et al. (2005) Cytokine production and hematopoiesis supporting activity of cord blood-derived unrestricted somatic stem cells. Exp Hematol 33: 573–583.

19

Administration of BMSCs with Muscone in Rats with Gentamicin-Induced AKI Improves Their Therapeutic Efficacy

Pengfei Liu[1,2], Yetong Feng[1,2], Chao Dong[1], Dandan Yang[1,2], Bo Li[1,2], Xin Chen[3], Zhongjun Zhang[4], Yi Wang[1]*, Yulai Zhou[1]*, Lei Zhao[4]*

1 Department of Regeneration Medicine, School of Pharmaceutical Science, Jilin University, Changchun, P.R. China, 2 Key Laboratory of Regenerative Biology, Guangdong Provincial Key Laboratory of Stem Cell and Regenerative Medicine, Guangzhou Institutes of Biomedicine and Health, Chinese Academy of Sciences, Guangzhou, P.R. China, 3 Department of Laboratory Medicine, Second Clinical Medical College of Jinan University, Shenzhen People's Hospital, Shenzhen, P.R. China, 4 Department of Anesthesiology, Second Clinical Medical College of Jinan University, Shenzhen People's Hospital, Shenzhen, P.R. China

Abstract

The therapeutic action of bone marrow-derived mesenchymal stem cells (BMSCs) in acute kidney injury (AKI) has been reported by several groups. However, recent studies indicated that BMSCs homed to kidney tissues at very low levels after transplantation. The lack of specific homing of exogenously infused cells limited the effective implementation of BMSC-based therapies. In this study, we provided evidence that the administration of BMSCs combined with muscone in rats with gentamicin-induced AKI intravenously, was a feasible strategy to drive BMSCs to damaged tissues and improve the BMSC-based therapeutic effect. The effect of muscone on BMSC bioactivity was analyzed in vitro and in vivo. The results indicated that muscone could promote BMSC migration and proliferation. Some secretory capacity of BMSC still could be improved in some degree. The BMSC-based therapeutic action was ameliorated by promoting the recovery of biochemical variables in urine or blood, as well as the inhibition of cell apoptosis and inflammation. In addition, the up-regulation of CXCR4 and CXCR7 expression in BMSCs could be the possible mechanism of muscone amelioration. Thus, our study indicated that enhancement of BMSCs bioactivities with muscone could increase the BMSC therapeutic potential and further developed a new therapeutic strategy for the treatment of AKI.

Editor: Pranela Rameshwar, Rutgers - New Jersey Medical School, United States of America

Funding: This study was supported by the grants from the Basic Research Project of Knowledge Innovation Program of Shenzhen City (No. 201404163000377), Graduate Innovation Fund of Jilin University, the Science and Technology Project of Shenzhen City (No. 200702120, 200602035), the Scientific Research of Health Department of Jilin Province (No. 2009Z043) and Science and Technology Development Project of Jilin Province (No. 20120955). The funders had no role in study design, data collection and analysis, decision to publish, or preparation of the manuscript.

Competing Interests: The authors have declared that no competing interests exist.

* E-mail: zhaoleicx@126.com (LZ); zhouyl@jlu.edu.cn (YZ); wangyi@jlu.ecu.cn (YW)

Introduction

Acute kidney injury (AKI) is the rapid deterioration of renal function. It can be induced by numerous insults, and has high morbidity and mortality rates [1–3]. The mortality rate of hospital-acquired AKI currently ranges from 30% to 80%, and recent dialysis techniques, such as pharmacological therapy and continuous renal replacement therapy, have no obvious effects on the overall mortality [4–6]. Thus, a novel therapeutic strategy should be developed to improve the survival outcomes of patients with AKI.

In recent years, great interest has shifted to stem cell-based therapy in the treatment of many diseases, such as diabetes [7], neural disease [8], and so on [9–12]. In the treatment of AKI, the therapeutic actions of stem cells in preventing and repairing damaged renal cells have been demonstrated by previous studies [13–15]. Different types of stem cells, such as amniotic fluid stem cells [16], hematopoietic progenitor cells [17], and kidney-derived mesenchymal stem cells[18], have been investigated, and their therapeutic effects in AKI treatment have been determined. Especially for the action of bone marrow derived mesenchymal

stem cells (BMSCs), several studies have used BMSCs to treat AKI in animal models and their results showed that renal function and structure could be improved with the infusion of BMSCs [19–21]. BMSCs can be isolated from the bone marrow of patients. Compared with other stem cells, BMSCs are feasible in autologous treatment because of their source, number, and safety [21,22].

The therapeutic mechanism of BMSCs comprises both differentiation-dependent mechanism [23,24] and differentiation-independent mechanism [25]. However, which of these two mechanisms is more significant during the improvement in kidney function and structure remains unclear.

Besides further exploring the accurate mechanism, researchers are also interested in further refining the therapeutic action of BMSCs. Liu et al. found that BMSC transplantation combined with erythropoietin (EPO) injection was a novel and effective approach for AKI repairing [19]. They showed that the AKI microenvironment had a direct chemotactic effect on BMSCs, which could be further enhanced by EPO treatment. The activation of PI3K/AKT and MAPK in BMSCs and the increase in stromal cell-derived factor (SDF-1) levels in the AKI microenvironment were the possible mechanisms for the effect of EPO.

However, the experimental events in their study were carried out *in vitro*, the function of EPO *in vivo* remained unknown. To further analyze the enhancement of EPO on the effects of stem cell-based therapy, Eliopoulos et al. reported that BMSCs, which were genetically enhanced to secrete EPO, could produce significant beneficial effects in AKI therapy [26]. They created mouse EPO-secreting BMSCs, which were implanted by intraperitoneal injection in allogeneic mice that were previously administered with cisplatin to induce AKI. Their results showed that EPO-BMSCs significantly improved the survival rate of implanted mice than normal BMSCs. Liu et al. investigated the effect of CXCR4 overexpression on BMSC migration to the kidney in AKI treatment [27]. CXCR4 gene-modified BMSCs (CXCR4-BMSCs) and normal BMSCs were prepared and transplanted into AKI mice. Their results showed that overexpression of the CXCR4 gene enhanced BMSC migration to the kidney after AKI. However, Gheisari et al. reported a different conclusion [28]. In their study, CXCR4 and CXCR7 were separately and simultaneously overexpressed in BMSCs with a lentiviral vector system, and the homing and renoprotective potentials of these cells were evaluated in the mouse model of cisplatin-induced AKI. They concluded that the overexpression of CXCR4 and CXCR7 receptors in BMSCs could not improve the homing and therapeutic potentials of these cells, and it could be due to severe chromosomal abnormalities in these cells during expansion *in vivo*. Therefore, gene-modified BMSCs are unstable during treatment, and they have yet to be used in clinical studies because of the associated laws and regulations. Thus, a novel strategy for the improvement of stem cell-based therapy should be developed.

In this study, muscone was used to optimize the therapeutic action of BMSCs by enhancing cell migration, proliferation, and secretory capacity. Natural muscone is obtained from musk, which is a ventral glandular secretion of the male musk deer, and is regarded as the main active ingredient of musk [29,30]. Musk has been extensively used in Chinese medicine for thousands of years, and muscone is believed to hold less toxicity and side effects compared with musk, and also hold the function as a refreshing agent, promoting blood flow and detumescence [30,31]. Therefore, muscone is widely used in clinical studies. Xie et al. first investigated the effect of muscone on BMSC migration *in vivo* [32], in which a rat model of skull defect was established through dental surgery. BMSCs combined with muscone were injected into skull-defect rats from the tail vein. Their results indicated that muscone could accelerate the migration of BMSCs to the injured area *in vivo*. However, whether muscone promotes cell migration in AKI model remains unclear. Our study showed that BMSCs combined with muscone was a feasible strategy in promoting BMSC migration, proliferation, and secretory capacity to repair kidney tissues in the rat model of gentamicin-induced AKI. This strategy could be used to develop a novel treatment mode for the preclinical study of AKI.

Results

Characterization of BMSCs and RTECs

The BMSCs isolated from Wistar rats were characterized by FACS for CD14, CD29, CD34, CD44, CD45, CD73, CD90, CD105, and CD166. The BMSCs used in our study were positive for CD29 (95.6%), CD44 (92.1%), CD73 (96.6%), CD90 (94.7%), CD105 (97.3%), and CD166 (88.2%), and nearly negative for CD14 (0.79%), CD34 (0.95%) and CD45 (1.64%), which were regared as the specific markers of hematopoietic cells (Figure 1A). The BMSCs displayed a spindle-shaped "fibroblast" appearance. They were successfully differentiated into adipocytes and osteo-

blasts, as demonstrated by Oil Red O staining and positive staining with alkaline phosphatase (AP), respectively (Figure 1B). In the negative control group, the BMSCs cultured with normal medium were used for each stain, and could not be stained with Oil Red O or AP kit (Figures 1B1 and 1B2). In the process of RTECs isolation, kidney tubules were first separated from kidney tissues (Figure 1C). With further culture *in vitro*, the kidney tubules gradually attached to the plastic surface. The RTECs grew out from the tubules and continued to pave stone, resembling growth (Figure 1C1). After one passage, the RTECs were characterized by FACS and immunohistochemistry for CK-18, and a positive stain was observed (Figure 1C). These results confirm the purity and differentiation ability of the cells used in our culture system.

Effect of muscone on BMSC and RTEC proliferation and secretion

The treatment of muscone at different concentrations (0.3, 1.0, and 3.0 mg/L) enhanced the proliferative ability of BMSCs to some degree (Figure 2A). The absolute values in proliferation indexes of each group were shown in Table S1, and the fold change can be counted based on those values, showing up to 1.37 fold increases with the treatment of muscone compared with the normal cells. The function of BMSCs in secreting renal protective cell factors (VEGF, HGF, BMP-7, and IL-10) was further analyzed. Real-time qPCR showed that BMP-7 expression in BMSCs was upregulated by muscone than that in normal BMSCs (Figure 2B). These results were further confirmed using ELISA. However, other cell factors secreted by BMSCs in ELISA still demonstrated a modest increasing tendency (Figure 2B). The effect on RTEC proliferation and secretion was also evaluated in our study. No obvious variation was observed in RTEC proliferative activity during 3 days of culture *in vitro* (Figure S1A). qPCR and ELISA showed that the expression of renal protective cell factors (HGF, TGF-β, TIMP-1, and ET-1) in RTECs and the function of RTECs to secrete these cell factors in each treatment group were similar to those of the normal group (Figure S1B).

Evaluation of rat AKI model

In our study, the AKI model was induced by gentamicin in adult male Wistar rats. To confirm the rat AKI model, urine and serum samples were collected on day 8 before treatment. The levels of N-acetyl-beta-glucosaminidase (NAG) and lysozyme (LZM) in urine normalized by urinary creatinine (Table S2), as well as creatinine and urea nitrogen in serum, were measured in the normal and AKI model groups. These indices in the model group presented significantly higher levels than those in the normal group (NAG: 63.1 ± 6.3 U/L vs. 13.8 ± 2.7 U/L; LZM: 302.4 ± 55.5 U/ml vs. 32.7 ± 2.7 U/ml; creatinine: 65.2 ± 9.9 μM vs. 21.2 ± 7.7 μM; and urea nitrogen: 26.46 ± 4.1 mM vs. 9.5 ± 2.1 mM, $P<0.05$). These findings demonstrated that the rat model displayed characteristics of AKI disease (Table 1).

Effect of treatment on kidney weight coefficient

To evaluate the therapeutic action on the accrementition and enlargement of the entire kidney, the kidney weight coefficient was determined in each group. We found that the kidney weight coefficient was below 0.01 in the normal group, whereas that of the model group exceeded 0.02. The coefficients of the other groups decreased in varying degrees with therapy (Figure 3). Both the positive drug and BMSCs had an effect to release the enlargement of kidney, and the kidney weight coefficient of the BMSC group was similar to that of the positive drug group. In the combined group, the BMSCs and muscone were given simulta-

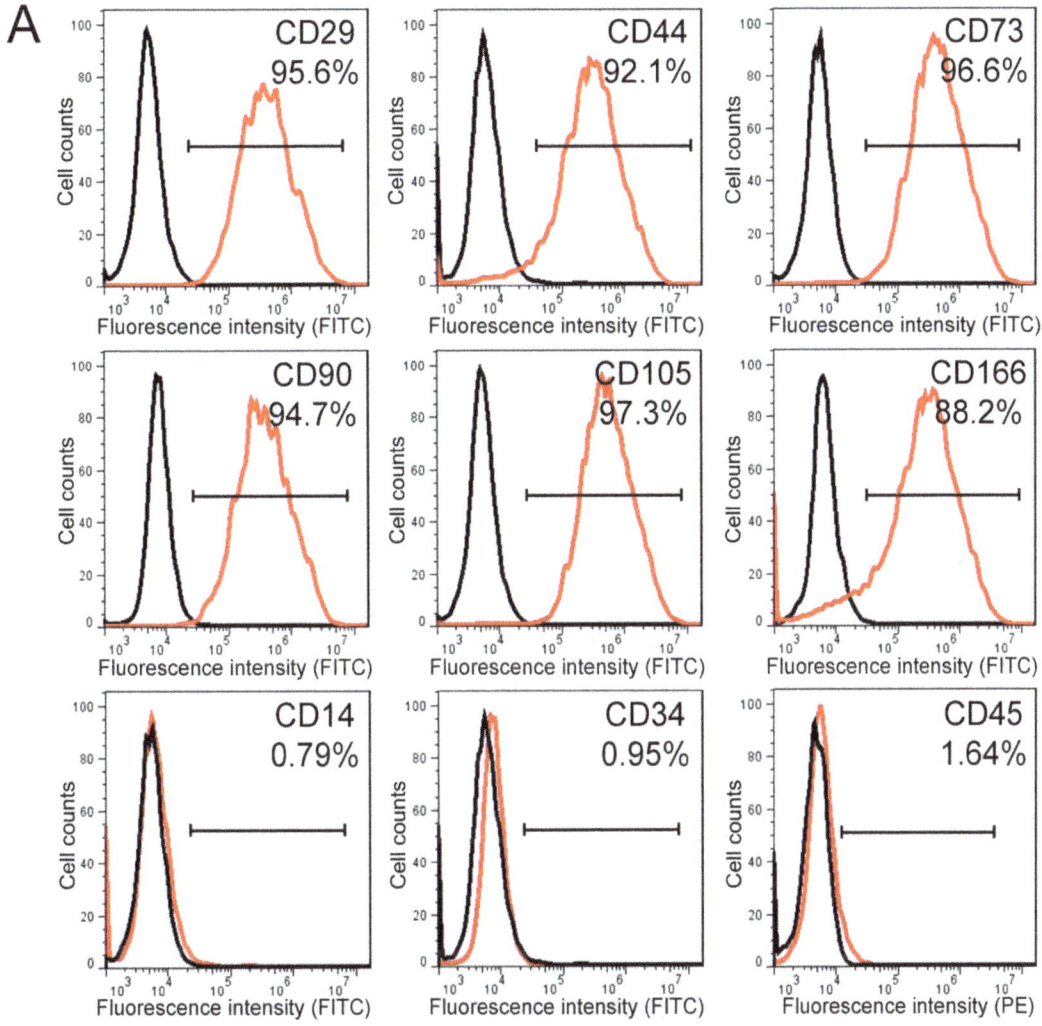

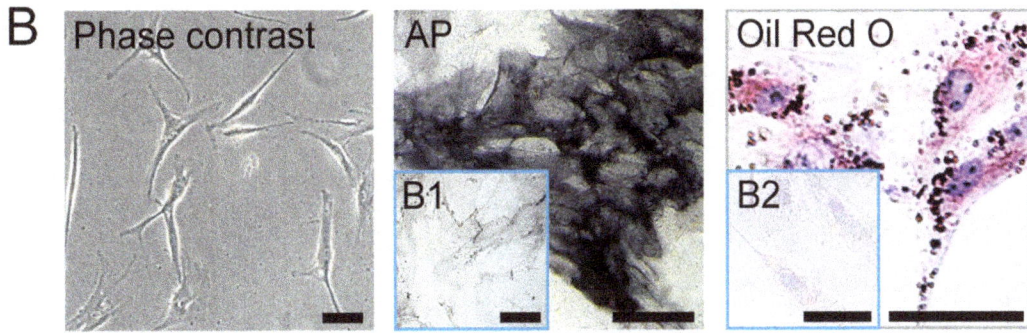

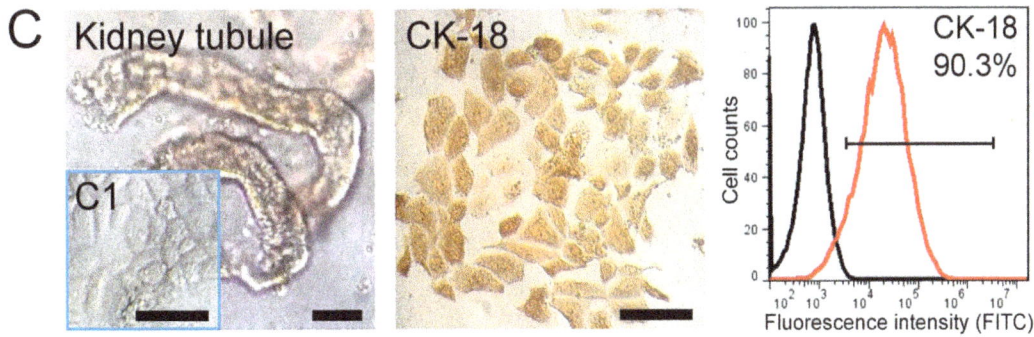

Figure 1. Characterization of BMSCs and RTECs. A: Immunophenotype of isolated BMSCs. Isolated rat BMSCs were characterized by FACS. BMSCs were positive for CD29, CD44, CD73, CD90, CD105, and CD166, and nearly negative for CD14, CD34, and CD45. B: Differentiation characteristics of BMSCs. The phase contrast of BMSCs is shown on the left. Osteogenic differentiation was detected by AP staining (middle), and adipogenic differentiation was visualized by Oil Red O staining of the lipid vesicles (right). BMSCs cultured with normal medium were used as the negative control group for each stain (B1 and B2). C: Immunophenotype of isolated RTECs. The appearance of kidney tubules is shown on the left and the confluent RTECs in culture were shown in C1. The isolated rat RTECs were characterized by immunohistochemistry (middle) and FACS (right), and they were positive for CK-18. All scale bars correspond to 200 μm.

neously, and the therapeutic effect of BMSCs was further enhanced with muscone treatment, and an optimal protective function in inducing the accrementition of kidney tissues was observed in the combined group. However, no significant therapeutic effect was observed in the muscone group, which indicated that muscone did not have an obvious therapeutic function in AKI.

Effect of treatment on biochemical variables in urine and serum

The levels of urea nitrogen, NAG, and LZM in urine normalized by urinary creatinine (Table S2), as well as creatinine, urea nitrogen, and NAG in serum, were measured using a Biochemistry Autoanalyzer. LZM showed a statistically significant difference between the model and treatment groups. The levels of NAG and urea nitrogen in the treatment groups were similar to the normal levels, especially urea nitrogen in the combined group (Table 1). The injection of BMSCs and muscone significantly decreased NAG and creatinine in serum ($P<0.05$). The creatinine level in the combined group was the most similar to that of the normal group. However, no amelioration was found in urea nitrogen in serum (Table 1).

Effect of treatment on renal histology

To evaluate the therapeutic effect of BMSCs and muscone in the AKI model, the pathological changes in the kidney tubules, kidney glomeruli, and collecting tubules in each group were observed by H&E staining. The typical pathological changes in AKI induced by gentamicin were mainly reflected in the kidney tubules and collecting tubules. The kidney glomeruli in all groups were similar to one another. Notable damage, including tubular necrosis, dilatation, and effusion in the kidney tubules and collecting tubules, was observed in the model group compared with the normal group. Amelioration of various degrees was observed in the treatment group, and both the BMSC and combined groups showed better restoration than the muscone group, and there is no difference between the combined group (the BMSCs and muscone were given simultaneously) and the BMSCs group in terms of histopathology. The BMSC and combined groups exhibited fewer necrotic and dilated tubules and less effusion in the tubules, especially in the recovery of collecting tubules. The muscone group still showed therapeutic effects than the model group (Figure 4A). A histological scoring system was used to further evaluate kidney tissue morphology. The histological score of normal group was below 1, whereas that of the model group exceeded 3. The score of other groups could be decreased with the treatment, and there was no significant difference between the BMSC and the combined groups (Figure 4B).

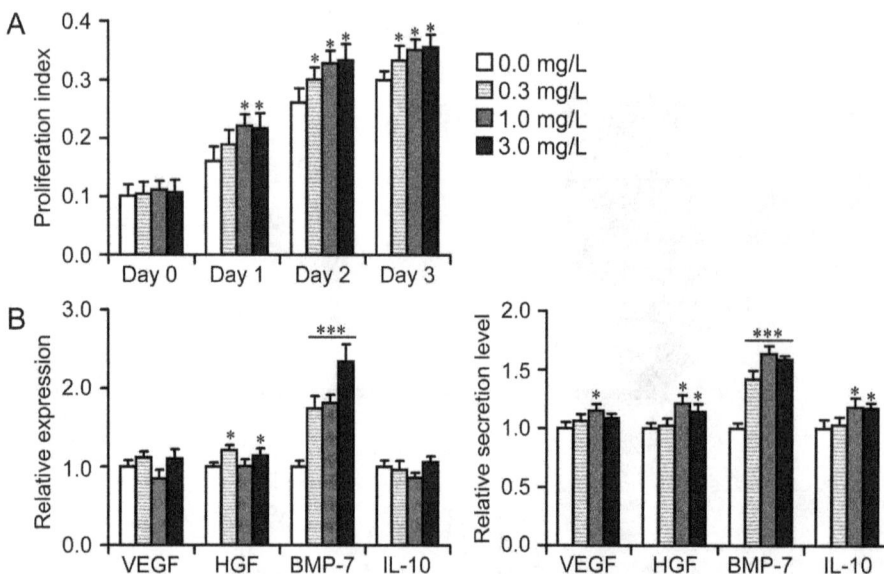

Figure 2. Effect of muscone on BMSC bioactivity *in vitro.* A: Effect of muscone on BMSC proliferation. Proliferation index (the absorbance of experimental group − the absorbance of blank group) was measured using CCK-8. B: Effect of muscone on BMSC secretion. To evaluate the secretory function, the secretion or cytokine expression level of normal BMSCs without any muscone treatment was 1.0 for each cytokine. Expression of the cytokines in BMSCs detected with qPCR is shown on the left, and the BMSC secretory function evaluated using ELISA is shown on the right. Similar results were obtained in at least three independent experiments. Results were expressed as mean ± SEM. A *t*-test was used to compare the various groups, and $P<0.05$ was considered statistically significant. *$P<0.05$ compared with normal BMSC group without any muscone treatment; ***$P<0.001$ compared with normal BMSC group without any muscone treatment.

Table 1. Change of biochemical variables in serum or urine during the therapy process.

	Characteristics of AKI model in urine or serum				Biochemical variables in urine after therapy			Biochemical variables in serum after therapy		
	NAG(U/L)	LZM(U/ml)	Creatine(μM)	UN(mM)	NAG(U/L)	LZM(U/ml)	UN(mM)	NAG(U/L)	Creatine(μM)	UN(mM)
normal group	13.8±2.7	32.7±2.7	21.2±7.7	9.5±2.1	17.4±3.2	34.2±2.4	33.1±3.1	17.9±3.1	19.9±6.7	7.7±2.1
model group	63.1±6.3*	302.4±55.5*	65.2±9.9*	26.5±4.1*	62.7±17.2	406.7±71.4	15.7±5.6	58.3±5.6	78.4±6.5	28.1±4.6
positive drug group					19.3±5.6*	249.1±32.6*	24.3±4.9*	16.3±3.2*	22.1±4.3*	26.3±4.4
muscone group					23.0±5.8*	278.1±66.6*	16.1±4.2	17.8±5.7*	42.2±3.2*	27.3±6.2
BMSCs group					20.1±7.1*	273.5±82.3*	25.1±2.6*	19.7±4.3*	33.2±3.8*	26.4±5.3
combined group					17.8±5.3*	254.2±59.4*	30.8±3.5*#	19.5±5.3*	26.6±4.3*#	28.7±4.8

To assay the characteristics of rat AKI model, NAZ and LAM in urine, as well as creatinine and urea nitrogen (UN) in serum, were detected on day 7. *$P<0.05$ compared with the normal group. To analyze the therapeutic effect, urine and serum samples of each group were collected from each group on day 15 after animals were sacrificed. The levels of UN, NAZ, and LAM in urine, as well as creatinine, UN, and NAG in serum, were measured. *$P<0.05$ compared with the model group; #$P<0.05$ compared with the muscone and BMSC groups.

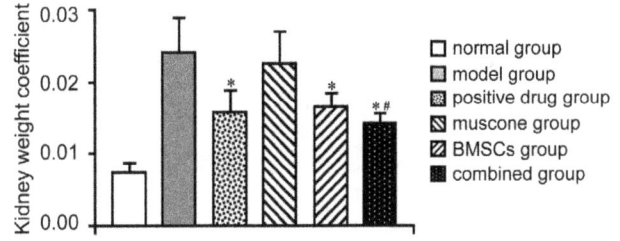

Figure 3. Level of kidney weight coefficients in each group. *$P<0.05$ compared with the model group; #$P<0.05$ compared with the muscone and BMSC groups.

Samples of each group were observed under TEM to analyze the ultrastructures of RTECs. The RTECs in the normal group had clear ultrastructures with minimal heterochromatin and numerous organelles in the cytoplasm. In the model group, the RTECs exhibited noticeable apoptosis with fewer organelles, deformed nuclei, and markedly more heterochromatin. The ultrastructures of RTECs in the treatment groups were improved in some degree. Abundant mitochondria and some cytolysosomes were observed in the cytoplasm of the positive drug group. After treatment with BMSCs or muscone, the RTEC structures exhibited better restoration with abundant organelles and some cytoplasmic vacuoles. However, no significant difference was found between the BMSC and combined groups (Figure 4A).

Cell apoptosis assay

TUNEL assay was used to detect apoptosis in renal cells in the kidney sections. The number of TUNEL-positive cells in both the kidney tubules and collecting tubules markedly increased in the model group but decreased in the treatment group (Figure 5). Motic Image Advanced 3.2 software showed that the color intensity of the combined group (the BMSCs and muscone were given simultaneously) was lighter than that of the BMSC or muscone group in the kidney tubules. However, the combined group did not show a superior improvement in the therapeutic effect in the collecting tubules. TUNEL assay detected no significant difference in cell apoptosis of the kidney glomeruli between the model and other groups (Figure 5).

The expression levels of apoptotic genes (Caspase 3, Fas, and Bax) and anti-apoptotic gene (Bcl-2) in the kidney tissue of each group were detected. Real-time qPCR showed that the expression levels of apoptotic genes in the model group were much higher than those in the normal group. Most treatments were efficient in inducing anti-apoptosis. By contrast, muscone did not significantly downregulate Bax expression, and the combined group showed reduced expression levels of Caspase 3 and Fas (Figure 6A). The expression of anti-apoptotic gene (Bcl-2) was much higher in the combined group than that in the single therapy and model groups (Figure 6A).

The expression levels of apoptotic genes (Caspase 3, Fas, and Bax) and anti-apoptotic gene (Bcl-2) in kidney tissues were also detected using immunohistochemistry. A representative set close to the average level of each group is shown in Figure S2, and most of the results were similar to those of qPCR. Gene expression levels in the kidney tubules, kidney glomeruli, and collecting tubules were analyzed. Motic Image Advanced 3.2 software was used to measure the color intensity of each group. No noticeable difference in gene expression was observed between the model and other groups in the kidney glomeruli. Both the positive drug and combined groups showed a modest therapeutic effect on cell apoptosis in the collecting tubules. The differences among each

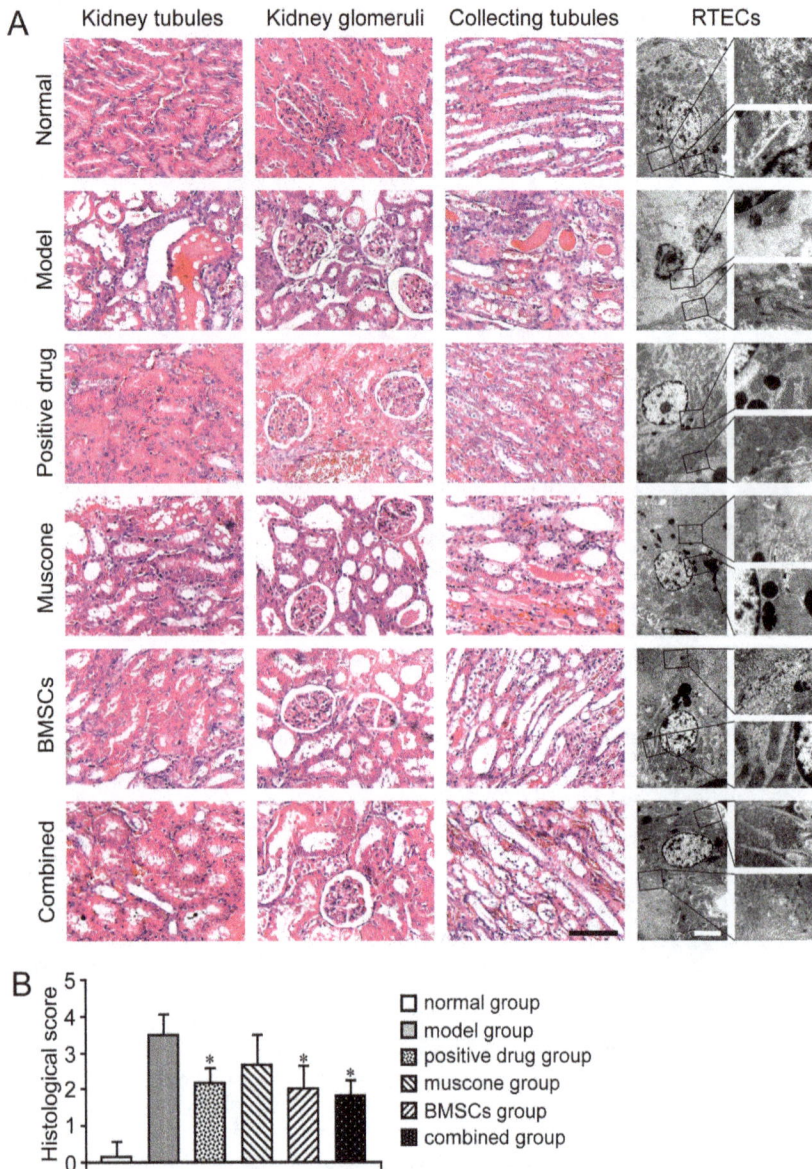

Figure 4. Detection of pathological changes in each group. A: Pathological changes of each group. Pathological changes in the kidney tubules, kidney glomeruli, and collecting tubules were observed under a light microscope using H&E staining. Dilatation and necrosis in the kidney tubules and collecting tubules of the model group were observed in various degrees. In the combined and BMSC groups, nearly no effusion was observed in the kidney tubules or collecting tubules, and both tubules remained slightly expanded compared with the normal group. Ultrastructures of RTECs were further observed under TEM. The RTECs in the normal group had minimal heterochromatin and numerous organelles in the cytoplasm. In the AKI model, the RTECs showed an apoptotic appearance with deformed nuclei, more heterochromatin, and fewer organelles. Positive drug or muscone treatment resulted in numerous mitochondria and few cytolysosomes in the RTEC cytoplasm. The RTECs of both the BMSC and combined groups were restored well, and showed no significant difference. Scale bar of the phase observed through H&E staining corresponds to 100 μm, and scale bar of the phase observed under TEM corresponds to 4 μm. B: Histological score of each group. Results are expressed as mean ± SEM. A t-test was used to compare the various groups. *$P<0.05$ compared with the model group.

group were mainly reflected in the kidney tubules. All apoptotic genes were downregulated in the treatment groups, except the muscone group. The combined group exhibited lower expression of apoptotic genes (Caspase 3 and Fas) and higher expression of anti-apoptotic gene (Bcl-2) than those in the BMSC or muscone group (Figure 6B). In addition, the activated Caspase3 in the kidney tissue was also determined by the kit. The result was similar to that of qPCR and the combined group showed better therapeutic action on inhibiting cell apoptosis, compared with the single therapy (Figure 6C).

Inflammatory components assay

Expression of MCP-1, IL-10, RANTES and MIP-2 in RNA level and protein level were examined in our study. All of the inflammatory components were down-regulated with the therapy in different degree (Figure 7A, 7B). Both the positive drug and muscone showed therapeutic effect on the regulation of some inflammatory components. Compared with other group, qPCR results showed the combined group (the BMSCs and muscone were given simultaneously) hold better therapeutic action on down-regulating RANTES and MIP-2 (Figure 7A), and the effect

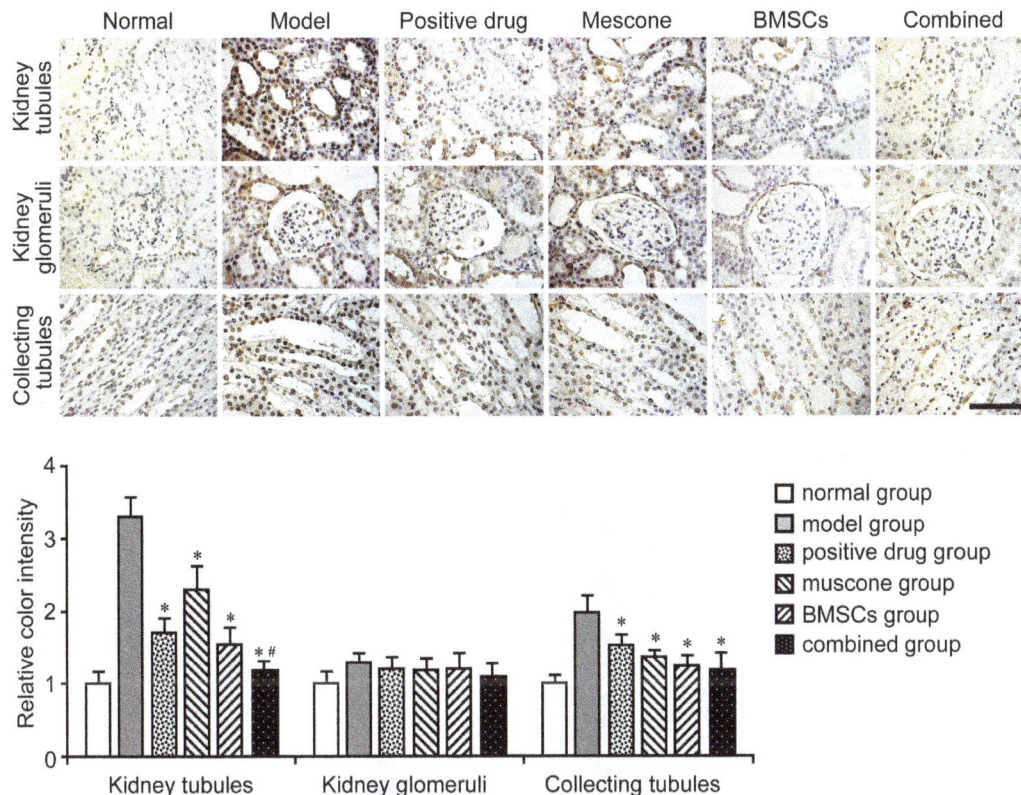

Figure 5. Evaluation of cell apoptosis in each group using TUNEL staining. Images of TUNEL staining in the kidney tubules, kidney glomeruli, and collecting tubules are shown for each group. The color intensity was measured using Motic Image Advanced 3.2. The color intensity of the normal group was 1.0, and the relative color intensity of the other groups was evaluated. Similar results were obtained in at least three independent experiments. Results are expressed as mean ± SEM. A t-test was used to compare the various groups, and P<0.05 was considered statistically significant. *P<0.05 compared with the model group; #P<0.05 compared with the muscone and BMSC groups. Scale bar corresponds to 100 μm.

on RANTES could be further confirmed in the ELISA assay (Figure 7B).

Effect of muscone on BMSC migration

To investigate the effect of muscone on BMSCs and explain the improved therapeutic effect in the combined group, the migratory ability of BMSCs was detected *in vitro* and *in vivo*. BMSC healing capacity was quantified using a scrape-healing assay in which cells were scratched. Muscone treatment could enhance the migration of BMSCs toward the injured area and reduce the scratch surface area. The number of BMSCs migrating to the injured area was counted in the observed field, and more cells were observed in the injured area with muscone treatment (Figure 8A). An *in vitro* injury-migration model was then applied as a reference [33] based on a transwell system consisting of BMSCs co-cultured with cisplatin-injured RTECs. BMSC migration from the upper chamber across the membrane to the cisplatin-damaged RTECs could be enhanced with muscone treatment compared with that without any treatment (Figure 8A). Preconditioned cells co-cultured with untreated RTECs or without RTECs did not exhibit significant migratory activity (Figure S3A). To evaluate migration *in vivo*, BrdU-labeled BMSCs were detected using immunohistochemistry, and the combined group had more labeled BMSCs than the BMSC group (Figure 8B).

To investigate the molecular mechanism by which cell migration is promoted, we detected the expression of CXCR4 and CXCR7 in the treated BMSCs using qPCR and western blot, respectively. CXCR4 and CXCR7 had pivotal functions in cell migration and proliferation. The qPCR results showed that muscone increased the expression of both CXCR4 and CXCR7 (Figure 8C). These results were similar to those of western blot. However, low muscone concentration could not increase the expression of those genes at a protein level (Figure 8D). To further define the biological significance of muscone-mediated CXCR4 and CXCR7 upregulation, we performed scrape-healing assay and transfilter assay in the presence of a neutralizing antibody. Our results showed that the functional blocking of CXCR4 or CXCR 7 could abrogate the promigratory effect of muscone on BMSC (Figure 8A).

The transfilter assay was also tried to be performed after a shorter incubation time (8 h, 12 h and 24 h) to avoid the obvious cell proliferation. However, no significant difference existed among each group under the circumstances (8 h and 12 h), and the obvious difference could be observed after 24 h (Figure S3B). This result indicated that enough time for the incubation could be necessary in the transfilter assay. Thus, to further investigate the influence of cell proliferation on the migration, BMSCs treated with mitomycin C were used for transfilter assay in our study. However, the results showed that there was no significant difference between the normal BMSCs and the mitomycin C treated BMSCs in the migration assay (Figure S3C). This indicated that the modest function to promote cell proliferation of muscone or cell proliferation by themselves, would not affect the

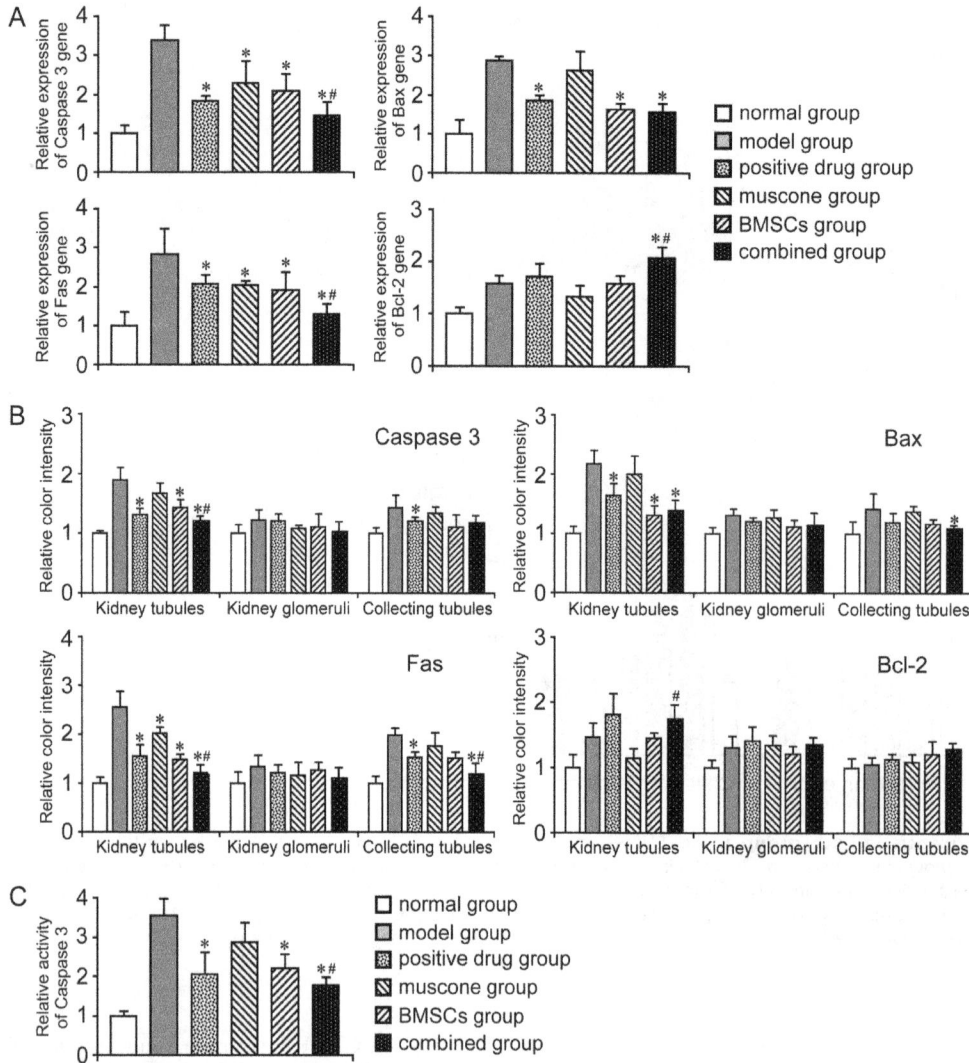

Figure 6. Expression of apoptotic genes and anti-apoptotic gene in the kidney tissues of each group. A: Detection of apoptotic genes (Caspase 3, Fas, and Bax) and anti-apoptotic gene (Bcl-2) expression using qPCR. The gene expression level in the normal group was 1.0, and the relative gene expression level of each group was further evaluated. B: Detection of apoptotic genes and anti-apoptotic gene expression using immunohistochemistry. The color intensity was further measured using Motic Image Advanced 3.2. The color intensity of the normal group was 1.0, and the relative color intensity of the other groups was further evaluated. C: Detection of activated Caspase3 in kidney tissues. Similar results were obtained in at least three independent experiments. Results are expressed as mean ± SEM. A t-test was used to compare the various groups, and $P < 0.05$ was considered statistically significant. *$P < 0.05$ compared with the model group; #$P < 0.05$ compared with the muscone and BMSC groups.

migration in our transfilter assay system which only contained 2% FBS, though more cells would affect the migratory assay results.

Discussion

BMSCs have been reported to preserve renal function in various AKI models [27,34,35]. Both the differentiation-dependent and differentiation-independent mechanisms have functions in the therapy process, but increasing reports have indicated that the differentiation-independent mechanism has a more important function during treatment *in vivo* [23,24,35]. Chen et al. [35] evaluated the organ bio-distributions of transplanted BMSCs, and correlated the survival of transplanted cells. They found that BMSCs were largely localized in pulmonary capillaries after intravenous administration. Moreover, they reported that only a minute fraction of BMSCs could enter the kidneys, exhibiting

transient survival. The development of novel strategies in enhancing cell homing to target tissues is regarded as a prerequisite for the success of BMSC-based systemic therapies. Xinaris et al. showed that the preconditioning of BMSCs with IGF-1 before infusion could improve cell migration and restore normal renal function after AKI. They demonstrated that promoting BMSC migration could increase their therapeutic potential, and indicated a novel therapeutic paradigm for organ repair [33].

The present study reported a novel approach to optimize BMSC engraftment efficiency and increased its therapeutic effects based on the combination with muscone in systemic infusion. Muscone improved BMSC engraftment in the injured kidney, as well as other bioactivities, such as cell proliferation and secretion. This function possibly improved the therapeutic effect of the combined group in some respects, including kidney weight

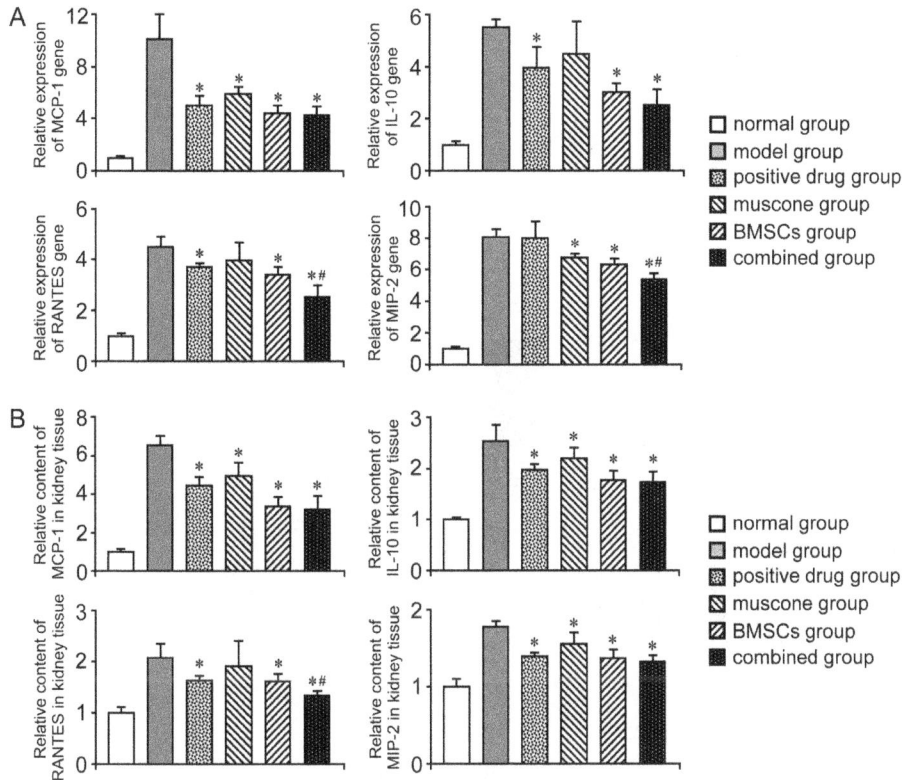

Figure 7. Evaluation of inflammatory reaction in the kidney tissues of each group. A: Detection of inflammatory component using qPCR. B: Detection of inflammatory component using ELISA. Similar results were obtained in at least three independent experiments. Results are expressed as mean ± SEM. A t-test was used to compare the various groups, and P<0.05 was considered statistically significant. *P<0.05 compared with the model group; #P<0.05 compared with the muscone and BMSC groups.

coefficient, some biochemical variables, cell apoptosis and inflammation. Muscone also showed a therapeutic effect in the AKI model through anti-apoptosis, anti-inflammation and amelioration in pathological changes. However, the kidney weight coefficient did not significantly decrease with muscone treatment. This result may be attributed to the amount of effusion in kidney tissues, which was observed by H&E staining.

Furthermore, dexamethasone were used as the positive drug in our study and also showed a therapeutic function to treat the nephrotoxic injury. As a kind of glucocorticoids, dexamethasone was widely used in the clinical studies for the treatment of kidney injury [36–39]. In succession, its therapeutic function for different kidney injury animal models (cyclosporine A nephrotoxicity injury, ischemia/reperfusion injury, uranyl nitrate induced injury, endotoxin induced injury and gentamicin nephrotoxic injury) has been demonstrated, and associated mechanisms, such as ameliorating microvascular oxygenation and stabilizing TRPC6 expression and distribution, were also investigated by several groups [34,40–45]. Therefore, it was chosen as the positive drug in our study. Even though dexamethasone showed some therapeutic action against AKI, it was still not as effective as BMSCs or the combined group in some respects.

The pathological changes in the kidney tubules, kidney glomeruli, and collecting tubules in each group were observed. Typical changes in AKI induced by gentamicin were mainly reflected in the kidney tubules and collecting tubules. However, the degree of cell apoptosis in the kidney tubules was more serious than that in the collecting tubules. Thus, the kidney tubules were the main damaged area in the AKI model induced by gentamicin.

The combined group showed superior effects, but no significant difference in pathological changes was observed compared with the BMSC group. This result was similar to our previous study [34]. This phenomenon might be due to the unsuitable dosage of the muscone and BMSCs, and only one dosage was tested in our study after all. Therefore, further investigations on the dosage or other factors are needed to improve the therapeutic effect on pathological changes. Our previous study investigated the therapeutic effect of BMSCs combined with vitamin E against AKI in rats, and the combined treatment was better than BMSCs or vitamin E alone. However, the two treatments may function independently from each other. Muscone, which has been used in clinical studies for years, was used in this study to improve the therapeutic effect of BMSCs by enhancing cell migration and proliferation. This feasible therapeutic approach may produce better effects in future clinical studies.

Herein, several inflammatory components (MCP-1, IL-10, RANTES and MIP-2) in the kidney tissues of each group were analyzed with qPCR and ELISA. In these inflammatory components, IL-10 is capable of inhibiting synthesis of pro-inflammatory cytokines. For the compensation in vivo, IL-10 was up-regulated together with other pro-inflammatory cytokines, and down-regulated when The inflammatory reaction was relieved gradually. Our study had showed that muscone had no ability to increase the expression or secretion of IL-10 in BMSCs (Figure 2). The detection of IL-10 in vivo further confirmed no obvious difference existed between the BMSCs group and combined group (Figure 7). Even though our results showed that the combined group hold better therapeutic action on down-regulating

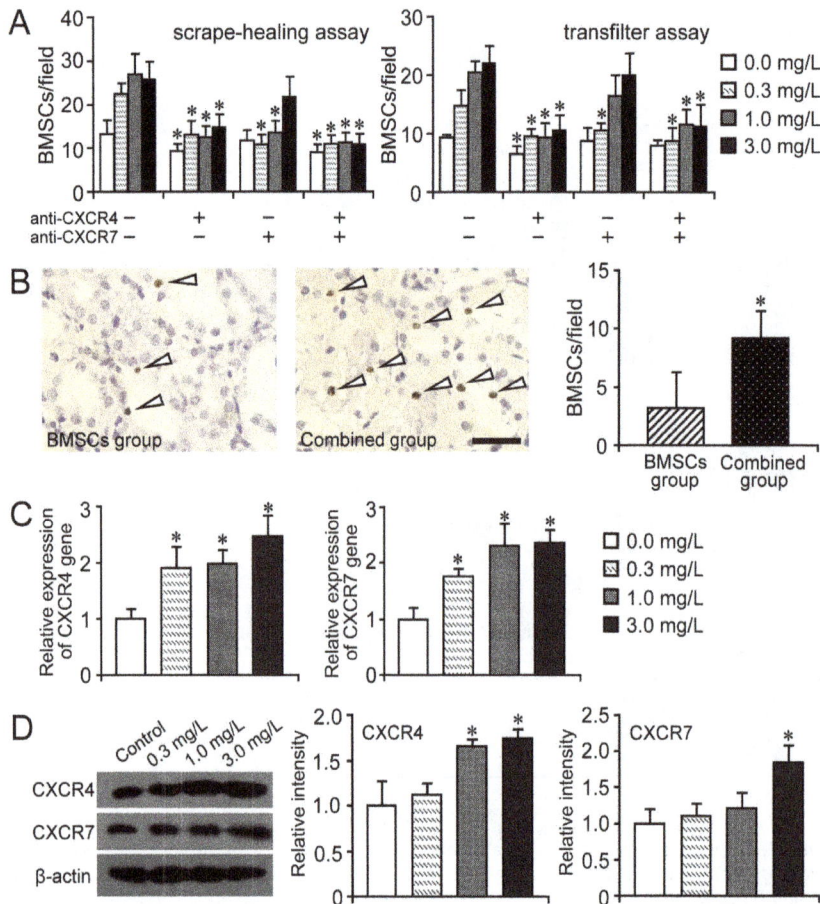

Figure 8. Effect of muscone on BMSC migration capacity. A: Quantification of the BMSC migration capacity using scrape-healing assay and transfilter assay. The number of migrating BMSCs was counted in each field. BMSCs exposed to CXCR4 and CXCR7 neutralizing antibodies were used for the both assay, and the promigratory effect of muscone on BMSC was abrogated in different degree. *$P<0.05$ compared with the corresponding group without the treatment of CXCR4 or CXCR7 neutralizing antibody. B: Enhancement of BMSC migration *in vitro*. Representative images of BrdU-positive cells in the BMSC and combined groups. The number of BrdU-positive cells was further counted in each field. Scale bar corresponds to 30 μm. C and D. Expression of CXCR4 and CXCR7 in BMSCs with or without muscone treatment detected using qPCR and western blot, respectively. Similar results were obtained in at least three independent experiments. Results are expressed as mean ± SEM. A *t*-test was used to compare the various groups, and $P<0.05$ was considered statistically significant. *$P<0.05$ compared with the normal BMSC group without any muscone treatment.

RANTES and MIP-2 than the BMSCs group, the mechanism still need further exploration.

To further analyze the mechanism by which muscone promotes BMSC migration and proliferation, we detected the expression of CXCR4 and CXCR7 in BMSCs treated with muscone. Several groups have reported that the SDF-1/CXCR4 axis is a pivotal mediator of migration, proliferation, and survival of BMSCs [33,46–49]. For most cell therapy experiments, expanding the cells *in vitro* is unavoidable. However, some reports showed that CXCR4 expression declined after several passages in the culture process [46,50], which possibly decreased the homing and engraftment potentials of stem cells. Thus, the upregulation of CXCR4 expression on the stem cell surface should be an effective strategy to overcome this limitation. Although SDF-1 was originally assumed to signal exclusively through the chemokine receptor, CXCR4, another research identified CXCR7 as a second SDF-1 receptor [51], which interacted with CXCR4 and modulated its functions. Mazzinghi et al. investigated the function of CXCR4 and CXCR7 in renal progenitor cells [52]. They reported that both receptors were crucial for the homing and therapeutic potentials of these cells, and showed that CXCR7 was

more involved in cell survival and adhesion to endothelium, whereas CXCR4 was involved in cell chemotaxis. In the present study, both CXCR4 and CXCR7 expression were upregulated in the BMSCs with muscone treatment, which could explain the function of muscone in promoting cell migration and proliferation.

In this study, the BMSCs and muscone were given simultaneously in rats with gentamicin-induced AKI. BMSCs preconditioned with muscone (3.0 mg/L, for 36 h) were also used to treat AKI rats in our previous research, but the therapeutic effect was not enhanced compared with the normal BMSC group, such as the improvement of kidney weight coefficient, biochemical variables in urine and serum, and cell apoptosis (data not shown). This result indicates the necessity of long-term treatment with muscone to enhance the therapeutic effect of BMSCs against AKI *in vivo*.

Conclusions

In our study, muscone enhanced the therapeutic action of BMSCs by promoting cell proliferation, secretion, and migration. This finding could be used as a novel therapeutic approach for AKI or other diseases in the field of regenerative medicine through anti-apoptosis, anti-inflammation and amelioration in some biochemical variables. The mechanism was preliminarily investigated, and the expression of CXCR4 and CXCR7 was upregulated in BMSCs after muscone treatment. However, the combined group showed no increase in pathological changes or some physiological indices. Thus, the optimal medication used in stem cell-based therapy for AKI should be further explored in different animal models of AKI, even in clinical studies.

Methods and Materials

This study was carried out in strict accordance with the recommendations in the Guide for the Care and Use of Laboratory Animals of the National Institutes of Health. The protocol was approved by the Committee on the Ethics of Animal Experiments of Jilin University. All operations were performed under sodium pentobarbital anesthesia, and all efforts were made to minimize suffering.

Isolation and characterization of BMSCs and renal tubular epithelial cells (RTECs)

Whole bone marrow was collected from the tibia and femur of a male Wistar rat weighing 120 g to 150 g. BMSCs were isolated from the whole bone marrow through density gradient centrifugation. The isolated cells were cultured in DMEM/F12 medium (Gibco, USA) containing 10% fetal bovine serum (FBS; Gibco), penicillin/streptomycin (100 U/mL to 0.1 mg/mL; Hyclone, USA), and 2 mM L-glutamine (Gibco, USA). The cells were incubated in a humidified incubator with 5% CO_2 at 37°C. The media were changed every other day. The adherent cells were passaged by a dilution of 1:3 once every 4 days or 5 days until passage 4, when a portion of the resulting BMSCs was prepared for phenotypic analysis by flow cytometry. The cells were then screened for CD14, CD29, CD34, CD44, CD45, CD73, CD90, CD105, and CD166. BMSCs were differentiated into adipocytes and osteocytes as previously described [53]. The rat RTECs were collected and cultured as described [34,54], and screened for CK-18 after one passage with flow cytometry and immunohistochemistry. This operation was similar to that in our previous research [34].

Effect of muscone on BMSC and RTEC proliferation and secretion

BMSCs and RTECs were planted in 96-well plates at 1×10^3 and 3×10^3 cells per well, respectively. The muscone (Santa, USA) concentrations used in the culture system were 0.3, 1.0, and 3.0 mg/L. Cell proliferation was detected at day 0, day 1, day 2, and day 3. The DMEM/F12 medium containing 10% FBS was used as the blank control group, and the proliferation index of each group was determined using the CCK-8 method (Dojindo, Japan) according to the manufacturer's instructions [34], which allowed sensitive colorimetric assays for the determination of the number of viable cells in cell proliferation. In brief, 10 μL of CCK-8 solution was added into each well (containing 100 μL of medium), and cultured for 1 h to 2 h at 37°C. The absorbance of each group at 450 nm was detected (n = 4) and it was directly proportional to the number of living cells. The proliferation index = the absorbance of experimental group − the absorbance of blank group, was used to measure cell proliferation in our study. After 2 days of culture, the medium of each group was collected for secretory cell function analysis. Vascular endothelial growth factor (VEGF), hepatocyte growth factor (HGF), bone morphogenetic protein-7 (BMP-7), and interleukin-10 (IL-10), which were secreted by BMSCs and relieved kidney injury, were measured

using an ELISA Kit (Abcam, United Kingdom) following the manufacturer's instructions. The cells were collected for real-time quantitative PCR (qPCR) analysis. A similar operation was used to evaluate the effect of muscone on RTEC secretion [HGF, transforming growth factor-β (TGF-β), tissue inhibitor of metalloproteinase-1 (TIMP-1), and endothelin-1 (ET-1)].

Flow cytometry

BMSCs or RTECs were dissociated into single cells with 0.25% trypsin, further fixed and permeated with Fixation Buffer (BD) and Perm/Wash Buffer (BD), respectively, and prepared at a concentration of 1.0×10^5 cells in 100 μL of PBS. The antibodies, including CD14 conjugated to FITC (Biorbyt, UK), CD29 conjugated to FITC (BD, USA), CD44 conjugated to FITC (BD), CD45 conjugated to PE (BD), and CD90 conjugated to FITC (BioLegend, USA), were added and incubated for 30 min at 4°C. After two washes in PBS, cells were acquired and analyzed by FACScalibur (BD Bioscience). The antibodies, including CK-18 (Bioss, China), CD73 (BD), CD105 (Boster, China), CD34 (Abcam, United Kingdom), and CD166 (Santa), were added. After 30 min of incubation at 37°C, the cells were washed three times with PBS and incubated with FITC-conjugated goat anti-rabbit or donkey anti-goat IgG (Invitrogen, USA) for 30 min at 37°C. After two washes in PBS, cells were obtained and analyzed.

Preparation and treatment of AKI model

Male Wistar rats, weighing 250 g to 300 g, were used in the experiment. All rats were housed in a room with constant temperature room with a 12 h dark-12 h light cycle and fed a standard diet. The method used to prepare or treat the AKI model is shown in Figure S4. In brief, animals were divided into six groups (six rats per group), namely, normal group, AKI model group, positive drug group, muscone group, BMSC group, and BMSCs combined with muscone group (combined group). To generate a rat model of AKI for testing the BMSCs and muscone *in vivo*, the model was induced by gentamicin (140 mg/kg/day for 7 days, i.p.) in rats. At day 8, BMSCs were administered by intravenous injection of 3.3×10^6 cells/kg combined with muscone (75 mg/kg), whereas dexamethasone (0.08 mg/kg) was used as a positive control drug for 7 d. To trace the cells *in vivo*, BMSCs were labeled with 10 μM BrdU (Sigma). The normal group did not receive any treatment. All research experiment protocols adhered to the Principles of Laboratory Animal Care, and were approved by the Institutional Animal Care and Use Committee of Jilin University.

Evaluation of kidney function

In our study, all animals were sacrificed on day 15. To determine the biochemical variables (e.g., urea nitrogen) using a Biochemistry Autoanalyzer, blood and urine were collected on days 8 (before treatment) and 15, respectively. Urinary levels of NAG and LZM were normalized by urinary creatinine, since it corrects for the filtered contribution of a given marker. To evaluate the effect of BMSCs and muscone on the accrementition and enlargement of tissue damage induced by gentamicin in the kidney, the kidney weight coefficient of each group was determined using the following algorithm: kidney weight coefficient = bilateral kidney weight (g) / body weight (g). Our previous study had showed that the kidney weight coefficient of normal rat was below 0.01 while average coefficient was over 0.02 in model group [34]. Therefore, this coefficient could be used for the evaluation of AKI model induced by gentamicin. For histological analysis, kidney tissue samples of each group were fixed with formalin, embedded in paraffin, sectioned to 5 μm-thick, and processed by hematoxylin and eosin (H&E) staining. Then the slides were reviewed blindly and scored with a semiquantitative scale evaluating changes found in AKI as the reference [55]. For each kidney, 100 cortical tubules from at least 10 different areas were scored, and avoid repeated scoring of different convolutions of the same tubule. The average histological score was used to valuate kidney tissue morphology. Higher scores represented more severe damage. On the other side, the samples fixed in glutaraldehyde were dehydrated in graded ethyl alcohols, and further embedded with Epon812. An ultrathin section was cut with an Ultracut E ultramicrotome, stained with uranyl acetate and lead citrate, and observed under transmission electron microscopy (TEM) as previously described [56,57].

Analysis of cell apoptosis

Apoptosis in the kidney tissue was evaluated by enzymatic labeling of DNA strand breaks using terminal deoxynucleotidyl transferase-mediated deoxyuridine triphosphate nick end-labeling (TUNEL) assay kit (KeyGEN) according to the manufacturer's instructions. The expression levels of apoptotic genes (Caspase 3, Fas, and Bax) and anti-apoptotic gene (Bcl-2) in the kidney tissues of each group were further detected using real-time qPCR and immunohistochemistry. The activated Caspase3 in kidney tissues was further detected with Caspase-3 colorimetric assay kit (KeyGEN) according to its manufacturer's instructions

Real-time qPCR

For real-time qPCR analysis, kidney tissues (40 mg) or 3×10^6 cells were homogenized in 1 mL of Trizol reagent (Invitrogen), and total RNA was extracted. RNA (2 µg) was reverse-transcribed using an RT–PCR kit (Takara, Japan), and qPCR was performed using a Thermal Cycler Dice^{TM} Real-Time System and SYBR Green Premix EX Taq^{TM} (Takara). GAPDH was used for qPCR normalization and all items were measured in triplicate. Primer sequences were obtained from the references [11,34,58–64].

VEGF forward (F) 5'-ACAGCTTTTTGCCTTCGAGCTA-3'
reverse (R) 5'-CATCAAAGCCCTTGTCGGGATA-3'
HGF forward (F) 5'-GGCTTTACTGCTGTACCTCC-3'
reverse (R) 5'-CAAATGCTTTCTCCGCTCT-3'
BMP-7 forward (F) 5'-AGACGCCAAAGAACCAAGAG-3'
reverse (R) 5'-GCTGTCGTCGAAGTAGAGGA-3'
IL-10 forward (F) 5'-GCAGGTGTCCCAAAGAAG-3'
reverse (R) 5'-TCAAAGGTGCTGAAGTCC-3'
TGF-β forward (F) 5'-CTGCTGACCCCCACTGATAC-3'
reverse (R) 5'-GTGAGCACTGAAGCGAAAGC-3'
TIMP-1 forward (F) 5'-ATTTGCACATCACTGCCTGC-3'
reverse (R) 5'-GGGATGGCTGAACAGGGAAA-3'
ET-1 forward (F) 5'-TCAGAGCAACCAGACACCGT-3'
reverse (R) 5'-CTTGGAAAGCCACAAACAGC-3'
Caspase 3 forward (F) 5'-GACAGTGGCATCGAGACAGA-3'
reverse (R) 5'-GAAAAGTGGCATCAAGGGAA-3'
Fas forward (F) 5'-CTCTGGAAGTGCATGCTGTAAGA-3'
reverse (R) 5'-GGTAGATGTCATTTGCGAAAGGT-3'
Bax forward (F) 5'-CTGCCAACCCACCCTGGT-3'
reverse (R) 5'-TGGCAGCTGACATGTTTTCTG-3'
Bcl-2 forward (F) 5'-TCTGTGGATGACTGAGTACCT-GAAC-3'
reverse (R) 5'-AGAGACAGCCAGGAGAAATCAAAC-3'
MCP-1 forward (F) 5'-CAGAAACCAGCCAACTCTCA-3'
reverse (R) 5'-GTGGGGCATTAACTGCATCT-3'
RANTES forward (F) 5'-GGGCAGATGATTCTGAGA-CAAC-3'
reverse (R) 5'-CCAGGAATGAGTGGGGAGTAGG-3'
MIP-2 forward (F) 5'-TTGTCTCAACCCTGAAGCCC-3'

reverse (R) 5'-TGCCCGTTGAGGTACAGGAG-3'
CXCR4 forward (F) 5'-TAGTGGGCAATGGGTTGG-TAATC-3'
reverse (R) 5'-CTGCTGTAAAGGTTGACGGTGTA-3'
CXCR7 forward (F) 5'-TCACCTACTTCACCAGCACC-3'
reverse (R) 5'-ACATGGCTCTGGCGAGCAGG-3'
GAPDH forward (F) 5'-GCCAGCCTCGTCTCATAGACA-3'
reverse (R) 5'-AGAGAAGGCAGCCCTGGTAAC-3'

Western blot

In our experiment, cells were harvested at the indicated times with RIPA lysis buffer (50 mM Tris–HCl, pH 7.4; 1% TritonX-100; 150 mM NaCl; 1% sodium deoxycholate; 0.1% SDS), including the phosphatase inhibitors (sodium orthovanadate, 2 mM) and protease inhibitor (0.5 µg/mL leupeptin, 0.1 µg/mL aprotinin, 0.6 µg/mL pepstatin A), for 30 min. After centrifugation at 12,000 rpm for 15 min, the protein content of cell lysates was determined using a BCA protein estimation kit (Pierce, USA), and bovine serum albumin was used as the standard. Equal amounts (15 µg) of protein were loaded per lane and electrophoresed in a 12% acrylamide gel, which was run at 120 V for 1 h. Protein transfer was performed using nitrocellulose for 1 h at 100 V. The primary antibodies used were anti-CXCR7 (1:1000; Santa) and anti-CXCR4 (1:1000; Abcam). Anti-rabbit or goat HRP and an Amersham ECL kit (GE Healthcare, USA) were used to detect protein. The band densities were quantified by densitometry (Quantity One v4.62).

Immunohistochemistry

The cells or kidney tissues were fixed in 4% paraformaldehyde in PBS. The kidney tissues of each group were embedded in paraffin and sliced into 5 µm-thick sections. For immunohistochemistry, the primary antibodies used were anti-CK-18 (1:50, Bioss), anti-Caspase 3 (1:100, Abcam), anti-Fas (1:50, Santa), anti-Bax (1:50, Santa), anti-Bcl-2 (1:100, Abcam), and anti-BrDU (1:100, Abcam) polyclonal antibodies. After 12 h of incubation at 4°C, the samples were washed three times with PBS and processed using an SABC kit and DAB solution. Finally, the sections were observed using Axio Scope A1 (Zeiss, Germany) with AxioCAM MRc5 (Zeiss), and processed with AxioVision software (Zeiss). The color intensity was measured using Motic Image Advanced 3.2.

Evaluation of inflammatory components

After the therapy, the inflammatory components in the kidney tissues of each group were detected with qPCR and ELISA. For the ELISA assay, kidney tissues (200 mg) were homogenized with PBS solution and the homogenate was centrifuged at 3000 rpm for 20 min. The supernatant was collected for ELISA. Herein, monocyte/macrophage chemotactic protein-1 (MCP-1), IL-10, regulated upon activation normal T cell expressed and presumably secreted (RANTES) and Macrophage inflammatory protein 2 (MIP-2) were examined for evaluation of inflammatory reaction.

Cell migration assays

Both the scrape-healing assay and transfilter assay were performed as described in the literature [33]. In the scrape-healing assay, BMSCs were seeded at 8×10^4 cells/well in 24-well plates and allowed to reach confluence. The cells were then switched to DMEM/F12 medium containing 2% FBS and incubated with muscone (0.3, 1.0, and 3.0 mg/L) for 24 h. The monolayers were then scratched by the tip of a 1ml pipette, washed with PBS, then observed under the microscope to determine the wound distance used for counting, and further incubated in the medium containing 2% FBS for 24 h. Finally, the images of cell samples were obtained for further analysis, and four random fields were counted for each group. For transfilter assay, BMSCs were treated as scrape-healing assay and seeded on the upper side of a porous polycarbonate membrane (pore size: 8 μm; Euroclone SPA, Italy) in co-culture with RTECs. RTECs were seeded at 9×10^4 cells/well in a 24-well plate. After 24 h, cells were incubated with DMEM/F12 plus 2% FBS alone or in the presence of 5 μM cisplatin for 6 h. After cisplatin removal, the RTECs were washed and co-cultured with BMSCs. After 36 h, the cells at the upper side of the filter (unmigrated cells) were mechanically removed. Cells that had migrated to the lower side of the filter were fixed for 30 min in 4% paraformaldehyde and further stained with crystal violet. The number of cells in six random fields was counted for each filter. We also tried to induce the RTECs damage with gentamicin (10 U/ml, 30 U/ml and 100 U/ml). However, the damaged RTECs could not induce the cell migration effectively. Therefore, cisplatin was used in our study as the reference [33]. To investigate the influence of cell proliferation on the migration, mitomycin C was used to block proliferation. In our study, BMSCs incubated with 10 mg/mL mitomycin C (Sigma) for 60 min, were used for transfilter assay as the reference [65]. In addition, for further exploring the regulation of muscone on cell migration, BMSCs exposed to CXCR4 and CXCR7 neutralizing antibody (10 μg/ml), were used for scrape-healing assay, as well as transfilter assay.

Statistical analyses

The results are expressed as means ± SEM, and statistical analysis was performed using SPSS 17.0. The differences among groups were analyzed by one-way ANOVA followed by t-test. $P < 0.05$ was considered statistically significant.

Supporting Information

Figure S1 Effect of muscone on RTEC bioactivity *in vitro*. A: Effect of muscone on RTEC proliferation. Proliferation index (the absorbance of experimental group − the absorbance of blank group) was measured using CCK-8. B: Effect of muscone on RTEC secretion. To evaluate the secretory function, the secretion or cytokine expression level of normal BMSCs without muscone treatment was 1.0 for each cytokine. Cytokine expression in RTECs detected using qPCR is shown on the left, and the RTEC secretory function evaluated using ELISA is shown on the right. Similar results were obtained in at least three independent experiments. Results are expressed as mean ± SEM. A *t*-test was used to compare the various groups, and $P < 0.05$ was considered statistically significant. *$P < 0.05$ compared with the normal RTEC group without any muscone treatment.

Figure S2 Expression of apoptotic genes and anti-apoptotic gene in the kidney tissues of each group. The expression levels of apoptotic genes (Caspase 3, Fas, and Bax) and anti-apoptotic gene (Bcl-2) in the kidney tissue were also detected using immunohistochemistry. A representative in the kidney tubules, kidney glomeruli, and collecting tubules close to the average level of each group is shown for each group. Scale bar corresponds to 100 μm.

Figure S3 A: Transmigration of muscone-treated BMSCs toward the cisplatin-injured RTECs. BMSCs co-cultured with healthy RTECs or without RTECs were used as the control group. *$P < 0.05$ compared with the normal BMSC group without any muscone treatment. B: Effect of incubation time on cell migration in transfilter assay. *$P < 0.05$ compared with the normal BMSC group without any muscone treatment after same incubation time. C: Effect of cell proliferation on cell migration in transfilter assay. BMSCs treated with mitomycin C were used for transfilter assay, and no significant difference existed between the normal BMSCs and the mitomycin C treated BMSCs.

Figure S4 Preparation and treatment of rat AKI model. The rat AKI model was induced with gentamicin and further treated with dexamethasone, muscone, and stem cells.

Table S1 The absolute values in proliferation indexes of each group. BMSCs were treated with muscone at different concentrations (0.3, 1.0, and 3.0 mg/L) Proliferation index (the absorbance of experimental group − the absorbance of blank group) on day 0, day 1, day 2 and day 3 was measured using CCK-8.

Table S2 Urinary creatinine during the therapy process. I: urinary creatinine for characteristics of AKI model; II: urinary creatinine after the therapy.

Acknowledgments

We thank Mr. Bo Sun, Mrs. Yi Shi and Dr. Jinglei Cai for their technical assistance.

Author Contributions

Conceived and designed the experiments: PL YF YZ ZZ. Performed the experiments: PL YF CD DY BL XC. Analyzed the data: PL YW LZ. Contributed reagents/materials/analysis tools: PL YW. Wrote the paper: PL LZ.

References

1. Chertow GM, Burdick E, Honour M, Bonventre JV, Bates DW (2005) Acute kidney injury, mortality, length of stay, and costs in hospitalized patients. J Am Soc Nephrol 16: 3365–3370.
2. Bagshaw SM (2008) Short- and long-term survival after acute kidney injury. Nephrol Dial Transplant 23: 2126–2128.
3. Kellum JA, Hoste EA (2008) Acute kidney injury: epidemiology and assessment. Scand J Clin Lab Invest Suppl 241: 6–11.
4. Nash K, Hafeez A, Hou S (2002) Hospital-acquired renal insufficiency. American Journal of Kidney Diseases 39: 930–936.
5. Bellomo R, Kellum JA, Ronco C (2012) Acute kidney injury. Lancet 380: 756–766.
6. Bonventre JV, Yang L (2011) Cellular pathophysiology of ischemic acute kidney injury. J Clin Invest 121: 4210–4221.
7. Wang L, Zhao S, Mao H, Zhou L, Wang ZJ, et al. (2011) Autologous bone marrow stem cell transplantation for the treatment of type 2 diabetes mellitus. Chin Med J (Engl) 124: 3622–3628.

8. Kanno H (2013) Regenerative therapy for neuronal diseases with transplantation of somatic stem cells. World J Stem Cells 5: 163–171.
9. Daley GQ (2012) The promise and perils of stem cell therapeutics. Cell Stem Cell 10: 740–749.
10. Bilousova G, Roop DR (2013) Generation of functional multipotent keratinocytes from mouse induced pluripotent stem cells. Methods Mol Biol 961: 337–350.
11. Li T, Zhu J, Ma K, Liu N, Feng K, et al. (2013) Autologous bone marrow-derived mesenchymal stem cell transplantation promotes liver regeneration after portal vein embolization in cirrhotic rats. J Surg Res 184: 1161–1173.
12. Fujimoto Y, Abematsu M, Falk A, Tsujimura K, Sanosaka T, et al. (2012) Treatment of a mouse model of spinal cord injury by transplantation of human induced pluripotent stem cell-derived long-term self-renewing neuroepithelial-like stem cells. Stem Cells 30: 1163–1173.

13. Raja N, Miller WE, McMillan R, Mason JR (1998) Ciprofloxacin-associated acute renal failure in patients undergoing high-dose chemotherapy and autologous stem cell rescue. Bone Marrow Transplant 21: 1283–1284.

14. Togel F, Hu ZM, Weiss K, Isaac J, Lange C, et al. (2005) Administered mesenchymal stem cells protect against ischemic acute renal failure through differentiation-independent mechanisms. American Journal of Physiology-Renal Physiology 289: F31–F42.

15. Iwasaki M, Adachi Y, Minamino K, Suzuki Y, Zhang Y, et al. (2005) Mobilization of bone marrow cells by G-CSF rescues mice from cisplatin-induced renal failure, and M-CSF enhances the effects of G-CSF. J Am Soc Nephrol 16: 658–666.

16. Rota C, Imberti B, Pozzobon M, Piccoli M, De Coppi P, et al. (2012) Human amniotic fluid stem cell preconditioning improves their regenerative potential. Stem Cells Dev 21: 1911–1923.

17. Li L, Black R, Ma Z, Yang Q, Wang A, et al. (2012) Use of mouse hematopoietic stem and progenitor cells to treat acute kidney injury. Am J Physiol Renal Physiol 302: F9–F19.

18. Choi HY, Moon SJ, Ratliff BB, Ahn SH, Jung A, et al. (2014) Microparticles from kidney-derived mesenchymal stem cells act as carriers of proangiogenic signals and contribute to recovery from acute kidney injury. PLoS One 9: e87853.

19. Liu N, Tian J, Cheng J, Zhang J (2013) Effect of erythropoietin on the migration of bone marrow-derived mesenchymal stem cells to the acute kidney injury microenvironment. Exp Cell Res 319: 2019–2027.

20. Reis LA, Borges FT, Simoes MJ, Borges AA, Sinigaglia-Coimbra R, et al. (2012) Bone marrow-derived mesenchymal stem cells repaired but did not prevent gentamicin-induced acute kidney injury through paracrine effects in rats. PLoS One 7: e44092.

21. Qi S, Wu D (2013) Bone marrow-derived mesenchymal stem cells protect against cisplatin-induced acute kidney injury in rats by inhibiting cell apoptosis. Int J Mol Med 32: 1262–1272.

22. Horwitz EM, Le Blanc K, Dominici M, Mueller I, Slaper-Cortenbach I, et al. (2005) Clarification of the nomenclature for MSC: The International Society for Cellular Therapy position statement. Cytotherapy 7: 393–395.

23. Zarjou A, Kim J, Traylor AM, Sanders PW, Balla J, et al. (2011) Paracrine effects of mesenchymal stem cells in cisplatin-induced renal injury require heme oxygenase-1. Am J Physiol Renal Physiol 300: F254–262.

24. Gatti S, Bruno S, Deregibus MC, Sordi A, Cantaluppi V, et al. (2011) Microvesicles derived from human adult mesenchymal stem cells protect against ischaemia-reperfusion-induced acute and chronic kidney injury. Nephrol Dial Transplant 26: 1474–1483.

25. Jia X, Xie X, Feng G, Lu H, Zhao Q, et al. (2012) Bone marrow-derived cells can acquire renal stem cells properties and ameliorate ischemia-reperfusion induced acute renal injury. BMC Nephrol 13: 105.

26. Eliopoulos N, Zhao J, Forner K, Birman E, Young YK, et al. (2011) Erythropoietin gene-enhanced marrow mesenchymal stromal cells decrease cisplatin-induced kidney injury and improve survival of allogeneic mice. Mol Ther 19: 2072–2083.

27. Liu N, Tian J, Cheng J, Zhang J (2013) Migration of CXCR4 gene-modified bone marrow-derived mesenchymal stem cells to the acute injured kidney. J Cell Biochem 114: 2677–2689.

28. Gheisari Y, Azadmanesh K, Ahmadbeigi N, Nassiri SM, Golestaneh AF, et al. (2012) Genetic modification of mesenchymal stem cells to overexpress CXCR4 and CXCR7 does not improve the homing and therapeutic potentials of these cells in experimental acute kidney injury. Stem Cells Dev 21: 2969–2980.

29. Lin DL, Chang HC, Huang SH (2004) Characterization of allegedly musk-containing medicinal products in Taiwan. J Forensic Sci 49: 1187–1193.

30. Wu Q, Li H, Wu Y, Shen W, Zeng L, et al. (2011) Protective effects of muscone on ischemia-reperfusion injury in cardiac myocytes. J Ethnopharmacol 138: 34–39.

31. Ford RA, Api AM, Newberne PM (1990) 90-day dermal toxicity study and neurotoxicity evaluation of nitromusks in the albino rat. Food Chem Toxicol 28: 55–61.

32. Xie XW, Hou FW, Li N (2012) Effects of musk ketone at different concentrations on in vivo migration of exogenous rat bone marrow mesenchymal stem cells. Zhongguo Zhong Xi Yi Jie He Za Zhi 32: 980–985.

33. Xinaris C, Morigi M, Benedetti V, Imberti B, Fabricio AS, et al. (2013) A novel strategy to enhance mesenchymal stem cell migration capacity and promote tissue repair in an injury specific fashion. Cell Transplant 22: 423–436.

34. Liu P, Feng Y, Dong C, Liu D, Wu X, et al. (2013) Study on therapeutic action of bone marrow derived mesenchymal stem cell combined with vitamin E against acute kidney injury in rats. Life Sci 92: 829–837.

35. Cheng K, Rai P, Plagov A, Lan X, Kumar D, et al. (2013) Transplantation of bone marrow-derived MSCs improves cisplatinum-induced renal injury through paracrine mechanisms. Exp Mol Pathol 94: 466–473.

36. Hanf W, Guillaume C, Jolivot A, Chapuis-Cellier C, Guebre-Egziabher F, et al. (2010) Prolonged hemodialysis for acute kidney injury in myeloma patients. Clin Nephrol 74: 319–322.

37. Hasegawa M, Kondo F, Yamamoto K, Murakami K, Tomita M, et al. (2010) Evaluation of blood purification and bortezomib plus dexamethasone therapy for the treatment of acute renal failure due to myeloma cast nephropathy. Ther Apher Dial 14: 451–456.

38. Shi SF, Zhou FD, Zou WZ, Wang HY (2012) Acute kidney injury and bilateral symmetrical enlargement of the kidneys as first presentation of B-cell lymphoblastic lymphoma. Am J Kidney Dis 60: 1044–1048.

39. Wadhwa NK, Kamra A, Skopicki HA, Richard S, Miller F, et al. (2012) Reversible left ventricular dysfunction and acute kidney injury in a patient with nonamyloid light chain deposition disease. Clin Nephrol 78: 501–505.

40. Johannes T, Mik EG, Klingel K, Dieterich HJ, Unertl KE, et al. (2009) Low-dose dexamethasone-supplemented fluid resuscitation reverses endotoxin-induced acute renal failure and prevents cortical microvascular hypoxia. Shock 31: 521–528.

41. Kumar S, Allen DA, Kieswich JE, Patel NS, Harwood S, et al. (2009) Dexamethasone ameliorates renal ischemia-reperfusion injury. J Am Soc Nephrol 20: 2412–2425.

42. Lee AK, Lee JH, Kwon JW, Kim WB, Kim SG, et al. (2004) Pharmacokinetics of clarithromycin in rats with acute renal failure induced by uranyl nitrate. Biopharm Drug Dispos 25: 273–282.

43. Sun QL, Chen YP, Rui HL (2010) Establishment of a new rat model of chronic cyclosporine A nephrotoxicity. Zhongguo Yi Xue Ke Xue Yuan Xue Bao 32: 205–209.

44. Yu S, Yu L (2012) Dexamethasone Resisted Podocyte Injury via Stabilizing TRPC6 Expression and Distribution. Evid Based Complement Alternat Med 2012: 652059.

45. Zhou L, Yao X, Chen Y (2012) Dexamethasone pretreatment attenuates lung and kidney injury in cholestatic rats induced by hepatic ischemia/reperfusion. Inflammation 35: 289–296.

46. Honczarenko M, Le Y, Swierkowski M, Ghiran I, Glodek AM, et al. (2006) Human bone marrow stromal cells express a distinct set of biologically functional chemokine receptors. Stem Cells 24: 1030–1041.

47. Chamberlain G, Wright K, Rot A, Ashton B, Middleton J (2008) Murine mesenchymal stem cells exhibit a restricted repertoire of functional chemokine receptors: comparison with human. PLoS One 3: e2934.

48. Peled A, Petit I, Kollet O, Magid M, Ponomaryov T, et al. (1999) Dependence of human stem cell engraftment and repopulation of NOD/SCID mice on CXCR4. Science 283: 845–848.

49. Herberg S, Shi X, Johnson MH, Hamrick MW, Isales CM, et al. (2013) Stromal cell-derived factor-1beta mediates cell survival through enhancing autophagy in bone marrow-derived mesenchymal stem cells. PLoS One 8: e58207.

50. Ahmadbeigi N, Seyedjafari E, Gheisari Y, Atashi A, Omidkhoda A, et al. (2010) Surface expression of CXCR4 in unrestricted somatic stem cells and its regulation by growth factors. Cell Biol Int 34: 687–692.

51. Balabanian K, Lagane B, Infantino S, Chow KY, Harriague J, et al. (2005) The chemokine SDF-1/CXCL12 binds to and signals through the orphan receptor RDC1 in T lymphocytes. J Biol Chem 280: 35760–35766.

52. Mazzinghi B, Ronconi E, Lazzeri E, Sagrinati C, Ballerini L, et al. (2008) Essential but differential role for CXCR4 and CXCR7 in the therapeutic homing of human renal progenitor cells. J Exp Med 205: 479–490.

53. Hwang NS, Varghese S, Lee HJ, Zhang Z, Ye Z, et al. (2008) In vivo commitment and functional tissue regeneration using human embryonic stem cell-derived mesenchymal cells. Proc Natl Acad Sci U S A 105: 20641–20646.

54. Hauser PV, De Fazio R, Bruno S, Sdei S, Grange C, et al. (2010) Stem cells derived from human amniotic fluid contribute to acute kidney injury recovery. Am J Pathol 177: 2011–2021.

55. Paller MS, Hoidal JR, Ferris TF (1984) Oxygen free radicals in ischemic acute renal failure in the rat. J Clin Invest 74: 1156–1164.

56. Ebenezer PJ, Mariappan N, Elks CM, Haque M, Soltani Z, et al. (2009) Effects of pyrrolidine dithiocarbamate on high-fat diet-induced metabolic and renal alterations in rats. Life Sci 85: 357–364.

57. Cai J, Zhang Y, Liu P, Chen S, Wu X, et al. (2013) Generation of tooth-like structures from integration-free human urine induced pluripotent stem cells. Cell Regeneration 2: 6.

58. Yamada Y, Watanabe Y, Zhang J, Haraoka J, Ito H (2002) Changes in cortical and cerebellar bcl-2 mRNA levels in the developing hydrocephalic rat (LEW-HYR) as measured by a real time quantified RT-PCR. Neuroscience 114: 165–171.

59. He B, Xiao J, Ren AJ, Zhang YF, Zhang H, et al. (2011) Role of miR-1 and miR-133a in myocardial ischemic postconditioning. J Biomed Sci 18: 22.

60. Lipfert J, Odemis V, Wagner DC, Boltze J, Engele J (2013) CXCR4 and CXCR7 form a functional receptor unit for SDF-1/CXCL12 in primary rodent microglia. Neuropathol Appl Neurobiol 39: 667–680.

61. Krzysiek-Maczka G, Targosz A, Ptak-Belowska A, Korbut E, Szczyrk U, et al. (2013) Molecular alterations in fibroblasts exposed to Helicobacter pylori: a missing link in bacterial inflammation progressing into gastric carcinogenesis? J Physiol Pharmacol 64: 77–87.

62. Smirkin A, Matsumoto H, Takahashi H, Inoue A, Tagawa M, et al. (2010) Iba1(+)/NG2(+) macrophage-like cells expressing a variety of neuroprotective factors ameliorate ischemic damage of the brain. J Cereb Blood Flow Metab 30: 603–615.

63. Luo Z, Liu H, Sun X, Guo R, Cui R, et al. (2013) RNA interference against discoidin domain receptor 2 ameliorates alcoholic liver disease in rats. PLoS One 8: e55860.

64. Zhang Z, Zhong W, Hall MJ, Kurre P, Spencer D, et al. (2009) CXCR4 but not CXCR7 is mainly implicated in ocular leukocyte trafficking during ovalbumin-induced acute uveitis. Exp Eye Res 89: 522–531.

65. Kolambkar YM, Bajin M, Wojtowicz A, Hutmacher DW, Garcia AJ, et al. (2014) Nanofiber orientation and surface functionalization modulate human mesenchymal stem cell behavior in vitro. Tissue Eng Part A 20: 398–409.

20

Combinatorial G-CSF/AMD3100 Treatment in Cardiac Repair after Myocardial Infarction

Constantin Rüder[1,2*⑨], Tobias Haase[1,2⑨], Annalena Krost[1], Nicole Langwieser[1¤], Jan Peter[1], Stefanie Kamann[1], Dietlind Zohlnhöfer[1,2]

1 Berlin Brandenburg Center for Regenerative Therapies (BCRT), Berlin, Germany, 2 Department of Cardiology, Campus Virchow Klinikum, Charité Berlin, Germany

Abstract

Aims: Several studies suggest that circulating bone marrow derived stem cells promote the regeneration of ischemic tissues. For hematopoietic stem cell transplantation combinatorial granulocyte-colony stimulating factor (G-CSF)/Plerixafor (AMD3100) administration was shown to enhance mobilization of bone marrow derived stem cells compared to G-CSF monotherapy. Here we tested the hypothesis whether combinatorial G-CSF/AMD3100 therapy has beneficial effects in cardiac recovery in a mouse model of myocardial infarction.

Methods: We analyzed the effect of single G-CSF (250 µg/kg/day) and combinatorial G-CSF/AMD3100 (100 µg/kg/day) treatment on cardiac morphology, vascularization, and hemodynamics 28 days after permanent ligation of the left anterior descending artery (LAD). G-CSF treatment started directly after induction of myocardial infarction (MI) for 3 consecutive days followed by a single AMD3100 application on day three after MI in the G-CSF/AMD3100 group. Cell mobilization was assessed by flow cytometry of blood samples drawn from tail vein on day 0, 7, and 14.

Results: Peripheral blood analysis 7 days after MI showed enhanced mobilization of white blood cells (WBC) and endothelial progenitor cells (EPC) upon G-CSF and combinatorial G-CSF/AMD3100 treatment. However, single or combinatorial treatment showed no improvement in survival, left ventricular function, and infarction size compared to the saline treated control group 28 days after MI. Furthermore, no differences in histology and vascularization of infarcted hearts could be observed.

Conclusion: Although the implemented treatment regimen caused no adverse effects, our data show that combinatorial G-CSF/AMD therapy does not promote myocardial regeneration after permanent LAD occlusion.

Editor: Marie Jose Goumans, Leiden University Medical Center, Netherlands

Funding: Work was funded by the Berlin-Brandenburg Center for Regenerative Therapies (BCRT). The funder had no role in study design, data collection and analysis, decision to publish, or preparation of the manuscript.

Competing Interests: The authors have declared that no competing interests exist.

* Email: constantinrueder@hotmail.com

¤ Current address: Lilly Deutschland GmbH, München, Germany

⑨ These authors contributed equally to this work.

Introduction

Cytokine mediated mobilization of peripheral blood stem cells for autologous stem cell transplantation is a generally accepted therapeutic option for the hematopoietic reconstitution after myoablative chemotherapy. The clinically used cytokine granulocyte-colony stimulating factor (G-CSF) is known to mobilize various subsets of hematopoietic stem and progenitor cells (HSPC) into blood circulation that may contribute to tissue repair. Additionally G-CSF was shown to have anti-apoptotic, anti-inflammatory and antioxidant effects [1,2,3]. These findings raised expectations of G-CSF as a promising therapeutic avenue in tissue regeneration.

Especially in the field of ischemic heart disease numerous studies investigated the efficacy of G-CSF induced stem cell mobilization in myocardial regeneration. While early animal studies and small clinical trials indicated beneficial effects on cardiac regeneration,

these results were later challenged by studies that could not confirm these positive effects or even reported deleterious effects of G-CSF therapy on cardiac recovery (for review see [4,5,6]). The missing benefit of G-CSF induced mobilization of progenitor cells might be due to a reduced homing capacity as G-CSF treatment results in significant downregulation of important adhesion molecules on mobilized cells [7].

Besides G-CSF, the CXCR4 antagonist AMD3100 (AMD) was shown to rapidly mobilize stem cells by reversibly disrupting the interaction between CXCR4 and SDF-1α that tethers stem cells to the bone marrow (BM) environment [8]. In patients that do not respond to single G-CSF treatment a combination of G-CSF and AMD has shown to effectively mobilize hematopoietic stem cells (HSC) from the BM [9]. Moreover combinatorial G-CSF/AMD therapy was shown to be superior to single G-CSF therapy with respect to HSC numbers and is clinically approved for autologous HSC mobilization [10]. Preclinical studies on AMD in tissue regeneration showed that acute application leads to enhanced vascularization of ischemic tissues [11,12] while continuous AMD treatment has deleterious effects on tissue regeneration [13,14]. This effect was attributed to the crucial role of the CXCR4/SDF-1α axis in stem cell homing towards injured tissues [7,15].

On the basis of these results we explored possible beneficial effects of combinatorial G-CSF/AMD therapy in myocardial regeneration in a mouse model of MI. We applied a treatment regimen were G-CSF administration started directly after induction of MI for 3 consecutive days followed by a single dose of AMD in order to attain positive effects on stem cell mobilization while avoiding negative effects on stem cell homing.

Methods

Surgical induction of myocardial infarction and study design

Eight to ten weeks old male FVB/NJ mice (Charles River) were anaesthetized with an intraperitoneal injection of midazolam (5.0 mg/kg), fentanyl (0.05 mg/kg), and medetomidin (0.5 mg/kg). The animals were intubated and ventilated using a rodent ventilator (MiniVent, Hugo Sachs) with a stroke volume of 0.2 ml and respiration rate of 200 strokes/min. Inhalation anesthesia was maintained with 1.5% isoflurane through a vaporizer with 100% oxygen. After left lateral thoracotomy at the left third intercostal space, the left anterior descending coronary (LAD) was ligated with 7-0 prolene sutures (Ethicon) 1 mm below the tip of the left atrial auricle. The chest and skin were closed with 6-0 vicryl (Ethicon) sutures. The sham group underwent the same procedure except for the ligation of the LAD. After induction of MI animals were randomly divided into 3 groups and stem cell mobilization was induced by the following dosing regimen:

1) G-CSF (250 µg/kg/day; Amgen GmbH) subcutaneously (s.c.) starting 1 h post-MI and then daily on days 1 and 2; 2) Combination of G-CSF s.c. starting 1 h post-MI and then daily on days 1 and 2+ AMD3100 (5 mg/kg/day; Sigma-Aldrich) s.c. as single dose on day 3 post-MI; 3) the control MI group received at the same time points equal volumes of saline (0.9% NaCl). Data acquired from control MI and sham groups contributed in parallel to another study but were pooled with new animals [16]. Postoperative, mice were housed singly in enriched standard cages with free access to food and water. Mice were monitored three times per day during the first three days and two times per day from day 4–7. After this acute phase, mice were monitored one time per day. During the first 7 days after MI analgesia was maintained by buprenorphine application (0.1 mg/kg). The state of health of mice was recorded on a score sheet. Animals that died in this study after induction of MI deceased due to acute heart failure or heart rupture as a result of the intervention. Euthanasia on the basis of humane endpoints was not done, but humane endpoints were included in the applied animal care guidelines approved by the local authorities Landesamt für Gesundheit und Soziales (LAGeSo) and Gesellschaft für Versuchstierkunde (GV-SOLAS). Humane endpoints were: automutilation, sepsis, local infection, dyspnea/apnea, apathy, dehydration, weight loss over 20% and drastic worsening of the general health condition on the basis of the score sheet rating. After 28 days mice were sacrificed in deep isoflurane anesthesia by cervical dislocation by trained personnel. All animal procedures were performed in accordance with institutional and federal animal care guidelines and approved by the ethics committee of the LAGeSo (Permit Number: G003109).

Flow cytometry analysis of peripheral blood

Whole blood samples were drawn from the tail vein 0, 7 and 14 days after MI and circulating white blood cells (WBC) were counted with an animal blood counter (Scil Vet abc). For fluorescence activated cell sorting (FACS) analysis, blood mononuclear cells were separated via gradient-density centrifugation using Histopaque-1083 (Sigma-Aldrich). Cells were blocked with normal rat serum and anti-CD16/32 monoclonal antibody (mAb) (clone 93) in FACS buffer (PBS, 0.5% BSA, 0.05% NaN₃) and incubated with Alexa Fluor 647-labeled anti-Flk-1/VEGFR2 (clone 89B3A5), Phycoerythrin (PE)-conjugated anti-Ly-6A/E (Sca-1) (clone D7), and allophycocyanin(APC)-labeled anti-CD117 (c-kit) (clone 2B8) (all purchased from Biolegend). Appropriate isotype controls were always included. Cells were analyzed on FACSCanto II flow cytometer using FACSDiva software (BD Biosciences) and analyzed with FlowJo software (TreeStar).

Hemodynamic measurements

Evaluation of ventricular pressure–volume relationships was done 28 days after surgical induction of MI in isoflurane anesthetized ventilated mice as described above. A 1.4F polyimide pressure-conductance catheter (Millar Instruments) was inserted through the right carotid artery into the left ventricle to record baseline pressure-volume loops in the closed chest. Conversion of raw conductance data to calibrated volumes was performed by determination of parallel conductance (Vp) using hypertonic saline dilution method [17,18]. Afterward, mice were euthanized in deep anesthesia and hearts were excised. Measurements and data analysis were performed by a blinded person using LabChart® Software (AD Instruments).

Histology and Immunofluorescence

At day 28, hearts were excised, fixed overnight with 4% formalin/PBS-buffered and embedded in paraffin. Transversal sections of a thickness of 3 µm were cut from apex to base and mounted on glass slides for histological and immunhistochemical staining. Masson trichrome (MT) staining was done according to standard protocols. Infarction size was determined using midline length measurement on MT stained sections from the surgical LAD occlusion to the base [19]. For the quantification of fibrosis, blue stained areas of sequential MT stained heart sections were determined and correlated to the whole heart section area using ImageJ software. Additionally, infarction size was determined by staining with 2% tri-phenyltetrazolium chloride (TTC). Therefore, hearts were frozen at −20°C and cut in semi frozen state into five equally thick sections. Slices were then incubated in TTC solution for 15 min at 37°C and fixed in 10% formalin. Viable myocardium stained red while the infarcted area appeared pale-white. The area of infarction was measured in each slice with ImageJ (1.44; National Institutes of Health) software and expressed as percentage of the entire left ventricular area (including septum). Vascularization was evaluated by immunostaining with CD31/PECAM-1 (sc-1506-R M-20 clone, Santa Cruz Biotechnology) and α-smooth muscle actin (clone 1A4, Sigma-Aldrich) primary antibody followed by incubation with respective AlexaFluor-labeled secondary antibodies (Invitrogen). Images were acquired with a Zeiss Axioskop microscope. Capillary density was determined by counting CD31/PECAM-1 positive vascular structures in three randomly chosen high-power fields (each 40000 µm²) in 10 sections per heart (n = 5–8 for each group) within the border zone, infarcted and remote area. Alpha-smooth muscle actin positive vascular structures were counted per area

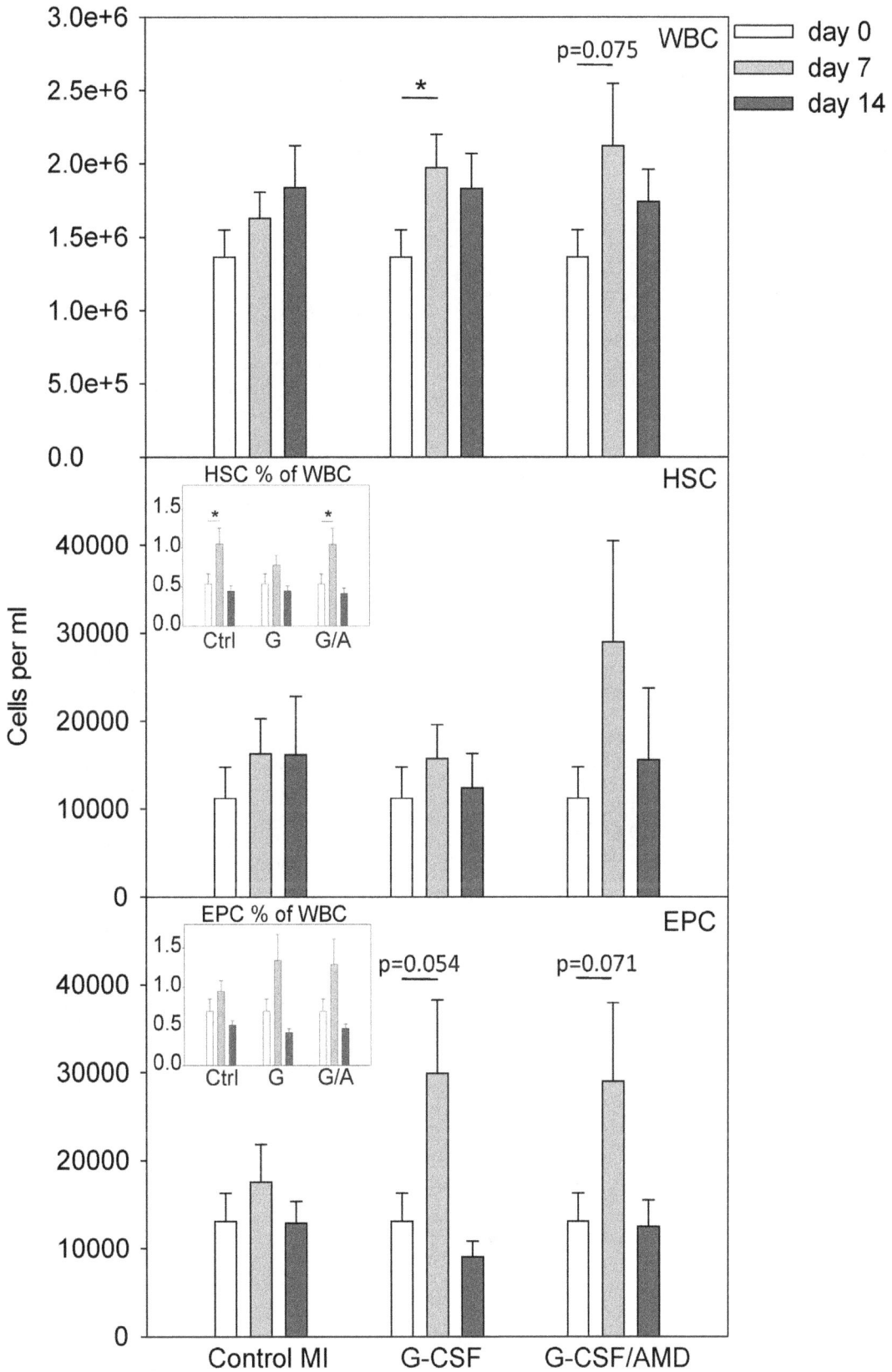

Figure 1. Mobilized peripheral blood cells after MI. Circulating white blood cells were counted before (day 0), 7 and 14 days after MI. **(upper panel)** G-CSF and combinatorial G-CSF/AMD treatment enhances white blood cell numbers 7 days after MI (*p<0.05 G-CSF day 7 vs. day 0, p<0.075 G-CSF/AMD day 7 vs. day 0). Flow cytometry analysis on peripheral blood mononuclear cells was done before (day 0) and 7 and 14 days after induction of MI in control MI, G-CSF and G-CSF/AMD treated mice. **(middle panel)** The absolute numbers of circulating c-Kit$^+$Sca-1$^+$ double positive

HPC were not different between control MI and drug treated groups. The HSC fraction was significantly increased in the control and G-CSF/AMD group (*p<0.05 vs. day 0), but not in G-CSF treated mice **(middle panel inset)**. **(lower panel)** Flk-1[+]Sca-1[+] double positive EPC mobilization peaked 7 days after MI in drug treated mice (p<0.054 G-CSF day 7 vs. day 0; p<0.071 G-CSF/AMD day 7 vs. day 0). EPC fractions were increased upon drug treatment, but did not reach statistical significance **(lower panel inset)**. Data represent means ± SEM. (n>10 per group).*p<0.05.

assessed with ImageJ software. Both parameters are expressed as positive stained vascular structures per mm².

Statistics

Data are reported as mean value ± standard error of the mean (SEM), and were analyzed by two-tailed unpaired Student's t-test. A p-value of less than 0.1 was considered as a trend a p-value less than 0.05 was considered significant. Group comparisons were performed using One-way ANOVA followed by the Tukey's test. Survival analysis was assessed by Kaplan-Meier method.

Results

Mobilized peripheral blood cells

Circulating white blood cells (WBC) and progenitor cells were determined before (0 days) and after MI in drug treated mice and untreated controls. Treatment with G-CSF (*p<0.05 vs. day 0) and G-CSF/AMD (p = 0.075 vs. day 0) enhanced the MI induced mobilization of circulating WBC, without reaching statistical significance compared to the untreated group 7 days after MI (Figure 1 upper panel). Mobilization of HSC and endothelial progenitor cells (EPC) was analyzed by co-expression of receptor tyrosin kinase c-kit and stem cell antigen-1 (Sca-1) or fetal-liver kinase-1 (Flk-1) on peripheral blood mononuclear cells, respectively (Figure S1). Although not statistically significant, G-CSF/AMD3100 treatment led to elevated absolute numbers of c-Kit/Sca-1 positive HSC into circulation compared to single G-CSF administration and untreated control mice 7 days after MI (Figure 1 middle panel). The percentage of HSC was significantly increased in the control and G-CSF/AMD group (*p<0.05 vs. day 0), but not in G-CSF treated mice (middle panel inset). Absolute numbers of EPC were elevated by single treatment with G-CSF (p = 0.054 vs. day 0) and combinatorial G-CSF/AMD treatment (p = 0.071 vs. day 0) 7 days after MI (Figure 1 lower

panel). Furthermore, EPC percentages were increased upon drug treatment, but did not reach statistical significance (lower panel inset). No synergistic augmentation of circulating cell numbers could be observed when G-CSF treatment was combined with AMD after MI.

Survival

A total of 106 mice (38 control MI, 36 G-CSF, 32 G-CSF/AMD) were included into cumulative Kaplan-Meier survival analysis. There were no statistical differences in overall mortality between saline and drug treated groups (Figure 2 A). In order to prevent that early deaths after surgery has masked beneficial effects, the 70 (27 control MI, 23 G-CSF, 20 G-CSF/AMD) animals that survived for the first 4 days after MI were included in a modified Kaplan-Meier survival analysis. However, even after exclusion of mice that died early after MI there were no significant differences in survival rates between saline and drug treated animals (Figure 2 B). Post-mortem examination confirmed that all dead mice suffered from MI.

Heart function and infarction size

Surgical induction of MI severely impaired cardiac function compared to sham-operated animals as assessed by hemodynamic measurements 28 days after surgery. In sham-operated mice the left ventricular ejection fraction (EF) was 66.2% ±5.2 and declined to 23.8% ±2.9 (p<0.001) in control MI mice. G-CSF and G-CSF/AMD treatment did not improve EF 28 days after MI (Table 1). Detailed examination of P-V loop derived hemodynamic parameters confirmed a severely decreased cardiac function in all MI groups (see Table 1). The parameters stroke work (SW), stroke volume (SV), cardiac output (CO) and the rate in fall of ventricular pressure (dP/dt min) were significantly reduced, while end-systolic and end-diastolic volumes (Ves, Ved) increased in control MI compared to sham-operated mice. Drug treated animals showed

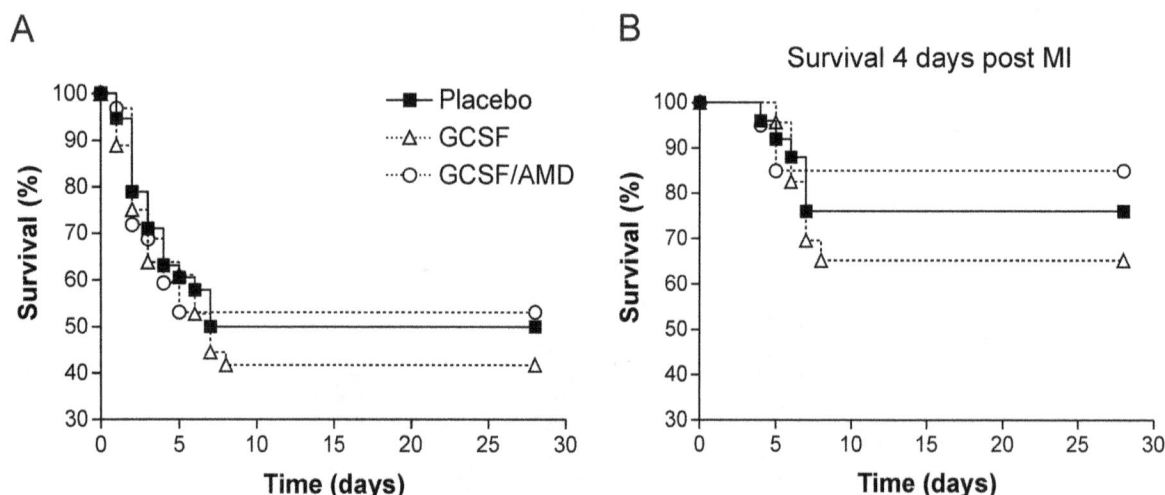

Figure 2. Cumulative Kaplan-Meier survival analysis. Kaplan-Meier survival curve of control MI and drug-treated mice during the observation period of 28 days after MI. Treatment of mice with G-CSF or G-CSF/AMD did not improve the (**A**) overall survival and did not alter (**B**) the mortality of mice that survived the first 4 days after MI.

Table 1. Left ventricular hemodynamics recorded by pressure-volume catheterization in the closed chest 28 days after LAD ligation.

	Sham (n = 11)	Control MI (n = 11)	GCSF (n = 5)	GCSF/AMD (n = 9)
HR (bpm)	443.9±27.6	475.7±17.8	492.6±48.7	465.2±13.6
EF (%)	66.2±5.2	23.8±2.9#	34.8±7.7#	32.9±4.8#
SW (mmHg* μl)	761.8±57.9	310.8±33.6#	354.7±60.6#	413±49.9#
dP/dt max (mmHg/s)	6337.5±593.6	5053.4±343.5	6272.7±631.4	5540.4±321.1
dP/dt min (mmHg/s)	−5961.9±629.6	−3961.7±286.3#	−4807.3±388.4	−4701.1±297.2
SV (μl)	11.9±0.9	6.8±0.6#	6.5±1#	7.5±0.9#
CO (μl/min)	5245.9±417.4	3231±307#	3346±711.5#	3541.6±514.5#
Ves (μl)	7.8±1	25.7±2.5#	16.1±3.3	19.2±2.9#
Ved (μl)	18.7±1.3	31.2±2.2#	21.6±3.8	25.3±3.1

Values are means ± SEM. HR, heart rate; EF, ejection fraction; SW, stroke work; dP/dt max, maximum first derivative of change in pressure rise with respect to time; dP/dt min, maximum first derivative of change in pressure fall with respect to time; SV, Stroke volume; CO, cardiac output; Ves, end-systolic volume; Ved, end-diastolic volume; One-way ANOVA post hoc Tukey's Multiple Comparison Test # $p < 0.05$ vs. sham; no significant differences vs. control MI.

significantly reduced EF, SW, SV and CO compared to sham-operated animals. The rates in rise and fall of ventricular pressure (dP/dt max, dP/dt min) and Ves, Ved were not significantly altered in drug treated compared to sham-operated mice. However, in comparison to control MI animals, drug treatment did not significantly improve any of the recorded parameters of left ventricular function (see Table 1).

Determination of infarction size 28 days post MI either by TTC staining or midline infarct length method on Masson trichrome stained sections revealed no significant difference between drug treated animals and control MI mice (Figure 3 A, B, C). Quantification of collagen rich fibrotic areas of the infarcted hearts on sequential transversal sections of Masson trichrome stained sections revealed no differences in fibrosis between control MI mice and drug treated groups (Figure 3 D, E). In summary, neither single G-CSF nor combinatorial G-CSF/AMD therapy significantly altered left ventricular hemodynamics and infarction size.

Cardiac histology and vascularization

Twenty-eight days after LAD ligation characteristic signs of late phase postinfarction remodelling of the heart were visible. Infarcted left ventricles showed typical loss of myocardium, left ventricular wall thinning and fibrous scar formation. Masson trichrome staining of infarcted regions revealed a viable subendocardial layer followed by collagen rich fibrous tissue reaching into border zone myocardium (Figure 4). No obvious differences in quantity of viable myocardium as well as interstitial collagen deposition in border zone or remote areas between control MI and drug treated mice hearts were visible.

EPC as well as certain subsets of WBC are known to exert angiogenic properties in ischemic tissues. Using antibodies against endothelial (CD31/PECAM-1) and smooth muscle cells (α-smooth muscle actin), the abundance of capillaries and arterioles in remote area (RA), border zone (BZ) and infarct zone (IF) was analyzed (Figure 5 A). Treatment with G-CSF and G-CSF/AMD had no significant effect on the vessel density in any of the designated regions compared to the control MI group (see Figure 5 B, C).

Discussion

In the present study, the potential regenerative properties of a combinatorial G-CSF/AMD therapy were tested in a model of permanent LAD occlusion. During the last decade numerous studies investigated the hypothesis that cytokine mediated mobilization of stem cells contribute to cardiac regeneration after MI. Implementing different cytokines and mobilization protocols these studies yielded controversial results ranging from beneficial effects to even deleterious effects on cardiac regeneration [20,21,22,23,24,25,26,27,28]. Moreover, the precise mode of action of stem cells in cardiac repair is still a matter of debate. While some studies showed direct differentiation of stem cells into functional cardiomyocytes [29,30,31], these results could not be reproduced by others [32,33] supposing paracrine effects on surrounding cells to be the cause of regeneration in ischemic tissues [34,35]. In addition, a number of studies proposed that the applied cytokine itself directly influences survival of myocytes and endothelial cells [22,23,24,25,26,36], thereby promoting myocardial recovery and neovascularization. In view of the fact that cardiac regeneration certainly involves interplay of complex protective mechanisms this study was aimed to optimize the mobilization of progenitor cells and combine it with possible cytoprotective effects of the most widely used mobilizing cytokine G-CSF. Therefore, we implemented a mobilization scheme in which G-CSF was applied in a relatively high dosage (250 μg/kg) for a short period of 3 days starting directly after induction of MI. This experimental setting was directed to support early cytoprotective actions [5] while avoiding long-term detrimental effects of G-CSF promoted inflammatory processes [20,37].

G-CSF as well as the CXCR4 inhibitor AMD3100 has been shown to mobilize HSPC and potential angiogenic cells from BM [38,39,40]. However, both agents exhibit different mobilization kinetics. While G-CSF leads to a delayed mobilization [41,42], AMD was shown to be a rapid mobilizer leading to a peak mobilization after 1–3 hours in mice [8,41]. Combination of both has been proven to synergistically enhance HSPC mobilization in mice [8] and humans [10] with the potential to promote neovascularization in a mouse model of hindlimb ischemia [11]. Although considered as a reliable mobilizing agent, continuous AMD application seems to attenuate positive effects of stem cell mobilization due to blockade of SDF-1α/CXCR4 mediated stem cell homing [13,14]. In view of these results, we combined G-CSF therapy with a single AMD administration at day 3 after MI to further enhance G-CSF mediated stem cell mobilization while avoiding negative effects of long term AMD application on stem cell homing.

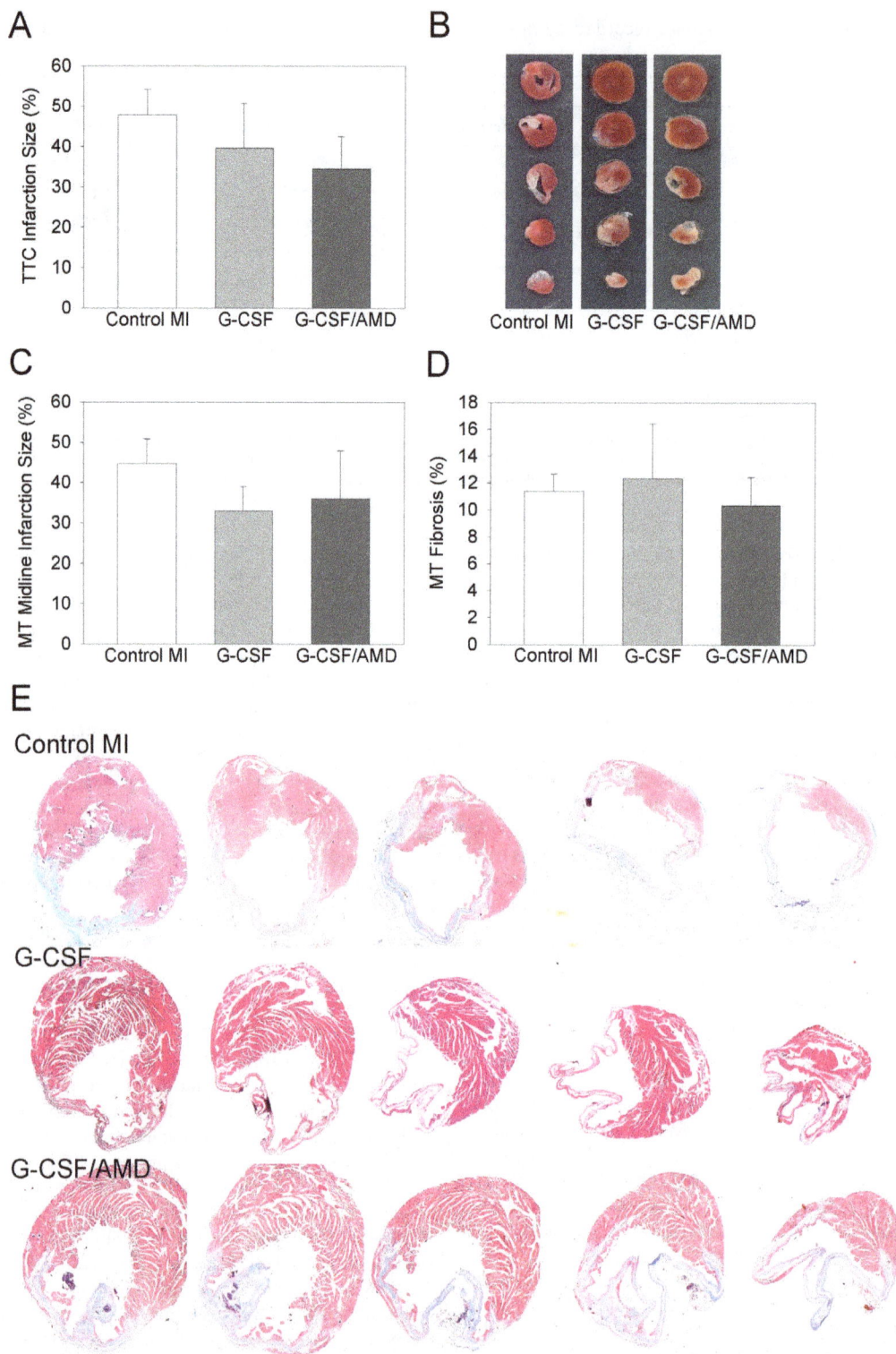

Figure 3. TTC and Masson trichrome staining of infarcted cardiac tissue for assessment of infarction size and fibrosis. Infarction size expressed as percentage of left ventricular area of control MI and drug treated mice assessed by TTC (**A, B**) and Masson trichrome (**C**) staining 28 days after MI. (**D, E**) Masson trichrome staining of sequential heart sections of control MI, G-CSF and G-CSF/AMD treated mice reveals no difference in left ventricular dilation, infarction size and fibrosis.

In the present study we found that G-CSF and G-CSF/AMD treatment promoted mobilization of WBC and EPC into peripheral blood. Mobilization tended to be higher in treatment groups compared to control mice without reaching statistical significance. This was due to large variations within treatment groups suggesting that not all animals responded to the treatment

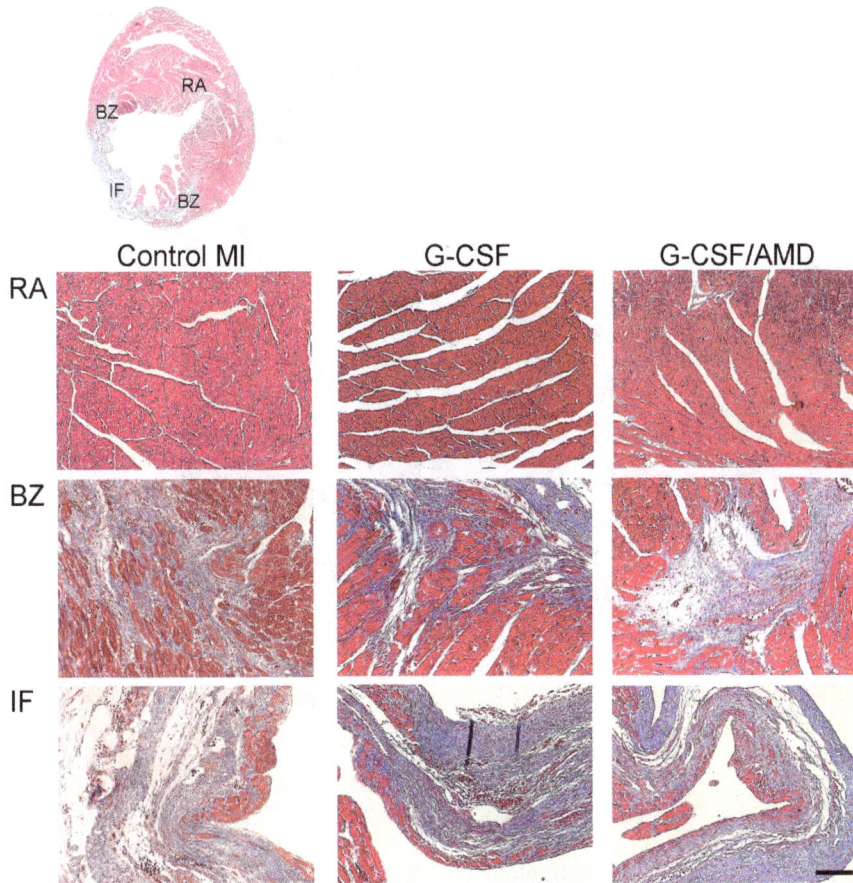

Figure 4. Cardiac histology of infarcted hearts 28 days after MI. Overview of Masson trichrome stained heart section (**upper panel**) and higher magnification images (**lower panels**) of border zone (BZ), infarcted region (IF) and remote area (RA). Images show no evident alteration of collagen deposition in designated areas between treatment groups. Bar: 100 μm.

in the same way. Inter-individual variations in G-CSF induced HSPC mobilization are also evident in humans [43]. A clear mobilization of CD34$^+$ cells, but with huge animal-to-animal variations was also seen in a rat model of MI after G-CSF treatment by Werneck-de-Castro et al. [27]. Nevertheless, these inter-individual variations in stem cell mobilization reflect the practical circumstances that a regenerative therapy has to face.

Combination of G-CSF with AMD did not synergistically increase WBC counts, HPC or EPC numbers at day 7 after MI (4 days after AMD treatment). The rapid AMD mediated mobilization might be an explanation for this, however, significantly increased EPC mobilization could be detected even 7 days after single AMD injection in a mouse MI model [12]. Furthermore, the rather moderate mobilization in the present study could be due to the specification of HPC and EPC as c-Kit/Sca-1 or Flk-1/Sca-1 double positive cells that likely defines a more specific subtype than often used single CD34$^+$ or CD133$^+$ HSPC.

Survival analysis showed no significant differences in mortality between saline and drug treated groups. On the basis of the applied drug regimen, beneficial effects on cardiac recovery resulting from either direct or paracrine actions of mobilized stem cells would not be conceivable until day 4 after MI. However, exclusion of animals that died in the first 4 days after MI did not uncover a reduced mortality of drug treated animals. Furthermore, pressure volume relationships of control and drug treated animals were recorded to evaluate heart function 28 days after MI.

Drastically reduced heart function was evident in control MI mice compared to sham operated mice. Although G-CSF and G-CSF/AMD treatment led to a slight improvement of some hemodynamic parameters, no significant changes compared to the control MI group were observable. Moreover, there was no reduction in infarction size visible in drug-treated versus control MI animals. Since these basic parameters indicated no improvement of heart regeneration with respect to the applied therapy, histology and vascularization of infarcted hearts was evaluated. Histological analysis showed drastically reduced myocardium at the ischemic site of the left ventricle that was replaced by a thin collagen rich, fibrous tissue layer. In control MI as well as treatment groups similar histological pattern were visible showing no obvious signs of cardiac regeneration.

Besides postulated direct actions of cytokines or stem cells on myocyte regeneration, numerous studies linked HSPC mobilization to favorable angiogenic effects promoting neovascularization of ischemic tissues [11,14,25,44,45]. On that account, formation of capillaries and arterioles was determined in the remote area, border zone and infarcted area of control MI and drug treated animals 28 days after MI. There were no indications for significantly altered vascularization in any region of the heart among MI groups. These results indicated that the applied drug regimen did not provoke measurable vasculogenic properties. This is in conflict with studies showing G-CSF and AMD induced neovascularization [11,12,45]. However, absent effects on cardiac

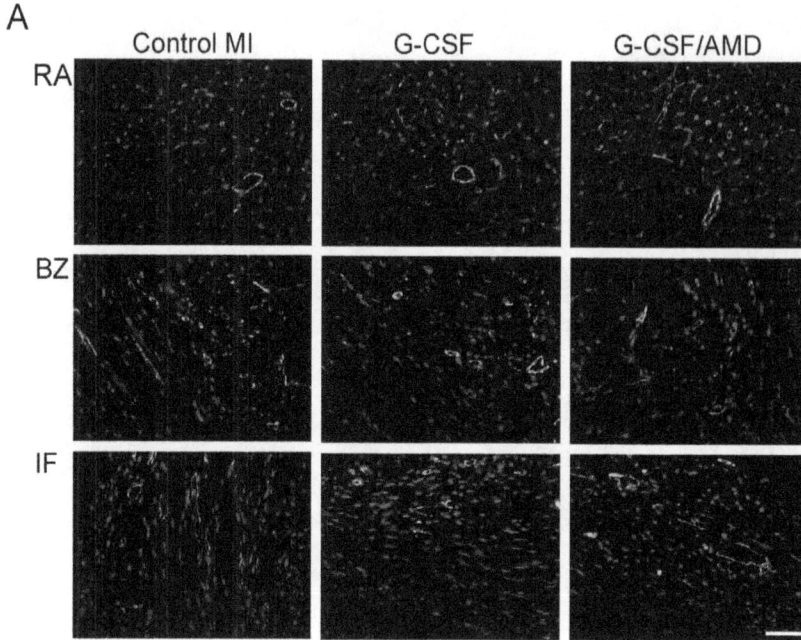

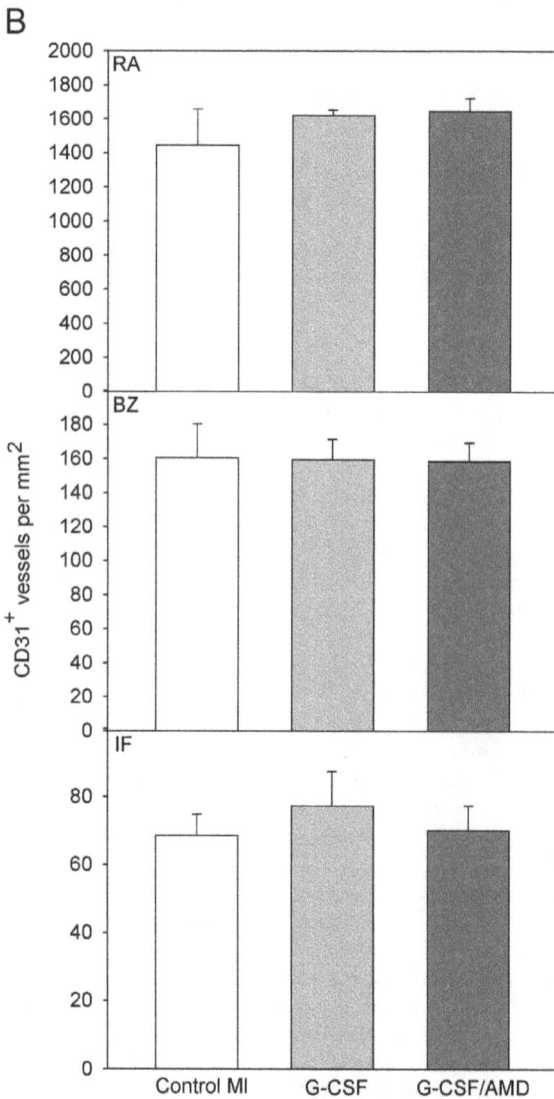

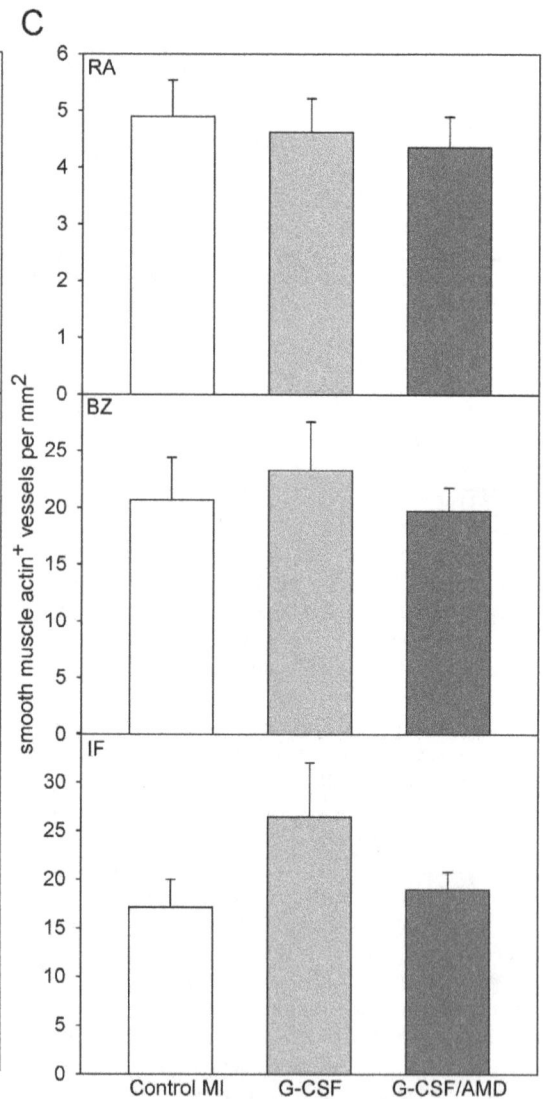

Figure 5. Cardiac vascularization 28 days after MI. Blood vessel formation was analyzed by immunfluorescence staining with specific antibodies against endothelial (CD31/PECAM-1) and smooth muscle (α-smooth muscle actin) cells (**A**). Comparable density of (**B**) CD31/PECAM-1 and (**C**) smooth muscle actin positive vascular structures in the remote area (RA, upper panel), border zone (BZ, middle panel) and infarcted area (IF, lower panel) among control MI and drug-treated groups. Data represent means ± SEM. (n = 5−8 hearts per group).

vascularization [20] and even inhibitory actions of G-CSF on vascular tubule formation and vascularization of subcutaneous sponges [46] have been reported by others. The inflammatory response after myocardial ischemia plays a pivotal role in heart regeneration being accountable for positive as well as adverse outcomes [47,48,49]. Elevated WBC numbers, reported in this study and by others [11,20,21] are capable to induce increased inflammation and adverse events after myocardial infarction [37,50]. In a study of Maekawa et al. [51] induction of the closely related cytokine granulocyte-macrophage stimulating factor (GM-CSF) led to increased macrophage infiltration into the infarcted myocardium. Moreover, expression of collagen and fibrogenic TGF-β1 was increased 14 days after MI. These effects resulted in infarct expansion, aggravated cardiac remodeling and increased mortality of treated rats after permanent LAD ligation. Cheng et al. [20] reported that G-CSF therapy affects expression of matrix-metalloproteinases (MMP) and their tissue inhibitors (TIMP) leading to increased fibrosis, mortality and left ventricular dysfunction after MI in the long term. Our implemented treatment regimen with G-CSF and G-CSF/AMD did not provoke negative effects on myocardial regeneration. This might be due to the short time period of drug therapy starting directly after MI. Numerous studies showed beneficial effects of mesenchymal stromal cell (MSC)- infusion on MI recovery in rodents and humans [52]. In a study of Pitchford et al. combinatorial G-CSF/AMD3100 treatment resulted in elevated peripheral EPC and HSC but not stromal progenitor cells (SPC) [53]. Furthermore single G-CSF therapy in patients after percutaneous intervention surprisingly led to decreased numbers of putative MSC in peripheral blood and had no effect on systolic performance [54]. From these observations it is tempting to speculate that missing MSC mobilization might be responsible for the poor regenerative properties of G-CSF or G-CSF/AMD therapy. However, the precise actions of G-CSF and cell based therapies in cardiac regeneration are uncertain and divergent results ranging from multi-level benefits to adverse effects on cardiac remodeling could be observed. It must be carefully taken into account that in addition to differences in experimental design, animal dependent factors such as genetic background, age, body temperature, and

even colony substrain differences critically influence the susceptibility to myocardial ischemia and cardiac healing [55,56].

In conclusion, although the applied drug regimen enhanced the mobilization of potentially regenerative cells, the combination of G-CSF and AMD did not significantly improve cardiac recovery after MI compared to control MI mice. On the other hand, no adverse effects of the applied drug treatment on cardiac function and remodeling could be observed. Further studies are needed to elucidate the complex mechanisms after cardiac injury to figure out treatment regimens that more specifically promote cardiac healing.

Study limitations

Beneficial effects of stem cells or cytokines were already observed after permanent LAD occlusion [25,57] however, this frequently used model has the limitation that it does not reflect the actual conditions in patients after angioplasty. The type of model (permanent vs. transient LAD occlusion) could influence the regenerative capacities of G-CSF therapy as suggested for experimental stroke models [58].

Supporting Information

Figure S1 FACS analysis of peripheral white blood cells of naïve (day 0) mice and 7 days post MI. (A) Mononuclear cells were gated on forward scatter (FSC) and side scatter (SSC) plot to exclude blood cells, debris and dead cells. Percentages of (B) c-Kit/Sca-1 double positive and (C) Flk1/Sca-1 double positive sub-populations were recorded. Representative dots plots are shown.

Acknowledgments

We thank Mrs Kunkel from the FACS core facility of the BCRT.

Author Contributions

Conceived and designed the experiments: CR TH AK NL DZ. Performed the experiments: CR TH AK JP NL. Analyzed the data: CR TH AK NL SK. Wrote the paper: CR TH DZ.

References

1. Guo Y, Graham-Evans B, Broxmeyer HE (2006) Murine embryonic stem cells secrete cytokines/growth modulators that enhance cell survival/anti-apoptosis and stimulate colony formation of murine hematopoietic progenitor cells. Stem Cells 24: 850–856.
2. Basu S, Broxmeyer HE (2005) Transforming growth factor-{beta}1 modulates responses of CD34+ cord blood cells to stromal cell-derived factor-1/CXCL12. Blood 106: 485–493.
3. Yu RY, Wang X, Pixley FJ, Yu JJ, Dent AL, et al. (2005) BCL-6 negatively regulates macrophage proliferation by suppressing autocrine IL-6 production. Blood 105: 1777–1784.
4. Moazzami K, Roohi A, Moazzami B (2013) Granulocyte colony stimulating factor therapy for acute myocardial infarction. Cochrane Database Syst Rev 5: CD008844.
5. Sanganalmath SK, Abdel-Latif A, Bolli R, Xuan YT, Dawn B (2011) Hematopoietic cytokines for cardiac repair: mobilization of bone marrow cells and beyond. Basic Res Cardiol 106: 709–733.
6. Zohlnhofer D (2008) G-CSF for left ventricular recovery after myocardial infarction: is it time to face reality? Cardiovasc Drugs Ther 22: 343–345.
7. Stein A, Zohlnhofer D, Pogatsa-Murray G, von Wedel J, Steppich BA, et al. (2010) Expression of CXCR4, VLA-1, LFA-3 and transducer of ERB in G-CSF-

mobilised progenitor cells in acute myocardial infarction. Thromb Haemost 103: 638–643.
8. Broxmeyer HE, Orschell CM, Clapp DW, Hangoc G, Cooper S, et al. (2005) Rapid mobilization of murine and human hematopoietic stem and progenitor cells with AMD3100, a CXCR4 antagonist. J Exp Med 201: 1307–1318.
9. Calandra G, McCarty J, McGuirk J, Tricot G, Crocker SA, et al. (2008) AMD3100 plus G-CSF can successfully mobilize CD34+ cells from non-Hodgkin's lymphoma, Hodgkin's disease and multiple myeloma patients previously failing mobilization with chemotherapy and/or cytokine treatment: compassionate use data. Bone Marrow Transplant 41: 331–338.
10. Brave M, Farrell A, Ching Lin S, Ocheltree T, Pope Miksinski S, et al. (2010) FDA review summary: Mozobil in combination with granulocyte colony-stimulating factor to mobilize hematopoietic stem cells to the peripheral blood for collection and subsequent autologous transplantation. Oncology 78: 282–288.
11. Capoccia BJ, Shepherd RM, Link DC (2006) G-CSF and AMD3100 mobilize monocytes into the blood that stimulate angiogenesis in vivo through a paracrine mechanism. Blood 108: 2438–2445.
12. Jujo K, Hamada H, Iwakura A, Thorne T, Sekiguchi H, et al. (2010) CXCR4 blockade augments bone marrow progenitor cell recruitment to the neovascu-

lature and reduces mortality after myocardial infarction. Proc Natl Acad Sci U S A 107: 11008–11013.

13. Misao Y, Takemura G, Arai M, Ohno T, Onogi H, et al. (2006) Importance of recruitment of bone marrow-derived CXCR4+ cells in post-infarct cardiac repair mediated by G-CSF. Cardiovasc Res 71: 455–465.

14. Theiss HD, Vallaster M, Rischpler C, Krieg L, Zaruba MM, et al. (2011) Dual stem cell therapy after myocardial infarction acts specifically by enhanced homing via the SDF-1/CXCR4 axis. Stem Cell Res 7: 244–255.

15. Abbott JD, Huang Y, Liu D, Hickey R, Krause DS, et al. (2004) Stromal cell-derived factor-1alpha plays a critical role in stem cell recruitment to the heart after myocardial infarction but is not sufficient to induce homing in the absence of injury. Circulation 110: 3300–3305.

16. Rüder C, Haase T, Krost A, Langwieser N, Böttcher S, et al. (2014) Combinatorial treatment with VEGF and AMD3100 promotes cardiac repair after myocardial infarction by enhanced neovascularization. In press.

17. Steendijk P, Staal E, Jukema JW, Baan J (2001) Hypertonic saline method accurately determines parallel conductance for dual-field conductance catheter. Am J Physiol Heart Circ Physiol 281: H755–763.

18. Zaruba MM, Huber BC, Brunner S, Deindl E, David R, et al. (2008) Parathyroid hormone treatment after myocardial infarction promotes cardiac repair by enhanced neovascularization and cell survival. Cardiovasc Res 77: 722–731.

19. Takagawa J, Zhang Y, Wong ML, Sievers RE, Kapasi NK, et al. (2007) Myocardial infarct size measurement in the mouse chronic infarction model: comparison of area- and length-based approaches. J Appl Physiol (1985) 102: 2104–2111.

20. Cheng Z, Ou L, Liu Y, Liu X, Li F, et al. (2008) Granulocyte colony-stimulating factor exacerbates cardiac fibrosis after myocardial infarction in a rat model of permanent occlusion. Cardiovasc Res 80: 425–434.

21. Deten A, Volz HC, Clamors S, Leiblein S, Briest W, et al. (2005) Hematopoietic stem cells do not repair the infarcted mouse heart. Cardiovasc Res 65: 52–63.

22. Harada M, Qin Y, Takano H, Minamino T, Zou Y, et al. (2005) G-CSF prevents cardiac remodeling after myocardial infarction by activating the Jak-Stat pathway in cardiomyocytes. Nat Med 11: 305–311.

23. Hasegawa H, Takano H, Iwanaga K, Ohtsuka M, Qin Y, et al. (2006) Cardioprotective effects of granulocyte colony-stimulating factor in swine with chronic myocardial ischemia. J Am Coll Cardiol 47: 842–849.

24. Miki T, Miura T, Nishino Y, Yano T, Sakamoto J, et al. (2004) Granulocyte colony stimulating factor/macrophage colony stimulating factor improves postinfarct ventricular function by suppression of border zone remodelling in rats. Clin Exp Pharmacol Physiol 31: 873–882.

25. Ohtsuka M, Takano H, Zou Y, Toko H, Akazawa H, et al. (2004) Cytokine therapy prevents left ventricular remodeling and dysfunction after myocardial infarction through neovascularization. FASEB J 18: 851–853.

26. Sugano Y, Anzai T, Yoshikawa T, Maekawa Y, Kohno T, et al. (2005) Granulocyte colony-stimulating factor attenuates early ventricular expansion after experimental myocardial infarction. Cardiovasc Res 65: 446–456.

27. Werneck-de-Castro JP, Costa ESRH, de Oliveira PF, Pinho-Ribeiro V, Mello DB, et al. (2006) G-CSF does not improve systolic function in a rat model of acute myocardial infarction. Basic Res Cardiol 101: 494–501.

28. Bocchi L, Savi M, Graiani G, Rossi S, Agnetti A, et al. (2011) Growth factor-induced mobilization of cardiac progenitor cells reduces the risk of arrhythmias, in a rat model of chronic myocardial infarction. PLoS One 6: e17750.

29. Badorff C, Brandes RP, Popp R, Rupp S, Urbich C, et al. (2003) Transdifferentiation of blood-derived human adult endothelial progenitor cells into functionally active cardiomyocytes. Circulation 107: 1024–1032.

30. Orlic D, Kajstura J, Chimenti S, Jakoniuk I, Anderson SM, et al. (2001) Bone marrow cells regenerate infarcted myocardium. Nature 410: 701–705.

31. Takamiya M, Haider KH, Ashraf M (2011) Identification and characterization of a novel multipotent sub-population of Sca-1(+) cardiac progenitor cells for myocardial regeneration. PLoS One 6: e25265.

32. Gruh I, Beilner J, Blomer U, Schmiedl A, Schmidt-Richter I, et al. (2006) No evidence of transdifferentiation of human endothelial progenitor cells into cardiomyocytes after coculture with neonatal rat cardiomyocytes. Circulation 113: 1326–1334.

33. Murry CE, Soonpaa MH, Reinecke H, Nakajima H, Nakajima HO, et al. (2004) Haematopoietic stem cells do not transdifferentiate into cardiac myocytes in myocardial infarcts. Nature 428: 664–668.

34. Gnecchi M, He H, Liang OD, Melo LG, Morello F, et al. (2005) Paracrine action accounts for marked protection of ischemic heart by Akt-modified mesenchymal stem cells. Nat Med 11: 367–368.

35. Huang C, Gu H, Yu Q, Manukyan MC, Poynter JA, et al. (2011) Sca-1+ cardiac stem cells mediate acute cardioprotection via paracrine factor SDF-1 following myocardial ischemia/reperfusion. PLoS One 6: e29246.

36. Minatoguchi S, Takemura G, Chen XH, Wang N, Uno Y, et al. (2004) Acceleration of the healing process and myocardial regeneration may be important as a mechanism of improvement of cardiac function and remodeling by postinfarction granulocyte colony-stimulating factor treatment. Circulation 109: 2572–2580.

37. Lian WS, Lin H, Cheng WT, Kikuchi T, Cheng CF (2011) Granulocyte-CSF induced inflammation-associated cardiac thrombosis in iron loading mouse heart and can be attenuated by statin therapy. J Biomed Sci 18: 26.

38. Hu J, Takatoku M, Sellers SE, Agricola BA, Metzger ME, et al. (2002) Analysis of origin and optimization of expansion and transduction of circulating peripheral blood endothelial progenitor cells in the rhesus macaque model. Hum Gene Ther 13: 2041–2050.

39. Powell TM, Paul JD, Hill JM, Thompson M, Benjamin M, et al. (2005) Granulocyte colony-stimulating factor mobilizes functional endothelial progenitor cells in patients with coronary artery disease. Arterioscler Thromb Vasc Biol 25: 296–301.

40. Zohlnhofer D, Ott I, Mehilli J, Schomig K, Michalk F, et al. (2006) Stem cell mobilization by granulocyte colony-stimulating factor in patients with acute myocardial infarction: a randomized controlled trial. JAMA 295: 1003–1010.

41. Shepherd RM, Capoccia BJ, Devine SM, Dipersio J, Trinkaus KM, et al. (2006) Angiogenic cells can be rapidly mobilized and efficiently harvested from the blood following treatment with AMD3100. Blood 108: 3662–3667.

42. Stroncek DF, Clay ME, Herr G, Smith J, Jaszcz WB, et al. (1997) The kinetics of G-CSF mobilization of CD34+ cells in healthy people. Transfus Med 7: 19–24.

43. Roberts AW, DeLuca E, Begley CG, Basser R, Grigg AP, et al. (1995) Broad inter-individual variations in circulating progenitor cell numbers induced by granulocyte colony-stimulating factor therapy. Stem Cells 13: 512–516.

44. Dubois C, Liu X, Claus P, Marsboom G, Pokreisz P, et al. (2010) Differential effects of progenitor cell populations on left ventricular remodeling and myocardial neovascularization after myocardial infarction. J Am Coll Cardiol 55: 2232–2243.

45. Takahashi T, Kalka C, Masuda H, Chen D, Silver M, et al. (1999) Ischemia- and cytokine-induced mobilization of bone marrow-derived endothelial progenitor cells for neovascularization. Nat Med 5: 434–438.

46. Tura O, Crawford J, Barclay GR, Samuel K, Hadoke PW, et al. (2010) Granulocyte colony-stimulating factor (G-CSF) depresses angiogenesis in vivo and in vitro: implications for sourcing cells for vascular regeneration therapy. J Thromb Haemost 8: 1614–1623.

47. Frangogiannis NG, Smith CW, Entman ML (2002) The inflammatory response in myocardial infarction. Cardiovasc Res 53: 31–47.

48. Lichtenauer M, Mildner M, Werba G, Beer L, Hoetzenecker K, et al. (2012) Anti-thymocyte globulin induces neoangiogenesis and preserves cardiac function after experimental myocardial infarction. PLoS One 7: e52101.

49. Roberts R, DeMello V, Sobel BE (1976) Deleterious effects of methylprednisolone in patients with myocardial infarction. Circulation 53: I204–206.

50. Barron HV, Cannon CP, Murphy SA, Braunwald E, Gibson CM (2000) Association between white blood cell count, epicardial blood flow, myocardial perfusion, and clinical outcomes in the setting of acute myocardial infarction: a thrombolysis in myocardial infarction 10 substudy. Circulation 102: 2329–2334.

51. Maekawa Y, Anzai T, Yoshikawa T, Sugano Y, Mahara K, et al. (2004) Effect of granulocyte-macrophage colony-stimulating factor inducer on left ventricular remodeling after acute myocardial infarction. J Am Coll Cardiol 44: 1510–1520.

52. Salem HK, Thiemermann C (2010) Mesenchymal stromal cells: current understanding and clinical status. Stem Cells 28: 585–596.

53. Pitchford SC, Furze RC, Jones CP, Wengner AM, Rankin SM (2009) Differential mobilization of subsets of progenitor cells from the bone marrow. Cell Stem Cell 4: 62–72.

54. Ripa RS, Haack-Sorensen M, Wang Y, Jorgensen E, Mortensen S, et al. (2007) Bone marrow derived mesenchymal cell mobilization by granulocyte-colony stimulating factor after acute myocardial infarction: results from the Stem Cells in Myocardial Infarction (STEMMI) trial. Circulation 116: I24–30.

55. Gorog DA, Tanno M, Kabir AM, Kanaganayagam GS, Bassi R, et al. (2003) Varying susceptibility to myocardial infarction among C57BL/6 mice of different genetic background. J Mol Cell Cardiol 35: 705–708.

56. Guo Y, Flaherty MP, Wu WJ, Tan W, Zhu X, et al. (2012) Genetic background, gender, age, body temperature, and arterial blood pH have a major impact on myocardial infarct size in the mouse and need to be carefully measured and/or taken into account: results of a comprehensive analysis of determinants of infarct size in 1,074 mice. Basic Res Cardiol 107: 288.

57. Orlic D, Kajstura J, Chimenti S, Limana F, Jakoniuk I, et al. (2001) Mobilized bone marrow cells repair the infarcted heart, improving function and survival. Proc Natl Acad Sci U S A 98: 10344–10349.

58. England TJ, Gibson CL, Bath PM (2009) Granulocyte-colony stimulating factor in experimental stroke and its effects on infarct size and functional outcome: A systematic review. Brain Res Rev 62: 71–82.

21

In Vivo Bioluminescence Imaging for Prolonged Survival of Transplanted Human Neural Stem Cells Using 3D Biocompatible Scaffold in Corticectomized Rat Model

Do Won Hwang[1,2❂], Yeona Jin[1❂], Do Hun Lee[3], Han Young Kim[1,2], Han Na Cho[4], Hye Jin Chung[4], Yunwoong Park[1], Hyewon Youn[1,5,6], Seung Jin Lee[4], Hong J. Lee[7], Seung U. Kim[7], Kyu-Chang Wang[8]*, Dong Soo Lee[1,2]*

1 Department of Nuclear Medicine, Seoul National University College of Medicine, Seoul, Korea, 2 Department of Molecular Medicine and Biopharmaceutical Science, WCU Graduate School of Convergence Science and Technology, Seoul National University, Seoul, Korea, 3 University of Miami School of Medicine, Miami Project to Cure Paralysis, Department of Neurological Surgery, Miami, Florida, United States of America, 4 College of Pharmacy, Ewha Womans University, Seoul, Korea, 5 Cancer Imaging Center, Seoul National University Cancer Hospital, Seoul, Korea, 6 Cancer Research Institute, Seoul National University College of Medicine, Seoul, Korea, 7 Medical Research Institute, Chung-Ang University College of Medicine, Seoul, Korea, 8 Division of Pediatric Neurosurgery, Seoul National University Children's Hospital, Seoul, Korea

Abstract

Stem cell-based treatment of traumatic brain injury has been limited in its capacity to bring about complete functional recovery, because of the poor survival rate of the implanted stem cells. It is known that biocompatible biomaterials play a critical role in enhancing survival and proliferation of transplanted stem cells via provision of mechanical support. In this study, we noninvasively monitored in vivo behavior of implanted neural stem cells embedded within poly-L-lactic acid (PLLA) scaffold, and showed that they survived over prolonged periods in corticectomized rat model. Corticectomized rat models were established by motor-cortex ablation of the rat. F3 cells expressing enhanced firefly luciferase (F3-effLuc) were established through retroviral infection. The F3-effLuc within PLLA was monitored using IVIS-100 imaging system 7 days after corticectomized surgery. F3-effLuc within PLLA robustly adhered, and gradually increased luciferase signals of F3-effLuc within PLLA were detected in a day dependent manner. The implantation of F3-effLuc cells/PLLA complex into corticectomized rats showed longer-lasting luciferase activity than F3-effLuc cells alone. The bioluminescence signals from the PLLA-encapsulated cells were maintained for 14 days, compared with 8 days for the non-encapsulated cells. Immunostaining results revealed expression of the early neuronal marker, Tuj-1, in PLLA-F3-effLuc cells in the motor-cortex-ablated area. We observed noninvasively that the mechanical support by PLLA scaffold increased the survival of implanted neural stem cells in the corticectomized rat. The image-guided approach easily proved that scaffolds could provide supportive effect to implanted cells, increasing their viability in terms of enhancing therapeutic efficacy of stem-cell therapy.

Editor: Ramasamy Paulmurugan, Stanford University School of Medicine, United States of America

Funding: This work was supported by Basic Science Research Program through the National Research Foundation of Korea (NRF) funded by the Ministry of Education, Science and Technology (2012R1A1A2008799), and a grant of the Korean Health Technology R&D Project, Ministry of Health & Welfare, Republic of Korea (HI13C1299). The funders had no role in study design, collection and analysis of data, decision to publish, or preparation of the manuscript.

Competing Interests: The authors have declared that no competing interests exist.

* Email: dsl@plaza.snu.ac.kr (DSL); kcwang@sun.ac.kr (K-CW)

❂ These authors contributed equally to this work.

Introduction

Traumatic brain injury (TBI), often defined as an acquired brain injury or simply a brain injury, is the leading cause of mortality and disability among young adults and elderly people, and it occurs when the brain is damaged by a sudden trauma such as those associated with falls, motor vehicle accidents, and surgical operations for epilepsy treatment [1,2]. Treatment of TBI has been largely dependent on use of various types of neuronal progenitors, or stem cells, to restore the lost brain tissue. Neural stem cells (NSCs) have drawn much attention because of their therapeutic potential for neurological disorders and because of their ability to differentiate into functional neuronal cell types [3–6]. Since the adult mammalian central nervous system (CNS) is limited in its capacity to utilize endogenous NSCs to repair neurologic deficits, cell replacement therapy can offer a potential means to recovery from the disability associated with neuronal loss. Much evidence suggests that transplanted NSCs can play a vital role in functional recovery in various animal models of CNS disorders including Parkinson's disease, Huntington's disease, stroke, and spinal cord injury [7–15]. In particular, NSC transplantation has recently been shown to restore brain function in animal models of TBI [16,17]. Despite intensive research, the severe conditions (oxidative stress, necrosis, inflammation) at the site of the injury are not favorable for the survival of grafted stem cells, thus limiting the effectiveness of stem cell therapy. To overcome this problem, a variety of methods for the introduction of neural stem cells that secrete growth factors, such as brain-derived neurotrophic factor (BDNF), have been investigated for the improvement of motor function in TBI models [18].

Gel- or solid-type biocompatible scaffolds have proven invaluable for therapy aimed at reconstitution of the injured brain tissue, since they not only provide the grafted stem cells with structural support and a three-dimensional (3D) environment for improved cell adhesion and proliferation, but also can directly induce stem cell differentiation in 3D cultures [19–23]. Commercially available scaffolds composed of extracellular matrix have been utilized for research and clinical purposes [24]. In this study, we used an electrospun-nanofibrous poly-L-lactic acid (PLLA) polymer scaffold. This biomaterial has proven to be biodegradable, biocompatible, and non-toxic, and is FDA-approved. Our previous research regarding PLLA scaffolds was conducted in the subcutaneously engrafted mouse model of cell/scaffold complexes, and the survival duration of the grafted stem cells was monitored in vivo [25].

Previously, a remarkable study examined extensively the in vivo behavior of polyglycolic acid (PGA)-encapsulated implanted neural stem cells and found effects such as enhanced NSC differentiation and reciprocal interactions with host cells in the injured brain [26]. This study aimed to provide fluorescence-based microscopic information to evaluate the in vivo characteristics of implanted neural stem cells within scaffold in an invasive manner, with the need for animal sacrifice. Therefore, the non-invasive monitoring system to be able to evaluate the supportive effect of biocompatible scaffold for viable grafted stem cells is required in brain injured condition.

For noninvasive monitoring, various imaging modalities, including positron emission tomography (PET), single-photon emission computed tomography (SPECT), magnetic resonance imaging (MRI), and bioluminescence imaging, are commonly applied to living animal models. In particular, bioluminescence imaging has been widely used for noninvasive and highly sensitive visualization of implanted stem cell localization, proliferation, and migration. Bioluminescence imaging based on the light-emitting firefly luciferase reporter gene continues to be popular because it is simple, cost-effective, and uses hypersensitive instrumentation especially free from background auto-luminescence. The luminescence observed is the light produced when luciferase catalyzes the conversion of D-luciferin to oxy-luciferin, in the presence of ATP and O_2 in living cells [27,28].

In this study, we used a corticectomized rat, with ablation of the motor cortex as a proper brain injury model [29]. Ablation models of the motor cortex have been applied to investigate brain plasticity and drug treatments for motor deficits induced by motor cortex injury [30]. The symptoms of the motor-cortex-resected corticectomy includes decreased consciousness, limb weakness, paralysis, seizures, and involuntary movement [29,31,32], indicating that the motor cortex is important for muscular and behavioral control. Physically, motor-cortex-ablated rats can be considered an ideal TBI model, since it allows easy implantation of solid-type scaffolds into the damaged brain cavity, and it provides a clearly abnormal behavioral pattern with minimal variation across individual animals.

In the present study, through in vivo bioluminescence imaging in motor-cortex-ablated rats, we investigated in vivo survival of human neural stem cells dependent on the mechanical support provided by biocompatible PLLA scaffolds.

Materials and Methods

Cell culture

HB1.F3 human neural stem cells were initially isolated from embryonic brains at 15 weeks gestation and immortalized by retroviral transduction with the v-myc oncogene. The F3 cells were maintained in Dulbecco's modified Eagle's medium (Invitrogen, Grand Island, NY) with L-glucose and L-glutamine, containing 10% (v/v) fetal bovine serum (Invitrogen) with 10 U/mL penicillin and 10 μg/mL streptomycin (Invitrogen) in a humidified incubator at 37°C and 5% CO_2. The cultures were passaged every 3 days by treating them with 3 mL 0.25% trypsin (w/v) and 1 mM EDTA (Invitrogen) per T75 flask for 1 min at 37°C. They were then harvested with the culture medium, centrifuged, and then resuspended in fresh flasks (Thermo Fischer Scientific, Roskilde, Denmark). The cell stock was supplemented with 10% DMSO, stored at −80°C in a freezer, and then transferred to liquid nitrogen.

Preparation of PLLA scaffold

The fibrous scaffolds were fabricated by the wet spinning method. The PLLA solution (6%) was prepared in methylene chloride/acetone (9:1 v/v). The polymer solution was loaded into a syringe, which was placed in a syringe pump. A blunt-tipped needle (27G) was used and the tip of the needle was immersed in a coagulation bath filled with methanol. The flow rate was between 0.9 and 1.1 mL/h and the polymer fiber was immediately formed in the bath. The collected PLLA fibers were freeze-dried to eliminate the organic solvent. Prior to cell seeding, PLLA-based scaffolds were cleaned with 70% isopropyl alcohol overnight, and were washed using phosphate buffered saline (PBS) three times. To help cell attachment to the microfibers of the PLLA scaffold, the cell-seeded scaffold was incubated for 2 h in a CO_2 incubator and complete medium was carefully added.

Luciferase reporter gene and retroviral modifications

To visualize the transplanted neural stem cells, F3 cells were engineered using retroviruses to express the enhanced firefly luciferase gene (effLuc) modified by codon optimization [33]. The DNA backbone in retroviral vector contained a Thy1.1 (CD90.1) marker and the effLuc gene linked to an internal ribosome entry site (IRES), under the control of the cytomegalovirus (CMV) promoter. To produce retroviruses, an effLuc viral vector and a DNA vector carrying major structural proteins (GAG, Pol, and Env) were transfected into a 293FT packaging cell line seeded in a 10-cm flask dish and 48 h after transfection, the supernatant containing retroviruses was collected. The produced retrovirus was added to F3-effLuc cells with 10 mM polybrene. The infected cells were separated into CD90.1$^+$ and CD90.1$^-$ by magnetic-activated cell sorting (MACS; Miltenyi Biotech Ltd., Bisley, Surrey, UK) using monoclonal anti-CD90.1 conjugated to magnetic microbeads. The purity of CD90.1$^+$ cells were identified by FACS (BD Immunocytometry System; Becton Dickinson, CA, USA) analysis using the monoclonal antibody, anti-CD90.1, conjugated to fluorescein isothiocyanate (FITC).

Flow cytometry

We incubated 1×10^6 F3 or F3-effLuc cells in FACS buffer (PBS with 5% fetal bovine serum and 0.05% sodium azide) for 30 min at 4°C. The cells were fixed using 2% paraformaldehyde fixation buffer in PBS, and suspended cells in 100 μL of permeabilization buffer (0.1% Triton X-100 in PBS), followed by incubation at room temperature for 30 min. F3 and F3-effLuc cells were stained with antibodies (Becton Dickinson) for the human proteins, CD44,

Nestin, Sox1, Sox2, GFAP, and Ki-67 in FACS buffer for 30 min. After staining, each cell-type was resuspended in FACS buffer and analyzed using a FACSCalibur system with CellQuest software (Becton Dickinson).

Scanning electron microscopy

The morphology of F3-effLuc cells seeded into PLLA scaffolds were characterized with field emission scanning electron microscopy (SEM; JEM-7401F, Joel Ltd., Tokyo, Japan) on 1 and 4 days. The samples were fixed using 1 mL of 2% glutaraldehyde (Sigma-Aldrich, MO, USA) at 4°C for 2 hr. To prepare dried and hydrated samples, they were washed with PBS and dehydrated by soaking in increasing concentrations of ethanol (30–100%). Specimens were covered with gold-palladium alloy on an aluminum stub after drying in hexamethyldisilazane (HMDS).

Corticectomy and the animal model

Animal were maintained without unnecessary pain or distress, and all animal experiments were approved by Seoul National University Hospital Institutional Animal Care and Use Committee (IACUC NO. 13-2010-005-0). Six-week-old adult male Sprague-Dawley (SD) albino rats weighing 180–200 g were provided by the Clinical Research Institute of Seoul National University Hospital. Rats were anesthetized with zoletil 50 (75 mg/kg, i.p.) and xylazine (10 mg/kg, i.p.). The anesthetized rats were placed in a stereotactic frame, and the height was adjusted between lambda and bregma. The midline scalp and temporal muscles were incised, and the exposed skull was stereotactically removed using a hand drill at the following coordinates: A) 4 mm rostral and 1 mm lateral, B) 2 mm caudal and 1 mm lateral, and C) 4 mm rostral and 6 mm lateral. After removing the skull, the exposed dura and motor cortices were ablated with a surgical blade, and the resected area was filled with Gelfoam (Pharmacia and Upjohn, Kalamazoo, MI, USA). Finally, the incised skin was sutured.

Limb placement behavior test in corticectomized rats

Two types of proprioceptive-response-related behavior tests were conducted to evaluate the corticectomized rat model: the whisker tactile and forelimb tests. Behavioral tests were carried out in semi-darkness in a silent room to minimize the impact of other environmental stimulation. The proprioceptive forelimb observation test was performed on an experimental table by gently pulling down the forelimb of corticectomized rats to examine the degree of forelimb retraction. Scores for the test were calculated as follows: 0 for normal retraction and 1–3 for abnormal retraction according to the amount of stretching. The whisker tactile test evaluated the level of sensory perception of whisker stimulation. The test examined whether the rat's forelimb reached out to the table when its whiskers approached within 2 mm of the table surface. Each animal was tested three times, and the scores were calculated as follows: 0 for normal reaching, and 1–3 for abnormal reaching according to the amount of stretching.

In vitro luciferase assay and in vivo bioluminescence imaging

In vitro luciferase assays were performed using luciferase assay kits (Applied Biosystems, Carlsbad, CA, USA). F3-effLuc cells were seeded onto 24- or 6-well plate and 24 h after plating, the cells were washed using PBS, and lysed using lysis buffer. The lysated F3-effLuc cells were transferred to a 96-well plate for detection of the bioluminescence signal. Luciferase intensity was measured using a Varioskan Flash (Thermo Fisher Scientific, Vantaa, Finland) at an acquisition time of 1 s. For *in vivo* imaging,

the wet-spinning PLLA scaffold was pre-wetted overnight using 70% v/v isopropyl alcohol, and washed three times with PBS. For implantation of cell/scaffold complex, 2×10^6 F3-effLuc cells were harvested using trypsin, and resuspended in PBS. Then, 40 µL of resuspened F3-effLuc cells were incorporated into the PLLA scaffold, and after the F3-effLuc/PLLA complex was incubated for 2 h, it was implanted into the cavity of the corticectomized rat brain (n = 3). For acquisition of the bioluminescence images, the rats were sedated with 2% isoflurane in 100% O_2 through a nose cone. D-Luciferin (Caliper Life Sciences, Hopkinton, MA, USA) was diluted to 3 mg/100 µL in normal saline and 0.6 mg of D-Luciferin was directly administrated into the brain on 0, 1, 3, 5, 8, 11, and 14 days post-transplantation. To suppress the innate immune response against the human F3-effLuc cells, cyclosporine A (5 mg/kg) was intraperitoneally administered every day after transplantation. An IVIS-100 imaging system equipped with a CCD camera (Caliper Life Sciences) was used for *in vivo* bioluminescence imaging. Images were acquired by integrating light for 5 min. The luminescence intensity in regions of interest from each image was quantified to examine the viability of the implanted cells.

Brain sectioning and histological analysis

Rats were anesthetized and transcardially perfused with normal saline containing heparin and 4% paraformaldehyde (PFA; Sigma-Aldrich). The rats' brains were removed, post-fixed in 4% paraformaldehyde for 24 h, and dehydrated in 10%, 20%, and 30% sucrose at 4°C. Specimens were frozen in OCT medium (Leica, IL, USA) and then, 14-µm-thick coronal serial sections were cut and mounted on gelatin-coated slides. One brain section in each group was processed with basic hematoxylin and eosin (H&E) staining, and six sets of sections were immunohistochemically stained. Sections on the glass slides were permeabilized in PBS containing 0.5% Triton-X for 5 min at 4°C, rinsed 3 times with PBS for 5 min, and then incubated in 1% normal horse serum for 1 h at room temperature. The slides were incubated overnight in a 1:500-diluted solution of anti-TuJ1 (Millipore Co., Billerica, MA, USA), anti-luciferase (Millipore Co.). Localization of transplanted NSCs was investigated via staining with anti-Thy1.1 and anti-luciferase antibodies. Immunohistochemistry analyses were performed using confocal laser microscopy (LSM 510; Carl Zeiss, Jena, Germany).

Statistical analysis

Data are expressed as means ± standard errors of means (SEM) from six biological replicates and were calculated using the Student's *t*-test. Statistical significance was accepted at P value of <0.005.

Results

Establishment of F3 cells stably expressing the effLuc transgene

To establish genetically engineered F3 human neural stem cells for visualizing *in vivo* characteristics of the implanted stem cells, we used F3 neural stem cells stably expressing the codon-optimized enhanced firefly luciferase gene and a Thy1.1 (CD90.1) marker linked with IRES under the control of the CMV promoter in the retroviral DNA backbone (Fig. 1A). CD90.1+ F3-effLuc cells in a heterogeneous cell population were collected via MACS. To measure the transduction efficiency of the infected cells, FACS analysis was conducted on the collected cells. The results indicated that these cells were 90.7% F3-effLuc

A

Retroviral vector carrying effLuc reporter gene

B

C

D

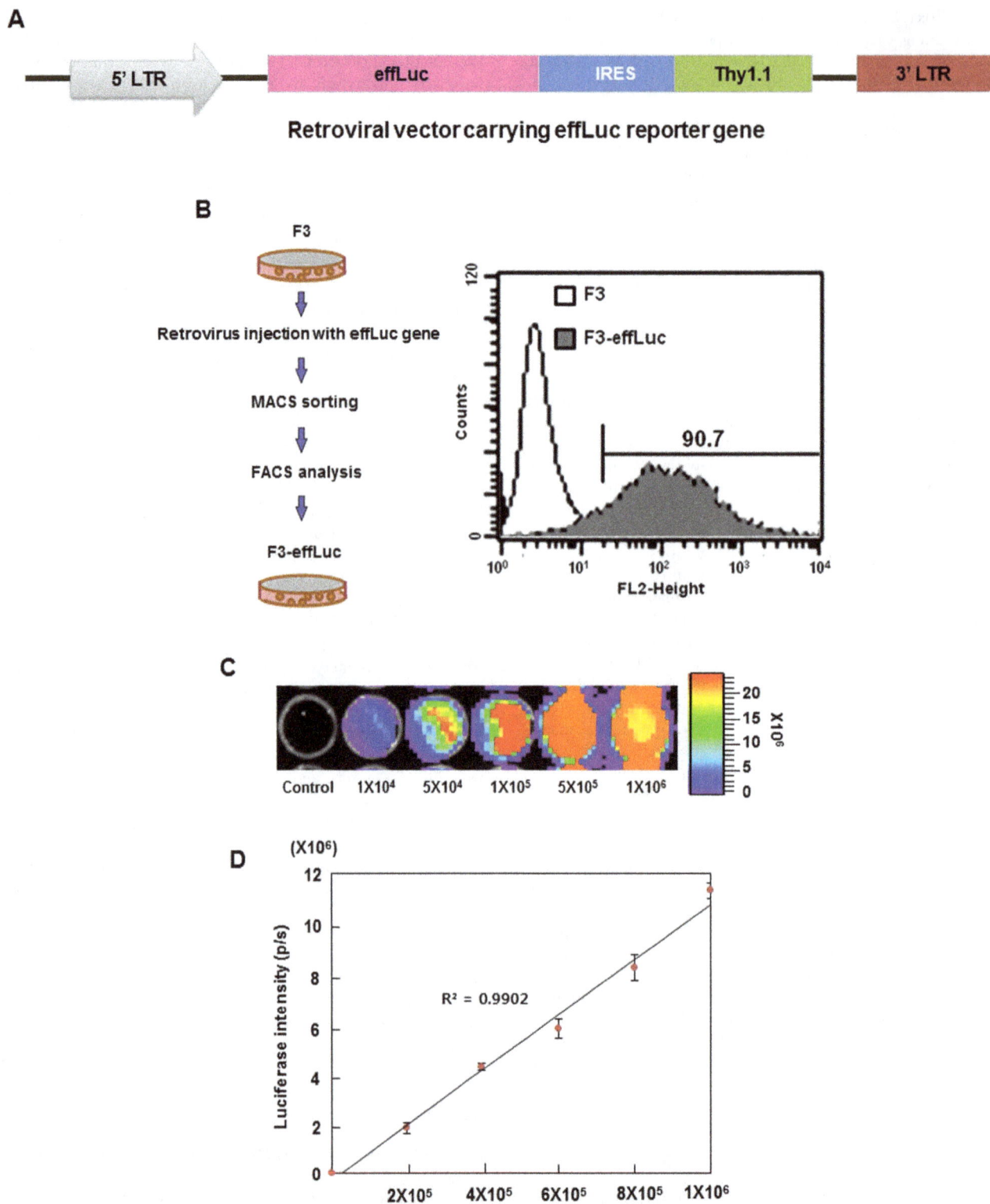

Figure 1. *In vitro* **luciferase reporter activity in F3-effLuc cells.** (A) Retroviral construct that contains the *effLuc* gene and Thy1.1 (CD90.1), linked with an IRES (internal ribosomal entry site). (B) Magnetic-activated cell sorting (MACS) was performed to collect the F3-effLuc cells. Flow activated cell sorting (FACS) analysis showed that more than 90% of cells were successively transfected with the *effLuc* vector. (C) The luciferase activity (n = 3) of F3-effLuc cells cultured in a 96-well plate were measured using an IVIS-100 optical imaging device. Firefly luciferase activity continuously increased in F3-effLuc cells in proportion to cell number, and (D) quantitative analysis showed a linear relationship between the cell number and bioluminescence signals.

(Fig. 1B). The luciferase intensity in the 96-well microplate showed the gradually increasing pattern as the number of F3-effLuc increased (Fig. 1C). The luciferase activity and F3-effLuc cell number were significantly correlated via a quantitative lumino-metric assay (Fig. 1D).

Examination of the characteristics of stem cells transfected with effLuc

To test whether the *effLuc* transgene influenced F3 stem cell characteristics, four different proteins involved in stem cell self-renewal were examined in non-transfected F3 cells and F3-EffLuc cells. FACS analysis showed that more than 90% of F3 cells were positive for the four stem cell markers (CD44, Nestin, SOX1, and SOX2). Expression levels of all four markers were similarly high (almost identical) in transfected cells, indicating that the transgene did not affect the stem cell properties (Fig. 2). No difference in Ki-67 (proliferation marker) expression was observed between F3 and F3-effLuc cells.

Spatial distribution and proliferation of F3-effLuc cells within the PLLA scaffold

To examine whether genetically modified F3-effLuc cells were stably attached to the synthetic PLLA microfiber scaffold, we examined the reciprocal interaction of F3-effLuc with the pre-wet PLLA scaffold. F3-effLuc cells (5×10^5) were incorporated within a sterile PLLA scaffold, and incubated for 2 h to allow the F3-effLuc cells to evenly distribute throughout the scaffold. The porous microstructure of the wetspun microcomposite PLLA scaffold (pore size distribution: 50–200 µm) was clearly observed under SEM [34]. F3-effLuc cells grown in PLLA scaffolds robustly adhered, and showed widespread cell attachment within the whole PLLA scaffold (Fig. 3A). To examine whether the engineered F3-effLuc cells proliferated well within the microfiber PLLA scaffolds, F3-effLuc cells were seeded into sterile PLLA scaffolds and incubated until 10 days in 24 well plate. An *In vitro* luciferase assay showed that F3-effLuc cell proliferation inside the PLLA scaffold gradually increased up to 2 days. Exponential proliferation was found to occur from 2 days (Fig. 3B).

Evaluation of the corticectomy rat model through behavioral testing

The motor cortex region in the rat brain was stereotactically removed at three coordinates from bregma (Fig. S1), and then F3-effLuc or F3-effLuc/PLLA cells were implanted in the corticecto-mized rat brain at 7 days after the operation. The complete experimental procedure is summarized in Fig. 4A. The implanted F3-effLuc cells were monitored at 0, 1, 3, 5, 8, 11, and 14 days using local administration of D-luciferin and a bioluminescence imaging device. The behavior tests indicated that the corticecto-mized rats showed abnormal behavior compared to control rats, exhibiting limb weakness and no response to whisker stimuli (Fig. 4B). The motor-cortex-ablated rats clearly displayed im-paired motor function, scoring highly for abnormality.

In vivo bioluminescence imaging of PLLA-encapsulated transplanted F3-effLuc cells in the corticectomized rat model

On post-operative 7 days, 1×10^6 F3-effLuc cells were harvested for seeding of the PLLA scaffold. After the harvested F3-effLuc cells were incubated for 2 h with the PLLA scaffold to keep F3-effLuc cells well attached within the scaffold, F3-effLuc cells were implanted into the ablated area with or without a PLLA scaffold. D-Luciferin was locally injected into the brain, and

bioluminescence images were acquired up to 14 days. Immediately after the cells were implanted, bioluminescence signals were clearly observed in both the F3-effLuc-only group and the F3-effLuc/PLLA group. Gradually increasing signals were detected up to 5 days in both groups, showing a significantly increased luciferase signal 3 days after transplantation. The luciferase activity in the cell-only group was decreased dramatically at 8 days and eventually disappeared at 11 days (Fig. 5A, n = 3). However, in the F3-effLuc/PLLA group, intense bioluminescence activity was detected in the implanted area at 8 days, and this was maintained up to 14 days (Fig. 5A). These results demonstrate that the supportive effect of the PLLA scaffold increases survival duration of implanted stem cells in the corticectomized rat model. For the quantitative analysis, photon counts in the region of interest (ROI) were measured from *in vivo* bioluminescence data. *In vivo* bioluminescence images from the corticectomized rat model revealed a continuously increasing luminescence signal of longer duration for the scaffold-encapsulated cells (Fig. 5B).

Histological analysis of PLLA-encapsulated F3-effLuc cells in the corticectomized rat

For the histological analyses, the whole brain was extracted from corticectomized rats (n = 3 per group) immediately after formalin perfusion via transcardial injection. The fixed brain tissue showed that the ablated area was filled with the F3-effLuc/PLLA scaffold (Fig. S2). To evaluate the characteristics of the transplanted F3-effLuc/PLLA complex, 14-µm-thick brain slices were stained with several antibodies including those for luciferase and the early neuronal marker beta III tubulin (TuJ1). H&E staining allowed us to observe the F3-effLuc cell morphology on the fibers of the PLLA scaffold in the implanted region (Fig. 6A). To examine the differentiation pattern of transplanted F3-effLuc cells within the PLLA scaffold, brain slices were stained with a TuJ1-specific antibody. Tuj1 and luciferase expression in the F3-effLuc/PLLA slice were partially co-localized (Fig. 6B). Lucif-erase expression in F3 cell only-implanted rat brain was not detected at 14 days after cell implantation. This result indicates that the F3-effLuc cells within the PLLA scaffold were differen-tiated into the neuronal lineage about 2 weeks after implantation.

Discussion

Newly developed scaffolds that are biochemically and physico-chemically suitable for clinical use have been explored for applications associated with cell-based therapies; these include synthetic polymers such as polylactic acid (PLA) [35], polyglycolic acid (PGA) [36], and poly(lactic-co-glycolic acid) (PLGA) [37], as well as natural biomaterials such as hydrogel [38], gelatin [39], chitosan, and collagen [40,41].

Numerous studies have investigated the use of stem cells for cell replacement therapy because stem cells have the potential to replace tissue deficits that occur in brain injury via their direct implantation into the damaged area. NSCs are especially suitable for implantation treatments of neurodegenerative diseases, and they have been widely used in neurological disease models. To improve the survival duration of implanted stem cells, biocom-patible scaffolds have been explored for use in tissue engineering. PLLA scaffold functions as a bioartificial niche, providing a 3D structure that enhances cell survival and proliferation *in vivo*. We have previously reported that the survival of implanted stem cells within scaffolds can be compared to that of scaffold-free cells through optical or radionuclide imaging [25]. In this study, we applied this image-based methodology to the brain injury model,

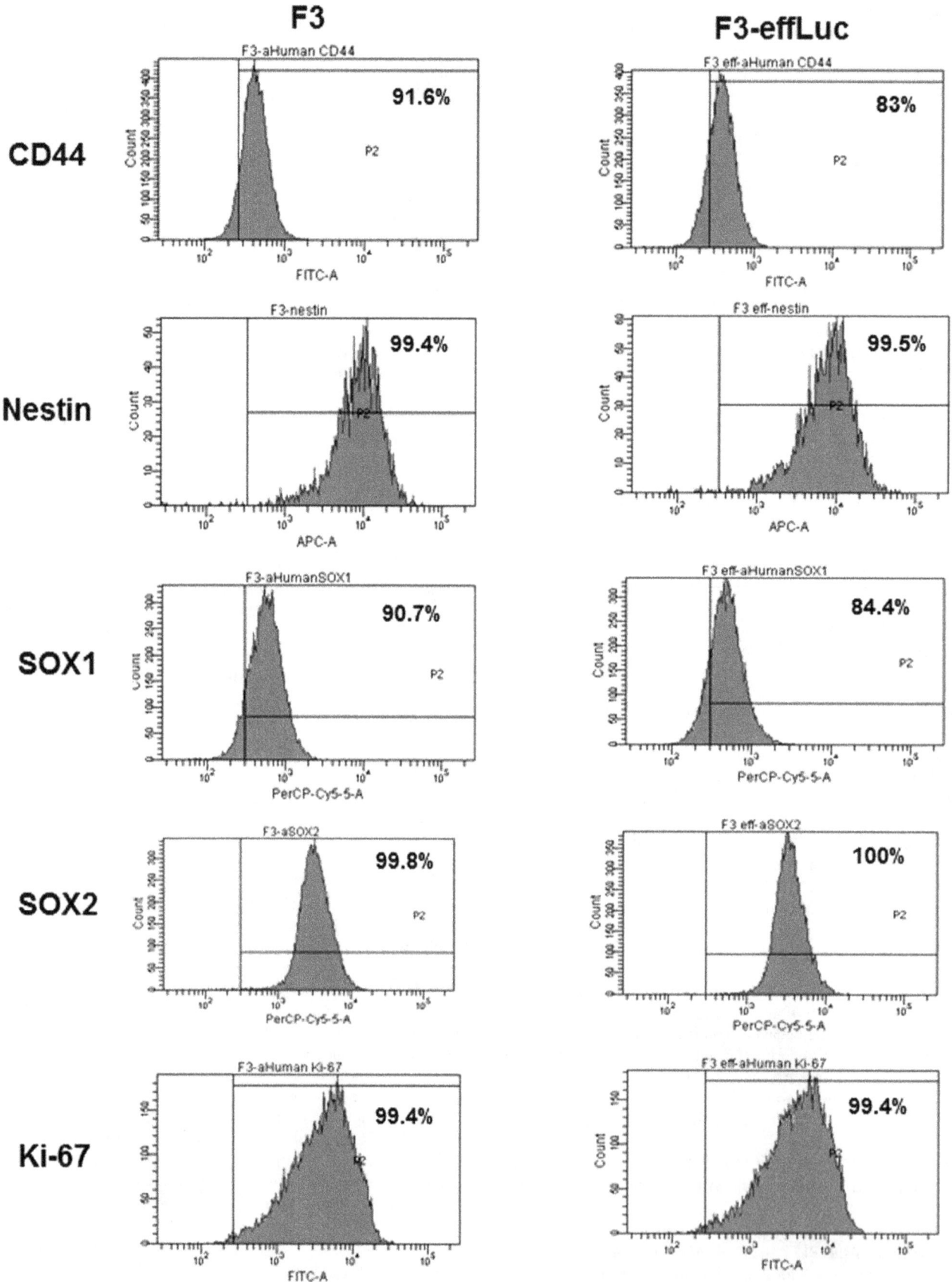

Figure 2. Validation of stem cell characteristics in F3 cells and F3-effLuc cells. Flow cytometric analysis showed F3 and F3-effLuc cells are positive for the stem cell surface marker, (a) CD44, and the intracellular marker s, (b) Nestin, (c) Ki67, (d) Sox1, and (e) Sox2.

A

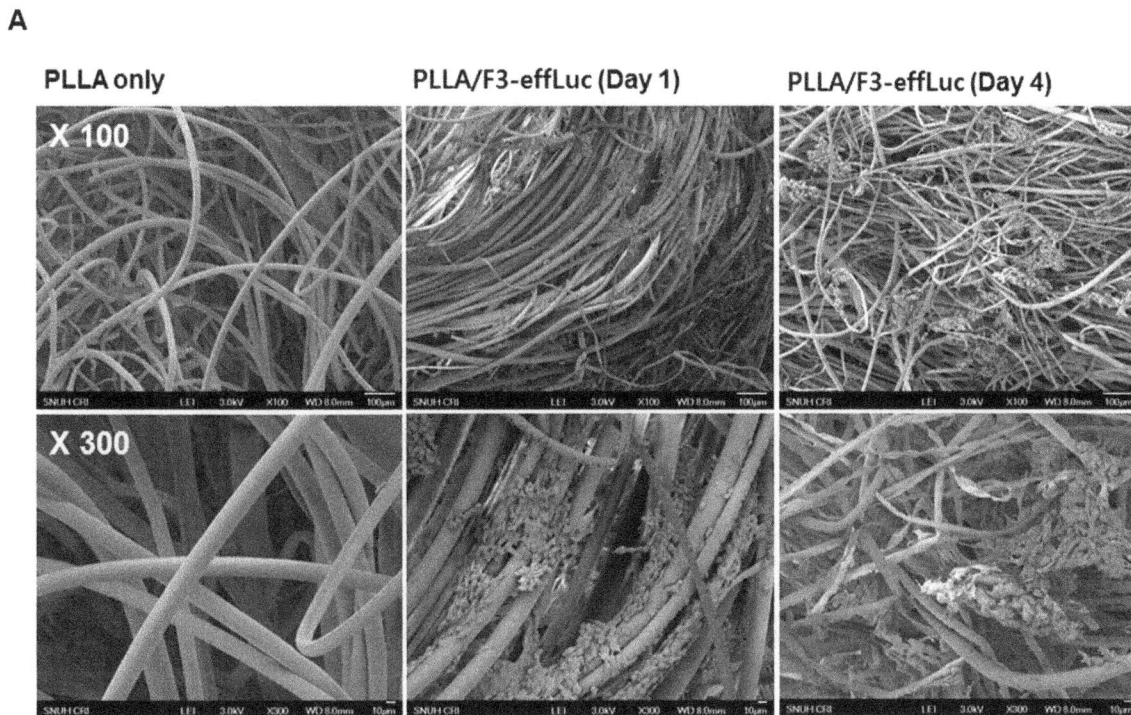

B

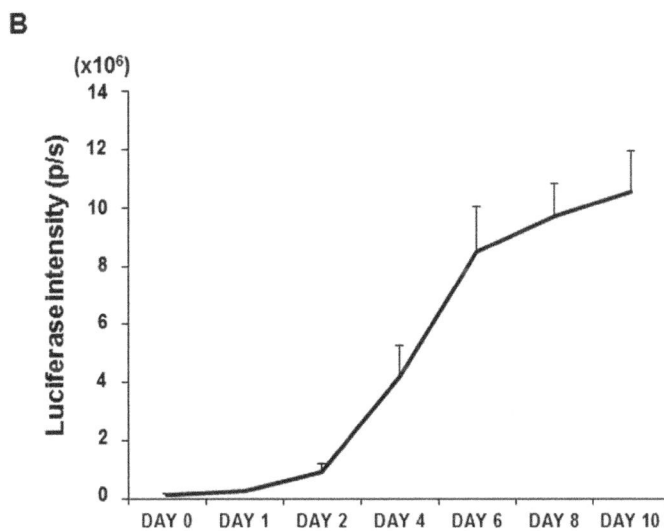

Figure 3. *In vitro* **proliferative effect in F3-effLuc cells within the PLLA scaffold.** (A) Scanning electron microscope (SEM) analysis was conducted to confirm adhesion of F3-effLuc cells to the PLLA scaffold. SEM images showed that F3-effLuc cells were stably attached onto the microfibers of the PLLA scaffold. (B) The luciferase intensity was quantified after F3-effLuc cells were incubated with the sterile PLLA scaffold. F3-effLuc cells incorporated within the PLLA scaffold were stably proliferated at 10 days.

demonstrating enhanced survival and proliferation rates for scaffold-encapsulated transplanted NSCs.

Our results confirmed the supportive effects of PLLA scaffolds for implanted stem cells in the corticectomized rat model, by evaluating the *in vivo* viability of transplanted stem cells using bioluminescence imaging. When F3-effLuc/PLLA scaffold complexes were transplanted into the ablated motor cortex, survival of transplanted stem cells was maintained for a longer time, and the proliferation rate was greater in the presence of the scaffold (Fig. 5A). Directly after F3-effLuc cell implantation, the prolifer-

ation rate of these cells increased gradually up to 5 days without the scaffold, and up to 8 days with the scaffold. A dramatic decline in cell survival occurred after 5 days without the scaffold, and after 11 days with the scaffold (despite daily immunosuppressant treatment). We suggest that the decline in survival was because of the environment around the cortical lesion, where implanted cells would be subjected to immunological and necrotic insults from the host cells. Despite the harsh environment, the survival duration of PLLA-encapsulated implanted F3-effLuc cells was

A

B

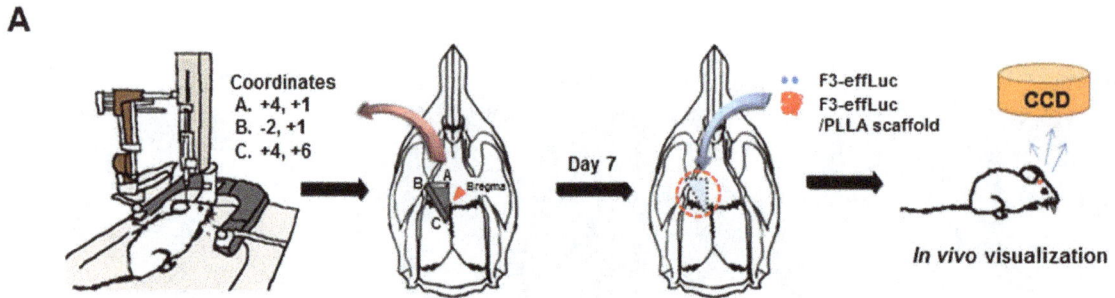

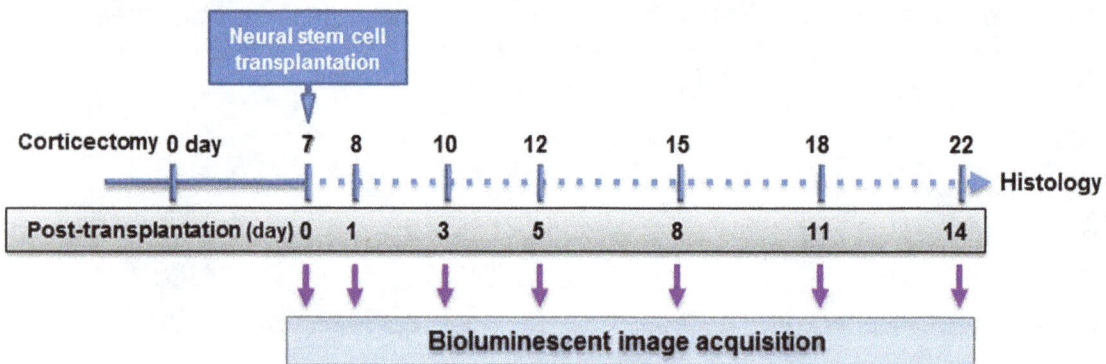

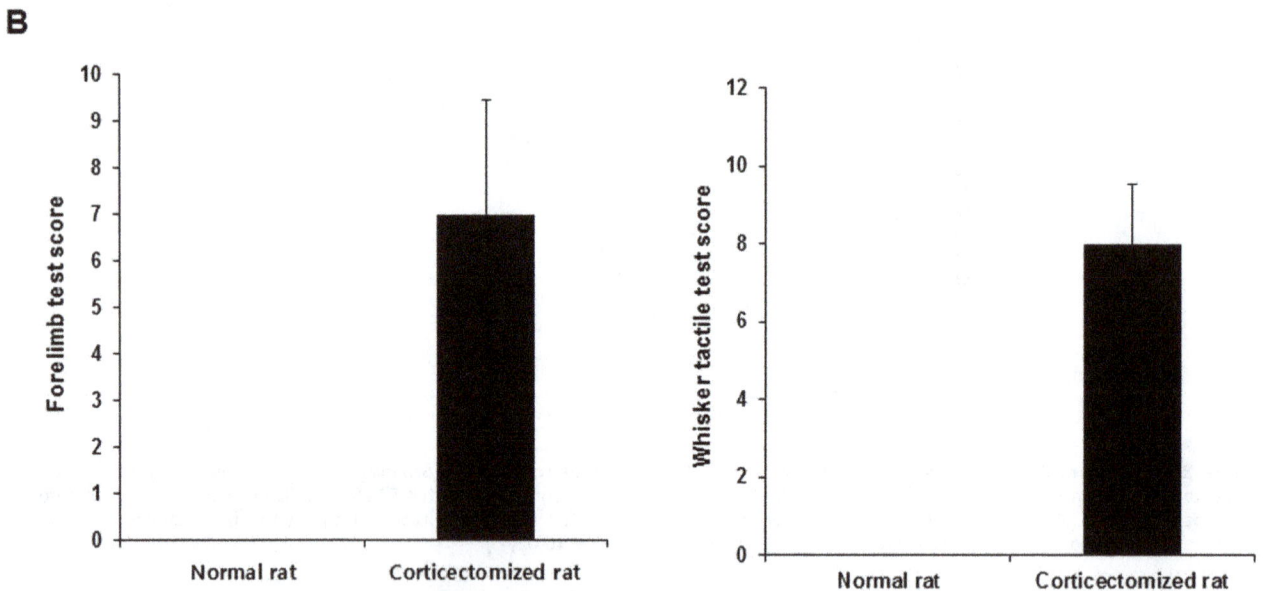

Figure 4. Schematic representation of the procedure for *in vivo* optical imaging. (A) The protocol is for *in vivo* monitoring of F3-effLuc cells implanted in a corticectomized rat model. The motor cortex region of the Sprague-Dawley rat brain was surgically removed at the given coordinates and after 7 days, the rats were transplanted with F3-effLuc cells alone or F3-effLuc/PLLA scaffold complexes, and then administered cyclosporine A everyday. The grafted cells were monitored at 0, 1, 3, 5, 8, 11, and 14 days using a bioluminescence-imaging device. At the end of the implant period, histological analyses were performed using hematoxylin and eosin (H&E) staining and immunohistochemistry. (B) Behavior tests were performed 7 days after motor cortex ablation. The traumatic brain injury (TBI) models were evaluated by forelimb placing tests and whisker tactile tests in normal and corticectomized rats (n = 10).

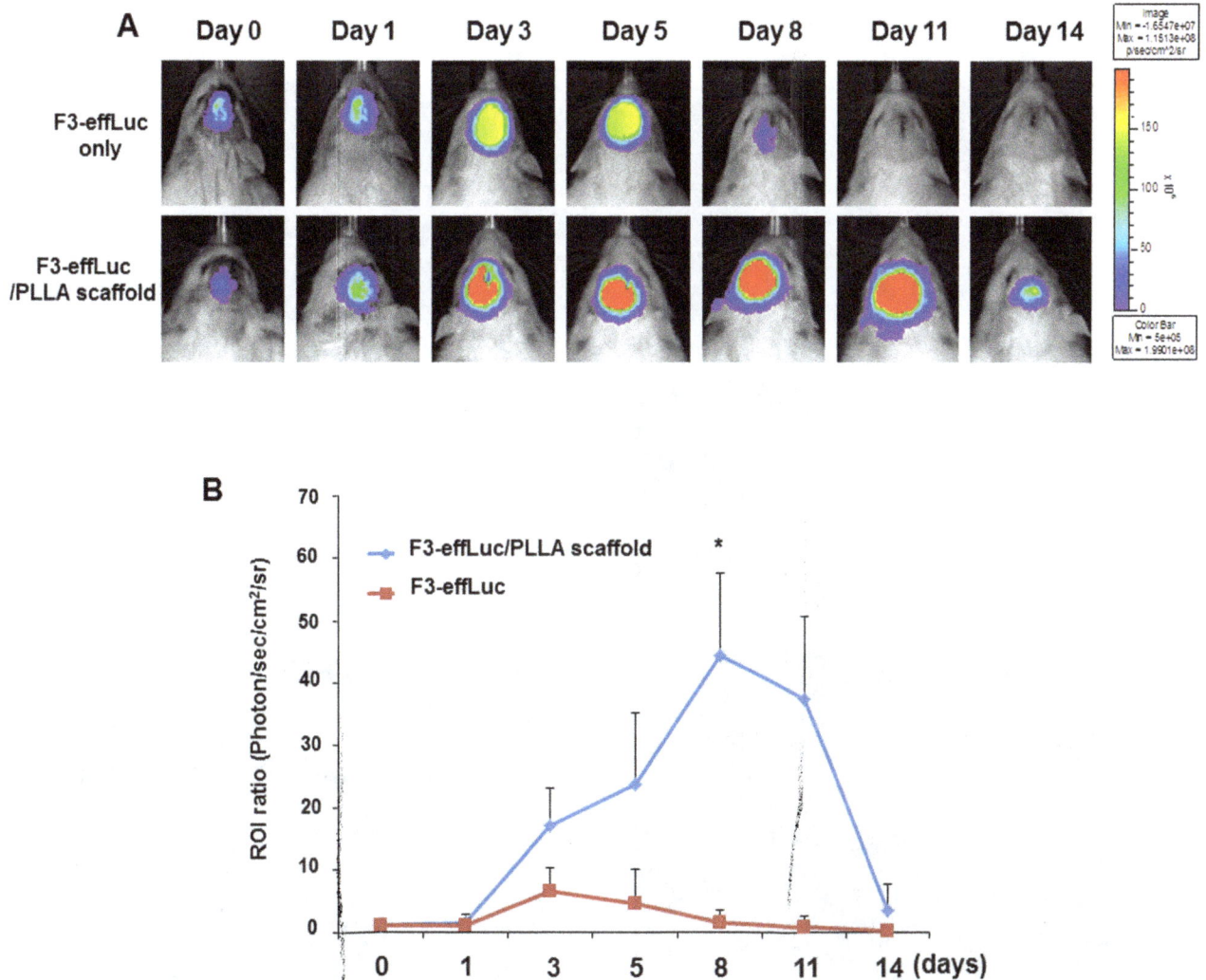

Figure 5. *In vivo* **bioluminescence imaging of the implanted F3-effLuc/PLLA scaffold in a corticectomized rat model.** (A) After F3-effLuc cells were incubated within the PLLA scaffold for 2 hr, the cell/scaffold complex was implanted into the ablated motor cortex area of the rat brain. Firefly luciferase bioluminescence imaging was performed over 14 days. The prolonged luminescence signals in F3-effLuc cells within the PLLA scaffold were clearly visualized in the ablated area. (B) Quantitative ROI analysis showed significantly enhanced survival duration for F3-effLuc cells within the PLLA scaffold (n = 6). P value, * <0.005.

significantly longer on account of the protection given by the fibrous scaffold.

Co-immunostaining results showed that the expression of the early neuronal marker Tuj1 was detected in the F3-effLuc/PLLA cells about 2 weeks after implantation in corticectomized region, which demonstrated that the cells within the PLLA scaffold had entered the neuronal lineage. This result indicates the possibility of improving functional recovery of brain tissue via enhanced neuronal differentiation resulting from the presence of the PLLA scaffold.

In the present study, we used a noninvasive reporter gene system to show that human NSCs transplanted into the ablated region exhibited increased viability thanks to the protection provided by a PLLA scaffold. We also confirmed that scaffold-encapsulated stem cells implanted into the injured sites began to be well differentiated along the neuronal lineage and had higher survival rates. It is important to note that long-term survival of transplanted cells is a fundamental requirement for their use in

improving recovery in animal models of brain injury. To further enhance survival duration, surface-engineered PLLA scaffolds coated with tropic factors such as fibroblast growth factor (FGF) could be used [34].

In the current study, we showed that implanted human NSCs incorporated within PLLA scaffolds were clearly visualized in motor-cortex-ablated rats. In particular, the supportive effect of a biocompatible PLLA scaffold, playing the role of a bio-bridge *in vivo*, was proven by using a noninvasive luminescence imaging technique in a traumatic injury model. Thus, the survival of stem cells implanted in a cortex-resected animal can be prolonged by using a polymeric scaffold to support implanted cells *in vivo*. *In vivo* imaging techniques for stem cell implantation in models of brain disorders are useful because they allow easy monitoring of implanted-cell survival. Scaffold-mediated implantation can play a supportive role in the treatment of trauma injury by enhancing implanted cell survival. Therefore, this study could provide useful

A

B

C

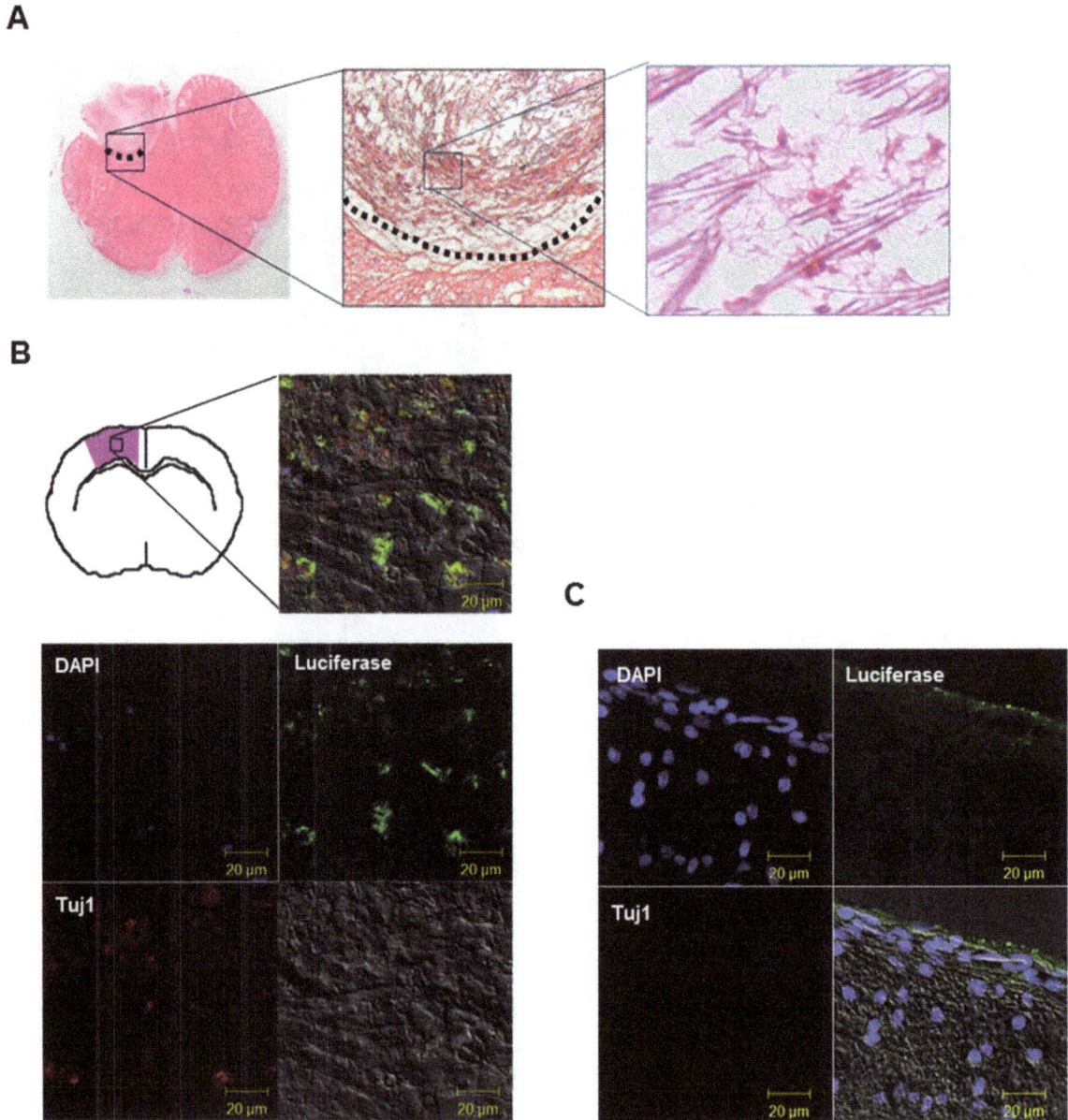

Figure 6. Immunohistochemistry results for PLLA-encapsulated F3-effLuc cells implanted in a corticectomized rat brain. (A) Hematoxylin and eosin (H&E) staining was performed on the corticectomized rat brains transplanted with the F3-effLuc/PLLA scaffold complex. Slices of the fixed brain were stained with hematoxylin and eosin to examine the presence of transplanted cells on the PLLA microfibers. (B) Confocal fluorescence images of the transplanted region (purple) revealed luciferase expression (green) in F3-effLuc cells was partially co-localized with the expression of Tuj1, a neuron-specific marker (red). Nuclei were visualized with DAPI (blue). Scale bars represent 20 μm. No luciferase expression was observed in F3 cell only-transplanted rat model.

insights to follow up stem-cell-based therapeutic research and for clinical applications.

Supporting Information

Figure S1 The corticectomized rat model. Surgical resection of the motor cortex was carried out at three coordinates (red circles). * bregma region.

Figure S2 Comparison of corticectomized rat brains bearing the (a) F3-effLuc cells and (b) F3-effLuc/PLLA scaffold complex.

Author Contributions

Conceived and designed the experiments: DWH YJ. Performed the experiments: DWH YJ HYK. Analyzed the data: HY K-CW DSL. Contributed reagents/materials/analysis tools: HNC HJC SJL SUK HJL YP DHL. Wrote the paper: DWH YJ DSL.

References

1. Zhang X, Zhang G, Yu T, Ni D, Cai L, et al. (2013) Surgical treatment for epilepsy involving language cortices: A combined process of electrical cortical stimulation mapping and intra-operative continuous language assessment. Seizure 9: 1059–1311.

2. Sitthinamsuwan B, Nunta-aree S (2013) Surgical treatment of epilepsy: principles and presurgical evaluation. J Med Assoc Thai 96: 121–131.

3. Gage FH, Coates PW, Palmer TD, Kuhn HG, Fisher LJ, et al. (1995) Survival and differentiation of adult neuronal progenitor cells transplanted to the adult brain. Proc Natl Acad Sci U S A 92: 11879–11883.

4. McKay R (1997) Stem cells in the central nervous system. Science 276: 66–71.

5. Brüstle O, Choudhary K, Karram K, Hüttner A, Murray K, et al. (1998) Chimeric brains generated by intraventricular transplantation of fetal human brain cells into embryonic rats. Nat Biotechnol 16: 1040–1044.

6. Uchida N, Buck DW, He D, Reitsma MJ, Masek M, et al. (2000) Direct isolation of human central nervous system stem cells. Proc Natl Acad Sci U S A 97: 14720–14725.

7. Rafuse VF, Soundararajan P, Leopold C, Robertson HA (2005) Neuroprotective properties of cultured neural progenitor cells are associated with the production of sonic hedgehog. Neuroscience 131: 899–916.

8. Richardson RM, Broaddus WC, Holloway KL, Fillmore HL (2005) Grafts of adult subependymal zone neuronal progenitor cells rescue hemiparkinsonian behavioral decline. Brain Res 1032: 11–22.

9. Ourednik J, Ourednik V, Lynch WP, Schachner M, Snyder EY (2002) Neural stem cells display an inherent mechanism for rescuing dysfunctional neurons. Nat Biotechnol 20: 1103–1110.

10. Ryu JK, Kim J, Cho SJ, Hatori K, Hatori K, et al. (2004) Proactive transplantation of human neural stem cells prevents degeneration of striatal neurons in a rat model of Huntington disease. Neurobiol Dis 16: 68–77.

11. McBride JL, Behrstock SP, Chen EY, Jakel RJ, Siegel I, et al. (2004) Human neural stem cell transplants improve motor function in a rat model of Huntington's disease. J Comp Neurol 475: 211–219.

12. Jeong SW, Chu K, Jung KH, Kim SU, Kim M, et al. (2003) Human neural stem cell transplantation promotes functional recovery in rats with experimental intracerebral hemorrhage. Stroke 34: 2258–2263.

13. Chu K, Kim M, Park KI, Jeong SW, Park HK, et al. (2004) Human neural stem cells improve sensorimotor deficits in the adult rat brain with experimental focal ischemia. Brain Res 1016: 145–153.

14. Cummings BJ, Uchida N, Tamaki SJ, et al. (2005) Human neural stem cells differentiate and promote locomotor recovery in spinal cord-injured mice. Proc Natl Acad Sci U S A 102: 14069–14074.

15. Hofstetter CP, Holmström NA, Lilja JA, Schweinhardt P, Hao J, et al. (2005) Allodynia limits the usefulness of intraspinal neural stem cell grafts; directed differentiation improves outcome. Nature Neurosci 8: 346–353.

16. Riess P, Zhang C, Saatman KE, Laurer HL, Longhi LG, et al. (2002) Transplanted neural stem cells survive, differentiate, and improve neurological motor function after experimental traumatic brain injury. Neurosurgery 51: 1043–1052.

17. Shear DA, Tate MC, Archer DR, Hoffman SW, Hulce VD, et al. (2004) Neural progenitor cell transplants promote long-term functional recovery after traumatic brain injury. Brain Res 1026: 11–22.

18. Ma H, Yu B, Kong L, Zhang Y, Shi Y (2012) Neural stem cells over-expressing brain-derived neurotrophic factor (BDNF) stimulate synaptic protein expression and promote functional recovery following transplantation in rat model of traumatic brain injury. Neurochem Res 37: 69–83.

19. Tabata Y (2004) Tissue regeneration based on tissue engineering technology. Congenit Anom 44: 111–124.

20. Srouji S, Kizhner T, Livne E (2006) 3D scaffolds for bone marrow stem cell support in bone repair. Regenerative Medicine 1: 519–528.

21. Chun TH, Hotary KB, Sabeh F, Saltiel AR, Allen ED, et al. (2006) A pericellular collagenase directs the 3-dimensional development of white adipose tissue. Cell 125: 577–591.

22. Horne MK, Nisbet DR, Forsythe JS, Parish CL (2010) Three-dimensional nanofibrous scaffolds incorporating immobilized BDNF promote proliferation and differentiation of cortical neural stem cells. Stem Cells Dev 19: 843–852.

23. Delcroix GJ, Schiller P, Benoit JP, Montero-Menei CN (2010) Adult cell therapy for brain neuronal damages and the role of tissue engineering. Biomaterials 31: 2105–2120.

24. Badylak SF, Freytes DO, Gilbert TW (2009) Extracellular matrix as a biological scaffold material: Structure and function. Acta Biomater 5: 1–13.

25. Hwang DW, Jang SJ, Kim YH, Kim HJ, Shim IK, et al. (2008) Real-time in vivo monitoring of viable stem cells implanted on biocompatible scaffolds. Eur J Nucl Med Mol Imaging 35: 1887–1898.

26. Park KI, Teng YD, Snyder EY (2005) The injured brain interacts reciprocally with neural stem cells supported by scaffolds to reconstitute lost tissue. Nat Biotechnol 20: 1111–1117.

27. Weissleder R, Mahmood U (2001) Molecular imaging Radiology 219: 316–33.

28. Wu JC, Sundaresan G, Iyer M, Gambhir SS (2001) Noninvasive optical imaging of firefly luciferase reporter gene expression in skeletal muscles of living mice. Mol Ther 4: 297–306.

29. Lee DH, Hong SH., Kim SK, Lee CS, Phi JH, et al. (2009) Reproducible and persistent weakness in adult rats after surgical resection of motor cortex: evaluation with limb placement test. Childs Nerv Syst 25: 1547–1553.

30. Dancause N, Barbay S, Frost SB, Plautz EJ, Chen D, et al. (2005) Extensive cortical rewiring after brain injury. J Neurosci 25: 10167–10179.

31. Biernaskie J, Szymanska A, Windle V, Corbett D (2005) Bihemispheric contribution to functional motor recovery of the affected forelimb following focal ischemic brain injury in rats. Eur J NeuroSci 21: 989–999.

32. Gonzalez CL, Gharbawie OA, Williams PT, Kleim JA, Kolb B, et al. (2004) Evidence for bilateral control of skilled movements: ipsilateral skilled forelimb reaching deficits and functional recovery in rats follow motor cortex and lateral frontal cortex lesions. Eur J NeuroSci 20: 3442–3452.

33. Rabinovich BA, Yang Y, Etto T, Chen JQ, Levitsky HI, et al. (2008) Visualizing fewer than 10 mouse T cells with an enhanced firefly luciferase in immunocompetent mouse models of cancer. Proc Natl Acad Sci 105: 14342–14346.

34. Jung MR, Shim IK, Kim ES, Park YJ, Yang YI, et al. (2011) Controlled release of cell-permeable gene complex from poly(L-lactide) scaffold for enhanced stem cell tissue engineering. J Control Release. 152: 294–302.

35. Saito N, Okada T, Horiuchi H, Ota H, Takahashi J, et al. (2003) Local bone formation by injection of recombinant human bone morphogenetic protein-2 contained in polymer carriers. Bone 32: 381–386.

36. Ameer GA, Mahmood TA, Langer R (2002) A biodegradable composite scaffold for cell transplantation. J Orthop Res 20: 16–9.

37. Fialkov JA, Holy CE, Shoichet MS, Davies JE (2003) In vivo bone engineering in a rabbit femur. J Craniofac Surg 14: 324–332.

38. Saim AB, Cao Y, Weng Y, Chang CN, Vacanti MA, et al. (2000) Engineering autogenous cartilage in the shape of a helix using an injectable hyrogel scaffold. Laryngoscope 110: 1694–1697.

39. Payne RG, McGonigle JS, Yaszemski MJ, Yasko AW, Mikos AG (2002) Development of an injectable, in situ crosslinkable, degradable polymeric carrier for osteogenic cell populations. Biomaterials 23: 4373–8430.

40. Hermanns S, Reiprich P, Müller HW (2001) A reliable method to reduce collagen scar formation in the lesioned rat spinal cord. J Neurosci Methods 110: 141–146.

41. O'Connor SM, Andreadis JD, Shaffer KM, Ma W, Pancrazio JJ, et al. (2000) Immobilization of neural cells in three-dimensional matrices for biosensor applications. Biosens Bioelectron 14: 871–881.

22

Tualang Honey Improves Human Corneal Epithelial Progenitor Cell Migration and Cellular Resistance to Oxidative Stress *In Vitro*

Jun Jie Tan[1]*, Siti Maisura Azmi[1], Yoke Keong Yong[2], Hong Leong Cheah[1], Vuanghao Lim[1], Doblin Sandai[1], Bakiah Shaharuddin[1]¤

1 Advanced Medical and Dental Institute, Universiti Sains Malaysia, Kepala Batas, Penang, Malaysia, 2 Department of Human Anatomy, Faculty of Medicine and Health Sciences, Serdang, Selangor Darul Ehsan, Malaysia

Abstract

Stem cells with enhanced resistance to oxidative stress after *in vitro* expansion have been shown to have improved engraftment and regenerative capacities. Such cells can be generated by preconditioning them with exposure to an antioxidant. In this study we evaluated the effects of Tualang honey (TH), an antioxidant-containing honey, on human corneal epithelial progenitor (HCEP) cells in culture. Cytotoxicity, gene expression, migration, and cellular resistance to oxidative stress were evaluated. Immunofluorescence staining revealed that HCEP cells were holoclonal and expressed epithelial stem cell marker p63 without corneal cytokeratin 3. Cell viability remained unchanged after cells were cultured with 0.004, 0.04, and 0.4% TH in the medium, but it was significantly reduced when the concentration was increased to 3.33%. Cell migration, tested using scratch migration assay, was significantly enhanced when cells were cultured with TH at 0.04% and 0.4%. We also found that TH has hydrogen peroxide (H_2O_2) scavenging ability, although a trace level of H_2O_2 was detected in the honey in its native form. Preconditioning HCEP cells with 0.4% TH for 48 h showed better survival following H_2O_2-induced oxidative stress at 50 µM than untreated group, with a significantly lower number of dead cells ($15.3 \pm 0.4\%$) were observed compared to the untreated population ($20.5 \pm 0.9\%$, $p < 0.01$). Both TH and ascorbic acid improved HCEP viability following induction of 100 µM H_2O_2, but the benefit was greater with TH treatment than with ascorbic acid. However, no significant advantage was demonstrated using 5-hydroxymethyl-2-furancarboxaldehyde, a compound that was found abundant in TH using GC/MS analysis. This suggests that the cellular anti-oxidative capacity in HCEP cells was augmented by native TH and was attributed to its antioxidant properties. In conclusion, TH possesses antioxidant properties and can improve cell migration and cellular resistance to oxidative stress in HCEP cells *in vitro*.

Editor: Xuefeng Liu, Georgetown University, United States of America

Funding: This project was supported by the Universiti Sains Malaysia Short Term Grant (304/CIPPT/6311032)(http://www.research.usm.my/?tag = 22) and Advanced Medical and Dental Institute. The funders had no role in study design, data collection and analysis, decision to publish, or preparation of the manuscript.

Competing Interests: The authors have declared that no competing interests exist.

* E-mail: jjtan@amdi.usm.edu.my

¤ Current address: Institute of Genetic Medicine, Newcastle University, International Centre for Life, Newcastle Upon Tyne, United Kingdom

Introduction

The cornea, which is a transparent and tough tissue that covers the anterior segment of the eye, is responsible for transmitting and reflecting light onto the retina. As the first layer of defence against external insults and oxidative insult from the sunlight, the corneal epithelium possesses mechanisms to maintain its cellular homeostasis. These mechanisms are orchestrated by a population of self-renewing stem cells that reside in the basal limbal epithelium, which is a niche area situated at the peripheral edge of the cornea called the limbus [1]. These cells express keratinocyte stem cell markers (p63, EGFR, K19, ABCG2 and integrin β1) and exhibit low expression of corneal differentiation markers (K3, involucrin, and connexin-43) [2], and they are responsible for sustaining the clarity and integrity of the corneal epithelium required for normal vision. Complete depletion or dysfunction of the corneal epithelial stem cells that occurs in some severe ocular surface abnormalities as a result of caustic injuries such as burns, inflammatory

conditions, or hereditary genetic disorders such as aniridia [3,4] will lead to conjunctivalisation, keratinisation, and opacification and result in impaired vision or blindness. Under these circumstances, visual acuity can only be restored by transplanting corneal epithelial stem cells onto the injured cornea. To minimise the risk of immune rejection, autologous stem cell transplantation may be the only option. Hence, culturing corneal epithelial stem cells and preserving their stemness and functions *in vitro* are pivotal for ensuring successful regeneration following transplantation.

Reactive oxygen species (ROS) are common metabolic by-products of aerobic metabolism, and their level is maintained through intrinsic antioxidant mechanisms in healthy cells. When maintained at the appropriate physiological level, ROS are vital in modulating several cellular signalling pathways that affect cell growth and function, including the phosphoinositide 3-kinase (PI3K) [5] and mitogen-activated protein kinase (MAPK) pathways [6]. In addition, ROS have the capability to dictate stem cell

Table 1. Dilution factor for Tualang honey supplemented medium.

Tualang honey concentrations (%)	20% Tualang honey: KSFM
0.004%	1:5000
0.04%	1:500
0.4%	1:50
3.33%	1:5

Abbreviation: KSFM, keratinocyte-serum free medium.

fate at physiological levels [7–9]. However, abnormal redox homeostasis involving ROS overproduction can induce oxidative stress, a physiological condition that renders cells susceptible to damage. Studies have confirmed that overproduction of ROS can compromise genomic stability [10] and trigger mutations and promote cancer growth [11]. High ROS levels also contribute to poor cell engraftment and viability, which impede regeneration after transplantation [12]. Although stem cells have greater antioxidant capacity compared to differentiated cells [13,14], they can exhibit telomere shortening-induced replicative senescence and reduced self-renewal capability under oxidative stress [15]. Hence, protecting stem cells from oxidative damage may help to promote cell survival, homing, and regeneration after transplantation. This protection could be achieved by maintaining a reduced environment at the site of transplantation through adjunctive therapy with dietary antioxidant [16] or by adding an antioxidant supplement to cells during *in vitro* expansion prior to transplantation. The efficacy of the latter strategy is supported by studies showing the potential for supplemental antioxidant in the culture medium to enhance the intracellular antioxidant activity of stem cells [15], prevent cellular damage, and salvage culture-induced loss of stemness [17].

Tualang honey is a medicinal honey that is collected from the honeycomb of *Apis dorsata*, the bees that built their hives on the Tualang tree (*Koompassia excels*) that is common in tropical forests in Malaysia. This honey has been extensively studied and has been found to have therapeutic benefits in treating various medical conditions, including protecting post-menopausal bone structure [18], promoting burn wound contraction with antibacterial effects [19,20], enhancing post-tonsillectomy healing [21], and inhibiting the growth of various cancer cells [22–24]. Generally, honey has been traditionally used to treat eye-related diseases [25,26].

Uwaydat et al. (2011) recently showed that raw honey accelerated healing of corneal abrasions and attenuated the inflammatory response and angiogenesis in endotoxin-induced keratitis [27]. Similar to other types of honey, previous studies of Tualang honey have also described its anti-inflammatory and antioxidant properties in treating alkali-induced eye injury *in vivo* [28] and its ability to save keratinocytes from inflammation and DNA damage as a result of ultraviolet radiation *in vitro* [29].

Although many *in vivo* studies of the effects of Tualang honey have been conducted, the potential for using Tualang honey in the cultivation of stem cells has not been investigated. To date, only one study described the use of Tualang honey to supplement the culture medium when cultivating a human osteoblast cell line (CRL1543) [30]. Although many studies have shown the therapeutic benefits of Tualang honey in treating cornea injury [28,29], its effects on corneal epithelial stem cells have yet to be evaluated. Herein we characterised the effects of Tualang honey on cytotoxicity, gene expression, and migration of human corneal epithelial progenitor (HCEP) cells and assessed its potential for improving cell resistance to oxidative stress.

Methodology

HCEP cell culture and expansion

HCEP cells were purchased from Gibco (Invitrogen Life Technologies Co., Carlsbad, CA, US) and ATCC (Manassas, VA, US). Cells were expanded in standard keratinocyte serum-free medium (KSFM, Gibco) that was supplemented with 5 ng/ml recombinant epidermal growth factor (rEGF) and 50 µg/ml bovine pituitary tissue extracts (Invitrogen Life Technologies Co., Carlsbad, CA, US). Passage 2–5 HCEP cells were used in all of the experiments.

Preparation of Tualang honey

Tualang honey used in this experiment was from Federal Agriculture Marketing Authorities of Malaysia (FAMA) and was a gift from Professor Siti Amrah Sulaiman, Universiti Sains Malaysia. Tualang honey was diluted to 20% in serum-free DMEM/F12 (Gibco, Invitrogen Life Technologies Co., Carlsbad, CA, US) and filtered through a 0.2 µm syringe filter (Pall Co., Port Washington, NY, US) prior to use in cell culture. Filtered Tualang honey was further diluted in KSFM according to the dilution factor described in Table 1.

Table 2. Sequences of primers used in this study.

Primers	Sequence (5′ – 3′)	Accession Number	PCR product (bp)
Beta-actin-F	GAGGCGTACAGGGATAGCA	NM_001101.3	302
Beta-actin-R	GTGGGCATGGGTCAGAAG		
ABCG2-F	GAGCTCGTCCCCTGGATGT	NM_004827.2	186
ABCG2-R	CGGAACCTTTTGAGTGGGCA		
Connexin43-F	CAAAATCGAATGGGGCAGGC	NM_000165.3	136
Connexin43-R	GCTGGTCCACAATGGCTAGT		
K12-F	CTCGCAGAGTGTGATAGGCA	NM_000223.3	146
K12-R	CCCCAAAGCCGGAACTAGAA		

Abbreviations: F, forward primer; R, reverse primer; PCR, polymerase chain reaction.

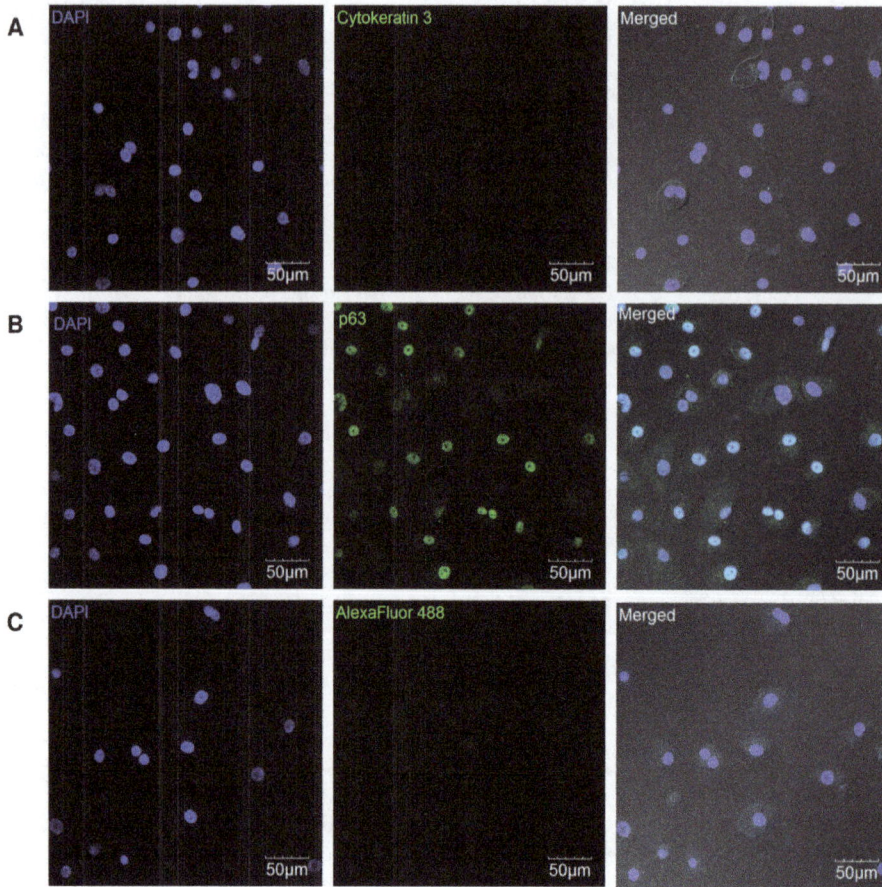

Figure 1. Representative confocal images of immunofluorescence-labelled HCEP cells. HCEP cells did not express the cornea-specific marker cytokeratin 3 (green) (A) but did express nuclear p63 transcription factor (green) (B). HCEP cells stained with AlexaFluor488 secondary antibody without any primary antibody were served as the negative control (C). Abbreviations: DAPI, 4′,6-diamidino-2-phenylindole.

Immunofluorescence labelling

HCEP cells were seeded onto Lab-Tek8 well-chamber slides (ThermoSci., Logan, UT, USA) using honey supplemented medium. Cells were fixed with 4% paraformaldehyde at 4°C for 10 min and then permeabilised with phosphate buffered saline containing 0.1% Triton-X and 0.1% Tween-20 (all from Sigma, St Louis, MO, USA). Cell samples were washed with PBS and blocked with 10% goat serum (Cedarlane Lab, Ontario, Canada) for 1 h. The cells then were incubated with rabbit anti-p63 (Santa Cruz Biotechnology, Heidelberg, Germany) and mouse anti-keratin 3 (Clone AE5, Millipore, Billerica, MA, USA) primary antibodies overnight at 4°C. Samples were washed thrice with PBS and incubated with AlexaFluor488 goat anti-rabbit secondary antibody at 37°C for 2 h. Samples were then counterstained with DAPI and imaged using a confocal microscope (Olympus, Japan).

RNA extraction and cDNA synthesis

The total RNA was isolated using the RNeasy Mini Kit (Qiagen, Hilden, Germany) according to the manufacturer's instructions. The isolated RNA was treated with DNase I (Sigma, St. Louis, MO, USA) to ensure the absence of genomic DNA contamination. Total RNA samples were quantitatively and qualitatively assessed using UV-spectrophotometric measurement and agarose gel electrophoresis, respectively. The cDNA was synthesised using the QuantiTect Reverse Transcription Kit (Qiagen, Hilden, Germany) according to manufacturer's protocol.

Primer design

Full-length exon spanning gene sequences were obtained from the NCBI GenBank Database (http://www.ncbi.nlm.nih.gov/genbank/). All primers were designed from the gene sequences using Primer-BLAST online software (NCBI, NIH, USA). All primers are listed in Table 2. The sequences of the primers were compared to the GenBank Database using BLAST in order to

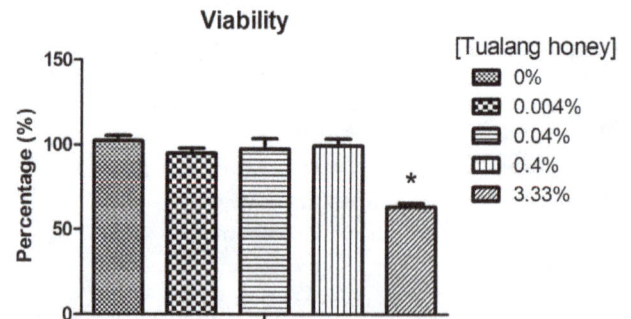

Figure 2. HCEP cell viability after treatment with 0, 0.004, 0.04, 0.4, and 3.33% Tualang honey for 48 h. Significant lower viability was observed in HCEP cells treated with 3.33% Tualang honey compared to the other treatments. $***p < 0.001$.

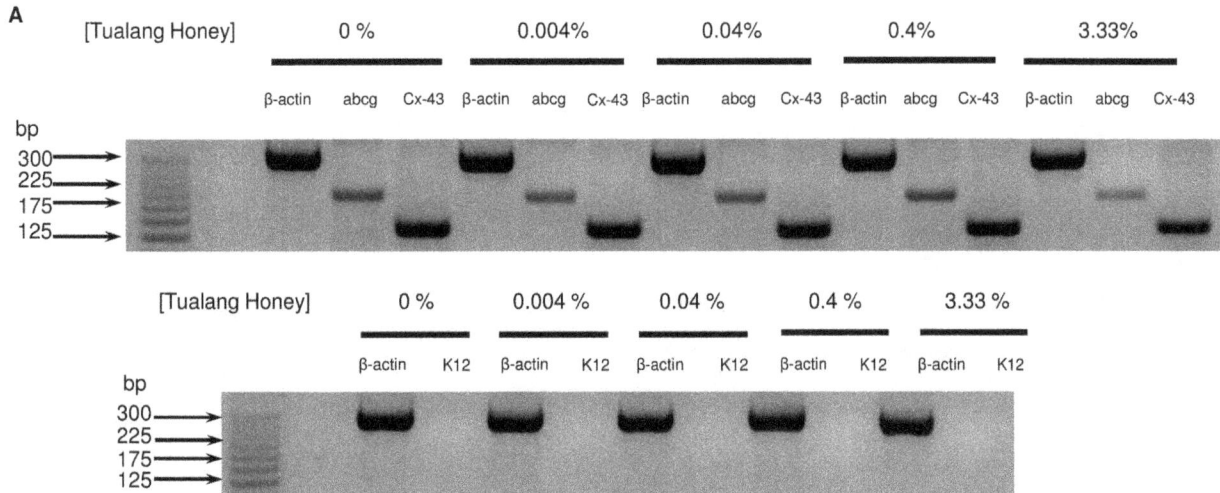

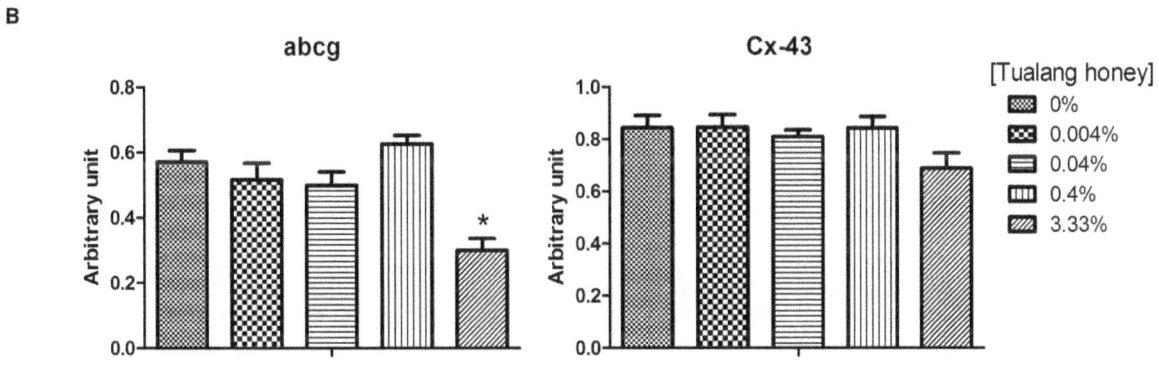

Figure 3. mRNA expression of Tualang honey treated HCEP cells. (A) RTPCR showed abcg and connexion-43 mRNA expressions (normalised to β-actin expression) but not cytokeratin-12 expression in HCEP cells. (B) abcg mRNA expression was down-regulated in HCEP cells treated with 3.33% Tualang honey, but the mRNA for connexin-43 remained unchanged in all groups. *$p<0.05$ compared to untreated control (0%). Abbreviations: abcg, ATP-binding cassette transporter; Cx-43, connexion-43.

determine their specificity. Primers that could have resulted in non-specific signals were excluded. All primers were purchased from BioBasic Inc. (Ontario, Canada).

Polymerase chain reactions (PCR)

PCR amplification was performed using the MyCycler Thermal Cycler (Bio-Rad, Hercules, CA, USA). All PCR reagents used in this study were purchased from Biotools B&M Labs (S.A., Madrid, Spain). Total cDNA used in each reaction was 1 μL unless specified otherwise. Reaction buffer (1 X), 1.5 mM $MgCl_2$, 1.75 mM of each dNTP, 0.5 μM of each primer, 100 ng of cDNA, and 1 U of DNA Taq Polymerase were added to all samples to reach a final volume of 20 μl in each reaction. A negative control (sample without template) was also included. PCR was performed at 95°C for 2 min as initial denaturation followed by 35 cycles of 30 sec at 95°C, 30 sec at 60°C, and 30 sec at 72°C, and then at 72°C for another 2 min. The PCR products were loaded with gel loading dye (Promega, Madison, WI, USA) onto 4% agarose gel in 1X TAE buffer and electrophoresed for 60 min at 80 V. A 25 base pair molecular ladder (Promega, Madison, WI, USA) was used. Subsequently, the gel was stained with ethidium bromide (Bio-Rad, Hercules, CA, USA) to visualize the presence of PCR amplicons.

Cytotoxicity assay

HCEP cells were seeded at 5000 cells/well onto a 96-well plate and left in an incubator with 5% CO_2 at 37°C overnight. The culture medium then was changed to test medium supplemented with Tualang honey at various concentrations (Table 1), and the cells were incubated for 48 h. Cell viability was tested using AlamarBlue (Molecular Probes, Invitrogen Life Technologies Co., Carlsbad, CA, USA) according to the manufacturer's protocol. After 3 h of incubation with AlamarBlue, the fluorescence intensity of each sample was read using a FLUOstar Omega multi-mode microplate reader (BMG Labtech, Germany) with an excitation wavelength of 570 nm and an emission wavelength of 585 nm. Controls were HCEP cells in regular medium without Tualang honey. Cell viability was expressed as a percentage using the following formula:

$$\text{Viability} = \frac{FI_{treated}}{FI_{non-treated}} \times 100\%$$

Where

$FI_{treated}$ = Fluorescence intensity of test samples (Tualang honey treated HCEP cells)

$FI_{non-treated}$ = Fluorescence intensity of control samples (non-treated HCEP cells)

[Tualang honey]   0%   0.004%   0.04%   0.4%   3.33%

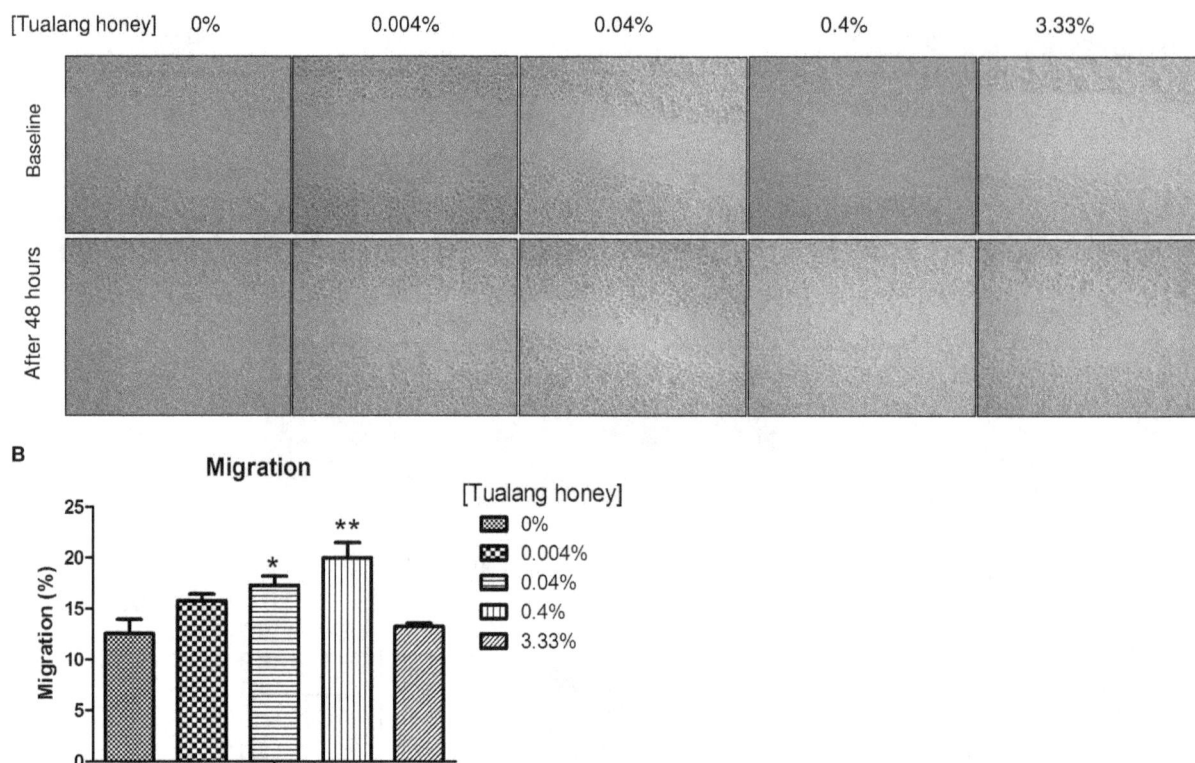

Figure 4. HCEP cell migration following treatment with Tualang honey. (A) Representative images of the HCEP cell scratch migration assay at baseline and after 48 h. (B) The gap size occupied by HCEP cells after 48 h was largest in the 0.4% Tualang honey treatment ($*p<0.05$; $**p<0.01$ compared to untreated control).

Scratch migration assay

HCEP cells were seeded onto a 6-well plate and allowed to expand and reach 90% confluence. A scratch was made onto each HCEP cell-containing well using a 200 μl pipette tip. The culture medium then was changed to standard KSFM supplemented with 0.004, 0.04, 0.4, or 3.33% (v/v) Tualang honey, and the cells were incubated at 37°C for 2 days. Ten representative images were taken using a charge-coupled device camera under an inverted light microscope pre- and post-treatment. Images obtained were analysed for wound closure using ImageJ software (NIH). All experimental samples were compared to a control (i.e., HCEP cells without Tualang honey treatment). The percentage of closure of the gap area was calculated using the following formula:

$$Gap_\Delta(\%) = Gap_{48}(\%) - Gap_0(\%)$$

Where
Gap_Δ = Occupied gap area after 48 h
Gap_{48} = Gap area at 48 h
Gap_0 = Gap area at baseline

Hydrogen peroxide inhibition assay

Tualang honey was prepared at 0.04, 0.4, and 4% and added to wells containing 5, 10, 20, 30, and 40 μM hydrogen peroxide (H_2O_2). The concentration of H_2O_2 was assessed using the Amplex Red hydrogen peroxide assay kit (Molecular Probes, Invitrogen Life Technologies Co., Carlsbad, CA, USA) after 30 min of incubation at room temperature according to the manufacturer's protocol. The excitation and emission wavelengths

were set at 544 and 590 nm, respectively, to obtain the fluorescence reading using FLUOstar Omega multi-mode microplate reader.

Oxidative stress assay

HCEP cells were cultured with 0.004, 0.04, 0.4, and 3.33% (v/v) Tualang honey in supplemented KSFM for 48 h. Treated cells then were exposed to 50 μM H_2O_2 for 24 h. The number of dead cells in treated groups after H_2O_2 exposure was measured using propidium iodide staining (BD Bioscience Franklin Lakes, NJ, USA) and quantified using a flow cytometer (FACS Canto, BD Bioscience Franklin Lakes, NJ, USA). Untreated HCEP cells served as the negative control. Ascorbic acid, a potent anti-H2O2 agent used as a positive control and 5-hydroxymethyl-2-furancarboxaldehyde were both purchased from Sigma, St Louis, MO, USA.

Gas Chromatography-Mass Spectrometry (GC-MS) analysis

GC-MS analysis of the honey was conducted using a Shimadzu system (Japan) consisting of a GC-MS-QP2010 gas chromatograph and a quadrupole mass spectrometer. The interface and source temperatures were 280°C and 250°C, respectively. Electron impact mass spectra were recorded in the 20–650 amu range at 70 eV ionization energy. Separation was performed on a fused-silica bonded-phase capillary column BP X5 (30 mx0.25 mm ID and 0.25 μm film thickness). The injector temperature was set at 250°C; and the samples were run in split mode with the ratio being adjusted to 25:1 and an injection volume of 1 μL. The temperature program was isothermal at

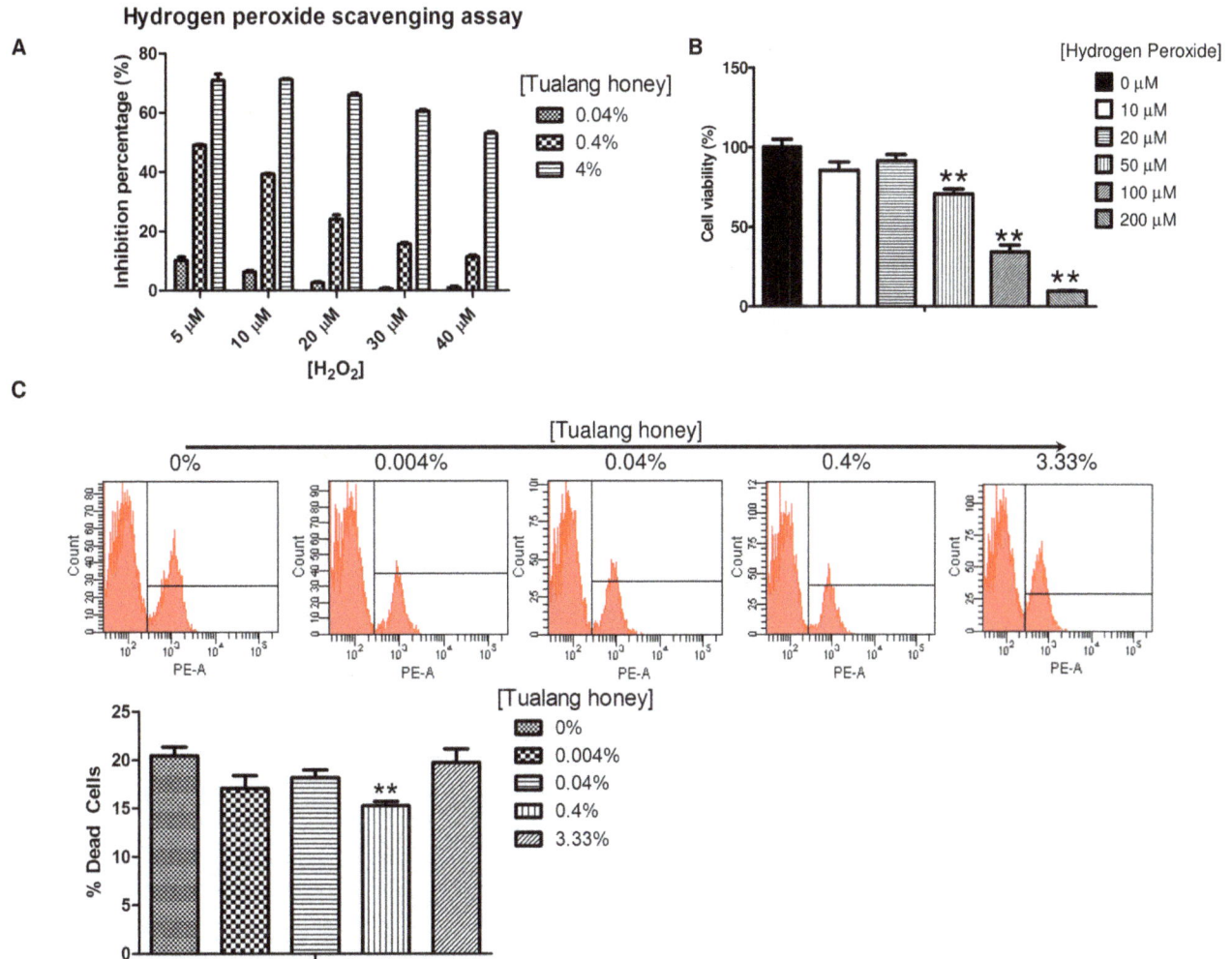

Figure 5. Effects of Tualang honey on hydrogen peroxide and HCEP cell resistance to oxidative stress. (A) *In vitro* inhibitory effects of Tualang honey at 0.04, 0.4, and 4% (v/v) against 5, 10, 20, 30, and 40 μM H_2O_2. (B) HCEP viability in response to H_2O_2 induction at 0, 10, 20, 50, 100 and 200 μM after 24 hours using AlamarBlue assay. (C) Flow cytometric analysis showed the total number of PI-stained dead cells in Tualang honey treated HCEP cells in response to exposure to 50 μM H_2O_2 after 24 h (**$p < 0.01$ compared to untreated control). Abbreviation: PI, propidium iodide.

35°C for 1 min, then it was raised to 280°C at 25°C/min and to 310°C at 10°C/min. This temperature was held for 2 min. Samples in the chromatograms were identified by comparing their mass spectra with NIST08 library data as well as the retention times against known standards.

Statistical analysis

All experiments were repeated three times and data were expressed as mean ± standard error (SEM). All statistical analyses were performed using Graphpad Prism (La Jolla, CA, USA). The differences between groups were analysed using one-way ANOVA with Dunnett post-hoc test. Differences were considered significant at $p < 0.05$.

Results

Cultured HCEP cells possessed stem cell characteristics

The cultured HCEP cells were highly proliferative and capable of forming holoclones. Immunofluorescence staining confirmed that the cultured HCEP cells used in this study expressed nuclear p63 but not cytokeratin 3/12 (Figure 1), suggesting a keratinocyte stem cell phenotype. However, the cultured HCEP cells exhibited signs of replicative senescence as soon as they reached passage 5 or 6.

Low levels of Tualang honey favoured HCEP cell viability, gene expression, and migration

Viability of HCEP cells was assessed after 24 h and no sign of cytotoxicity was observed in groups treated with 0.004, 0.04, or 0.4%Tualang honey (Figure 2). However, cell viability was significantly reduced to 63.1±2.1% at the concentration of 3.33% compared to the untreated control ($p < 0.001$). We also sought to investigate whether Tualang honey could affect HCEP cell stemness or trigger corneal epithelial differentiation. To do this, the mRNA level of ATP-binding cassette transporter 2 (abcg2), a marker which was found to express in human corneal epithelial stem cells, and corneal epithelial differentiation marker connexin-43 and cytokeratin-12, were examined. RTPCR revealed that the mRNA expression for connexin-43 and abcg2 were detected in all groups, but expression of cytokeratin-12 was not (Figure 3A). Semi-quantitative analysis of the expressed mRNA revealed that connexin-43 mRNA remained unaffected by

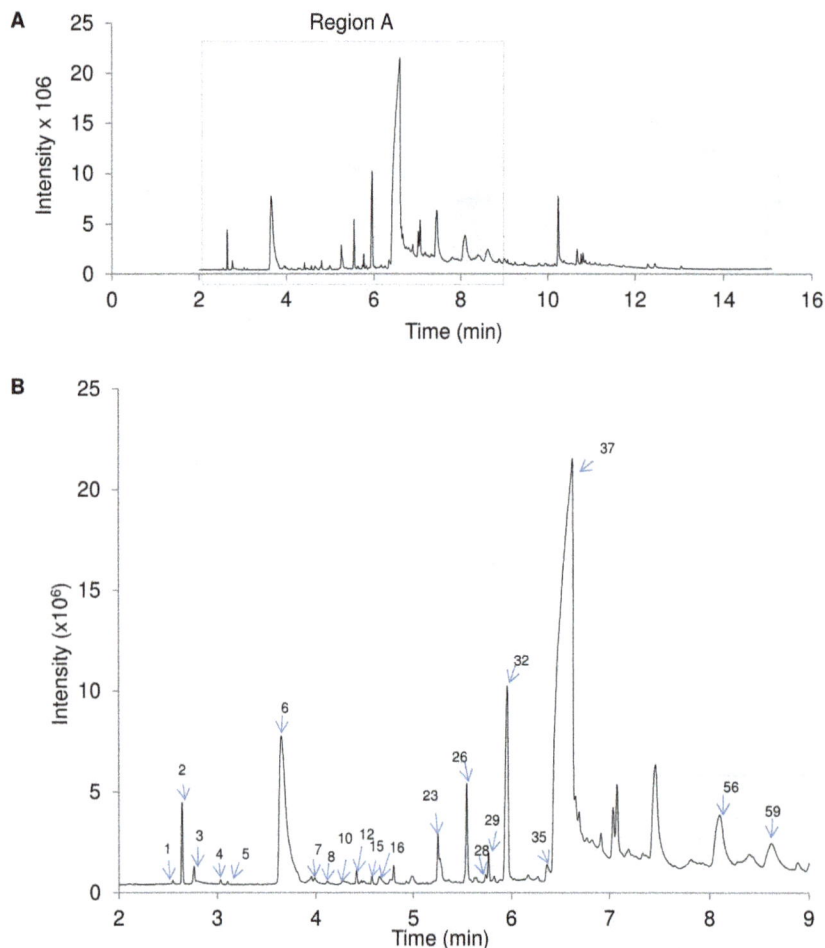

Figure 6. GC-MS analysis of Tualang honey. (A) Region A in the chromatogram of 20% Tualang honey diluted in distilled water. (B) Region A was enlarged and analysed further to identify the phytochemical constituents in Tualang Honey based on the NIST08 Mass Spectral Library.

Tualang honey in all groups, which also indicates that Tualang honey had minimal or no potential in triggering corneal epithelial differentiation. However, a significant down-regulation of abcg2 mRNA was found in HCEP cells treated with 3.33% Tualang honey; this represents a decrease of 47% when compared with the expression in untreated cells ($p < 0.05$) (Figure 3B). These results suggest that Tualang honey did not affect the HCEP stemness and differentiation at non-cytotoxic level.

The *in vitro* scratch migration assay showed that the percentage of gap area covered by migrated HCEP cells after 48 h of treatment increased in a dose-dependent manner (Figure 4). HCEP cells treated with 0.04 and 0.4% Tualang honey occupied $17.3 \pm 0.9\%$ ($p < 0.05$) and $20 \pm 1.5\%$ ($p < 0.01$) of the gap area, respectively, compared to $12 \pm 6\%$ in the untreated group. In the 3.33% Tualang honey treatment, only $13.3 \pm 0.3\%$ of the gap area was occupied after 48 h.

Tualang honey scavenged H_2O_2 in a dose-dependent manner and enhanced HCEP cell resistance to oxidative stress

When Tualang honey was tested for its H_2O_2 scavenging ability, the level of inhibition was found to increase with increasing Tualang honey concentration (Figure 5A). This result suggests that Tualang honey is a source of anti-H_2O_2 antioxidant. Notably, only negligible scavenging effects were observed at 30 and 40 μM

H_2O_2 when the Tualang honey was diluted to 0.04%. This indicates that over-dilution of Tualang honey may compromise its H_2O_2 scavenging ability. To identify the cytotoxicity of H_2O_2, HCEPs were exposed to 10, 20, 50, 100 and 200 H_2O_2 and the viability was assessed after 24 h using AlamarBlue assay. As shown in Figure 5B, HCEP viability was reduced by H_2O_2 in a dose-dependent manner, and the difference became apparent and statistically significant when H_2O_2 level was increased to ≥ 50 μM ($p < 0.01$).

This study was extended to investigate the potential for Tualang honey to improve HCEP cell resistance to H_2O_2-induced oxidative stress. Interestingly, HCEP cells treated with 0.4% honey had a higher survival rate than the untreated HCEP cells, with a significantly lower number of dead cells ($15.3 \pm 0.4\%$) compared to the control ($20.5 \pm 0.9\%$, $p < 0.01$) (Figure 5C).

Tualang honey contains active compounds with known antioxidant properties

The chromatogram of filtered 20% Tualang honey diluted in distilled water revealed the presence of several volatile compounds (Figure 6). Of those, only 20 chemical constituents were identified (Table 3). The major constituent in Tualang honey was 5-hydroxymethyl-2-furancarboxaldehyde (5HMF), a five carbon ring aromatic aldehyde antioxidant that accounted for a relative peak area of 36.21% [31]. Other known antioxidants, such as 3-

Table 3. Phytochemical constituents detected in Tualang honey.

Peak	Area (%)	Phytochemical	Activity
37	36.21	5-(hydroxymethyl) 2-Furancarboxaldehyde,	Antioxidant [31]
6	7.46	3-Furaldehyde	Antioxidant [32]
56	4.73	Beta.-D-Glucopyranose, 1,6-anhydro-	-
32	4.23	4H-Pyran-4-one, 2,3-dihydro-3,5-dihydroxy-6-methyl-	Antimicrobial, anti-inflammatory [60,61]; antioxidant [62–64]
59	3.06	1,6-Anhydro-.beta.-D-glucofuranose	-
26	1.18	Methyl 2-furoate	-
23	1.13	Phenylacetaldehyde	Antioxidant [65]
2	0.71	Formic acid	-
29	0.33	Levoglucosenone	Anticancer, treatment for autoimmune system and cardiovascular diseases [66]
16	0.2	2-Furancarboxaldehyde, 5-methyl-	-
3	0.16	Acetic acid	Antihistamine [67]
12	0.14	2(5H)-Furanone	-
7	0.13	2-Furanmethanol	Antioxidant [68–70]
28	0.12	Maltol	Antioxidant [71,72], anticonvulsant, depressant [73], anti-aging [74]
35	0.1	2(3H)-Furanone, dihydro-4-hydroxy-	-
8	0.09	Propanoic acid, 2-hydroxy-, ethyl ester	-
15	0.09	2(5H)-Furanone, 5-methyl-	-
10	0.08	2-Propanone, 1,3-dihydroxy-	-
4	0.04	2-Propanone, 1-hydroxy- (CAS) Acetol	-
1	0.03	Hydrogen chloride	-
5	0.02	Propanoic acid, 2-oxo- (CAS) Pyruvic acid	-

furaldehyde (7.46%), phenylacetaldehyde (1.13%), 2-furanmethanol (0.13%), and maltol (0.12%), were also present in Tualang honey [31,32], albeit at lower percentages.

The improvement in HCEP resistance to H_2O_2-induced oxidative stress was attributed to the antioxidant properties of Tualang honey at its native form

To test whether the observed improvement in HCEP resistance to H_2O_2-induced oxidative stress following the treatment with Tualang honey was attributed to 5-hydroxymethyl-2-furancarboxaldehyde (5HMF), the viability of HCEPs treated with 100 µM of 5HMF for 48 h were examined with AlamarBlue after exposed to 50 and 100 µM H_2O_2 for 24 h. As shown in Figure 7, HCEP viability was significantly decreased after treated with 5HMF for 48 h compared to the untreated group ($p<0.05$). However, the viability became comparable to that of untreated control after 24 h induction of H_2O_2 at 50 and 100 µM. This indicates that although 5HMF had little effects in improving HCEP resistance to H_2O_2, the benefit was nullified by its mild cytotoxicity.

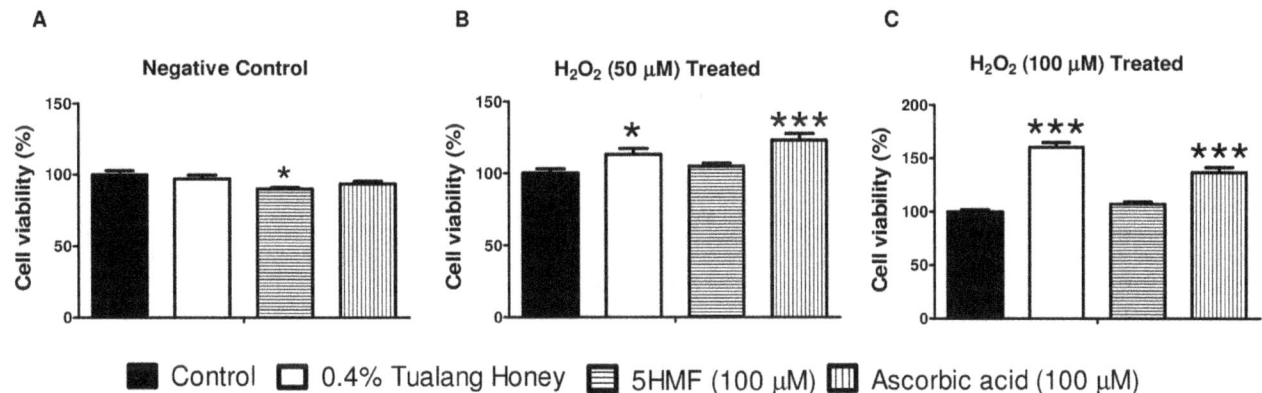

Figure 7. AlamarBlue cell viability assay of Tualang honey (0.4%), ascorbic acid (100 µM) and 5HMF(100 µM) treated HCEP cells after H_2O_2 induction. The resistance of treated cells against H_2O_2 was tested at (A) 0, (B) 50 and (C) 100 µM for 24 h. Significant difference in viability was found between Tualang honey and ascorbic acid-treated group using ANOVA with Tukey multiple comparisons test (*$p<0.05$; ***$p< 0.01$ compared to untreated control). Abbreviation: 5HMF, 5-hydroxymethyl-2-furancarboxaldehyde.

Table 4. pH values of HCEP cell culture medium after supplementation with Tualang honey at various concentrations.

Tualang Honey concentrations	0%	0.004%	0.04%	0.4%	3.33%	5.0%	10.0%
pH	7.65	7.67	7.58	7.59	7.17	5.83	4.81

To assess whether the antioxidant property of Tualang honey was responsible for the elevated resistance to H_2O_2-induced oxidative stress, HCEPs treated with 100 μM ascorbic acid, a known powerful antioxidant, was included as the positive control. AlamarBlue assay showed better HCEP survival following H_2O_2 insults at 50 μM and 100 μM after treated with ascorbic acid or Tualang honey as compared to the untreated group. More importantly, the resistance to 100 μM H_2O_2 was found significantly greater in Tualang honey-treated HCEP cells than that of ascorbic acid-treated (Figure 7C). Taken together, these results confirm that the enhanced HCEP resistance to oxidative stress is conferred by the antioxidant property of Tualang honey in its native form.

Discussion

Tualang honey has been used to treat bacterial infections [33], promote burn wound healing [19,34], and protect bone in post-menopausal women [18]. However, its potential as a supplement in cell culture medium for stem cell expansion has not been studied, and little is known about the effects of honey on cells at the cellular level. This is the first study to evaluate the use of Tualang honey in culturing HCEP cells and its effects in enhancing cell function in vitro. Thus, it is a pivotal study for exploring the potential of this honey and its application in corneal regeneration.

The ATP binding cassette transport 2 (abcg2) is known for its capability to efflux Hoechst 33342 and has been used as a marker to identify side population cells [35]. This marker was also found to present in clonogenic human limbal-derived epithelial stem cells [36]. A study by Kubota et al. (2010) suggested that abcg2 has a vital role in maintaining endogenous anti-oxidative capacity in HCEPs [37]. Here, the cultured human corneal epithelial cells used in this study expressed mRNA for abcg2 and nuclear p63 protein, the putative corneal epithelial stem cell markers [38,39]. However, significant downregulation of abcg2 mRNA was detected when Tualang honey was introduced to the cells at cytotoxic level, 3.33%. This may also suggest that the cytotoxicity of Tualang honey in HCEP cells is via the downregulation of abcg2 expression, reduces the cell anti-oxidative capacity which renders the cells susceptible to oxidative damage. Although no terminal corneal epithelial differentiation was detected (based on the lack of cytokeratin 3 protein and cytokeratin 12 mRNA expression), we found that the cells constantly expressed gap junction protein connexin-43 mRNA, a negative marker for corneal epithelial stem cells [40]. This suggests that the cells had partial commitment to the corneal epithelial lineage. Hence, in order to accurately describe the cell population, we called them human corneal epithelial progenitor cells (i.e., HCEPs).

Previous studies showed that the bactericidal effects of honey were partly, if not entirely, due to its acidity, high osmolality, and H_2O_2 content [41,42]. These characteristics could also be cytotoxic to cells. We addressed the cytotoxicity of Tualang honey to cultured HCEP cells by lowering the concentration to a level that favoured cell growth (≤0.4%). Our data show that the pH of the culture medium was not altered by Tualang honey at low concentration (0.004–0.4%), but the medium became acidic at high concentration (3.33%) (Table 4). Furthermore, a high level of honey supplementation could produce a hyperosmotic culture medium, which could induce ROS generation and apoptosis in corneal epithelial cells [43]. In contrast to our finding, Ghashm et al. (2010) reported that the pH of the culture medium remained suitable for cell culture after adding Tualang honey at high concentrations (3.5–20%) compared to the control medium without Tualang honey [23]. This discrepancy may be due to differences in preparation and storage conditions of the Tualang honey used in the experiments, as storage conditions can alter the content and properties of the honey [44].

We also found that Tualang honey diluted to 0.04% did not affect gene expression and was favourable to HCEP cells, as shown by the enhanced cell migration in vitro in this treatment. This result suggests that dilution of Tualang honey does not hamper its beneficial effects on HCEP cells. These data coincide with a past study that showed improvement in proliferation of human osteoblast cells (CRL 1543) when the culture medium was supplemented with 0.0195% Tualang honey [30]. The underlying mechanism for the observed improvements is unknown, but it could be related to the active compounds present in Tualang honey or to the presence of H_2O_2, a ROS that is essential for modulating stem cell behaviours at the physiological level [45–47]. Pan et al. (2011) recently suggested that low levels of H_2O_2 (10–50 μM) could promote rabbit corneal epithelial cell attachment, mobility, and wound repair [48]. This contradicts our findings using human cells that H_2O_2 showed significant cytotoxicity at 50 μM and no noticeable changes in cell number at 10 and 20 μM when compared with the untreated control. Nonetheless, this discrepancy may due to the difference in the species of origin of corneal epithelial cells used in the experiments.

Furthermore, we also found that H_2O_2 was present in Tualang honey, but only at a negligible level of <2 μM after dilution to ≤ 0.4%. This finding suggests that the H_2O_2 level may not be sufficient to exert significant changes in cell behaviours. Although H_2O_2 is present in naturally occurring Tualang honey, it also contains several phytochemical compounds with antioxidant properties that are capable of mitigating H_2O_2. The predominant phytocompound identified in Tualang honey was 5-hydroxy-methyl-2-furancarboxaldehyde, which is a five carbon ring aromatic aldehyde antioxidant that is commonly found in fruits [49] and marine products [50]. Compared to other types of honey, Tualang honey had been identified as a new source of antioxidant with superior activity that is attributed to its high phenolic content [51].

Oxidative stress has been shown to cause chromosomal instability, shortened telomeres, and cellular replicative senescence in stem cells in vitro [52,53], all of which hamper cell function and limit the therapeutic outcome of cell transplantation. Accumulating evidence supports the premise that cellular antioxidant levels in stem cells play a vital role in dictating their fate, regenerative capability, and therapeutic outcome after transplantation [54,55]. Overexpression of cellular redox regulators such as Cu/Zn superoxide dismutase enzymes and nuclear factor erythroid 2-

related factor-2 (Nrf-2) has been shown to improve stem cell functions [56]. The oxidative stress tolerance of stem cells can also be augmented by the addition of exogenous catalase [57] or antioxidants from natural products [16,58,59]. Herein, Tualang honey at 0.4% was found to improve HCEP cell resistance to oxidative stress, as shown by the significantly lower number of dead cells after treatment with 50 μM H_2O_2 in the treatment versus control group. This indicates that Tualang honey can be used as a source of antioxidant for preconditioning HCEP cells prior to transplantation. However, more evidence is needed to confirm the changes in HCEP cell antioxidant enzyme levels that likely account for the Tualang honey-induced enhancement of the oxidative stress tolerance of HCEP cells.

In summary, this study describes a novel approach to integrating the application of natural products to the process of corneal regeneration. Our data indicate that Tualang honey is an antioxidant and contains active phytocompounds that enhance HCEP cell migration and resistance to oxidative stress *in vitro* in a dose-dependent manner. The beneficial effects, however, are offset by the cytotoxicity of Tualang honey at high concentration. In-depth studies to isolate and identify the active components in Tualang honey and to identify the mechanism responsible for the observed benefits *in vitro* and *in vivo* are needed to enable use of this natural product for the treatment of corneal diseases.

Acknowledgments

The authors thank Dr. Gary Yam (Singapore Eye Research Institute), Dr. Fam Pei Shan, Nur Hikmah binti Ramli, and Noraida binti Zakaria (Universiti Sains Malaysia) for their technical help with this project, and special appreciation goes to David Lyn (Matrix Optics) for his assistance with confocal microscopy.

Author Contributions

Conceived and designed the experiments: JJT BS. Performed the experiments: JJT SMA HLC VL. Analyzed the data: JJT YKY VL. Contributed reagents/materials/analysis tools: JJT DS BS. Wrote the paper: JJT.

References

1. Chen Z, de Paiva CS, Luo L, Kretzer FL, Pflugfelder SC, et al. (2004) Characterization of putative stem cell phenotype in human limbal epithelia. Stem Cells 22: 355–366.
2. Kim HS, Jun Song X, de Paiva CS, Chen Z, Pflugfelder SC, et al. (2004) Phenotypic characterization of human corneal epithelial cells expanded ex vivo from limbal explant and single cell cultures. Exp Eye Res 79: 41–49.
3. He Y, Pan Z, Luo F (2012) A Novel PAX6 Mutation in Chinese Patients with Severe Congenital Aniridia. Current Eye Research 37: 879–883.
4. Smith W, Lange J, Sturm A, Tanner S, Mauger T (2012) Familial peripheral keratopathy without PAX6 mutation. Cornea 31: 130–133.
5. Le Belle JE, Orozco NM, Paucar AA, Saxe JP, Mottahedeh J, et al. (2011) Proliferative neural stem cells have high endogenous ROS levels that regulate self-renewal and neurogenesis in a PI3K/Akt-dependant manner. Cell Stem Cell 8: 59–71.
6. Ito K, Hirao A, Arai F, Takubo K, Matsuoka S, et al. (2006) Reactive oxygen species act through p38 MAPK to limit the lifespan of hematopoietic stem cells. Nat Med 12: 446–451.
7. Li TS, Marban E (2010) Physiological levels of reactive oxygen species are required to maintain genomic stability in stem cells. Stem Cells 28: 1178–1185.
8. Chaudhari P, Ye S, Jang YY (2012) Roles of Reactive Oxygen Species in the Fate of Stem Cells. Antioxid Redox Signal.
9. Pendergrass KD, Boopathy AV, Seshadri G, Maiellaro-Rafferty K, Che PL, et al. (2013) Acute Preconditioning of Cardiac Progenitor Cells with Hydrogen Peroxide Enhances Angiogenic Pathways Following Ischemia-Reperfusion Injury. Stem Cells Dev.
10. Limoli CL, Giedzinski E (2003) Induction of chromosomal instability by chronic oxidative stress. Neoplasia 5: 339–346.
11. Kryston TB, Georgiev AB, Pissis P, Georgakilas AG (2011) Role of oxidative stress and DNA damage in human carcinogenesis. Mutat Res 711: 193–201.
12. Song H, Cha MJ, Song BW, Kim IK, Chang W, et al. (2010) Reactive oxygen species inhibit adhesion of mesenchymal stem cells implanted into ischemic myocardium via interference of focal adhesion complex. Stem Cells 28: 555–563.
13. Dernbach E, Urbich C, Brandes RP, Hofmann WK, Zeiher AM, et al. (2004) Antioxidative stress-associated genes in circulating progenitor cells: evidence for enhanced resistance against oxidative stress. Blood 104: 3591–3597.
14. Urish KL, Vella JB, Okada M, Deasy BM, Tobita K, et al. (2009) Antioxidant levels represent a major determinant in the regenerative capacity of muscle stem cells. Mol Biol Cell 20: 509–520.
15. Ko E, Lee KY, Hwang DS (2012) Human umbilical cord blood-derived mesenchymal stem cells undergo cellular senescence in response to oxidative stress. Stem Cells Dev 21: 1877–1886.
16. Gurusamy N, Ray D, Lekli I, Das DK (2010) Red wine antioxidant resveratrol-modified cardiac stem cells regenerate infarcted myocardium. J Cell Mol Med 14: 2235–2239.
17. Alves H, Mentink A, Le B, van Blitterswijk CA, de Boer J (2013) Effect of antioxidant supplementation on the total yield, oxidative stress levels, and multipotency of bone marrow-derived human mesenchymal stromal cells. Tissue Eng Part A 19: 928–937.
18. Mohd Effendy N, Mohamed N, Muhammad N, Mohamad IN, Shuid AN (2012) The effects of tualang honey on bone metabolism of postmenopausal women. Evid Based Complement Alternat Med 2012: 938574.
19. Khoo YT, Halim AS, Singh KK, Mohamad NA (2010) Wound contraction effects and antibacterial properties of Tualang honey on full-thickness burn wounds in rats in comparison to hydrofibre. BMC Complement Altern Med 10: 48.
20. Nasir NA, Halim AS, Singh KK, Dorai AA, Haneef MN (2010) Antibacterial properties of tualang honey and its effect in burn wound management: a comparative study. BMC Complement Altern Med 10: 31.
21. Mat Lazim N, Abdullah B, Salim R (2013) The effect of Tualang honey in enhancing post tonsillectomy healing process. An open labelled prospective clinical trial. Int J Pediatr Otorhinolaryngol 77: 457–461.
22. Yaacob NS, Nengsih A, Norazmi MN (2013) Tualang honey promotes apoptotic cell death induced by tamoxifen in breast cancer cell lines. Evid Based Complement Alternat Med 2013: 989841.
23. Ghashm AA, Othman NH, Khattak MN, Ismail NM, Saini R (2010) Antiproliferative effect of Tualang honey on oral squamous cell carcinoma and osteosarcoma cell lines. BMC Complement Altern Med 10: 49.
24. Fauzi AN, Norazmi MN, Yaacob NS (2011) Tualang honey induces apoptosis and disrupts the mitochondrial membrane potential of human breast and cervical cancer cell lines. Food Chem Toxicol 49: 871–878.
25. Albietz JM, Lenton LM (2006) Effect of antibacterial honey on the ocular flora in tear deficiency and meibomian gland disease. Cornea 25: 1012–1019.
26. Mansour AM (2002) Epithelial corneal oedema treated with honey. Clinical & experimental ophthalmology 30: 149–150.
27. Uwaydat S, Jha P, Tytarenko R, Brown H, Wiggins M, et al. (2011) The use of topical honey in the treatment of corneal abrasions and endotoxin-induced keratitis in an animal model. Current eye research 36: 787–796.
28. Bashkaran K, Zunaina E, Bakiah S, Sulaiman SA, Sirajudeen K, et al. (2011) Anti-inflammatory and antioxidant effects of Tualang honey in alkali injury on the eyes of rabbits: experimental animal study. BMC Complement Altern Med 11: 90.
29. Ahmad I, Jimenez H, Yaacob NS, Yusuf N (2012) Tualang honey protects keratinocytes from ultraviolet radiation-induced inflammation and DNA damage. Photochem Photobiol 88: 1198–1204.
30. Kannan TP, Ali AQ, Abdullah SF, Ahmad A (2009) Evaluation of Tualang honey as a supplement to fetal bovine serum in cell culture. Food Chem Toxicol 47: 1696–1702.
31. Mohamed SA, Khan JA (2013) Antioxidant capacity of chewing stick miswak Salvadora persica. BMC Complementary and Alternative Medicine 13: 40.
32. Frankel EN (1985) Chemistry of autoxidation: mechanism, products and flavor significance. American Oil Chemists' Society, Champaign.
33. Tan HT, Rahman RA, Gan SH, Halim AS, Hassan SA, et al. (2009) The antibacterial properties of Malaysian tualang honey against wound and enteric microorganisms in comparison to manuka honey. BMC Complement Altern Med 9: 34.
34. Sukur SM, Halim AS, Singh KK (2011) Evaluations of bacterial contaminated full thickness burn wound healing in Sprague Dawley rats Treated with Tualang honey. Indian J Plast Surg 44: 112–117.
35. Zhou S, Schuetz JD, Bunting KD, Colapietro AM, Sampath J, et al. (2001) The ABC transporter Bcrp1/ABCG2 is expressed in a wide variety of stem cells and is a molecular determinant of the side-population phenotype. Nat Med 7: 1028–1034.
36. de Paiva CS, Chen Z, Corrales RM, Pflugfelder SC, Li DQ (2005) ABCG2 transporter identifies a population of clonogenic human limbal epithelial cells. Stem Cells 23: 63–73.
37. Kubota M, Shimmura S, Miyashita H, Kawashima M, Kawakita T, et al. (2010) The anti-oxidative role of ABCG2 in corneal epithelial cells. Invest Ophthalmol Vis Sci 51: 5617–5622.
38. Priya CG, Prasad T, Prajna NV, Muthukkaruppan V (2013) Identification of human corneal epithelial stem cells on the basis of high ABCG2 expression

combined with a large N/C ratio. Microscopy research and technique 76: 242–248.

39. Arpitha P, Prajna NV, Srinivasan M, Muthukkaruppan V (2005) High expression of p63 combined with a large N/C ratio defines a subset of human limbal epithelial cells: implications on epithelial stem cells. Investigative ophthalmology & visual science 46: 3631–3636.

40. Chen Z, Evans WH, Pflugfelder SC, Li DQ (2006) Gap junction protein connexin 43 serves as a negative marker for a stem cell-containing population of human limbal epithelial cells. Stem Cells 24: 1265–1273.

41. Mavric E, Wittmann S, Barth G, Henle T (2008) Identification and quantification of methylglyoxal as the dominant antibacterial constituent of Manuka (Leptospermum scoparium) honeys from New Zealand. Mol Nutr Food Res 52: 483–489.

42. Cheng G, He Y, Xie L, Nie Y, He B, et al. (2012) Development of a reduction-sensitive diselenide-conjugated oligoethylenimine nanoparticulate system as a gene carrier. Int J Nanomedicine 7: 3991–4006.

43. Chen Y, Li M, Li B, Wang W, Lin A, et al. (2013) Effect of reactive oxygen species generation in rabbit corneal epithelial cells on inflammatory and apoptotic signaling pathways in the presence of high osmotic pressure. PloS one 8: e72900.

44. Brudzynski K, Kim L (2011) Storage-induced chemical changes in active components of honey de-regulate its antibacterial activity. Food Chem 126: 1155–1163.

45. Daniels-Wells TR, Helguera G, Rodriguez JA, Leoh LS, Erb MA, et al. (2013) Insights into the mechanism of cell death induced by saporin delivered into cancer cells by an antibody fusion protein targeting the transferrin receptor 1. Toxicol In Vitro 27: 220–231.

46. Cho KA, Woo SY, Seoh JY, Han HS, Ryu KH (2012) Mesenchymal stem cells restore CCl4-induced liver injury by an antioxidative process. Cell Biol Int 36: 1267–1274.

47. Ding H, Keller KC, Martinez IK, Geransar RM, zur Nieden KO, et al. (2012) NO-beta-catenin crosstalk modulates primitive streak formation prior to embryonic stem cell osteogenic differentiation. J Cell Sci 125: 5564–5577.

48. Pan Q, Qiu WY, Huo YN, Yao YF, Lou MF (2011) Low levels of hydrogen peroxide stimulate corneal epithelial cell adhesion, migration, and wound healing. Investigative ophthalmology & visual science 52: 1723–1734.

49. Palma M, Taylor LT (2001) Supercritical fluid extraction of 5-hydroxymethyl-2-furaldehyde from raisins. J Agric Food Chem 49: 628–632.

50. Kulkarni A, Suzuki S, Etoh H (2008) Antioxidant compounds from Eucalyptus grandis biomass by subcritical liquid water extraction. J Wood Sci 54: 153–157.

51. Kishore RK, Halim AS, Syazana MS, Sirajudeen KN (2011) Tualang honey has higher phenolic content and greater radical scavenging activity compared with other honey sources. Nutr Res 31: 322–325.

52. Liu AM, Qu WW, Liu X, Qu CK (2012) Chromosomal instability in in vitro cultured mouse hematopoietic cells associated with oxidative stress. Am J Blood Res 2: 71–76.

53. Richter T, von Zglinicki T (2007) A continuous correlation between oxidative stress and telomere shortening in fibroblasts. Exp Gerontol 42: 1039–1042.

54. Abasi M, Massumi M, Riazi G, Amini H (2012) The synergistic effect of beta-boswellic acid and Nurr1 overexpression on dopaminergic programming of antioxidant glutathione peroxidase-1-expressing murine embryonic stem cells. Neuroscience 222: 404–416.

55. Sakata H, Niizuma K, Wakai T, Narasimhan P, Maier CM, et al. (2012) Neural stem cells genetically modified to overexpress cu/zn-superoxide dismutase enhance amelioration of ischemic stroke in mice. Stroke 43: 2423–2429.

56. Chatterjee S, Browning EA, Hong N, DeBolt K, Sorokina EM, et al. (2012) Membrane depolarization is the trigger for PI3K/Akt activation and leads to the generation of ROS. Am J Physiol Heart Circ Physiol 302: H105–114.

57. Urtasun R, Lopategi A, George J, Leung TM, Lu Y, et al. (2012) Osteopontin, an oxidant stress sensitive cytokine, up-regulates collagen-I via integrin alpha(V)beta(3) engagement and PI3K/pAkt/NFkappaB signaling. Hepatology 55: 594–608.

58. Schauen M, Spitkovsky D, Schubert J, Fischer JH, Hayashi J, et al. (2006) Respiratory chain deficiency slows down cell-cycle progression via reduced ROS generation and is associated with a reduction of p21CIP1/WAF1. J Cell Physiol 209: 103–112.

59. Esteban MA, Wang T, Qin B, Yang J, Qin D, et al. (2010) Vitamin C enhances the generation of mouse and human induced pluripotent stem cells. Cell Stem Cell 6: 71–79.

60. Ragupathi Raja Kannan R, Arumugam R, Anantharaman P (2012) Chemical composition and antibacterial activity of Indian seagrasses against urinary tract pathogens. Food chemistry 135: 2470–2473.

61. Kumar PP, Kumaravel S, Lalitha C (2010) Screening of antioxidant activity, total phenolics and GC-MS study of Vitex negundo. African Journal of Biochemistry Research 4: 191–195.

62. Xiangying Y, Zhao M, Fei L,Shitong Z,Jun H (2013) Identification of 2,3-dihydro-3,5-dihydroxy-6-methyl-4H-pyran-4-one as a strong antioxidant in glucose–histidine Maillard reaction products. Food Research International 51: 397–403.

63. Čechovská L, Cejpek K, Konečný M, Velíšek J (2011) On the role of 2,3-dihydro-3,5-dihydroxy-6-methyl-(4H)-pyran-4-one in antioxidant capacity of prunes. European Food Research and Technology 233: 367–376.

64. Osada Y, Shibamoto T (2006) Antioxidative activity of volatile extracts from Maillard model systems. Food chemistry 98: 522–528.

65. Nam S, Jang HW, Shibamoto T (2012) Antioxidant Activities of Extracts from Teas Prepared from Medicinal Plants, Morus alba L., Camellia sinensis L. and Cudrania tricuspidata and Their Volatile Components. Journal of Agricultural and Food Chemistry 60: 9097–9105.

66. Westman J, Wiman K, Mohell N (2007) Levoglucosenone derivatives for the treatment of disorders such as cancer, autoimmune diseases and heart diseases. In: W. I. P. . Organization, editor editors. Geneva Switzerland.

67. Ruiz CM, Gomes JC (2000) Effects of ethanol, acetaldehyde, and acetic acid on histamine secretion in guinea pig lung mast cells. Alcohol 20: 133–138.

68. Wei A, Mura K, Shibamoto T. (2001) Antioxidative activity of volatile chemicals extracted from beer. Journal of Agricultural and Food Chemistry 49: 4097–4101.

69. Fuster MD, Mitchell AE, Ochi H, Shibamoto T (2000) Antioxidative activities of heterocyclic compounds formed in brewed coffee. Journal of Agricultural and Food Chemistry 48: 5600–5603.

70. Yanagimoto K, Lee KG, Ochi H, Shibamoto T. (2002) Antioxidative activity of heterocyclic compounds found in coffee volatiles produced by Maillard reaction. Journal of Agricultural and Food Chemistry 50: 5480–5484.

71. Suh DY, Han YN, Han BH (1996) Maltol, an antioxidant component of Korean red ginseng, shows little prooxidant activity. Archives of Pharmacal Research 19: 112–115.

72. Lee KG, Shibamoto T (2000) Antioxidant properties of aroma compounds isolated from soybeans and mung beans. Journal of Agricultural and Food Chemistry 48: 4290–4293.

73. Aoyagi N, Kimura R, Murata T (1974) Studies on Passiflora incarnata dry extract. I. Isolation of maltol and pharmacological action of maltol and ethyl maltol. Chemical & pharmaceutical bulletin 22: 1008–1013.

74. Choi KT (2008) Botanical characteristics, pharmacological effects and medicinal components of Korean Panax ginseng CA Meyer. Acta Pharmacologica Sinica 29: 1109–1118.

Permissions

All chapters in this book were first published in PLOS ONE, by The Public Library of Science; hereby published with permission under the Creative Commons Attribution License or equivalent. Every chapter published in this book has been scrutinized by our experts. Their significance has been extensively debated. The topics covered herein carry significant findings which will fuel the growth of the discipline. They may even be implemented as practical applications or may be referred to as a beginning point for another development.

The contributors of this book come from diverse backgrounds, making this book a truly international effort. This book will bring forth new frontiers with its revolutionizing research information and detailed analysis of the nascent developments around the world.

We would like to thank all the contributing authors for lending their expertise to make the book truly unique. They have played a crucial role in the development of this book. Without their invaluable contributions this book wouldn't have been possible. They have made vital efforts to compile up to date information on the varied aspects of this subject to make this book a valuable addition to the collection of many professionals and students.

This book was conceptualized with the vision of imparting up-to-date information and advanced data in this field. To ensure the same, a matchless editorial board was set up. Every individual on the board went through rigorous rounds of assessment to prove their worth. After which they invested a large part of their time researching and compiling the most relevant data for our readers.

The editorial board has been involved in producing this book since its inception. They have spent rigorous hours researching and exploring the diverse topics which have resulted in the successful publishing of this book. They have passed on their knowledge of decades through this book. To expedite this challenging task, the publisher supported the team at every step. A small team of assistant editors was also appointed to further simplify the editing procedure and attain best results for the readers.

Apart from the editorial board, the designing team has also invested a significant amount of their time in understanding the subject and creating the most relevant covers. They scrutinized every image to scout for the most suitable representation of the subject and create an appropriate cover for the book.

The publishing team has been an ardent support to the editorial, designing and production team. Their endless efforts to recruit the best for this project, has resulted in the accomplishment of this book. They are a veteran in the field of academics and their pool of knowledge is as vast as their experience in printing. Their expertise and guidance has proved useful at every step. Their uncompromising quality standards have made this book an exceptional effort. Their encouragement from time to time has been an inspiration for everyone.

The publisher and the editorial board hope that this book will prove to be a valuable piece of knowledge for researchers, students, practitioners and scholars across the globe.

List of Contributors

Venkata Naga Srikanth Garikipati, Sachin Jadhav and Soniya Nityanand
Stem Cell Research Facility, Department of Hematology, Sanjay Gandhi Post-Graduate Institute of Medical Sciences, Lucknow, India

Lily Pal
Department of Pathology, Sanjay Gandhi Post-Graduate Institute of Medical Sciences, Lucknow, India

Prem Prakash and Madhu Dikshit
Cardio-vascular unit, Division of Pharmacology, Central Drug Research Institute, Lucknow, India

Chuan-Hang Yu
School of Dentistry, Chung Shan Medical University, Taichung, Taiwan
Department of Dentistry, Chung Shan Medical University Hospital, Taichung, Taiwan

Cheng-Chia Yu
School of Dentistry, Chung Shan Medical University, Taichung, Taiwan
Department of Dentistry, Chung Shan Medical University Hospital, Taichung, Taiwan
Institute of Oral Sciences, Chung Shan Medical University, Taichung, Taiwan

Michael Mildner, Maria Gschwandtner and Caterina Barresi
Department of Dermatology, Medical University Vienna, Vienna, Austria

Gregor Werba
Department of Surgery, Medical University Vienna, Vienna, Austria

Stefan Hacker
Department of Plastic Surgery, Medical University Vienna, Vienna, Austria
Christian Doppler Laboratory for Cardiac and Thoracic Diagnosis and Regeneration, Vienna, Austria

Thomas Haider, Matthias Zimmermann, Bahar Golabi and Jan Ankersmit
Christian Doppler Laboratory for Cardiac and Thoracic Diagnosis and Regeneration, Vienna, Austria
Department of Thoracic Surgery, Medical University Vienna, Vienna, Austria

Erwin Tschachler
Department of Dermatology, Medical University Vienna, Vienna, Austria
Centre de Recherches et dInvestigations Epidermiques et Sensorielles (CE.R.I.E.S.), Neuilly, France

Hua He
Department of Neurosurgery, Changzheng Hospital, Second Military Medical University, Shanghai, China

Ning Cheng
Department of Transfusion, Changhai Hospital, Second Military Medical University, Shanghai, China

Yanling Song, Weijin Ding, Yingfan Zhang, Jie Zhang and Hua Jiang
Department of Plastic Surgery, Changzheng Hospital, Second Military Medical University, Shanghai, China

Wenhao Zhang
Department of Hematology, XinHua Hospital, Affiliated to Shanghai Jiao Tong University (SJTU) School of Medicine, Shanghai, China

Heng Peng
Department of Mathematics, Hong Kong Baptist University, Kowloon, Hong Kong

Yi Tang
Department of Plastic Surgery, Changzheng Hospital, Second Military Medical University, Shanghai, China
Department of Plastic Surgery, No. 411 Hospital of CPLA, Shanghai, China

Mei Mei Wong, Xiaoke Yin, Claire Potter, Baoqi Yu, Hao Cai, Elisabetta Di Bernardini and Qingbo Xu
Cardiovascular Division, King's College London, British Heart Foundation Centre, London, United Kingdom

Viviana Meraviglia
Laboratorio di Biologia Vascolare e Medicina Rigenerativa, Centro Cardiologico Monzino, IRCCS, Milano, Italy

Simona Nanni
Istituto di Patologia Medica, Universitá Cattolica del Sacro Cuore, Roma, Italy

Andrea Barbuti, Angela Scavone and Dario DiFrancesco
Dipartimento di Bioscienze, The PaceLab, Universitá di Milano, Milano, Italy

Gualtiero I. Colombo
Laboratorio di Immunologia e Genomica Funzionale, Centro Cardiologico Monzino, IRCCS, Milano, Italy

Maurizio C. Capogrossi
Laboratorio di Patologia Vascolare, Istituto Dermopatico dell'Immacolata - IRCCS, Roma, Italy

Carlo Gaetano
Division of Cardiovascular Epigenetics, Department of Cardiology, Goethe University, Frankfurt am Main, Germany

Alessandra Rossini
Dipartimento di Scienze Cliniche e di Comunitá, Universitá di Milano, Milano, Italy

Matteo Vecellio and Giulio Pompilio
Dipartimento di Scienze Cliniche e di Comunitá, Universitá di Milano, Milano, Italy
Laboratorio di Biologia Vascolare e Medicina Rigenerativa, Centro Cardiologico Monzino, IRCCS, Milano, Italy

Antonella Farsetti
Istituto Nazionale Tumori Regina Elena, Roma, Italy
Istituto di Neurobiologia e Medicina Molecolare, Consiglio Nazionale delle Ricerche (CNR), Roma, Italy

Jianying Zhang and James H-C. Wang
MechanoBiology Laboratory, Departments of Orthopaedic Surgery, Bioengineering, Mechanical Engineering and Materials Science, and Physical Medicine and Rehabilitation, University of Pittsburgh, Pittsburgh, Pennsylvania, United States of America

Sayeh Khanjani, Manijeh Khanmohammadi, Mohammad-Mehdi Akhondi, Ali Ahani, Zahra Ghaempanah, Mohammad Mehdi Naderi and Somaieh Kazemnejad
Reproductive Biotechnology Research Center, Avicenna Research Institute, ACECR, Tehran, Iran

Saman Eghtesad
Department of Biochemistry and Molecular Biology, University of Maryland School of Medicine, Baltimore, Maryland, United States of America

Amir-Hassan Zarnani
Nanobiotechnology Research Center, Avicenna Research Institute, ACECR, Tehran, Iran
Immunology Research Center, Iran University of Medical Sciences, Tehran, Iran

Fang-Fang Huang, Deng-Shu Wu, Xiao-Yu Yuan, Fang-Ping Chen, Hui Zeng, Yan-Hui Yu and Xie-Lan Zhao
Department of Hematology, Xiang Ya Hospital, Central South University, Changsha, Hunan, China

Li Zhang
Department of Hematology, West China Hospital, Si Chuan University, Chengdu, Sichuan, China

Jianying Zhanga and James H.-C. Wang
MechanoBiology Laboratory, Departments of Orthopaedic Surgery, Bioengineering, Mechanical Engineering and Materials Science, and Physical Medicine and Rehabilitation, University of Pittsburgh, Pittsburgh, Pennsylvania, United States of America

Alexander M. Manya and Anthony M. C. Brown
Department of Cell & Developmental Biology, Weill Cornell Medical College, New York, New York, United States of America

Xiaoli Lan, Chunxia Qin, Xiaotian Xia, Hui Yuan, Zhiling Ding and Yongxue Zhang
Department of Nuclear Medicine, Union Hospital, Tongji Medical College of Huazhong University of Science and Technology, Hubei Province Key Laboratory of Molecular Imaging, Wuhan, China

Zhen Cheng
Molecular Imaging Program at Stanford and Bio-X Program, Stanford University, Stanford, California, United States of America

Zhijun Pei
Department of Nuclear Medicine, Union Hospital, Tongji Medical College of Huazhong University of Science and Technology, Hubei Province Key Laboratory of Molecular Imaging, Wuhan, China
Molecular Imaging Program at Stanford and Bio-X Program, Stanford University, Stanford, California, United States of America

Xiumei Liu, Xueming Wang, Lixia Li, Haiyan Wang and Xiaoling Jiao
Department of Pediatrics, Yuhuangding Hospital of Qingdao University, Yantai, Shandong, China

Yusuke Makino, Toshio Kukita and Takayoshi Yamaza
Department of Molecular Cell Biology and Oral Anatomy, Graduate School of Dental Science, Kyushu University, Fukuoka, Japan

Haruyoshi Yamaza, Yoshihiro Hoshino and Kazuaki Nonaka
Department of Pediatric Dentistry, Graduate School of Dental Science, Kyushu University, Fukuoka, Japan

Kentaro Akiyama and Songtao Shi
Center for Craniofacial Molecular Biology, Herman Ostrow School of Dentistry of USC, University of Southern California, Los Angeles, California, United States of America

Lan Ma
Department of Molecular Cell Biology and Oral Anatomy, Graduate School of Dental Science, Kyushu University, Fukuoka, Japan
Department of Pediatric Dentistry, Graduate School of Dental Science, Kyushu University, Fukuoka, Japan

Yixin Yao, Yinghua Lua and Wen-chi Chen
Department of Environmental Medicine, New York University Langone Medical Center, Tuxedo, New York, United States of America

Yongping Jiang
Biopharmaceutical Research Center, Chinese Academy of Medical Sciences & Peking Union Medical College, Suzhou, China

Tao Cheng
Institute of Hematology & Blood Disease Hospital, Chinese Academy of Medical Sciences & Peking Union Medical College, Tianjin, China

Yupo Ma
Yupo Ma, Department of Pathology, The State University of New York at Stony Brook, Stony Brook, New York, United States of America

Lou Lu
Department of Medicine, David Geffen School of Medicine, University of California Los Angeles, Torrance, California, United States of America

Wei Dai
Department of Environmental Medicine, New York University Langone Medical Center, Tuxedo, New York, United States of America
Department of Biochemistry and Molecular Pharmacology, New York University Langone Medical Center, Tuxedo, New York, United States of America

Simin Öz and Achim Breiling
Division of Epigenetics, DKFZ-ZMBH Alliance, German Cancer Research Center, Heidelberg, Germany

Christian Maercker
Mannheim University of Applied Sciences, Mannheim, Germany
Genomics and Proteomics Core Facilities, German Cancer Research Center, Heidelberg, Germany

Ryosuke Kondo, Ikiru Atsuta, Yasunori Ayukawa, Yuri Matsuura, Akihiro Furuhashi, Yoshihiro Tsukiyama and Kiyoshi Koyano
Section of Implant and Rehabilitative Dentistry, Division of Oral Rehabilitation, Faculty of Dental Science Kyushu University, Fukuoka, Japan

Takayoshi Yamaza
Department of Molecular Cell Biology and Oral Anatomy, Graduate School of Dental Science Kyushu University, Fukuoka, Japan

Lukas Wisgrill, Simone Schüller, Markus Bammer, Angelika Berger and Arnold Pollak
Dept. of Pediatrics and Adolescent Medicine, Division of Neonatology, Paediatric Intensive Care & Neuropaediatrics, Medical University of Vienna, Vienna, Austria

Teja Falk Radke and Gesine Kögler
Institute for Transplantation Diagnostics and Cell Therapeutics, Heinrich Heine University Medical Center, Duesseldorf, Germany

Andreas Spittler
Department of Surgery, Research Labs & Core Facility Flow Cytometry, Medical University of Vienna, Vienna, Austria

Hanns Helmer and Peter Husslein
Dept. of Obstetrics and Gynecology, Medical University of Vienna, Vienna, Austria

Ludwig Gortner
Dept. of Pediatrics and Neonatology, Saarland University, Homburg, Saar, Germany

Chao Dong, Yi Wang and Yulai Zhou
Department of Regeneration Medicine, School of Pharmaceutical Science, Jilin University, Changchun, P.R. China

Xin Chen
Department of Laboratory Medicine, Second Clinical Medical College of Jinan University, Shenzhen People's Hospital, Shenzhen, P.R. China

Zhongjun Zhang and Lei Zhao
Department of Anesthesiology, Second Clinical Medical College of Jinan University, Shenzhen People's Hospital, Shenzhen, P.R. China

Pengfei Liu, Yetong Feng, Dandan Yang and Bo Li
Department of Regeneration Medicine, School of Pharmaceutical Science, Jilin University, Changchun, P.R. China
Key Laboratory of Regenerative Biology, Guangdong Provincial Key Laboratory of Stem Cell and Regenerative Medicine, Guangzhou Institutes of Biomedicine and Health, Chinese Academy of Sciences, Guangzhou, P.R. China

Annalena Krost, Nicole Langwieser, Jan Peter and Stefanie Kamann
Berlin Brandenburg Center for Regenerative Therapies (BCRT), Berlin, Germany

Constantin Rüder and Dietlind Zohlnhöfer
Berlin Brandenburg Center for Regenerative Therapies (BCRT), Berlin, Germany
Department of Cardiology, Campus Virchow Klinikum, Charité Berlin, Germany

Yeona Jin and Yunwoong Park
Department of Nuclear Medicine, Seoul National University College of Medicine, Seoul, Korea

Do Hun Lee
University of Miami School of Medicine, Miami Project to Cure Paralysis, Department of Neurological Surgery, Miami, Florida, United States of America

Han Na Cho, Hye Jin Chung and Seung Jin Lee
College of Pharmacy, Ewha Womans University, Seoul, Korea

Hong J. Lee and Seung U. Kim
Medical Research Institute, Chung-Ang University College of Medicine, Seoul, Korea

Kyu-Chang Wang
Division of Pediatric Neurosurgery, Seoul National University Children's Hospital, Seoul, Korea

Do Won Hwang, Han Young Kim and Dong Soo Lee
Department of Nuclear Medicine, Seoul National University College of Medicine, Seoul, Korea
Department of Molecular Medicine and Biopharmaceutical Science, WCU Graduate School of Convergence Science and Technology, Seoul National University, Seoul, Korea

Hyewon Youn
Department of Nuclear Medicine, Seoul National University College of Medicine, Seoul, Korea
Cancer Imaging Center, Seoul National University Cancer Hospital, Seoul, Korea
Cancer Research Institute, Seoul National University College of Medicine, Seoul, Korea

Jun Jie Tan, Siti Maisura Azmi, Hong Leong Cheah, Vuanghao Lim, Doblin Sandai, Bakiah Shaharuddin
Advanced Medical and Dental Institute, Universiti Sains Malaysia, Kepala Batas, Penang, Malaysia

Yoke Keong Yong
Department of Human Anatomy, Faculty of Medicine and Health Sciences, Serdang, Selangor Darul Ehsan, Malaysia

Index

www.ingramcontent.com/pod-product-compliance
Lightning Source LLC
Chambersburg PA
CBHW061243190326
41458CB00011B/3568